ERRATA SHEET

Page number	Correction
xiii	Preston Harrison is *head* of the department.
12	Fuel Oil Transfer—Flash point should read 150°, not 110°.
18	Line 4 under *Combustible liquid* should read 26.7°C (80°F).
236	Last paragraph, first line should read: The port-side line (line 1 in Fig. 11.5) is advanced to the space.
238–244	On these pages, all *text* references to figure numbers are one digit behind. For example, in the third paragraph on page 238 (Fig. 11.5) should read (Fig. 11.6), and so on throughout this section.
331	In item number 3 under *Restowing* head: Figure 5 should read Figure 15.5. In subparagraph 1.a., under *Types of Breathing Apparatus,* sentence should read: These devices provide oxygen chemically or from a high pressure oxygen supply in a cylinder. In subparagraph 1.b., delete the words "or oxygen."
332	In the first column, last sentence, delete the word "Demand" and substitute the words "H.P. oxygen cylinder." Sentence should read: H.P. oxygen cylinder-type apparatus consist of a facepiece, regenerator, breathing bag, inhalation tube, exhalation tube, relief valve, high pressure oxygen cylinder, high pressure reducing valve and pressure gauge, cylinder control valve, and bypass valve. In the second column, delete entire paragraph (which begins with "Demand-type apparatus. . .") including two lines at top of page 333.
336	In the first column, under *Donning,* first sentence should read: The wearer can don the canister-type OBA without assistance as follows (Fig. 15.9 on pages 338 and 339):
344	For illustration of step 7 (donning the facepiece), see Figure 15.3 on page 330.
345	Two top photos should be labeled 2A and 2B.
346	Disregard note that follows photos.
347–355	Under *Minipack Unit* head: There is no illustration of a minipack unit (no Fig. 15.18). Therefore, all text references to figure numbers on these pages are one digit ahead of sequence. For example, in the text, Fig. 15.19 should read Fig. 15.18, etc.

Marine Fire Prevention, Firefighting and Fire Safety

Robert J. Brady Co. • Bowie, Maryland 20715
A PRENTICE-HALL PUBLISHING AND COMMUNICATIONS COMPANY

This publication was researched, developed, and produced by the Robert J. Brady Co. for the National Maritime Research Center, under Contract No. MA-2-4362.

ISBN 0-87618-994-X

Printed in the United States of America

80 81 82 83 84 85 86 10 9 8 7 6 5 4 3 2 1

Contents

Executive Producer: George D. Post
Project Coordinator: Joseph F. Connor
Technical Writer: Edward M. Millman
Production Editor: Marlise Reidenbach
Comprehensive Art Design: George D. Post
Text Design: Laura Lammens
Art Director: Don Sellers
Illustrators: Bruce Bollinger and Wayne R. Lewis
Photography: Rick Brady and George Dodson

foreword

Throughout history mariners have gone to sea in all types of watercraft, and, more often than not, with very limited protection against the threat of shipboard fires. In the event of fire, persons ashore often have available the immediate assistance of well-trained firefighting professionals. Mariners are alone aboard ship, and, when fires occur at sea they must remain onboard and cope with these incidents to the best of their own abilities. These efforts, often because of lack of knowledge, training, and experience, have produced less than satisfactory results and at times have resulted in tragedy. Because of the many technological advances in ship design and operation, today's mariner must possess more knowledge than his predecessors in many special areas. Fire prevention, control, and extinguishment is one of these areas.

While government agencies have through the years effected changes and promulgated regulations that have greatly reduced the ever-present danger of fire aboard vessels, fire tragedies have continued to occur. It therefore must be the mariner's responsibility to be as well-trained as possible and to understand the causes of fires so as best to prevent them. Furthermore, mariners must have a good working knowledge of the approaches that will best restrict the spread of fires and eventually extinguish them.

Maritime institutions have been doing their part in the training of individuals toward these ends. Until now, however, they have been at a disadvantage since no comprehensive marine firefighting textbook was available. With the publication of this book, the mariner will now have a comprehensive firefighting text for study and ready reference.

I salute all those who have given their time and effort to produce this manual, and I am certain that the men and women aboard our vessels, to whose protection this important work is dedicated, also will share in my appreciation.

SAMUEL B. NEMIROW
Acting Assistant Secretary
for Maritime Affairs
Department of Commerce

Preface

The objective of this manual is to fill a long-standing need for comprehensive source material in the specialized field of prevention, control, and extinguishment of fires aboard commercial vessels — in the safest and most expeditious manner.

Fire prevention, control, and extinguishment are similar for all types of water craft. However, so that no segment of the waterborne industry is neglected or left wanting for explicit instructions, special chapters have been included that deal directly with such other trades as the offshore oil drilling operations and their vessels and the tow boats on the inland waterways. This book, therefore, was written to provide detailed information concerning vessels of the deepsea oceans, the offshore drilling industry, and the boats in domestic waterborne trade — whether they are on the Great Lakes, inland waterways or domestic oceans.

This book should serve all present and future shipboard personnel by providing exhaustive source and reference material that may be used when dealing with the varied and complicated aspects of fire control and fire fighting. Furthermore, the book contains the most definitive information on fire, which can be used extensively by maritime training institutions throughout the country. As students prepare for service aboard merchant vessels, this manual will serve as a comprehensive study text. It will be useful to seamen, wiper or deck utility personnel, and Masters and Chief Engineers.

Because this book fills an important need within the industry, it is expected that it will be recommended reading for all personnel, kept in the reference libraries aboard ships, and available to all seamen. We further expect the manual to provide shoreside maritime executives with information that will assist them in their decisionmaking concerning fire fighting material, equipment, and requirements aboard vessels. It will also provide an understanding of the fire fighting training required for ship personnel to assure the greatest possible protection of crew, cargo and vessel.

So that this manual would fulfill these needs for a long period of time, the authors have researched the latest fire fighting equipment used aboard vessels today with an eye to future developments in the field. Countless hours were spent aboard all forms of water craft, in addition to time spent working with the manufacturers of fire detecting and fire fighting equipment, and fire fighters who have first hand knowledge of the subject. Finally, the manual was reviewed extensively by numerous people in the maritime industry who have made a career of teaching fire fighting to active shipboard personnel and to those aspiring to service within the industry.

In an ideal situation, it is most important to understand the nature of fires so that every effort may be made to *prevent* them. Failing this, important steps must be put into effect to control and thereby limit the spread of fires and finally to extinguish them. This three-fold approach is used effectively by the authors. They also point out the physical dangers caused by fires and the appropriate medical care that must be provided.

Much can be learned from the study of past fire tragedies that occurred aboard merchant vessels. These fires took a terrible toll in human lives and in material. Of course, these losses are irreversible, but the tragedies can be utilized positively — to change regulations in order to prevent similar accidents, and as a learning tool. The authors have skillfully included an in-depth study of past shipboard fires, dissecting and analyzing the events, to avoid repeating the mistakes of the past. These case studies provide a breakdown of the conditions that led to the problem — the inaction or untimely action taken, with an analysis of actions that should have been taken to limit the spread of the fire and to extinguish it. This manual is expected to be the "bible" on fire fighting and fire safety throughout the maritime industry, both ashore and afloat.

Captain Pasquale Nazzaro,
U.S. Maritime Service Master Mariner
Project Director
for the writing of the Firefighting Manual

Acknowledgments

The preparation of a manual of this magnitude and scope would have been beyond the capacities of any one person, particularly since it was necessary that it be produced without delays and available quickly for use throughout the industry. Therefore, this project involved not only the authors, but numerous individuals who gave unselfishly of their time and special expertise gained through years of dedicated maritime service. Their work included proofreading the manual and making whatever changes or corrections were necessary to assure technical accuracy. They gladly contributed their services because of a strong belief in this manual and a desire to bring to the maritime industry a monumental book that would help to save the lives of their fellow seafarers.

The original concept for publication of a firefighting manual was that of the Maritime Training Advisory Board (MTAB) — a nonprofit organization of maritime people from both labor and management, whose interest is the advancement and improvement of training among practicing mariners. To all members of the MTAB we extend special appreciation for providing the impetus for the writing of this text.

We acknowledge the National Maritime Research Center, Kings Point, New York, for their funding of this project and for their continued assistance and cooperation.

In addition, very special thanks go to the following members of the MTAB who reviewed the manual at various stages in its production.

Pasquale Nazzaro — Captain, USMS
Assistant Head and Professor
Department of Nautical Science
U.S. Merchant Marine Academy
Kings Point, NY 11024

Frank J. Boland — Training Director
N.M.U. Upgrading and Retaining Plan
346 West 17th St.
New York, NY 10011

Edwin M. Hackett — Training Specialist
Office of Maritime Labor and Training
Maritime Administration
Washington, DC 20230

Christopher E. Krusa — Manpower Development Specialist
Maritime Administration
Washington, DC 20230

Arthur Egle (deceased) — Maritime Institute of Technical and Graduate Studies
Linthicum Heights, MD 21090

Preston Harrison — Assistant Dept. Head, General Dept.
Calhoon MEBA Engineering School
Baltimore, MD 21202

R. T. Sommer — Captain, U.S. Coast Guard
MarAd Liaison Officer
Maritime Administration
Washington, DC 20230

Howard W. Patteson	Manpower Development Officer Maritime Administration Washington, DC 20230
William H. Sembler	Professor of Marine Transportation Maritime College — SUNY Fort Schuyler, NY 10465
Joseph Wall	Supervisor, Administrator Services Harry Lundeberg School Piney Point, MD 20674
William F. Fassler	Executive Director National River Academy of USA Helena, Arkansas
James Prunty	Mobil Oil Corporation New Orleans, LA

The Robert J. Brady Co. gratefully acknowledges the help of the following companies and organizations who allowed the use of printed matter and photographs from their publications or gave technical advice.

American District Telegraph
American Waterways Operators, Inc.
Ansul Company (dry chemical sk-3000)
Avondale Ship Building Corporation

Beckman Instruments, Inc. (gas detection system)
Bethlehem Steel Ship Repair Yard

City of New York Fire Department
Curtis Bay Towing Company

Delta Steamship Lines, Inc.
Detex (Newman portable watch clock)

Farrell Lines, Inc.
Fyrepel

Gaylord Industries, Inc. (galley duct washdown system)
Gulf Oil Corporation, Marine Division

C. J. Hendry Company (fireman's protective clothing)
Henschel Corp. (Henschel control unit)

Lykes Brothers Steamship Co., Inc.

Marine Chemists' Association
Merchant Marine Technical Division
Military Sealift Command
Mine Safety Appliances (gas detection system/breathing apparatus)
Mobil Oil Corporation (offshore oil drilling platforms)
Moore McCormack Steamship Lines

National Foam (foam systems and appliances)
Norris Industries

Offshore Marine Service Association
Offshore Operations Committee

Penniman and Browne, Inc.

Robertshaw Controls Company
Rockwood (foam systems and appliances)

Sea–Land Service, Inc.
Ships Operational Safety, Inc.

Walter Kidde and Company (CO_2 systems)
Walz and Krenzer, Inc. (watertight doors)

Contributing Authors

Gilbert W. O'Neill
Battalion Chief
Marine Division, Fire Department
New York City

Thomas J. Rush, Jr.
Deputy Chief in Charge
Marine Division, Fire Department
New York City

William J. Lanigan
Deputy Chief of Department
Fire Department, New York City

Francis P. McCormick
Deputy Chief of Dept. (Retired)
Fire Department, New York City
Assistant Professor
New York City Community College

Edwin J. Byrnes
Battalion Chief
Fire Department
New York City

Adolph S. Tortoriello
Chief in Charge, Fire Academy
Fire Department, New York City

Joseph F. Connor
Assistant Chief of Dept. (Retired)
Fire Prevention Consultant
Fire Department, New York City

George D. Post
Vice President
Fire Training Programs
Robert J. Brady Co.

J. David Bergeron
Educational Technologist
Robert J. Brady Co.

Dale E. Green
Beckman Instruments, Inc.
Somerset, New Jersey

Harvey D. Grant
Claymont, Delaware

Robert H. Murray, Jr.
Winchester, New Hampshire

Consultants

Walter M. Haessler, P.E.
Firefighting Specialist
Ocala, FLA

John Smith, Senior Instructor
Delaware State Fire School

Fire Prevention

Part I

Aboard ship as well as ashore, fire can be either a friend or an enemy. Harnessed and controlled, fire is so much a part of our everyday lives that we take it and its uses for granted. But uncontrolled fire brings disaster—loss of lives and millions of dollars in property damage. For example, in 1974 there were 198 fire incidents involving foreign and domestic vessels in American ports. These incidents resulted in an estimated fifty million dollars in losses. Moreover, the figures represent only incidents in which fire and explosion were the primary causes of the losses; fires resulting from collisions are not included. In some cases, fires that followed collisions have done much more damage than the collisions themselves.

Vessels are subject to all the fire hazards of land installations, and more. Passenger vessels may be likened to moving hotels, with spaces for sleeping, recreation, cooking and dining; these spaces and their occupants present as much of a potential fire problem at sea as they do on land. Tankers are mobile storage facilities for petroleum products and other hazardous fluids. Cargo, container, lash, and roll on–roll off ships are moving warehouses that often carry hazardous materials. Below deck are tons of fuel oil, engine rooms, boiler rooms and machinery spaces where many serious fires have originated.

The problems of fire prevention and firefighting become even more acute once a vessel leaves port. Then, rough seas and navigation difficulties may also increase the hazards. Assistance is far away, and the crew and vessel must provide their own fire protection. This lack of assistance makes shipboard fire prevention extremely important, a matter that must be of great concern to officers and crew alike.

This first part of the book on fire prevention contains three chapters. Chapter 1 deals with the major causes of fire aboard ship, and presents specific ways in which these causes can be eliminated. Chapter 2 deals with the organization and implementation of shipboard fire prevention programs. Chapter 3 presents the histories of a number of ship fires. All three chapters deal with reality, with actual experience and not with contrived situations. They show that fire prevention must be a continuing process on every vessel. There is no such thing as "It can't happen on this ship." Fires have occurred and probably will continue to occur on vessels that never had a fire before. It is up to the crew to minimize the possibility of fire and to minimize the damage that a fire can do if one occurs.

Causes & Prevention of Fire Aboard Ship

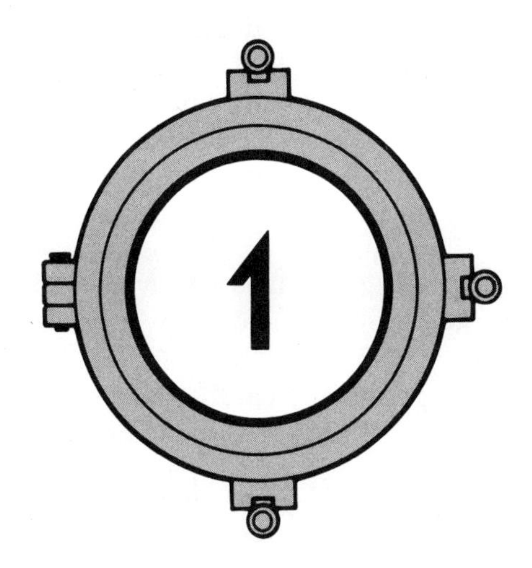

The major causes of shipboard fire are discussed in this chapter, along with actions that crew members can take to reduce the possibility of fire. These causes of fire—these situations and actions—are common to all vessels and are the responsibility of all crews.

Some fires may be purely accidental, and others may be caused by circumstances beyond control. But many fires have resulted from the acts or omissions of crew members. Carelessness and irresponsible or ill-advised actions have caused disastrous fires. And omissions—not taking the proper preventive measures when hazardous situations are discovered—have allowed many fires to "just happen."

No matter how a shipboard fire starts, it could result in the loss of the ship, and perhaps the loss of lives. It is therefore extremely important that crew members be constantly alert for situations that could cause fire aboard ship.

DESIGN SAFETY FEATURES

Before discussing the causes of shipboard fires, we should note that ships flying the U.S. flag are designed and built according to very detailed regulations. These regulations are, for the most part, based on maritime experience—in some cases tragic experience that resulted in loss of life and property. They provide uniform minimum requirements for the construction of vessels. The regulations, and the safety standards they represent, are continually being upgraded in the light of increased experience. Of course, the desire for absolute safety must be balanced against the cost of attaining it. Fire safety is well represented through the following design regulations.

1. Structural fire protection (hull, superstructure, bulkheads and decks)
2. Restrictions on the use of combustible materials
3. Insulation of exhaust systems
4. Venting of cargo spaces, fuel tanks and pump rooms
5. Means of escape
6. Minimum stairway sizes
7. Fire detection and alarm systems
8. Firemain systems
9. Fixed fire extinguishing systems
10. Portable and semiportable extinguisher requirements
11. Approved machinery, equipment and installation.

Each bulkhead, deck, hatch, ladder, and piece of machinery is built and located to serve a specific purpose or purposes including, wherever possible, fire safety. But good design is only the beginning; it must be combined with good construction and good workmanship to make a safe vessel. Then it is up to the crew to keep the vessel safe. Stated another way, safety begins on the drawing board and is completed only when the vessel is decommissioned.

CARELESS SMOKING

At the top of every list of fire causes—aboard ship or on land—is careless smoking and the careless disposal of lit cigarettes, cigars, pipe tobacco and matches.

Smoking is a habit. For some people it is so strong a habit that they "light up" without even realizing they are doing so. For others, nothing can quiet the urge to smoke; they will do so without regard for the circumstances or location. And some simply don't care or don't realize that smoking can be dangerous. Such people must be made aware of the risks of careless smoking.

Disposing of Butts and Matches

Glowing ashes and glowing tobacco contain enough heat to start a fire in such materials as dunnage, paper, cardboard, excelsior, rope and bedding. Therefore, matches, and ashes from cigarettes, cigars and pipes, butts and glowing pipe tobacco must be discarded in noncombustible receptacles. These receptacles should be placed throughout the vessel, wherever smoking is permitted. It is also a good idea to soak a cigarette or cigar butt with water before discarding it. The soaking provides added protection against fire.

Ashtrays should be emptied only when they contain no glowing embers. (A soaking under a faucet will ensure this.) Then they should be emptied into covered, noncombustible containers.

Smoking in Bed

Smoking in bed is dangerous at any time. After a busy day, when the smoker is tired, it can mean disaster (Fig. 1.1). A smoldering fire can be started just by touching the glowing tobacco to the bedding. The resulting smoke can cause drowsiness and possible asphyxiation before the fire is discovered.

Such fires can be prevented by following one simple but important rule: *Don't smoke in bed, under any circumstances.*

Smoking and Alcohol

A person who has been drinking alcohol tends to become careless. If that person is also smoking, he can be extremely dangerous. After one or two drinks, a few glowing embers that have dropped from a pipe may not seem important. Nor will a cigar butt that isn't quite extinguished, or a lit cigarette that someone has left on an ashtray. But these are actually small shipboard fires. If they come in contact with nearby flammable material, the fires will not stay small for very long (Fig. 1.2).

A smoker who is "under the influence" should be observed very carefully. Everyone should be responsible for seeing that the smoker's actions do not jeopardize the safety of the ship and its crew.

Figure 1.1. Smoking in bed is dangerous and unnecessary.

Figure 1.2. Drinking and smoking are a dangerous combination.

Figure 1.3. Smoking and careless disposal of smoking materials have caused many serious fires in cargo holds. Smoking must be prohibited in cargo spaces, and these spaces should be monitored during cargo handling.

No Smoking Areas

Open flames and glowing embers can be very dangerous in certain parts of a ship. Smoking must be prohibited in these spaces, and they should be clearly marked as No Smoking areas. Every crew member should know *where* smoking is prohibited and *why* it is prohibited there.

Visitors, longshoremen and other shoreside workers should be informed or reminded of smoking regulations whenever they come aboard. These people are not as concerned as crew members about fire safety. When the vessel sails, the shoreside people stay behind. If fire is discovered after the ship leaves port, only the crew is endangered, and the crew alone must fight the fire. For this reason, crew members must feel a strong responsibility to ensure that No Smoking regulations are followed by everyone on board their ship. Most people will comply with the regulations, and will not smoke in restricted areas. Those who persist in smoking after being told of the danger should be reported to the proper authorities.

Cargo Holds and Weather Deck. Smoking in the holds of cargo vessels, or on the weather deck when the hatches are open, is an invitation to disaster. Such smoking is strictly forbidden.

Break-bulk cargo vessels are especially vulnerable to cargo-hold fires during loading (Fig. 1.3). Such a fire may not be discovered for several days—after the vessel is well out to sea. By that time, much of the cargo may be involved in the fire, and the fire may be difficult to extinguish or control. To add to the problem, a number of port cities are reluctant to give refuge to a ship on fire. This is understandable: These cities do not have either the capability or the experience to combat ship fires.

The best way to deal with cargo-hold fires is to prevent them. This means *1)* smoking must be prohibited in cargo holds at all times; *2)* cargo holds should be posted as No Smoking areas; and *3)* holds should be monitored closely during loading and unloading operations.

Engine and Boiler Rooms. Engine rooms and boiler rooms contain relatively large amounts of petroleum products, such as fuel oil, lubricating oil and grease. Even the thickest of these products tends to vaporize and mix with the warm air of the engine room or boiler room. A lighted match or glowing tobacco can ignite this flammable air–vapor mixture. Carelessly discarded smoking materials can start fires in oily rags or other flammable materials.

Once ignited, an engine room fire is difficult to extinguish and very hazardous for the engine room crew. If the fire is serious enough, it could mean loss of propulsion and control of the vessel—an extremely dangerous situation. For these reasons, engine room and boiler room smoking regulations should be followed carefully.

Storage and Work Spaces. Smoking should be prohibited in storage rooms and work rooms, and this prohibition should be strictly enforced. These spaces—for example, paint and rope lockers and carpenter shops—contain large amounts of flammable materials. A stray ember or a hot match could easily ignite such materials.

SPONTANEOUS IGNITION

Spontaneous ignition is often overlooked as a cause of fire aboard ship. Yet many common materials are subject to this dangerous chemical phenomenon. They include materials that are carried as cargo and materials that are used in running the ship. An example of spontaneous ignition that could easily occur aboard a vessel might be a rag soaked with vegetable oil or paint that has been discarded in the corner of a workshop, storage area or engine room. The area is warm, and there is no ventilation (Fig. 1.4). The oil on the rag begins to oxidize—to react chemically with the oxygen in the warm air around it. Oxidation is a natural process that produces heat. The heat causes the remaining oil to oxidize faster and produce still more heat. Since the heat is not drawn away by ventilation, it builds up around the rag. After some time, the rag gets hot enough to burst into flames. It then can ignite any nearby flammable substances, perhaps other rags or stored materials, so that a major fire is very possible. All this can and does occur without any outside source of heat.

Figure 1.4. Careless disposal or storage of materials can lead to spontaneous ignition.

Materials Subject to Spontaneous Ignition

Ship's Materials. As noted in the previous section, oily rags and paint-soaked rags are subject to spontaneous ignition. In this case, fire prevention is simply a matter of good housekeeping (Chapter 2). However, some materials that are not usually subject to spontaneous ignition will ignite on their own under certain conditions. Wood is one such material.

Wood, like every other substance, must be heated to a certain temperature before it will ignite and burn. And most steam pipes do not get hot enough to ignite wood. Yet if a piece of wood is in constant contact with a steam pipe or a similar "low-temperature" heat source, it will ignite spontaneously. What happens is that wood is first changed to charcoal by the heat (Fig. 1.5). Then the charcoal, which burns at a lower temperature than wood, is ignited by the steam pipe. Even though the change from wood to charcoal may take several days to occur, it could easily go unnoticed. The first sign of a problem would be smoke or flames issuing from the wood.

To prevent such fires, combustible materials should be kept away from any heat source. If they cannot be moved, they can be protected with heat-insulating materials.

Cargo. Many materials that are carried as cargo are subject to spontaneous ignition. Ignition occurs through the chemical interaction of two or more substances, one of which is often air or water. Precautions for stowing many of these substances are included in the *Hazardous Ma-*

Figure 1.5. A hot steampipe can change wood to charcoal, ignite the charcoal and cause a fire.

terials Regulations of the Department of Transportation (DOT), which are enforced by the U.S. Coast Guard. These regulations may be found in Title 49 of the Code of Federal Regulations (CFR). Additionally, many items that may ignite spontaneously are mentioned in the current edition of the National Fire Protection Association's (NFPA) *Fire Protection Handbook*.

Types of Combustible Cargo. Chlorine produces a violent reaction when it combines with finely divided metals or certain organic materials, particularly acetylene, turpentine and gaseous ammonia. Title 49 CFR 172.101 cautions: "Stow in well-ventilated space. Stow away from organic materials."

The metals sodium and potassium react with water. Hence, 49 CFR 172.101 cautions: "Segregation same as for flammable solids labeled Dangerous When Wet."

Metal powders such as magnesium, titanium, calcium and zirconium oxidize rapidly (and produce heat) in the presence of air and moisture. Under certain conditions they can produce sufficient heat to ignite. The NFPA cautions "Moisture accelerates oxidation of most metal powders." In the DOT regulations, metallic aluminum powder is listed with the following requirements: "Keep dry. Segregation same as for flammable solids labeled Dangerous When Wet."

According to the NFPA, dry metal turnings do not tend to ignite spontaneously. However, piles of *oily* metal borings, shavings, turnings and cuttings have caused fires by igniting spontaneously. As in the case of oily rags, heat is produced by oxidation of the oil within the pile of shavings. Eventually enough heat is produced and held in the pile to ignite the most finely divided metal. Then the coarser shavings and other combustible materials, if present within the pile, ignite and compound the fire problem.

Soft coal may heat spontaneously, depending on several factors.

1. Geographic origin
2. Moisture content
3. Fineness of particles and ratio of fine particles to lump coal
4. Chemical makeup, including impurities
5. Whether or not the coal is newly crushed.

Both coal and metal shavings are regulated cargo, which means they must be handled and transported according to regulations in Title 49 CFR.

In addition, the following present a danger of fire through spontaneous heating: alfalfa meal, charcoal, codliver oil, colors in oil, cornmeal feeds, fish meal, fish oil, fish scrap, linseed oil, oiled and varnished fabrics of all kinds, redskin peanuts, and tung-nut meals. (Note the number of oils.)

A good rule of thumb in preventing the spontaneous ignition of cargo is to separate fibrous materials from oils. Other methods of preventing cargo fires are discussed under Cargo Stowage in this chapter as well as in Chapter 2.

FAULTY ELECTRIC CIRCUITS AND EQUIPMENT

For properly insulated and wired equipment, electricity is a safe and convenient source of power. However, when electrical equipment wears out, is misused or is poorly wired, it can convert electrical energy to heat. Then the equipment becomes a source of ignition and thus a fire hazard. For this reason, electrical equipment must be installed, maintained, tested and repaired in accordance with existing regulations, and only by qualified personnel.

Replacement Parts and Equipment

Standard residential or industrial electrical equipment does not last very long at sea. The salt air causes corrosion; the ship's vibration breaks down the equipment; and the steel hull can cause erratic operation or short-circuiting. As a result, the equipment or its wiring may overheat or arc, causing a fire if flammable materials are located nearby.

Approved electrical equipment is, however, specially designed and constructed for shipboard use. Given reasonable maintenance, it will withstand the strenuous conditions at sea. Thus, only approved replacement parts and equipment should be installed aboard ship—and only for the use for which they have been approved. The chief engineer should be consulted if there are any doubts concerning the installation, repair, use, or maintenance of this equipment.

Wiring and Fuses

The insulation on electrical wiring, particularly the type used for appliances, electric hand tools and cargo and drop lights, will not last forever. With age and use, it can become brittle and crack. It may be rubbed through or broken by abuse or by the vibration of the vessel. No matter how it happens, once the insulation is broken, the bare wire is dangerous. A single exposed wire can arc to any metal object. If both wires are exposed, they can touch and cause a short circuit. Either could produce enough heat to ignite the insulation on the wiring or some other nearby flammable material.

Further, if the fuse or circuit breaker in that particular circuit is too large, it will not break the circuit. Instead, an *increased* current will flow, and the entire circuit will overheat. Eventually the insulation will begin to burn and ignite combustible material in its vicinity.

This type of fire can be prevented by replacing wires that have bad insulation and by installing only fuses and circuit breakers of the proper size for their circuits.

Jury-Rigging

The "jury-rigging" of electrical outlets to serve additional appliances, particularly in crew's quarters and galleys, is a dangerous practice (Fig. 1.6). The wiring in every electrical circuit is designed to carry a certain maximum load. When this wiring is overloaded with too many operating appliances, it can overheat and burn its insulation. The hot wiring can also ignite flammable materials in the area. Cabins have been burned out by such fires, even though the need for jury-rigging can easily be avoided by planned use of appliances.

Exposed Light Bulbs

An exposed, lighted electric bulb can ignite combustible material by direct contact. A number of shipboard fires have started when a crew member left a lamp lit in unoccupied quarters. As the ship rolled, curtains or other flammable material came in direct contact with the hot bulb and ignited. The result in most cases was destruction of the crew member's quarters.

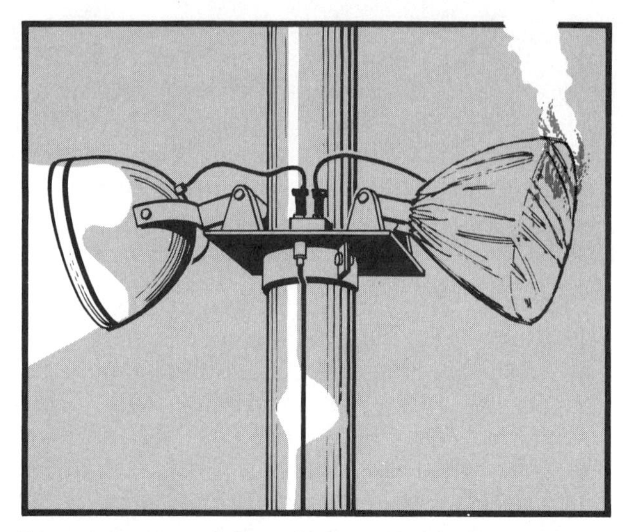

Figure 1.7. Covers left in place over floodlights can be ignited by the heat of the lamp.

On weather decks, high-intensity floodlights are usually protected from the elements by canvas or plastic covers. The covers are desirable when the lights are not in use. However, if a cover is left in place while the light is on, the heat of the lamp can ignite the material (Fig. 1.7).

Improperly protected drop-light or cargo-light bulbs could similarly ignite flammable materials, by contact or by breaking and arcing (Fig. 1.8). They should never be permitted to burn while unattended. What appears to be a safe situation in a calm sea could quickly become dangerous in a rough sea.

Vaportight Fixtures

Vaportight fixtures are protected against the effects of sea air. The vapor protection is designed to keep moisture out, but it also holds heat in.

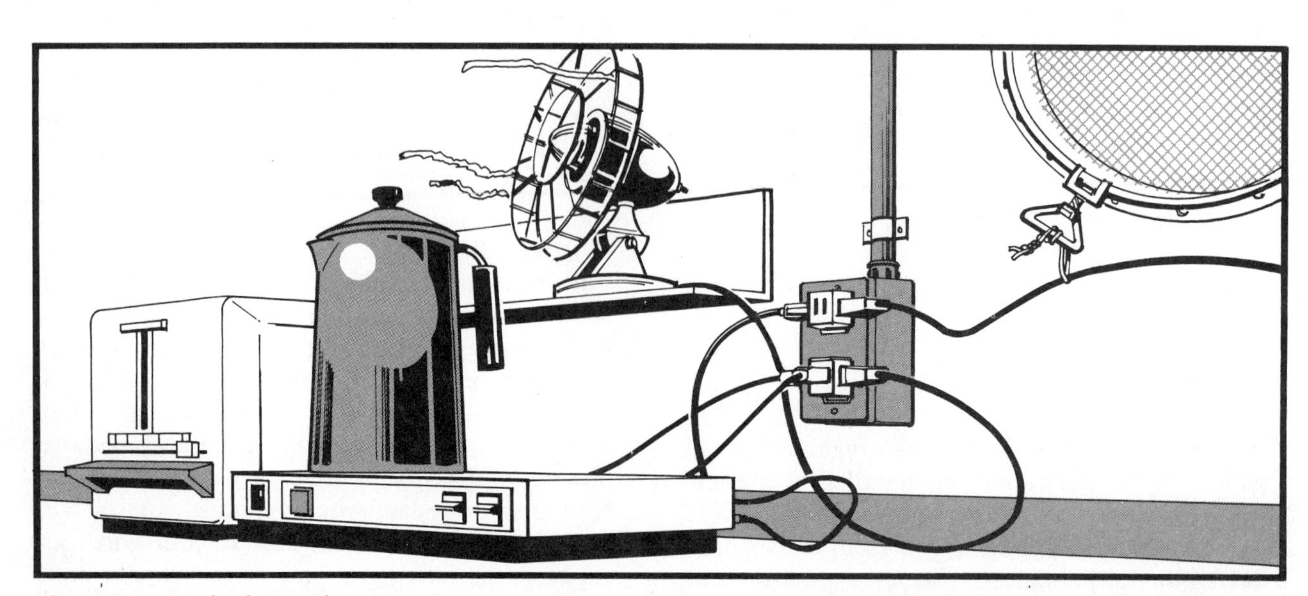

Figure 1.6. Overloading is dangerous. Only one appliance should be connected to each outlet in an electric circuit.

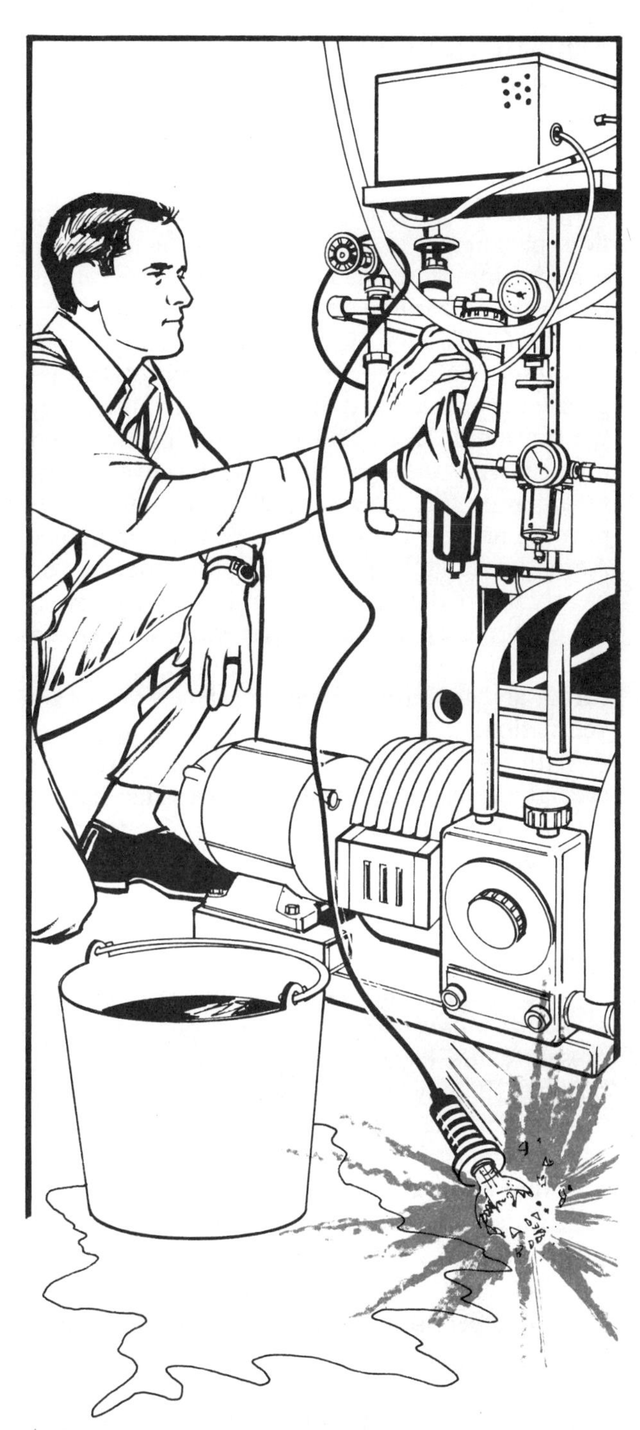

Figure 1.8. An unprotected drop-light bulb can easily break, allowing the live electric circuit to ignite nearby flammable material.

This causes the insulation to dry out and crack more rapidly than in standard fixtures. Thus, vaportight fixtures should be examined frequently and replaced as required, to prevent short circuits and possible ignition.

Electric Motors

Faulty electric motors are prime causes of fire. Problems may result when a motor isn't properly maintained or when it exceeds its useful life.

Motors require regular inspection, testing, lubrication and cleaning. Sparks and arcing may result if a winding becomes short-circuited or grounded, or if the brushes do not operate smoothly. If a spark or an arc is strong enough, it can ignite nearby combustible material. Lack of lubrication may cause the motor bearings to overheat, with the same result. (Lubrication is discussed further in Chapter 2.)

Engine Rooms

Engine rooms are particularly vulnerable to electrical hazards. Water dripping from ruptured sea-water lines can cause severe short-circuiting and arcing in electric motors, switchboards and other exposed electrical equipment. This, in turn, can ignite insulation and nearby combustible materials. Probably even more serious are ruptured fuel and lubrication lines above and near electrical equipment. The engineering staff must constantly monitor oil lines for leaks.

Charging Storage Batteries

When storage batteries are being charged, they emit hydrogen, a highly flammable gas. A mixture of air and 4.1% to 74.2% hydrogen by volume is potentially explosive. Hydrogen is lighter than air and consequently will rise as it is produced. If ventilation is not provided at the highest point in the battery charging room, the hydrogen will collect at the overhead. Then, any source of ignition will cause an explosion and fire.

To prevent hydrogen fires batteries should be charged in a well-ventilated area. Smoking and other sources of ignition should be strictly prohibited. The area should contain no machinery that might produce sparks.

UNAUTHORIZED CONSTRUCTION

Space for stowage is always at a premium aboard ship. There should be "a place for everything, and everything in its place." This in itself is a fire prevention measure, provided the stowage is safe to start with. But fires have resulted when stowed materials came loose and fell or slid across a deck in rough weather. Loose equipment can rupture fuel lines, damage essential machinery and smash electrical equipment, causing short circuits. In addition, it is difficult and dangerous to try to gain control of heavy equipment that has come loose during a heavy sea.

When unskilled personnel attempt to build stowage facilities, the results are usually less than satisfactory. In fact, jury-rigged stowage racks can be extremely dangerous. Generally, they are

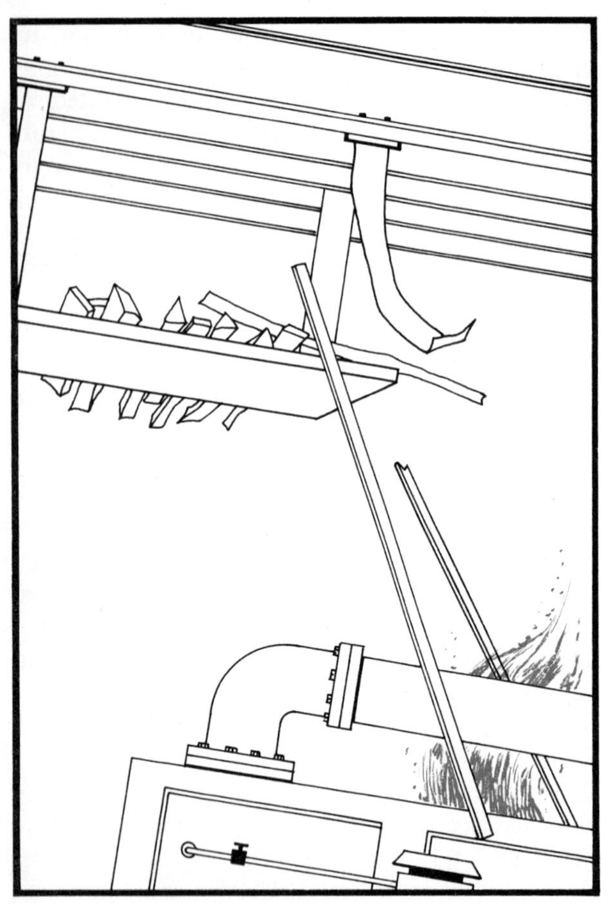

Figure 1.9. Unauthorized construction is usually poorly designed and engineered. Materials falling from a jury rigged stowage rack can damage equipment or cause a fire.

too weak to support the material to be stowed, or they are poorly designed, so they allow material to fall or slide from the structure (Fig. 1.9). The locations of unauthorized construction projects are usually chosen without regard for safety. For example, one of the worst places to stow angle iron is directly above a large item of electrical machinery, such as a generator; the dangers are obvious. Yet records show that serious fire was caused by falling materials in just such a situation.

CARGO STOWAGE

Even the most dangerous cargo can be transported safely if it is properly stowed. On the other hand, supposedly "safe" cargo can cause a fire if it is stowed carelessly. As noted earlier, shoreside personnel leave the ship after loading the holds. Only the crew remains to fight a fire that is discovered after their ship leaves port. For this reason, the master or his representative should always monitor the loading—even when stowage plans have been prepared in advance by port personnel prior to the ship's arrival.

Regulated Cargo

Materials carried as cargo aboard vessels can be divided into two general classifications—regulated cargo and nonregulated cargo. *Regulated cargo* is more generally referred to as *hazardous* cargo. Rules governing the classification, description, packaging, marking, labeling, handling and transporting of regulated cargo are given in detail in the Hazardous Materials Regulations, Subchapter C, Title 49, Code of Federal Regulations.

These regulations serve one purpose: to safeguard the carrier and its personnel. Ultimately, the master of the vessel is responsible for compliance with these regulations; however, every member of the crew should be aware of their purpose and the consequences of noncompliance. Among other things, the regulations define very clearly where dangerous materials may be stowed, on both passenger and cargo vessels. They include details concerning segregation from other cargo, and the proper humidity, temperature and ventilation.

With few exceptions, hazardous cargo requires special labels that define the particular hazard. The labels are discussed in Chapter 2 and shown in Figure 2.8.

Nonregulated Cargo

Cargo that is not specifically covered by Department of Transportation regulations is referred to as *nonregulated* cargo. Nonregulated cargo can present a fire hazard if it or its packing is combustible. It may be subject to spontaneous ignition, and it may be ignited by careless smoking or faulty electrical equipment. It could then act as a fuse if hazardous cargo is stowed nearby.

Loading and Unloading

Loading and unloading operations should be closely supervised by the ship's deck officers. Leaking cargo should be rejected immediately (Fig. 1.10); any liquid that has leaked into the hold should be removed or otherwise rendered harmless. (Remember, a vegetable oil that leaks onto baled cotton, rags or other fibrous material could cause spontaneous ignition.) When cargo is handled, it should not be allowed to bump hatch coamings or other cargo, or to land so heavily in the hold that the packaging is damaged. Such damage could go undetected and cause serious problems after the ship leaves port.

Even in home ports, loading and unloading should be carefully observed. In other ports, especially foreign, vigilance and close monitoring are of great importance.

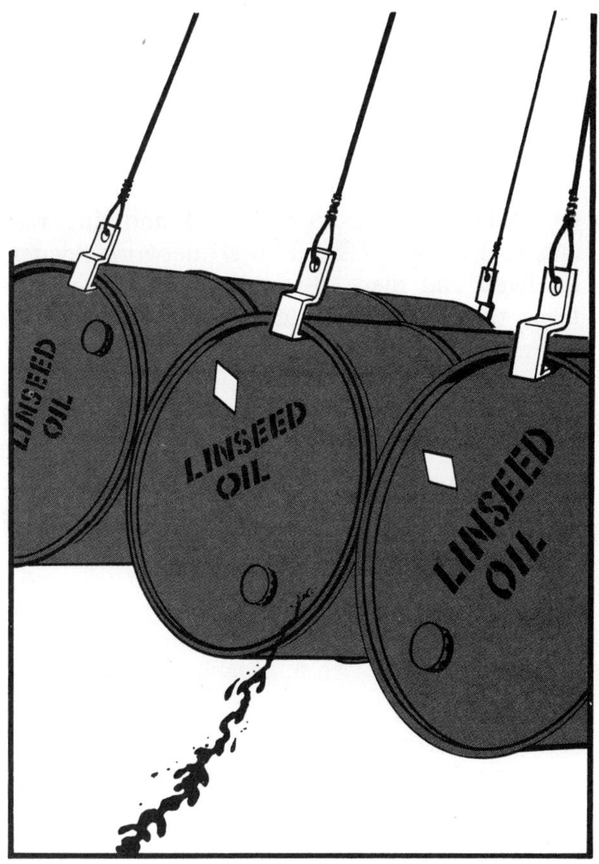

Figure 1.10. Leaking cargo should not be permitted aboard any vessel.

Shoring

At sea, a ship can move in many different directions. Proper shoring of cargo to keep it from shifting in rough seas is, of course, important for stability. It is also important from a fire safety standpoint. If stowed cargo is allowed to shift, hazardous materials that are incompatible can mix and ignite spontaneously or release flammable fumes. Further, metal bands on baled goods can produce sparks as they rub against each other—and one spark is enough to ignite some fumes. Heavy machinery, if not properly shored, can also produce sparks; or it can damage other packaging and thus release hazardous materials. As a precaution, hazardous material should be inspected frequently during the voyage for shifting, leakage and possible intermixing with other materials.

Bulk Cargo

Combustible bulk cargo such as grain can be extremely hazardous if required precautions are not followed. Title 46 CFR 97.55 outlines the master's responsibilities in connection with this type of cargo.

Before loading, the lighting circuits in the cargo compartments that are to be filled must be deenergized at the distribution panel or panel board. A sign warning against energizing these circuits must be posted at the panel. In addition, periodic inspections must be made to guard against reenergizing.

Containers

The loading of containers is, at present, receiving increased attention. Ship's personnel have little control over the contents, because they are usually stuffed many miles from the point where they are finally loaded aboard ship. This lack of control makes container safety a matter of great concern. The following precautions will reduce the chance of fire involving containers and their contents.

1. Containers with hazardous contents should be stowed aboard the vessel according to U.S. Coast Guard regulations.
2. If a container shows any sign of leakage or shifting cargo, it should not be allowed aboard the vessel.
3. If a container must be opened for any reason, extreme caution should be used, in case a potentially dangerous fire condition has developed inside.

GALLEY OPERATIONS

On a small harbor tug or a large passenger liner, a ship's galley is a busy place, and it can be a dangerous place. The intense activity, the many people, the long hours of operation, and the basic hazards—open flames, fuel lines, rubbish and grease accumulations—all add to the danger of fire due to galley operations. For these reasons, it is extremely important that the galley never be left unattended when it is in use.

Energy Sources

For cooking, the most common energy source is electricity. Diesel oil is used to a lesser degree, and liquefied petroleum gas (LPG) is used on some smaller vessels, such as harbor tugs. Electric ranges are subject to the same hazards as other electrical equipment. These include short circuits, brittle and cracked insulation on wiring, overloaded circuits and improper repairs.

When liquid fuels are used for cooking, extreme care should be taken to avoid accidental damage to fuel lines. All personnel should be alert to leaks in fuel lines and fittings. In the event of a leak, the proper valves should be closed at once; repairs should be made by competent personnel. Galley personnel should know the locations of fuel-line shutoff valves. These shutoff valves must be readily accessible.

Ranges

Ranges present a twofold fire danger: The heat of the range can cause a galley fire, and its fuel can be involved in one. Galley personnel should exercise extreme care when they are in the vicinity of an operating range. Clothing, towels, rags and other fabric or paper used in the galley can be ignited through carelessness. No material should be stowed above a range. At sea, the range battens should be used at all times (Fig. 1.11).

Pilot lights must be operative, and the main burners must light when they are turned on. Otherwise, fumes will leak into the galley, and any source of ignition will cause an explosion and fire. If a gas leak is discovered, all burners, pilot lights and other sources of ignition must be extinguished; then the emergency shutoff valves must be closed.

Deep Fryers

Deep fryers can also be a source of both heat and fuel for a galley fire. They must be used with caution and monitored carefully during operation. The fryer should be stationary, so that it cannot shift with vessel movement. Food that is too wet should not be placed in the fryer, and the basket should never be filled so full that the grease splatters or overflows. Once ignited, the grease will burn rapidly. Nothing should be stowed above the fryer. Most important, *the fryer should never be left unattended while it is operating.*

Housekeeping

The activities within a galley generate plenty of fuel for carelessly caused fires. Thus, good housekeeping is of the utmost importance. Used boxes, bags and paper, and even leftover food, should be placed in covered, noncombustible refuse cans where they cannot be ignited by a carelessly discarded butt or match.

Figure 1.11. When underway, range battens should be used to keep pots from sliding off the cooking surface.

Grease accumulations in and around the range, particularly in the hoods, filters and ductwork, can fuel a galley fire. If the ductwork becomes involved and there are heavy grease accumulations, the fire can extend to other areas and decks. Therefore, hoods, filters and ductwork should periodically be thoroughly cleaned. Fixed automatic extinguishing systems for ductwork are extremely valuable and most efficient in extinguishing grease fires. Some automatic duct cleaning systems are capable of protecting the galley ductwork from fire (Chapter 9).

FUEL OIL TRANSFER AND SERVICE OPERATIONS

Fuel oil for the ship's propulsion is stored in double-bottom tanks, deep tanks, and tanks in the vicinity of the engine room. The capacity of these tanks can be as high as 3,800,000 liters (4550 tonnes) (1,000,000 gal (5000 tons)), depending on the size of the vessel. The types of fuels most commonly used are No. 6 fuel oil, bunker C and diesel oil. Bunker C and No. 6 fuel oil are both heavy, tarry substances that require preheating before they can be transferred or burned. Both have flash points of approximately 65.6°C (150°F) and ignition temperatures of 368.3–407.2°C (695–765°F). (See Chapter 4 for the exact definitions of flash point and ignition temperature.) Double-bottom tanks and deep tanks are fitted with steampipe grids and coils near the suction pipe, to preheat the oil. Diesel oil does not require heating to be transferred and burned. Its flash point is 43.3°C (110°F), and its ignition temperature is 260°C (500°F).

Transfer of Fuel

When fuel is taken aboard, it is stored in double-bottom or deep tanks. If necessary, the fuel is heated, and then it is pumped to the service tanks or settling tanks. From there, it moves to a gravity or day tank, or to a fuel oil service pump, from which it is pumped to the fuel oil burners or diesel engines.

During this transfer of fuel under pressure, the liquid fuel itself is not a fire hazard if there are no mistakes. However, the fuel vapors that may be given off are very hazardous. Both the overfilling of fuel tanks and leaks in the transfer system can increase the danger of fire.

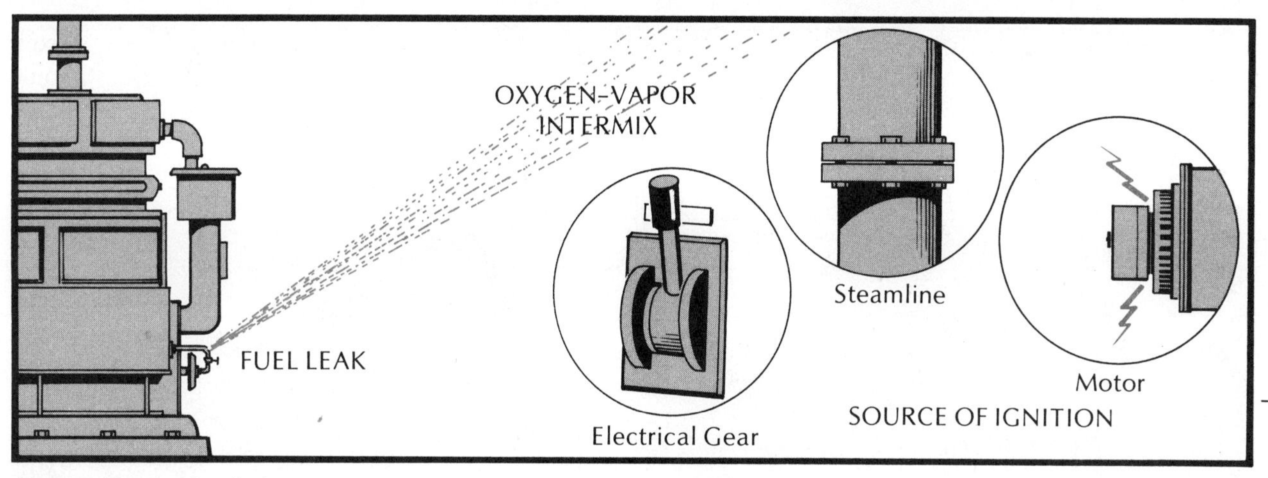

Figure 1.12. Fuel line leaks can spray vaporized fuel far enough to be ignited by steam lines or electrical equipment.

Overfilling. If a tank is overfilled, the fuel will rise through the overflow pipe, and eventually through the vent pipe that terminates topside.

The engine room crew should monitor the transfer process carefully and constantly, to prevent overfilling. However, if a tank is overfilled, strict control of flames, sparks and smoking should be put into effect until the danger of fire has passed.

Leaks in the Transfer System. If there is a leak in the transfer piping, the pressurized fuel will be sprayed out through the break. Spraying tends to vaporize the fuel, and the vapors are easily ignited. Thus, line breaks can be very hazardous if there are steam pipes, electric motors, electric panel boards and so forth in the area (Fig. 1.12). (This is also true of lubricating oil leaks near steam pipes). A diesel oil line break resulted in a serious engine room fire in a passenger liner in New York harbor several years ago. (*See* SS *Hanseatic,* Chapter 3.) It spread upward from the engine room and involved every deck of the vessel.

Before fuel is transferred, the system should be checked to ensure that strainers are in place and all flanged joints are properly tightened. During the pumping, the system should be continuously checked for leaks.

Oil Burner Maintenance

For proper atomization and operation, oil burner tips require regular cleaning and maintenance. An improperly operating oil burner tip can cause incomplete burning of the fuel and a buildup of unburned fuel in the windbox of the boiler. This fuel will eventually ignite. If sufficient fuel is present, the flames can spread away from the boiler and involve other materials and equipment.

Oil burner tips should therefore be cleaned and maintained regularly. They should be installed with care, since improper installation can also cause fuel buildup and ignition.

Bilge Area

Fires occur in bilge areas because of excess accumulations of oil. Most often, the oil leaks into the bilge from an undetected break in a fuel or lube-oil line. The oil vaporizes, and the flammable vapors build up in and around the bilge area. Once these vapors are mixed with air in the right proportion, a carelessly discarded cigarette or cigar butt, a match or a spark can ignite them and cause a fire (Fig. 1.13). Bilge fires can move very quickly around machinery and piping, and for this reason they are not easily controlled. They are more difficult to extinguish than most engine and boiler room fires.

Bilge areas should be watched closely. Excess oil almost always indicates a leak, and the oil lines should be checked until it is found. Oil/water bilge separators should also be checked frequently to prevent overflow, which can also be a source of large accumulations of oil in the bilges.

WELDING AND BURNING OPERATIONS

Welding and burning operations are hazardous by their very nature. This can best be appreciated by noting that the flame from an oxyacetylene torch can reach a temperature of 3315.5°C (6000°F).

Welding temperatures are reached either by burning a mixture of gas and oxygen or by using electricity. The most common welding gas is acetylene; others include hydrogen, LPG and natural gas. In electric welding, commonly called

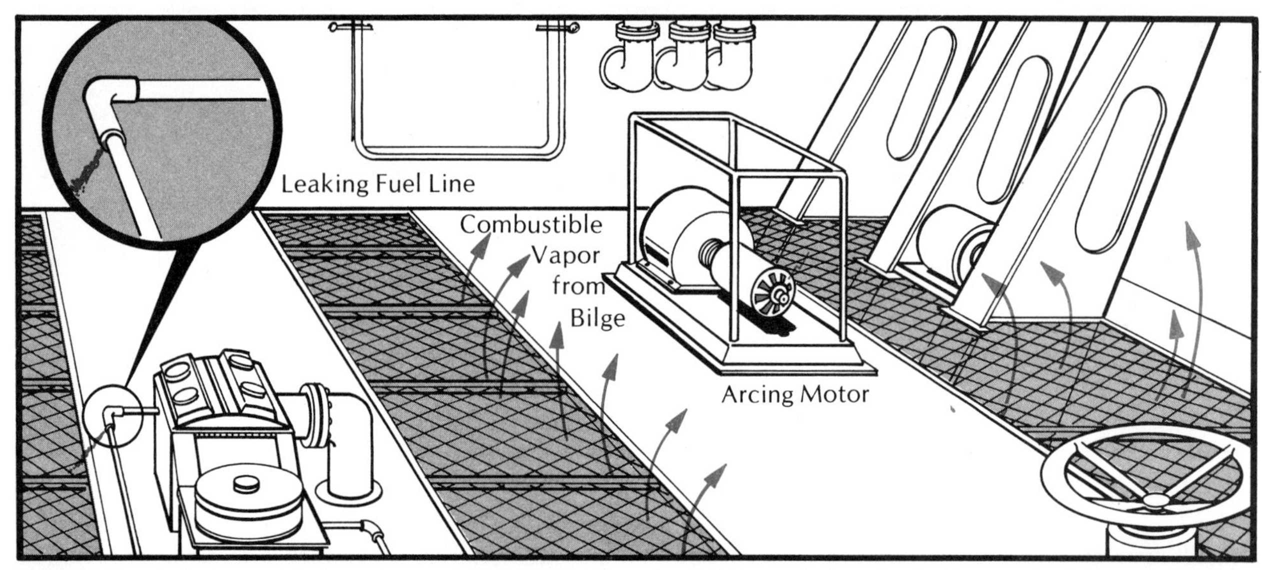

Figure 1.13. A disastrous combination not uncommon on vessels: Fuel from a leaking line collects in the bilge. Combustible fuel vapors from the bilge mix with air as they move toward the arcing motor. Ignition of the fumes by the motor can cause an explosion followed by fire.

arc welding, the required heat is produced by an electric arc that is formed at the workpiece. In either type of welding, dangerous, high-temperature sparks and slag are thrown off.

Burning is a gas-fueled operation and is more hazardous than welding. In gas burning, or gas cutting, the temperature of the metal is raised to the ignition point, and a jet of oxygen is introduced. This forms a metal oxide which melts, and the oxygen jet removes the molten metal.

Unsafe Burning and Welding Practices

The high temperatures, molten metal and sparks produced in welding and burning operations can be an extremely serious fire hazard. During these operations, shipboard fires may be caused by the following.

1. *Failure to provide a competent fire watch* in the immediate work area, below the work area and on the opposite side of a bulkhead that is being welded or burned. The fire watch should have no other duties and should inspect and reinspect the area for at least one-half hour after the operation. This is crucial, as hot metal and slag retain heat for a long time.
2. *Failure to move combustible materials* (or to protect them if they cannot be moved). Materials in the work area and the areas below deck and on the opposite side of a bulkhead should all be protected (Fig. 1.14). Hot sparks and slag can travel great distances, and heat moves quickly through metal decks and bulkheads.

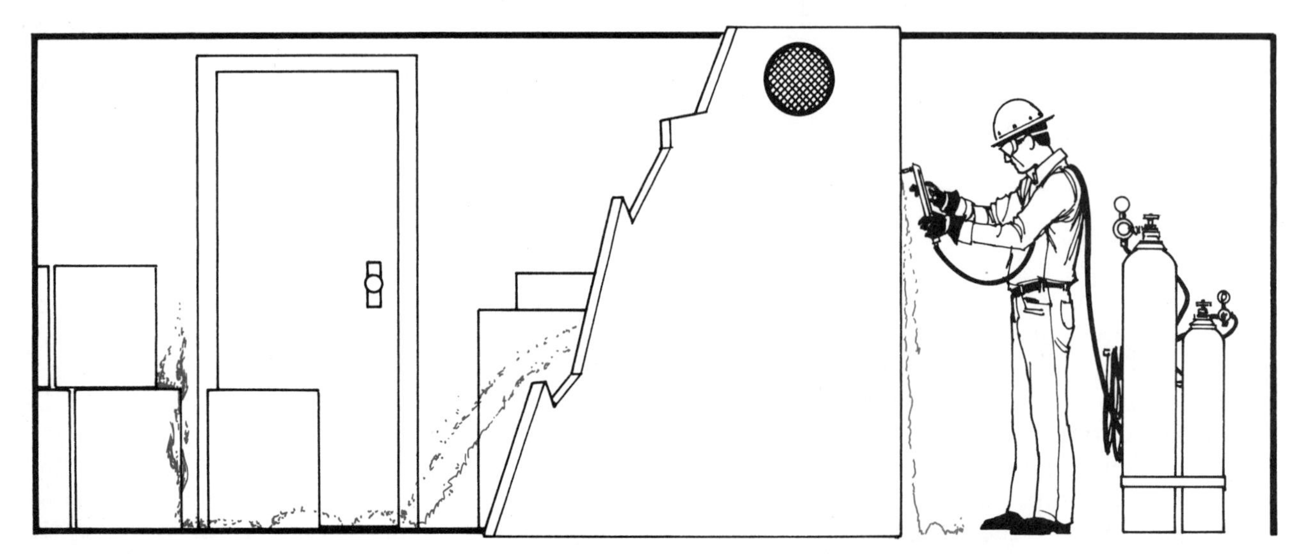

Figure 1.14. Failure to remove combustible materials or to establish a fire watch are major causes of fires during burning and welding operations.

3. *Burning near heavy concentration of dust or combustible vapors* such as those emitted by fuel oil, lubricating oil and other flammable liquids.
4. *Failure to remove flammable vapors, liquids or solids* from a container, pipe or similar workpiece and to obtain the proper clearance (including a certificate) from an NFPA-certified marine chemist or an officially designated "competent person."
5. *Failure to have the proper type of fire extinguisher at the scene,* along with a hoseline charged with water to the nozzle and ready for immediate use.
6. *Failure to secure oxygen and gas cylinders in an upright position.*
7. *Failure to protect the gas and oxygen hoses from mechanical damage,* or damage from the the flying sparks, slag and hot metal resulting from the operation.
8. *Failure to provide a shutoff valve* for gases outside a confined space.
9. *Failure to remove hoses* from confined spaces when the torches have been disconnected.

Safety Measures

Equipment and Personnel. Welding and burning equipment should be of an approved type and in good repair. Oxygen and gas cylinders are equipped with regulators that prevent excess pressures and provide for proper mixing of oxygen and gas. These regulators should, therefore, be handled with extreme care. Only standard gas hoses (oxygen is green, and acetylene, red) and fittings should be used. Repairs should not be improvised (as in the use of tape to seal a leak in a line), and all gas line connections should be tight.

Because welding and burning are hazardous operations, only well-trained, qualified operators should be permitted to handle the equipment. Before allowing the work to begin, the master or his representative should ensure that the person using the equipment—either crewman or shoreside worker—has the proper knowledge and experience. Welding-permit laws and other local regulations can be of aid in determining whether an operator is qualified. In any case, it is important to remember that a safe welding or burning operation begins with a qualified operator and properly maintained equipment.

Federal Regulations. Welding and burning operations are well regulated, but regulations do not ensure safety unless they are complied with. Title 33 CFR 126.15(c), Subchapter L, states that:

> Oxyacetylene or similar welding or burning or other hot work including electric welding or the operation of equipment is prohibited on waterfront facilities or on vessels moored thereto, during the handling, storing, stowing, loading, discharging, or transporting of explosives. Such work may not be conducted on waterfront facilities or vessels moored thereto while either the facility or vessel is handling, storing, stowing, loading, discharging, or transporting dangerous cargo without the specific approval of the Captain of the Port.

Approval of the U.S. Coast Guard is granted by the issuance of a Welding and Hot-Work Permit. form CG-4201 (Fig. 1.15).

Additional safety and fire prevention requirements for welding and burning are included in Title 29 of the Safety and Health Regulations for Maritime Employment of the Occupational Safety and Health Administration (OSHA) of the Department of Labor, Section 1915.32. These regulations are primarily for the protection of personnel. However, they include many of the restrictions and precautions that are part of the U.S. Coast Guard Welding and Hot-Work Permit.

Local Regulations. Many port communities in the United States have strict regulations governing welding and burning. The requirements of some communities are more severe than federal regulations. In some cities, workmen must pass qualifying examinations before they are licensed to operate burning equipment. Therefore, in the interest of safety, it is a good idea to check local regulations before permitting shoreside workers aboard for welding or burning operations.

The issuance of a permit means only that the hot work may be performed, not that it will be done safely. For that, the operator and his assistants must comply with all the safety requirements that are part of the permit. Fire prevention procedures and common sense must be an integral part of every welding or burning operation. No welding or burning should be performed by shoreside workers or crew members without the knowledge and approval of the master or his representative, who should ensure that safety regulations are followed.

DEPARTMENT OF TRANSPORTATION U. S. COAST GUARD CG-4201 (Rev. 11-70)	**WELDING AND HOT-WORK PERMIT** *(Electric Welding, Oxyacetylene Welding, Burning and Other Hot-Work)*	PERMIT NUMBER
FROM CAPTAIN OF THE PORT, U.S. COAST GUARD	TO	

This permit, issued in accordance with 33 CFR 126.15(c), authorizes the below described welding, burning or other hot-work to be performed on the waterfront facility or on vessels moored thereto while dangerous cargo other than explosives is being handled, stored, stowed, loaded, or discharged there, subject to the requirements listed hereafter.

DESCRIPTION OF WORK	LOCATION	VALID FROM	VALID TO

PERSONNEL	A qualified operator shall be in charge. All persons using any welding or hot-work equipment shall be fully qualified in its use and associated safety procedures.
EQUIPMENT	All equipment shall be in good condition. Oxygen and acetylene cylinders shall be placed in an upright position and properly secured. Hose shall be free of leaks. In confined spaces mechanical ventilating equipment shall be supplied to exhaust fumes to the outer atmosphere. Each electric welding machine shall be properly grounded to prevent arcing. Welding machines driven by liquid fuel shall be equipped with drip pans and shall be fueled off the pier.
SCENE	All flooring in the area of operation shall be swept clean. Wooden planking shall be wet down. All combustible material shall be removed 30 feet from the work area or protected with an approved covering such as, asbestos, baffles, metal guards, or flameproof tarpaulins. The welding foreman and the pier superintendent or his authorized representative shall make a joint inspection of the area and adjoining areas before any hot-work is started. The pier superintendent or his authorized representative shall notify the welder of all fire hazards in the area.
FIRE PROTECTION	A competent fire watch shall be maintained and, if the hot-work is on the boundary of a compartment (i.e., bulkhead, wall or deck) an additional fire watch shall be stationed in the adjoining compartment. The fire watch shall have immediately available at least one UL approved fire extinguisher containing an agent appropriate to existing conditions. The minimum rating of the extinguisher shall be not less than 4A or 4B as defined by National Fire Code 10. A fire hose shall be available, already led out, and with pressure at the nozzle. Fire watches shall have no duties except to watch for the presence of fire and to prevent the development of hazardous conditions. The fire watch shall be maintained for at least one half hour after completion of the hot-work. Flammable vapors, liquids or solids must first be completely removed from any container, pipe or transfer line subject to hot-work. Tanks used for storage of volatile substances must be tested and certified gas free prior to starting hot-work. Proper safety precautions in relation to purging, inerting, or venting shall be followed for hot-work on containers.
GENERAL	All safe practices, local laws and ordinances shall be observed. See NATIONAL FIRE CODE 51B. In case of fire or other hazard, all equipment shall be completely secured.

ADDITIONAL SPECIAL REQUIREMENTS

DATE	ISSUED BY

I acknowledge receipt of this welding and hot-work permit and agree to comply and to require my employees to comply with its requirements. I understand that failure to comply with these requirements will result in cancellation of the permit and may subject the responsible person to the penalties prescribed by Section 2, Title II of the Act of June 15, 1917 as amended (50 USC 192). I FURTHER UNDERSTAND THAT THIS PERMIT IS NOT VALID AND THAT NO WELDING, BURNING, OR OTHER HOT-WORK MAY BE UNDERTAKEN WHEN ANY EXPLOSIVES ARE BEING HANDLED, LOADED, DISCHARGED, OR STORED ON THE WATERFRONT FACILITY OR ON VESSELS MOORED THERETO.

DATE	TITLE	SIGNATURE

PREVIOUS EDITION MAY BE USED

GPO 947-574

Figure 1.15. U.S. Coast Guard Welding and Hot-Work Permit, form CG-4201.

SHORESIDE WORKERS ABOARD FOR CARGO MOVEMENT, REPAIR AND MAINTENANCE

Generally, shoreside personnel do not have as much concern for, or interest in, the vessel as do members of the crew. This is perhaps understandable, because many shoreside workers do not fully realize the dangers involved in a fire at sea, but it is not excusable. Their indifferent attitude and lack of interest in fire prevention measures can result in shipboard fires. This must be compensated for by extremely close supervision and extraordinary alertness on the part of the crew.

Cargo Movement

Because of the frequency with which they come aboard, the nature of their duties, their access to the ship's holds and the materials they handle, longshoremen require the closest supervision. The hazards involved in cargo handling have already been discussed, but they are important enough to be repeated here:

1. Careless and illegal smoking in the hold or on deck during loading and unloading.
2. Careless discarding of butts and matches.
3. Careless handling of cargo and the loading of damaged cargo.
4. Improper stowage of cargo so as to cause shifting under rough sea conditions. This is particularly dangerous if two types of cargo are incompatible and can ignite spontaneously when mixed.

Repairs and Maintenance

Contractors who come aboard to do repair work, particularly welding and burning, also require close and intensive supervision. The safety measures discussed in the previous section should be followed, and crew members should be assigned to watch for and report any of the unsafe practices listed there.

Any repair contractors or individuals who come aboard should be suspect, wherever their work is to be done. A member of the crew should be assigned to accompany every work party. The following are among the general safety precautions the crew should take.

1. Monitor the observance of No Smoking regulations.
2. Thoroughly test any machinery or equipment that has been worked on by a contractor. Improperly repaired equipment, particularly electrical equipment, can be a source of later trouble.
3. Check handheld power tools for the proper type of grounding plug and for frayed wiring.
4. Ensure that the work area is free of combustible rubbish and waste material when the job is completed.
5. If any fixed fire extinguishing system has been repaired check that the repairs were done properly.

Shoreside personnel who are working aboard should never be left unsupervised. Again we note that it is the ship's crew that is endangered by a fire at sea. Therefore, it is the ship's crew that must assume the responsibility for seeing that fire prevention procedures are followed by shipboard workers.

SHIPYARD OPERATIONS

The hazards of shipyard operations are closely related to the hazards of repair operations performed by shoreside workers, but on a much larger scale. A vessel is normally placed in a shipyard for major repairs, refitting or conversion—operations beyond the capability of the crew. Thus, the ship may be swarmed with shoreside workers, whose poor housekeeping habits and indifferent attitude can contribute to the fire hazards. In addition, shipyard work usually implies that:

- Welding and burning operations are being performed throughout the ship.
- Fire detection and extinguishing systems may be temporarily shut down for modification or to allow other repair operations.
- Very few crew members remain on board to monitor the observance of safety precautions by workers.

Coast Guard and OSHA Regulations

All this adds up to a different situation in terms of fire prevention, both in the shipyard and at sea. However, the picture is not completely bleak. United States Coast Guard and OSHA regulations provide some protection during shipyard operations.

The U.S. Coast Guard regulations are aimed at ensuring the safety of the ship, as far as is possible. They require that the Coast Guard also be notified before anyone makes any repairs that affect the ship's safety, and drawings must be submitted before any alteration work is begun. No exception is made for emergency repairs, even if they are performed in a foreign shipyard.

The OSHA regulations are primarily for the protection of employees. Fire safety is only one part of the overall safety picture, but an important one. The regulations include safety requirements for shipyards and personnel engaged in ship repairs. They make excellent reference material for officers.

Hazardous Practices

As mentioned earlier, the existence of regulations does not ensure that they will be followed. Through carelessness, indifference, lack of knowledge, oversight or deliberate violation of the regulations, hazardous conditions can be created at shipyards. Among the practices that can lead to such conditions are:

1. Drydocking the vessel or making major alterations without prior U.S. Coast Guard approval, or not requesting an inspection following repairs or alterations.
2. Installation of unapproved or substandard equipment, not designed for use aboard ship.
3. Improper or poor workmanship on bulkheads and decking, which destroys their resistance to fire.
4. Concealing poor repairs to tanks, bulkheads and so forth by conducting inadequate pressure tests.
5. Failure to complete repairs on the firemain system or CO_2 system before the vessel leaves the shipyard.
6. Failure to replace watertight doors following repairs, or making openings in bulkheads, in violation of fire safety standards.
7. Failure to take the proper precautions to free tanks and piping of flammable gas before welding or burning.
8. Dismantling fuel pipes that are under pressure.
9. Improper electrical wiring practices, such as
   a. Using wire of a gauge insufficient to carry the intended load.
   b. Bypassing overload protection.
   c. Running wires through bilges or other areas in the vicinity of water piping.

If this list makes you feel uncomfortable, then it has served at least part of its purpose. The crew's only defense against such practices is vigilance—while the work is being done, immediately after it is completed, and for as long as the ship is in service.

TANKER LOADING AND DISCHARGING OPERATIONS

Each year billions of gallons of flammable and combustible liquid cargo move into our ports and on the waters of the United States. They represent the most important commodity moved by vessel. For example, no other cargo has as much financial impact or is so vital to our national economy as petroleum products. Yet, these products are extremely hazardous to transport and transfer.

Responsibilities

The movement of combustible or flammable cargo from ship to shore, shore to ship, or ship to ship is an awesome responsibility. Carelessness, neglect, inattention to duties, poor equipment or violation of the regulations can have dire consequences. Tanker accidents have led to the destruction of vessels and the loss of lives; the resulting fires and explosions have been of such severity that shoreside installations have also been seriously affected. The licensed officer or tankerman as the person-in-charge must, therefore, know his duties and responsibilities and discharge them to the letter.

The rules and regulations governing the operation of tank vessels are contained in Title 46 CFR, Parts 30–40 inclusive, Subchapter D, Chapter I. For convenience, they have been extracted and published in Coast Guard manual CG-123, which also includes Parts 154–156 of Title 33 CFR (Pollution Prevention Regulations). Other excellent sources of information (required reading for tanker officers) are listed in the bibliography at the end of this chapter. (In reading this material, it is important to note that the terms "flammable" and "inflammable" are used interchangeably in the U.S. Coast Guard regulations.)

Title 33 CFR 156.150 requires that the persons-in-charge jointly and independently inspect both the vessel and the shoreside facility before any combustible or flammable liquid or other hazardous products are transferred. This is a very formal inspection during which a form containing 22 items must be completed and signed by both parties.

Hazardous Liquids

Over 500 types of flammable and combustible liquids and liquefied gases are carried as cargo on inland and ocean-going vessels. The regulations define these products as follows.

Combustible Liquid: Any liquid having a flash point above 26.7°C (80°F). There are two grades within this category: *1)* Grade D—any combustible liquid having a flash point above 26.7°F (80°F) and below 65.5°C (150°F), and *2)* Grade E—any combustible liquid having a flash point of 65.6°C (150°F) or above.

Flammable Liquids: Any liquid that gives off flammable vapors at or below a temperature of 26.7°C (80°F) as determined by flash point with an open-cup tester, used for testing burning oils. There are three grades within this category: *1)* Grade A—any flammable liquid having a Reid vapor pressure* of 96.5 kilopascals (14 psia) or more; *2)* Grade B—any flammable liquid having a Reid vapor pressure above 58.6 kilopascals (8.5 psia) and below 96.5 kilopascals (14 psia); and *3)* Grade C—any flammable liquid having a Reid vapor pressure of 58.6 kilopascals (8.5 psia) or less and a flash point of 26.6°C (80°F) or below.

Causes of Tanker Fires

The following errors or omissions could result in fire and explosion during the movement of combustible and flammable liquids.

Improper Fendering. Improper or inadequate fendering can generate sparks. This is particularly true during vessel-to-vessel operations. Since the vapors given off by petroleum products are heavier than air, they tend to drift down to the water. There they can be ignited by sparks caused by metal-to-metal contact.

Lack of Coordination During Transfers. Every transfer should be well planned, with close coordination throughout the operation. No transfer operation can begin without a person-in-charge at each end of the operation. Emergency shutdowns and the means of communication between persons-in-charge must be tested and found in order before the transfer is started. The person-in-charge on the vessel must be able to shut down the flow or request shutdown through a communication system that is used for no other purpose. Emergency shutdowns must be provided on the vessel.

Transfer systems are only as effective as the people who are charged with the responsibility for using them. Even a momentary lapse can permit an overflow with resultant spill on the vessel, at the terminal, in the water or at all three locations.

Cargo Expansion. Another cause of overflows is failure to allow for expansion of the product caused by temperature increases. There are tables that can be checked for the proper fill levels, when a vessel is headed for a warmer climate. These tables should be consulted to ensure that tanks are not overfilled.

Pump Room Hazards. Because it is subject to vapor accumulation, the pump room is the most hazardous area on a tank vessel. To ensure that vapors are removed during loading or unloading, the vent systems in pump rooms should be operated continuously. As a safety precaution, be sure the vent system is working before entering a pump room. No repair work should be permitted in the pump room unless absolutely necessary. In fact, proper maintenance will help to avoid both repairs and vapor accumulation through leaks in piping and pump seals. Any piece of equipment that might cause a spark should, of course, be prohibited, because of the possibility of ignition of vapor accumulations. This includes spark-producing tools, unapproved electrical equipment, and flashlights. Bilge areas should be well maintained and kept free of flammable materials. Smoking in the pump room would be inviting disaster.

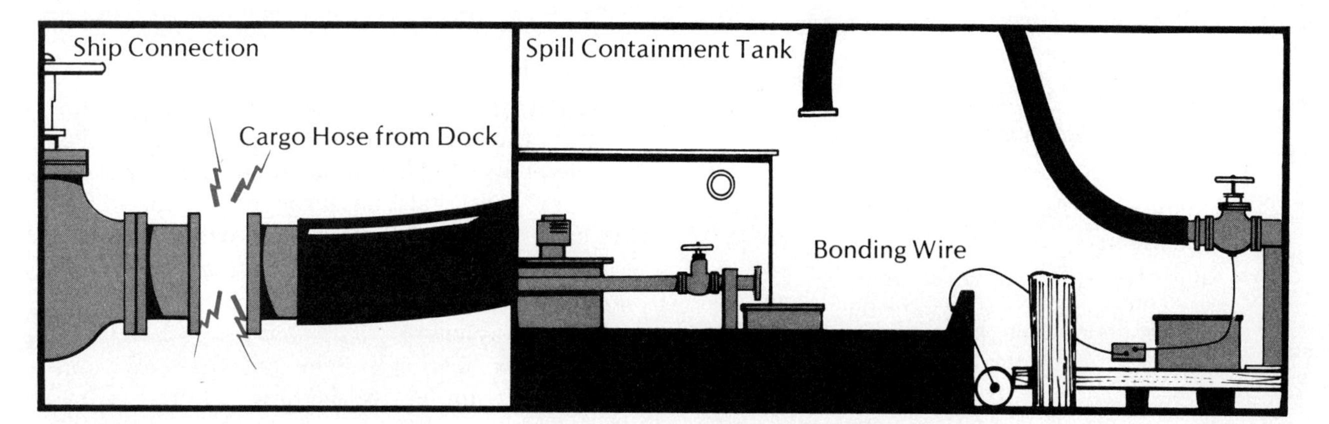

Figure 1.16. An electrical bond between the vessel and the shoreside facility *(right)* prevents sparks caused by static electricity *(left)*. The bonding should be completed before the cargo hose is connected to the shore connector.

*The Reid vapor pressure is a measure of the volatility, or tendency to vaporize, of a liquid. A small amount of the liquid is placed in a container that has a pressure gauge. The container is closed tightly, the liquid is heated to 37.8°C (100°F), and the pressure in the vapor above the liquid is read on the gauge.

Static Electricity. Static electricity is not an obvious cause of fire, but it is dangerous. Thus, precautions must be taken to prevent the generation of static sparks. During transfers, the usual method is to provide an electrical bond between the vessel and the shoreside facility (Fig. 1.16). This can be done in several ways, and the persons-in-charge are responsible for ensuring that it is done properly.

Certain cargos such as kerosene jet fuels and distillate oils can generate static electricity as they are moved. Water suspended in these cargos increases the possibility of static spark generation. To reduce the hazard, the operation should be started with a low loading rate. This permits the water to settle to the bottom of a tank more easily. The use of steel ullage tapes, metal sampling cans and metal sounding rods should be avoided when these static-producing cargos are being loaded. Only nonconductive devices should be used until the tanks have been topped off for at least 30 minutes. This waiting time permits suspended water to settle and static electricity to dissipate.

Oil that is splashed or sprayed may become electrostatically charged. For this reason, oil should never be loaded into a tank through an open hose. Oil splashing about the opening could generate static electricity.

Open Flames or Sparks. Ignition of flammable vapors by an open flame or a spark is the most obvious fire hazard during transfer operations. Some sources of sparks and flames are:

- Smoking and matches
- Boiler and galley fires
- Ship's radio equipment
- Welding and burning
- Machinery operation
- Electrical equipment in living quarters
- Unapproved flashlights or portable electrical equipment
- Sharp abrasion of ferrous metal.

The ship's person-in-charge is responsible for ensuring the safety of the transfer operation. This responsibility extends to areas around the ship, as well as to shore installations and other vessels. It includes the posting of signs indicating when radio equipment and boiler and galley fires may operate; the control of, and granting of, permission for hot work and other repair work; the securing of ventilation and air-conditioning intakes; and the securing of doors and ports on or facing cargo-tank areas. The entire crew is required to cooperate with the person-in-charge during transfer procedures.

Improper Use of Cargo Hose. Either the vessel or the shoreside facility may supply the hose used in a transfer. However, both persons-in-charge must inspect it to ensure its quality and stability. The importance of this inspection is obvious. If the hose is in good condition, the following precautions should be taken to prevent its rupturing during the transfer.

1. Position the hose so it cannot be pinched between the vessel and the dock.
2. Allow sufficient slack for tide conditions and lightening of the load.
3. Do not place the hose near a hot surface.
4. Support the hose properly, to prevent chafing.
5. Inspect the hose for leaks frequently during the transfer, and be prepared to shut down if necessary.

Vessel-to-Vessel Transfers

When vessel-to-vessel transfers are under way, several additional precautions must be taken.

1. Adequate fendering must be provided.
2. Changes in weather, sea and current conditions must be anticipated.
3. There must be a clear understanding as to which vessel is in charge of the operation.
4. The effect of drifting vapors on *both* vessels must be considered.

Cargo Heating System

High-viscosity cargos become so thick at low temperatures that heating is required before they can be pumped. The liquids are heated by steam pipes or coils that run through the bottoms of the tanks. The temperature to which they are heated is critical. Overheating can be hazardous, since dangerous flammable gases can be generated and released.

The tank heating system must be well maintained. A steam leak at the tank bottom can lead to the same problem as overheating—chemical reactions and the production of dangerous flammable gases. The cargo could also leak into the steam coils, with equally dangerous results. Title 46 CFR limits the heating of fuel oil in storage tanks to a maximum of 48.9°C (120°F).

COLLISIONS

Fires caused by collisions, particularly when tankers were involved, have resulted in serious damage and great losses of property and life. Some of these incidents were beyond the capacity of the crew to prevent. However, many important lessons have been learned; as a result of these lessons, it is expected that the incidence of these casualties will be reduced in the future.

No incident aboard ship emphasizes the importance of training and organization more than a collision followed by fire. The crew is faced with multiple problems: control of the vessel, control of the fire, and institution of damage control procedures after determination of the most immediate danger to the vessel. Ship's officers must ensure that all crew members know their duties in accordance with the station bill and know how to perform these duties.

If the fire cannot be controlled but stability is not a problem, it may be possible to take refuge aboard the ship, especially if assistance is not too far away. Previous drills and training will reduce the hazards of abandoning ship, if that procedure becomes necessary.

BIBLIOGRAPHY

CG–174, A Manual for the Safe Handling of Inflammable and Combustible Liquids and Other Hazardous Products

CG–329, Fire Fighting Manual for Tank Vessels

CG–388, Chemical Data Guide for Bulk Shipment by Water

CG–446–1, Chemical Hazards Response Information System (CHRIS), A condensed guide to chemical hazards

Fire Prevention Programs

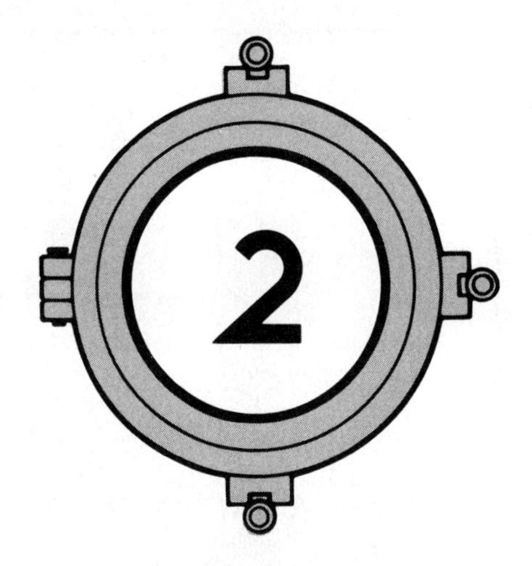

If most shipboard fires can be prevented, then who is responsible for preventing them? The answer is that fire prevention is the shared duty of *each and every member* of the crew—not just the master, or the chief engineer, or any particular individual or group of individuals. No fire prevention effort or program can be successful unless it involves everyone aboard ship.

Fire prevention is not easily defined, perhaps because it is primarily a matter of attitude, and its benefits are not easy to measure until after they are lost. For these reasons fire prevention is difficult to sell, and it requires continuing effort and strong guidance and leadership.

Every seaman probably fears the consequences of a serious fire at sea, but, unfortunately, awareness of the possibility of fire does not always lead to the attitudes and actions necessary to prevent it. Some individuals may be sensitive to the hazards of fire and to the means of preventing it. Others may be completely irresponsible, perhaps because of indifference; only good luck keeps these people from becoming victims of their own carelessness. Somewhere between these extremes is the majority who are in some respects very careful—in others, foolishly careless—perhaps from lack of knowledge.

Each member of the crew should analyze his own attitude toward safety, and toward fire prevention in particular. This may only require the answers to two simple questions: "Do I know the causes of shipboard fires?" "Have I considered the damage and loss of life that can result from a fire?" A carefully planned and conducted fire prevention program can help ensure that both questions are answered with a strong, unconditional "yes."

RESPONSIBILITY FOR THE PROGRAM
(WHAT WE OWE EACH OTHER)

We have noted that every crew member is responsible for the prevention of fire aboard ship. Similarly, every crew member has a role in the ship's fire prevention program. Because attitude is so much a part of fire prevention, it is also a most important part of the fire prevention program. To a great extent, the attitude of the crew will reflect that of the master.

The Responsibilities of the Master

The master of a vessel is responsible for developing the attitudes and cooperation required for the best operation of his ship. This responsibility obviously extends to fire safety. During formal meetings, informal discussions, casual conversations and training sessions, the master should convey his concern for fire prevention. There should be no doubt that he wants a fire-safe ship, and that he expects every member of the crew to assist in reaching this goal.

In most cases, the ship's safety committee will develop and implement the formal aspects of the fire prevention program. The master should participate in the management of this committee. He should contribute to its agenda and approve its programs. Most important, he must exhibit his continued interest, without which no program can be effective.

The Role of Supervisory Personnel

Department heads should take an active part in the work of the safety committee since they are responsible for the actions of the personnel under their supervision. Initially, they should evaluate

Figure 2.1. On-the-job training gives supervisory personnel the opportunity to teach safe practices. It ensures that crewmen get correct information, and it establishes an avenue of communication.

each subordinate for fire prevention attitude and level of training. This evaluation is especially important when there is a high turnover of crew personnel.

Daily on-the-job training and supervision will help develop good attitudes and habits in subordinates. These are probably the most effective means of communicating the details of the fire prevention program. Informal instructional sessions, given during the actual performance of routine duties, can be valuable learning experiences (Fig. 2.1). Unsafe actions may be corrected immediately, so that they do not become unsafe habits. Where needed, repeated corrections will reinforce the learning process.

Day-to-day training also provides an excellent opportunity for teaching (and learning) how regulations are developed to minimize fire hazards, and why they must be a part of the ship's fire prevention program. Often a crewman's duties will include the operation of equipment. Competence in its use and handling should be encouraged and checked, particularly as it relates to fire safety.

Supervisors should try to instill in subordinates a sense of pride in earning and maintaining a fire-safe record. This can develop unity among the crew and help bring a new crewman "aboard." At the same time, it tends to motivate individual crewmen to think in terms of fire safety and to incorporate this thinking in their actions.

It is important for department heads to keep up to date on the causes of recent fires aboard vessels. Then, if a crewman is observed doing something that once resulted in a fire, the supervisor can immediately call this fact to his attention. This is practical fire prevention at its best.

Responsibilities of Crewmen

The details of shipboard operation are the responsibility of the crew, and fire prevention duties are no exception. Every crewman is responsible for eliminating or reporting hazardous conditions in his quarters, his work area and wherever else he may find them. Crewmen are responsible for safely operating the ship's equipment. When a crewman is assigned to operate machinery with which he is not familiar, he should ask for instruction in its use; his inability to operate the equipment could cause an accident resulting in fire or injuries.

Perhaps the most important responsibility of crewmen in a fire prevention program is to develop and maintain the proper attitude. The crewmen who are interested and involved in the program will make it work. They will learn well and operate well, individually and as a crew, to prevent fires aboard their ship. The crewmen who are indifferent to the program will soon be wondering why so many fires "happen" on their ship.

ELEMENTS OF EFFECTIVE PROGRAMS

To be successful a fire prevention program must be carefully planned and structured. The details of the program should be tailored to the ship for which it is developed. Thus the fire prevention program for a tug would be much less formal than that for a tank vessel; but each program would reflect the master's concern for fire safety, be developed by the safety committee, be conducted by the master and department heads and receive the high priority that it merits.

On any vessel, the fire prevention program should include the following elements:

1. Formal and informal training
2. Periodic inspections
3. Preventive maintenance and repair
4. Recognition of effort.

These are discussed in some detail in the remainder of this chapter. First, however, it is important to emphasize that the fire prevention program itself should be the subject of continual review by the master and the safety committee. Both the scope and the conduct of the program should be modified as necessary to improve fire safety. That is, the safety committee cannot relax once they have developed and implemented a fire prevention program. They should question their program every time an unsafe situation is discovered, and extend the program to ensure that such a situation cannot occur again. To wait until a fire breaks out is to await disaster.

FORMAL AND INFORMAL TRAINING

The education of crew members may be difficult and, at times, frustrating, but it is a most important factor in any fire prevention program. It must be a continuing process that includes both formal training sessions and informal discussions. No opportunity should be missed and no effort spared to develop an awareness of fire safety. The objective of this training should be to teach every crew member to *think* fire prevention, before, during and after every action. Each crewman must ask himself, "Is it safe? Could it cause a fire?" This attitude toward fire prevention might be called "taking one second for safety."

Figure 2.2. Formal training sessions are the foundation of a fire prevention program.

Formal Training Sessions

Formal training sessions should be conducted on a regular basis during each voyage (Fig. 2.2). For the benefit of new crew members, it is essential that these training sessions be started as soon as possible. Until the first session can be held, department heads should convey the master's attitude toward fire safety to their subordinates.

The safety committee should plan and schedule the formal training sessions. (This book could serve as the basis for the fire prevention curriculum.) In addition, each vessel should build its own fire prevention and firefighting library, and crew members should be encouraged to use it. The library should be kept current to promote its use. Some of the publications listed in the bibliographies in this book would be excellent additions to such a library.

Training aids, films, slides and the new video tape cassettes (when available) should be used to add interest to the sessions. While repetition is sometimes necessary, it should be avoided whenever possible. People lose interest in (and pay little attention to) material that is presented over and over again in the same way. The sessions should vary as to topics, presentation and approach as much as possible. Practice sessions (equipment maintenance and inspections, for example) should be scheduled along with the required drills. They will help relieve the sameness of "sit-down" training sessions.

Schedules should be posted in advance. Sessions should be held at different times of day (for example, morning and afternoon) so that all watches can be accommodated. Total participation is just as important to these training sessions as it is to the overall fire prevention program.

Informal Training

Informal training can be a very effective teaching tool. When crewmen talk things over in a relaxed atmosphere, everyone gets a chance to speak and to listen. There is a free interchange of information and ideas (Fig. 2.3). This can lead to a better understanding of the responsibilities of crew members relative to their specific skills, the general safety of the ship and fire prevention in particular.

Visual reminders, posters, warning signs and personal messages to the crew can also be effective informal education media. Here again, it is important to vary the message and the media. The same posters or messages left in the same locations week after week indicate a lack of interest on the part of the program's planners. This lack of interest can easily become contagious.

Training Curriculum

The training should be focused primarily on the prevention of fires. A secondary goal should be to teach the crew how to isolate and then extinguish small fires. Toward these ends, the curriculum should include the following eight topics.

Theory of Fire. When they understand what fire is, crewman are better equipped to prevent it. (*See* Chapter 4 for a discussion on fire theory.)

Classes of Fires. The importance of this topic stems from the fact that different classes of fires (that is, different flammable materials) require different extinguishing agents. (*See* Chapter 5 for a discussion of classes of fires.)

Maintenance and Use of Portable Fire Extinguishers. Portable fire extinguishers can control a fairly large fire if they are used promptly and properly. Through training, crewmen should develop confidence in these appliances. They should check to see that fire extinguishers are in their proper places, in good condition and ready for use. Additionally, every crew member should be absolutely certain about the proper use of the different types of extinguishers. (*See* Chapter 7 for a discussion of portable fire extinguishers. *See* Chapter 8 for a discussion of extinguishing agents used in portable appliances.)

Figure 2.3. Informal discussions provide an opportunity for crew members to learn from each other, to stimulate interest in fire prevention, and help to establish the proper attitude toward safety.

Good Housekeeping. Basically this means *cleanliness.* However, from the fire prevention standpoint it means the elimination of sources of fuel for fires, that is, the elimination of fire "breeding grounds." (*See* Chapter 1 for a discussion of some potential fire hazards.) These and other housekeeping problem areas are listed below. Almost every one of them can be eliminated with a minimum of effort.

1. Cleaning rags and waste should be stored in covered metal containers.

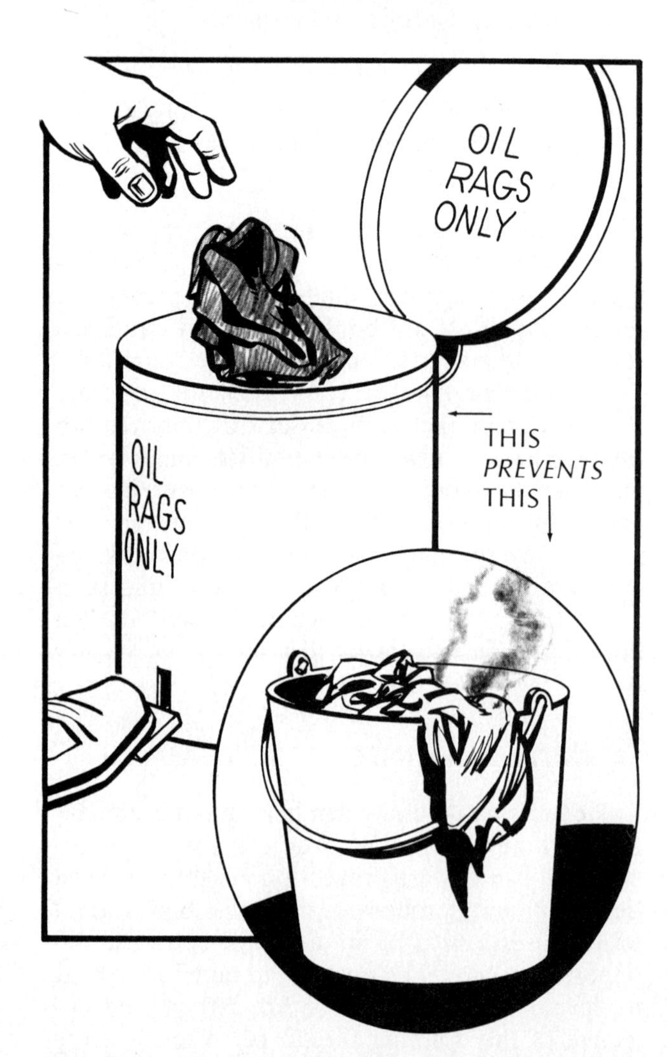

Figure 2.4. Oily rags should be placed in covered metal containers to prevent fires by spontaneous ignition.

2. Accumulations of oily rags should be placed in covered metal containers (Fig. 2.4) and discarded as soon as possible.
3. Accumulations of packaging materials should be disposed of immediately.
4. Dunnage should only be stored in the proper area.
5. Accumulations of sawdust (especially oil- or chemical-soaked sawdust), wood chips or shavings should be disposed of properly.
6. Accumulations of flammables in crew or passenger quarters should be avoided.
7. Oil-soaked clothing or other flammables should never be stored in crew lockers.
8. Paints, varnish and so forth should be stored in the paint locker when not in use—even overnight.
9. Leaks in product, fuel-oil or lubricating-oil piping and spilled oil or grease should be cleaned up; also oil in bilges or on tank tops and floor plates.
10. Kerosene and solvents should be stored in appropriate containers and in approved locations.
11. Oil-burner cleaning substances should not be left in open containers in the boiler room.
12. Oil-soaked clothing should not be worn by crew members.
13. .Grease filters and hoods over galley ranges should be cleaned regularly.
14. Avoid accumulations of dust in holds and on ledges in holds, and accumulations of lint and dust on light bulbs.
15. Avoid soot accumulations in boiler upstakes and air heaters.

Elimination and Control of Ignition Sources. The safety committee should be aware of the causes of recent fires on other vessels. "Proceedings of the Marine Safety Council," published by the U.S. Coast Guard, maritime-oriented publications and newspapers are excellent sources of such information.) Discussions of actual ship fires have the most impact and help crewmen to realize that fires still can and do occur aboard vessels (*see* Chapter 3).

As was pointed out, cleanliness can eliminate sources of shipboard fires. Good training, a good attitude and alertness can assist immeasurably in eliminating another necessary ingredient of fires, namely, the source of heat or ignition. (The major sources of ignition aboard ship are discussed in Chapter 1.) These can be eliminated by:

Figure 2.5. "Smokes" that are discarded carefully cannot become sources of ignition.

1. Not smoking in restricted areas; discarding ashes, butts and matches carefully (Fig. 2.5); using only saftey matches on tank vessels; closely observing longshoremen working in holds
2. Not overloading electrical circuits; protecting circuits with the proper fuses or circuit breakers; proper maintenance and repair of electrical equipment; following instructions and regulations for wiring (Fig. 2.6)

Figure 2.6. Inspection, maintenance and use of approved components reduce the possibility of fire in and around electrical equipment.

3. Keeping flammable materials clear of steam pipes, light bulbs and other sources of ignition.
4. Thoroughly cleaning cargo holds before any cargo is loaded. (Otherwise, there is a possibility of mixing incompatible cargos such as vegetable oil and fibers and causing a fire through spontaneous ignition.)
5. Careful loading operations (particularly the loading of baled fibers), so that bales do not strike coamings, machinery or other steel structures; care by longshoremen not to strike bands with their hooks (Fig. 2.7)
6. Removing cargo lights from holds when loading is completed; replacing receptacle watertight caps after portable lights are unplugged
7. Observing all precautions when welding or burning—including the posting of a fire watch—or seeing that shoreside workers do so. (Welding and burning are among the most hazardous operations performed aboard ship.)
8. Eliminating the causes of static electricity. (This is extremely important on tank vessels, especially when butterworthing.)
9. Awareness of the possibilities of spontaneous ignition, and how to avoid it. (Again, this is basically good housekeeping.)
10. Using approved flashlights and portable lights and nonsparking tools on tank vessels
11. Not using electric tools where a fire hazard may exist, and using only tools in good condition
12. Following the instructions of the senior deck officer on tank vessels when loading or discharging cargo, especially regarding smoking, boiler and galley fires and other possible sources of ignition; proper bonding of the vessel; ensuring the integrity of cargo hose and couplings
13. Continually observing cargo pumps during transfer operations. (Loss of suction or prolonged operation when tanks are empty may overheat the pump and result in explosion and fire.)

Preparation for Emergencies. Except for the knowledge and experience gained in actually fighting a fire, no training is as effective as live, well-conducted fire drills. The experience gained through drills can help prevent a major tragedy. This same experience can reduce the possibility of injuries during an actual emergency, and equipment deficiencies that show up during drills can be corrected before they become problems (*see also* Chapter 10).

The station bill is very important to the conduct of fire drills. The master and department heads

Figure 2.7. Cargo that is damaged during loading can leak and cause a fire several days later.

should ensure that it is up to date. All crew members should be aware of and familiar with their duties and responsibilities.

Fire drills should be conducted weekly (at least), and at irregular intervals to avoid expectations. Fire conditions should be staged in different parts of the vessel to add interest and create challenges. Each drill should begin with the sounding of the alarm and end with a constructive discussion and analysis.

Respiratory Protection Devices. The proper use of respiratory protection devices is a most important part of the rescue and firefighting education of every crewman. Masks are designed for different purposes, and each has certain limitations. It is important that the proper mask be chosen for the task to be performed. Manufacturers' instructions make excellent guidelines (*see also* Chapter 15).

Crewmen should practice donning masks; facepieces in particular must fit properly. Constant practice and training are required to develop proficiency, and breathing with a mask in place will develop confidence in its use. However, overdependence on a mask can be dangerous and can jeopardize the wearer. Close supervision and lifelines are essential for safety during operations in which respiratory protection devices are used.

Knowledge of Cargo. The crew should be familiar with the types of cargo carried on their ship. The crew is the ship's firefighting force, and knowledge of potential fire hazards is important information. An item-by-item review of the cargo manifest will provide the crew with information on the amounts of each cargo on board. To acquaint the crew with particular characteristics of hazardous cargos, they should refer to "Chemical Data Guide for Bulk Shipment by Water," CG-388. They should also review classifications of fires and the types of extinguishing agents they require.

Comparison of the flash points, ignition temperatures and explosive ranges of dangerous liquids and gases is of help in understanding their relative hazards. The *Hazardous Materials Regulations* of the U.S. Coast Guard are the best source of such information. They are contained in Title 49 CFR (Transportation); Parts 100–199. Where necessary, the information in those regulations can be supplemented with data supplied by the National Fire Protection Association (NFPA).

Title 49 CFR 172.101 contains a list of the materials classed as hazardous by the U.S. Coast Guard. It is important that these substances be easily recognized when they are being transported. For this reason, they are marked with distinctive labels to indicate their particular hazard.

These hazardous material warning labels (Fig. 2.8) are authorized by the U.S. Department of Transportation (DOT). United States Coast Guard regulations require that they be placed on the outside of every container in which a hazardous material is to be transported by ship. Placards similar to the labels are required for trailer-type shipments. In addition, hazardous material *class numbers* are required by some foreign governments. These numbers are also referred to as UN class numbers and are endorsed by the Inter-Governmental Maritime Consultative Organization (IMCO). The class number is located at the bottom corner of the DOT label. Most of the hazards indicated by the labels in Figure 2.8 are obvious. Two that are less obvious are: *1) oxidizer:* A substance that gives off oxygen readily to aid in the combustion of organic matter; and *2) organic peroxide:* a flammable solid or liquid that will increase the intensity of the fire. Many peroxides can be broken down by heat, shock or friction. They are widely used in the chemical and drug industries.

The labels are, of course, visible during loading. However, during a fire, they may be obscured by smoke or destroyed by flames. Here again, the ship's dangerous cargo manifest is extremely important. It is the only positive indicator of the type of materials involved in the fire.

PERIODIC INSPECTIONS

Inspection is one of the most important parts of the shipboard fire prevention program. Its purpose is to find and eliminate fuels and ignition sources that could cause fires. A number of these possible fire causes were listed earlier in this chapter. As noted, the elimination of these sources is not a technical matter, but mainly common sense and "good housekeeping."

Because vessels are large and complex, the responsibility for inspection cannot rest with any individual or group of individuals. Instead, every crewman should be an informal inspector, checking for fire hazards at all times, on and off duty, wherever he may be on the ship. This is a matter of attitude, and an extension of the idea of "one second for safety."

In addition, the master, chief officer and chief engineer should make a joint formal inspection of the entire vessel at least once each week. This should be a complete inspection, from bow to stern and bilge to bridge. The formal inspection should be systematic; a checklist should be used

SPONTANEOUSLY COMBUSTIBLE MATERIAL

UN CLASS 4

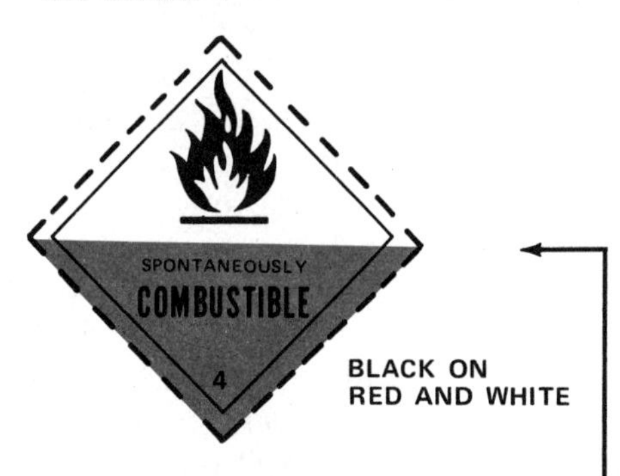

WATER-REACTIVE MATERIAL

UN CLASS 4

NOTE: May be used in addition to other required labels.

DANGEROUS WHEN WET

4

BLACK ON BLUE

CORROSIVE MATERIAL

UN CLASS 8

RADIOACTIVE MATERIALS

UN CLASS 7

POISONOUS MATERIAL

UN CLASS 2 or 6

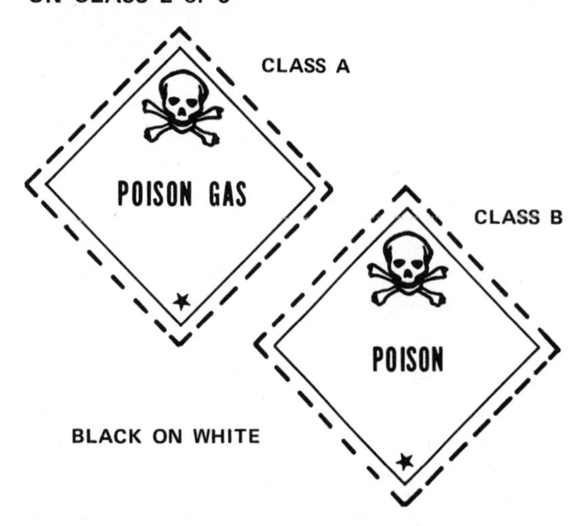

ETIOLOGIC AGENT

Required for domestic shipments including the domestic portion of import and export movements.

Note: A Poison Label may be used on import/export shipments in addition to this label.

RED ON WHITE

Figure 2.8. Hazardous material warning labels. Note the UN class number on each label.

Hazardous Materials Warning Labels

EXPLOSIVES

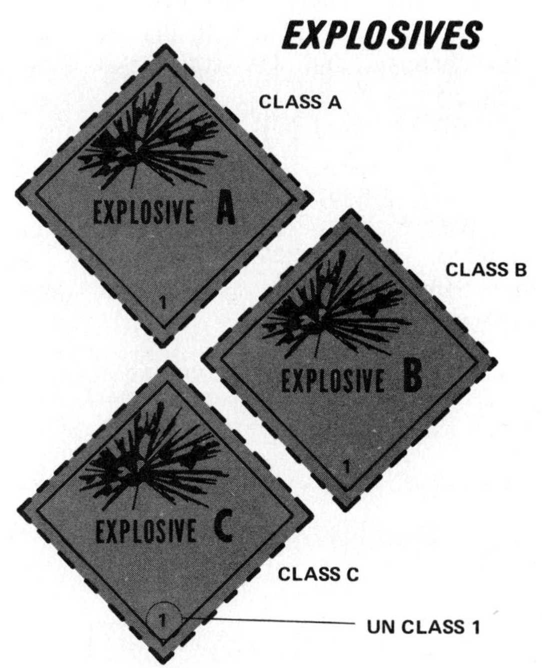

BLACK ON ORANGE BACKGROUND

COMPRESSED GASES

UN CLASS 2

OXIDIZING MATERIAL

UN CLASS 5

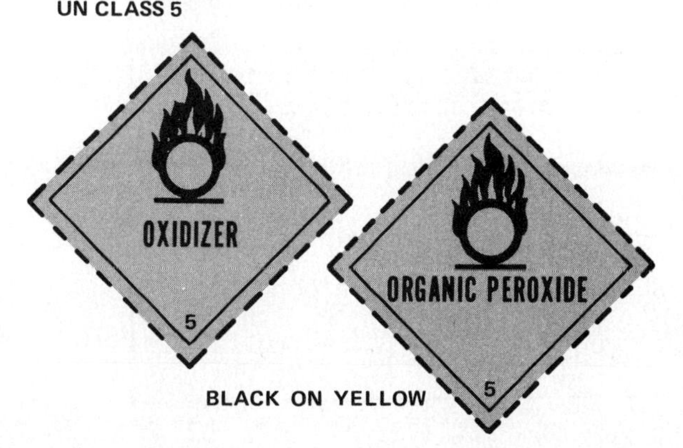

IRRITATING MATERIAL

UN CLASS 6

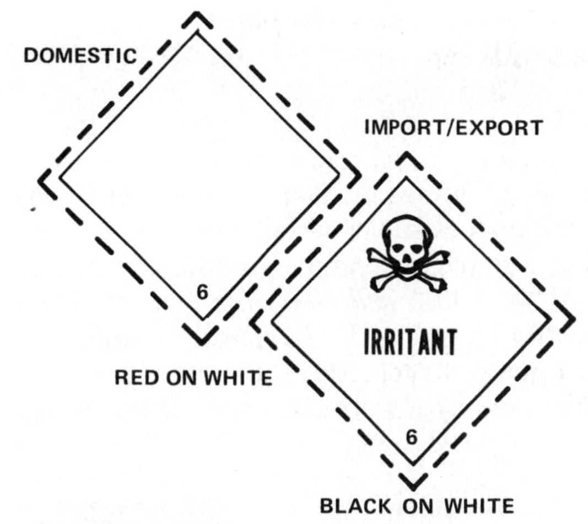

BLACK ON WHITE

FLAMMABLE LIQUID

UN CLASS 3

BLACK ON RED

FLAMMABLE SOLID

UN CLASS 4

EMPTY

EMPTY

BLACK ON WHITE

to assure that no area is overlooked. A sample checklist, included at the end of this chapter, can be used as a guide for informal inspections as well.

Requirements Prior to Repairs and Alterations

United States Coast Guard regulations, in Title 46 CFR (Shipping) require inspections before riveting, welding, burning or such fire-producing operations are undertaken in certain portions of a vessel. They also require that the provisions of NFPA standard No. 306, *Standard for the Control of Gas Hazards on Vessels to Be Repaired,* be used as a certified guide. Regulations state that the inspections be made by a marine chemist. If one is not available, then consideration will be given to other persons. The marine chemist makes a crucial judgment, based on his findings, as to whether or not burning, welding and other hot work may be performed. The master, chief officer and chief engineer should be familiar with the services of marine chemists and the certificates they issue (Fig. 2.9).

NFPA standard No. 306 requires the use of special designations to describe the conditions found during marine chemists' inspections. These are "Safe for Men" or "Not Safe for Men" and "Safe for Fire" or "Not Safe for Fire." Briefly, the "Safe for Men" conditions are defined as follows:

1. The compartment contains at least 18% oxygen by volume.

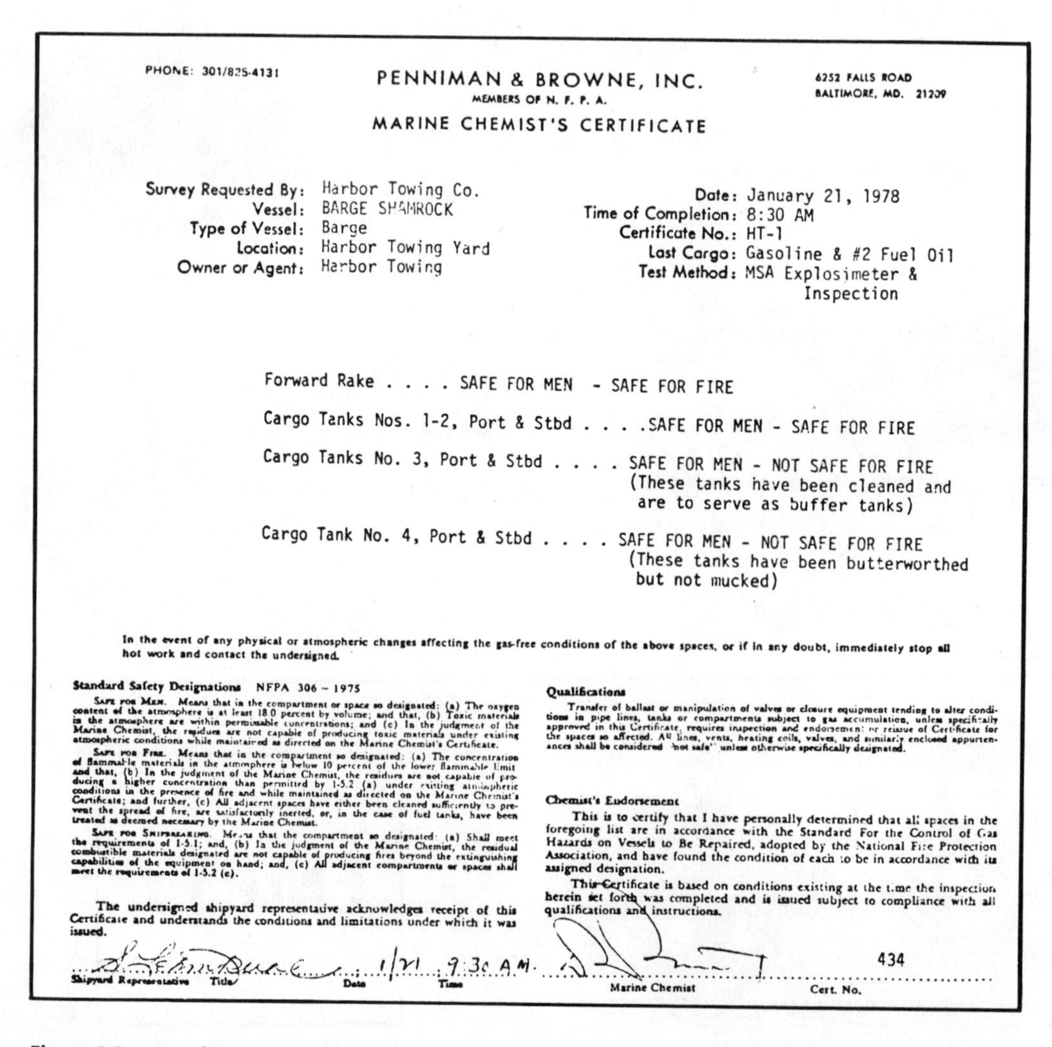
PHONE: 301/825-4131

PENNIMAN & BROWNE, INC.
MEMBERS OF N. F. P. A.

6252 FALLS ROAD
BALTIMORE, MD. 21209

MARINE CHEMIST'S CERTIFICATE

Survey Requested By: Harbor Towing Co.
Vessel: BARGE SHAMROCK
Type of Vessel: Barge
Location: Harbor Towing Yard
Owner or Agent: Harbor Towing

Date: January 21, 1978
Time of Completion: 8:30 AM
Certificate No.: HT-1
Last Cargo: Gasoline & #2 Fuel Oil
Test Method: MSA Explosimeter & Inspection

Forward Rake . . . . SAFE FOR MEN - SAFE FOR FIRE

Cargo Tanks Nos. 1-2, Port & Stbd . . . .SAFE FOR MEN - SAFE FOR FIRE

Cargo Tanks No. 3, Port & Stbd . . . . SAFE FOR MEN - NOT SAFE FOR FIRE
(These tanks have been cleaned and are to serve as buffer tanks)

Cargo Tank No. 4, Port & Stbd . . . . SAFE FOR MEN - NOT SAFE FOR FIRE
(These tanks have been butterworthed but not mucked)

In the event of any physical or atmospheric changes affecting the gas-free conditions of the above spaces, or if in any doubt, immediately stop all hot work and contact the undersigned.

Standard Safety Designations NFPA 306 – 1975

SAFE FOR MEN. Means that in the compartment or space so designated: (a) The oxygen content of the atmosphere is at least 18.0 percent by volume; and that, (b) Toxic materials in the atmosphere are within permissable concentrations; and (c) In the judgment of the Marine Chemist, the residues are not capable of producing toxic materials under existing atmospheric conditions while maintained as directed on the Marine Chemist's Certificate.

SAFE FOR FIRE. Means that in the compartment so designated: (a) The concentration of flammable materials in the atmosphere is below 10 percent of the lower flammable limit and that, (b) In the judgment of the Marine Chemist, the residues are not capable of producing a higher concentration than permitted by 1-5.2 (a) under existing atmospheric conditions in the presence of fire and while maintained as directed on the Marine Chemist's Certificate; and further, (c) All adjacent spaces have either been cleaned sufficiently to prevent the spread of fire, are satisfactorily inerted, or, in the case of fuel tanks, have been treated as deemed necessary by the Marine Chemist.

SAFE FOR SHIPBREAKING. Means that the compartment so designated: (a) Shall meet the requirements of 1-5.1; and, (b) In the judgment of the Marine Chemist, the residual combustible materials designated are not capable of producing fires beyond the extinguishing capabilities of the equipment on hand; and, (c) All adjacent compartments or spaces shall meet the requirements of 1-5.2 (c).

The undersigned shipyard representative acknowledges receipt of this Certificate and understands the conditions and limitations under which it was issued.

1/21 ; 9:30 A.M.
Shipyard Representative Title Date Time

Qualifications

Transfer of ballast or manipulation of valves or closure equipment tending to alter conditions in pipe lines, tanks or compartments subject to gas accumulation, unless specifically approved in this Certificate, requires inspection and endorsement or reissue of Certificate for the spaces so affected. All lines, vents, heating coils, valves, and similarly enclosed appurtenances shall be considered "not safe" unless otherwise specifically designated.

Chemist's Endorsement

This is to certify that I have personally determined that all spaces in the foregoing list are in accordance with the Standard For the Control of Gas Hazards on Vessels to Be Repaired, adopted by the National Fire Protection Association, and have found the condition of each to be in accordance with its assigned designation.

This Certificate is based on conditions existing at the time the inspection herein set forth was completed and is issued subject to compliance with all qualifications and instructions.

434
Marine Chemist Cert. No.

Figure 2.9. Certificate issued by marine chemists after an inspection according to NFPA standard No. 306.

2. Toxic materials are within permissible concentrations. The permissible limits are given in the current Table of Threshold Limit Values of the American Conference of Governmental Industrial Hygienists.
3. If the area is maintained as directed, residues will not produce toxic materials.

"Safe for Fire" conditions are defined as:

1. The concentration of flammable gas in the compartment is less than 10% of the lower flammable limit. (The lower flammable limit of a gas is the minimum flammable concentration of that gas in air.)
2. If the area is maintained as directed, the the residues are not capable of producing concentrations greater than 10% of the lower flammable limit.
3. Adjacent spaces have been properly cleaned or made inactive to prevent the spread of fire.

The meanings of the "Not Safe" categories are obvious from these definitions.

> Regulations provide that, if a marine chemist is not available, The Officer in Charge, Marine Inspection Office, upon the recommendation of the vessel owner and his contractor or their representative, shall select a person who, in the case of an individual vessel, shall be authorized to make the inspection.

If this authorized person is not available, the inspection is made by the senior officer present. For this purpose, several types of testing instruments are carried on board.

Combustible-Gas Indicator. Under Title 46 CFR 35.30-15, U.S. flag vessels (manned tank barges and tank ships) authorized to carry flammable and combustible liquid cargos at any temperature are required to carry combustible-gas indicators (*See* Chapter 1 for a description of these hazardous liquids.)

A combustible-gas indicator (sometimes referred to as an *explosimeter*) is used to determine whether there is a flammable atmosphere in any area of a vessel (*see* Chapter 16). The instrument should be a type that is approved by Underwriters Laboratories, Factory Mutual Engineering Division or some other agency acceptable to the U.S. Coast Guard. The manufacturer's operating and maintenance instructions should remain with the instrument. All officers should be thoroughly familiar with the instrument and know how to use it. The combustible-gas indicator is a safeguard against fire; if it is used improperly, an unsafe environment may appear to be safe. If it is not used at all, there is no way to know whether an environment is safe.

Flame Safety Lamp. United States Coast Guard regulations require that all cargo, miscellaneous and passenger vessels on international voyages carry flame safety lamps as part of their fireman's outfits.

The flame safety lamp is used to test only for oxygen deficiency. Its operating principle is simple: If there is enough oxygen in the surrounding atmosphere to keep the flame burning, then there is enough oxygen to support life. It is important to remember that this instrument contains a source of ignition and must be in good condition. Like the combustible-gas indicator, it must be used correctly, or it might give false results (*see* Chapter 16).

Oxygen Indicator. Some vessels carry oxygen indicators, although these instruments are not required by U.S. Coast Guard regulations. There are several different types, but all serve the same purpose as the flame safety lamp. That is, they are used to determine whether the atmosphere contains sufficient oxygen (15% or more) to sustain life. This is particularly important in spaces that have been closed for a long time, for example, cofferdams, deep tanks, double bottoms and chain lockers.

The oxygen indicator is preferred over the flame safety lamp. It has the advantage of being equipped with a meter so that the actual amount of oxygen in the atmosphere can be determined. The instrument can also be used to determine whether a "nominally inert" gas is free of oxygen, that is, contains less than 5% oxygen.

Like the combustible-gas indicator and the flame safety lamp, the oxygen indicator must be used and cared for properly. An incorrect reading on the meter can provide a false sense of security that may result in asphyxiation and death, or an incorrect reading can result in a fire if, for example, hot work is allowed in an atmosphere that is believed to be inert but actually isn't. Operators should carefully check the manufacturer's manual supplied with the instrument for proper operation and maintenance procedures.

PREVENTIVE MAINTENANCE AND REPAIR

The collision and resulting fire of the *SS C. V. Sea Witch* and *SS Esso Brussels* in New York harbor on June 2, 1973, was a major tragedy resulting in loss of life and in damages totaling approximately $23 million. In its marine casualty report released

March 2, 1976, the Department of Transportation listed the following two contributory causes:

> The modification to the differential gear mechanism stub shaft and connecting universal . . . on 23 April 1973, approximately six weeks prior to the collision, was improper. The milling of the stub shaft for the fitting of a square key to replace the originally designed captured or locked-in Woodruff key without a provision for securing the key allowed the new square key to slip out of position and permit free rotaton of the shaft.*

> The extensive loss of life of the crew on the *SS Esso Brussels* may not have occurred or may have been greatly reduced had there been no delay in releasing the lifeboat falls and had the hand cranked lifeboat engine immediately started.

On September 26, 1974, the *SS Transhuron* stranded at Kiltan Island, and the vessel had to be left for salvors. The stranding followed a loss of propulsion as the result of a fire in the engine room. The Department of Transportation, in its marine casualty report released December 30, 1976, stated:

> The loss of main propulsion power was due to a fire in the main propulsion control desk caused by the action of sea water directed onto high voltage components in the control circuitry. Contributing to the fire were:
> a. Failure of a pipe nipple in a gauge connection in the circulating water header of the freon condenser of the air conditioning unit.
> b. Wasting of the material of the pipe nipple due to the connection of dissimilar metals in a saltwater environment.

Firefighting efforts with semiportable CO_2 extinguishers were reported in part as follows: "The hose burst in way of its connection to the shut-off valve at the horn and the horn separated from its threaded connection at the valve and blew off." The report concludes: "Cause of the failure of the hose of the B-V semiportable CO_2 extinguisher is unknown."

The March 1977 issue of the *Proceedings of the Marine Safety Council* recounts the death of the chief engineer of a U.S. tanker, the *SS Thomas Q*. There was no fire, although all the ingredients were present. The tanker was undergoing ballasting after completing discharge of a cargo of naphtha. The pumpman entered the pump room against orders and was overcome. The chief engineer attempted a rescue and was also overcome by the naphtha. Both were finally removed, but it was too late for the chief engineer. Five causes contributed to the unfortunate incident, the first of which was equipment failure. According to the report, "The cargo pump seals were faulty and leaked naphtha."

These documented incidents clearly indicate the relationship between poor preventive maintenance and fire. Consideration of the possible losses in both human life and dollars leaves no doubt that preventive maintenance and repair programs can return dividends well beyond their cost.

Programs Require Supervision

Strong leadership and the backing of management are necessary ingredients of preventive maintenance programs. Information should be channeled from the master through department heads to the members of each department. A preventive maintenance and repair program is a form of discipline; to be effective, it must be carefully supervised and controlled.

It would be beyond the scope of this book to outline complete programs for preventive maintenance on typical vessels. (Such programs should already exist on all ships.) Instead, we shall discuss the basic elements of a program for the care of machinery and equipment, and its relation to fire prevention. A well-run program can become the first line of defense against fire.

Elements of a Preventive Maintenance Program

The four basic elements of a preventive maintenance program are *1)* lubrication and care, *2)* testing and inspection, *3)* repair or replacement, and *4)* record keeping.

The first three should be performed according to definite schedules that depend on the equipment in question. For example, some equipment might be serviced at various intervals during each watch. Other equipment might require maintenance once each watch, or daily or weekly, on up to annually or at even longer intervals. The manufacturer's manual is the best guide for establishing the schedules for periodic maintenance procedures.

This is by no means a new approach to preventive maintenance. However, it does imply that maintenance schedules must provide the answers to such questions as: What controls have been established to ensure that the schedule is being followed? Have provisions been made for turnover in both supervisory and other personnel? Many existing schedules have left such questions unanswered. Standardized maintenance schedules

*This refers to the *C.V. Sea Witch*.

are absolutely necessary, but they are effective only when they are implemented.

Lubrication and Care

Machinery and Equipment. Probably the most basic element in a preventive maintenance program for machinery and equipment is regular and proper lubrication. Scheduling alone is not sufficient to ensure this, because personnel may tend to neglect machinery that is difficult to reach. Controls must be instituted to ensure that manufacturers' lubrication recommendations are followed, and the lubrication schedule should be watched closely by supervisory personnel.

Machinery should be lubricated carefully to avoid spillage because most lubricants are flammable. A spark or other source of ignition could quickly transform some spilled lubricant into a fire problem.

Boilers and Appurtenances. Title 46 CFR (Shipping) requires that boilers, pressure vessels, piping and other machinery be inspected and tested at regular intervals. If emergency repairs are required between inspections, the nearest Officer in Charge of Marine Inspection, U.S. Coast Guard, must be notified of this fact as soon as possible after the repairs are completed.

Because they involve heat and high pressure, boilers and appurtenances require very special care. Perhaps no other type of equipment pays higher dividends for proper preventive maintenance. On the other hand, neglect of this equipment can result in poor operation, explosion and fire.

Chemical treatment and testing of water and fuel systems are recommended to reduce corrosion, prevent slag, scale and sludge buildups in boilers and protect diesel engines from fuel contaminants. Excessive corrosion can cause tube failure and an explosion. Then, if the fuel supply is not shut down, a fuel oil fire, which can be difficult to control, may follow. Fuel contaminants can eventually lead to inefficient burning with resultant soot buildup on tubes and in the stack. A heated piece of carbon could then ignite the soot.

Proper maintenance of burning equipment will prevent fuel oil from collecting in the furnace. Burner tips (atomizers) should be cleaned and the assembly adjusted periodically, for efficient operation and proper combustion. When not in use, a burner should *not* be left in place; it should be removed completely to prevent oil from dripping and collecting in the fire box. Any burner that has been shut off should be completely removed to preventing anyone from inadvertently attempting to light it.

As a precaution, the boiler should be inspected regularly when it is out of service. Oil accumulations can be an indication of a malfunction or a leak somewhere in the system.

Piping and Fittings. Piping and fittings that carry fuel, chemicals, flammable products, water or steam should not be abused or misused. They should not be used for handholds or footholds or for securing chain falls. The results of such misuse may not be evident immediately, but continued misuse can only weaken the equipment. It can lead to a slow leak or a sudden rupture.

Leaks in piping and fittings should be repaired immediately. In some cases, it is only necessary to tighten a gasket or some screw threads. In others, a section of piping may have to be replaced. Whatever the repair, care should be taken to ensure that it is done properly. The repair should leave the piping properly aligned and supported.

Bearings. Overheated bearings have caused a number of shipboard fires. Such fires can be prevented by following a few simple rules:

1. Bearings should be lubricated with the appropriate amount of the proper lubricant, using the correct pressure.
2. No piece of machinery should be started unless the operator is sure that its bearings have been lubricated with the proper lubricant.
3. Unless absolutely necessary, no piece of machinery should be used if its bearings are in poor condition.
4. The operator should know the approximate normal running temperature of the bearings and should check during operation to determine if they are running too hot.

Testing and Inspection

Fire Protection Equipment. Coast Guard regulations require owners, masters or persons-in-charge to ensure that portable fire extinguishers, semiportable fire extinguishing systems and fixed fire extinguishers are tested and inspected "at least once in every 12 months." The required tests are described in the appropriate USCG regulations corresponding to the service of the vessel. Records of such tests should be maintained in or with the logbook.

No voyage should begin unless all fixed systems are known to be in working order, and all portable extinguishers are usable and in their proper places. United States Coast Guard regulations require that the master conduct drills and inspections to familiarize the crew with the operation of all emergency equipment. The crew is the ship's firefighting team, but they can't be any more effective than the tools they are given. The crew should therefore share the concern of the master for the maintenance of this equipment. There are many case histories in which extinguishing systems and portable fire extinguishers failed when they were needed. In many respects, fire protection equipment that fails is worse than no protection at all.

Fixed Systems. Frequent testing and inspection are the only means of detecting the need for repairs to fixed systems. As is true for other equipment, these preventive maintenance procedures must be scheduled at definite intervals. The fire and boat drills required by USCG regulations provide an excellent opportunity for testing and inspection. The following checks are suggested; any problems that are found should be corrected immediately.

1. Check the capacity of the pumps by charging and utilizing a sufficient number of hoses. Check for proper volume and pressure and the integrity of the piping and fittings.
2. Inspect the hoses for cuts and abrasions, proper stowage and marking. Test them at 45 kg (100 lb) or the highest pressure to which each hose will be subjected in service, whichever is greater.
3. Inspect all threads and clean them with a wire brush if necessary. Keep the threads lubricated; replace gaskets when necessary.
4. Operate the all purpose nozzle, and clean it when necessary. Check the holes for clogging and corrosion.
5. Operate the hydrant valves to ensure that they are ready for use.
6. Check the stowage of applicators and clear the holes and internal strainer where necessary. Lubricate the threads and make sure the applicator fits the nozzle. Do not use lubricant on the heads.
7. Ensure that nothing is connected to the system that shouldn't be.
8. Check for proper operation of the relief valve, remote control pump starting and the pressure alarm where these controls are required.
9. In foam systems, check the quantity and quality of the foam; operate the foam proportioners and driving equipment.
10. Visually check all CO_2 lines and discharge outlet heads.

Every piece of equipment aboard ship should be inspected and tested before the vessel leaves port. No aircraft is taken from its hangar before all systems are checked, and before takeoff the pilot runs through an extensive checklist. The status of the equipment on a vessel prior to its sailing is just as critical as that on an aircraft. Vessel management should require a careful check, and the master should ensure that it is performed. The results of this testing and inspection should be the subject of a formal report to management.

Repair or Replacement

It is important that repairs be performed by competent and knowledgeable people. Whether these are ship's personnel or shoreside contractors aboard or in a shipyard, controls should be established to ensure that repairs are done properly. An improper repair to an electric range in the galley, to a leaky joint in a fuel line or to a defective boiler can have the same results—fire at sea.

Regulations requiring that the U.S. Coast Guard be notified when equipment is repaired or replaced should be followed. Replacements should be only approved types of machinery and equipment. Approval is based on past performance, and safety is an important criterion.

Record Keeping

The history of each major piece of machinery should be recorded. The record should include all tests, inspections, malfunctions, repairs, adjustments, readings and casualties. A card file with a separate card for each piece of equipment has worked well on many ships. Such a file can provide new personnel with the history of each piece of equipment from the day it was installed. It can be of great help in diagnosing problems and in deciding when to replace machinery.

RECOGNITION OF EFFORT

Vessel owners and operators cannot expect the people who operate their ships to participate in a continuing fire prevention program unless they, the owners and operators, demonstrate their interest in the program. Active participation, obvious concern for fire safety and recognition of effort will demonstrate such interest.

One way in which owners can recognize fire prevention efforts is by awarding a plaque to each vessel that achieves a certain number of "fire-free" years. The plaque might first be awarded for a five-year period. Then it would be updated for each successive fire-free year. The plaque would be mounted prominently on the vessel, where it could easily be viewed and admired.

Crew members who complete firefighting and fire prevention training courses should receive recognition from both ship owners and unions. This recognition could be in the form of a certificate, along with a writeup in company and union publications.

If a vessel is unfortunate enough to have a fire, its crew should be rewarded for noteworthy firefighting efforts. The particular situation will usually indicate the most effective method of recognition.

As noted at the beginning of this section, fire prevention is difficult to sell. Recognizing effort on an individual basis, by ship or crew member, will help provide the incentive to maintain a good record.

BIBLIOGRAPHY

U.S. Coast Guard. CG–115, Marine Engineering Regulations

U.S. Coast Guard. CG–123, Rules & Regulations for Tank Vessels

U.S. Coast Guard. CG–174, A Manual for the Safe Handling of Inflammable and Combustible Liquids and Other Hazardous Products

U.S. Coast Guard. CG–190, Equipment Lists

U.S. Coast Guard. CG–239, Security of Vessels & Waterfront Facilities

U.S. Coast Guard. CG–257, Rules & Regulations for Cargo and Miscellaneous Vessels

U.S. Coast Guard. CG–329, Fire Fighting Manual for Tank Vessels

U.S. Coast Guard. CG–388, Chemical Data Guide for Bulk Shipment by Water

U.S. Coast Guard. CG–466–1, Chemical Hazards Response Information System (CHRIS)

U.S. Coast Guard. Hazardous Materials Regulations, Title 49 CFR, Parts 171-177

U.S. Coast Guard. Marine Casualty Report—MAR-76-2—SS Transhuron

U.S. Coast Guard. Marine Casualty Report—MAR-75-6—SS CV Sea Witch—SS Esso Brussels

U.S. Department of Labor. Safety & Health Regulations for Maritime Employment, OSHA, Title 29

Preventive Maintenance in the Boiler Room Helps Prevent Breakdowns at Sea. Marine Engineering/Log, June 1977

Mine Safety Appliance Co.—Supplementary Inspections, MSA Explosimeter.

NFPA. Fire Protection Handbook. 14th ed.

Control of Gas Hazards on Vessels to be Repaired. Bulletin No. 306, National Fire Protection Association, Boston.

MASTER'S INSPECTION CHECKLIST

Accommodation Areas

Crew Quarters	*Yes*	*No*
1. Direct and uncluttered means of escape	____	____
2. General alarm system in good order	____	____
3. Area free of combustible rubbish	____	____
4. Area free of combustibles close to sources of heat	____	____
5. Area free of overloaded electric circuits	____	____
6. Area free of unauthorized repairs to electrical wiring	____	____
7. Area free of jury-rigged electrical wiring	____	____
8. Electrical equipment properly grounded	____	____
9. a. Extinguishers in place and unobstructed	____	____
b. Extinguishers of proper type and size	____	____
c. Extinguishers properly charged	____	____
d. Date of last examination noted	____	____
10. Noncombustible ashtrays, adequate in number and size, and properly placed	____	____

Galley	*Yes*	*No*
1. Area free of combustible rubbish	____	____
2. Noncombustible receptacles with covers provided	____	____
3. a. Oven hood and ducts clear and free of grease	____	____
b. Date of last cleaning recorded	____	____
4. Extinguishing system properly marked	____	____
5. a. Extinguishers in place and unobstructed	____	____
b. Extinguishers of proper type and size	____	____
c. Extinguishers properly charged	____	____
d. Date of last examination noted on inspection tag	____	____
6. Area free of leaking pipes and fittings	____	____
7. Area free of overloaded electrical outlets	____	____
8. Electrical appliances in good repair	____	____
9. Electric oven and ranges dry	____	____
10. Oven free of cracks or crevices	____	____
11. Oven burners secured	____	____
12. Noncombustible ashtrays, of adequate number and size, and properly placed	____	____

Mess Rooms and Lounges	*Yes*	*No*
1. Noncombustible ashtrays	____	____
2. a. Extinguishers in place and unobstructed	____	____
b. Extinguishers of proper type and size	____	____
c. Extinguishers properly charged	____	____
d. Date of last examination noted	____	____
3. Fireproof containers with covers provided	____	____

Deck Department	*Yes*	*No*
1. Decks free of combustible rubbish	____	____
2. Decks free of oil and grease	____	____
3. Decks free of leaking pipes and fittings	____	____
4. Electrical deck machinery in good repair	____	____
5. Holds clean and dry before loading	____	____
6. Cargo lights removed after loading	____	____
7. Dangerous cargo properly stowed	____	____
8. Cargo stowed to avoid shifting	____	____
9. Decks free of damaged or leaking containers	____	____
10. Dangerous-cargo manifest and cargo stowage plan in order	____	____
11. Fuel for lifeboats properly stored	____	____

12. No Smoking signs posted
13. Paints and flammables properly stowed
14. Bos'n stores properly secured

Tankers

1. Pump room free of leaks
2. Expansion trunks properly secured
3. Cargo valves properly marked
4. Venting system and screens

Engineering Department

1. Clear of rubbish, waste, oily rags
2. Noncombustible receptacles with covers provided
3. Noncombustible ashtrays of adequate number and size properly placed
4. Decks, tank tops, clear of oil and grease
5. Area free of leaking pipes and fittings
6. Out-of-service boilers free of oil accumulations
7. No combustible liquids in open containers
8. Paints and varnishes in proper storage room
9. Lumber in proper storage room
10. Area free of unapproved or jury-rigged wiring
11. Area free of unsafe or homemade stowage
12. No unapproved electrical fixtures in paint lockers, battery rooms, etc.
13. Warning signs (High Voltage–Keep Clear) posted
14. Switchboard area clear and free of obstructions
15. Area free of improper fusing or bridging
16. Motors free of lint and dust
17. Motors clear of combustible material
18. Ladders clear and unobstructed
19. No combustible runners on deck plates
20. a. Extinguishers in place and unobstructed
    b. Extinguishers of proper type and size
    c. Extinguishers properly charged
    d. Date of last examination noted

Fire Protection Equipment

Firemain System — *Yes* *No*

1. Hose in place and free of cuts and abrasions
2. Nozzle in place, and applicator provided (if required)
3. Valves unobstructed and easily operated
4. Hose spanner in place
5. Station properly marked

CO_2 System

1. CO_2 room clear of debris and improper stowage
2. Operating control valves unobstructed
3. Alarms and indicators in good order
4. Operating controls set for proper operation
5. Required number of CO_2 cylinders on hand and connected
6. Pipes and fittings in good condition
7. Discharge outlets in good condition
8. Operating instructions posted
9. Signs posted at all CO_2 alarms

Foam System

1. Hose in place and free of cuts and abrasions ____ ____
2. Nozzles and equipment ready for use ____ ____
3. Sufficient supplies of solution and/or powder ____ ____
4. Foam containers free of leaks ____ ____
5. No leaking pipes or fittings ____ ____
6. Valves in operating condition ____ ____
7. Valves and controls properly marked ____ ____
8. Operating instructions posted ____ ____
9. Monitor stations properly marked ____ ____

Steam Smothering System

1. Pipes and fittings in good condition ____ ____
2. Operating controls properly set ____ ____
3. Operating instructions properly posted ____ ____
4. Valves marked to indicate the protected compartments ____ ____
5. Discharge outlets in good condition ____ ____

Water Spray System

1. Apparatus marked ____ ____
2. Control valves marked to indicate the protected compartments ____ ____
3. Spray heads in place and unobstructed ____ ____

Emergency Equipment

1. Storage space properly marked ____ ____
2. Gas mask or self-contained breathing apparatus properly located outside refrigeration equipment space ____ ____
3. Self-contained breathing apparatus in good condition ____ ____
4. Firefighter's outfits in good condition and stored in widely separated, accessible locations ____ ____
   a. Self-contained breathing apparatus in good condition ____ ____
   b. Lifeline free of tangles and ready for immediate use ____ ____
   c. Explosion-proof flashlight with spare batteries ____ ____
   d. Flame safety lamp (except tank vessels) ____ ____
   e. Helmet ____ ____
   f. Boots—electrically nonconducting material ____ ____
   g. Gloves—electrically nonconducting material ____ ____
   h. Protective clothing ____ ____
   i. Fire axe ____ ____

NOTE: A *no* answer to any of the above items requires positive action.

Submitted by

____
Signature

____
Title Date

Case Histories of Shipboard Fires

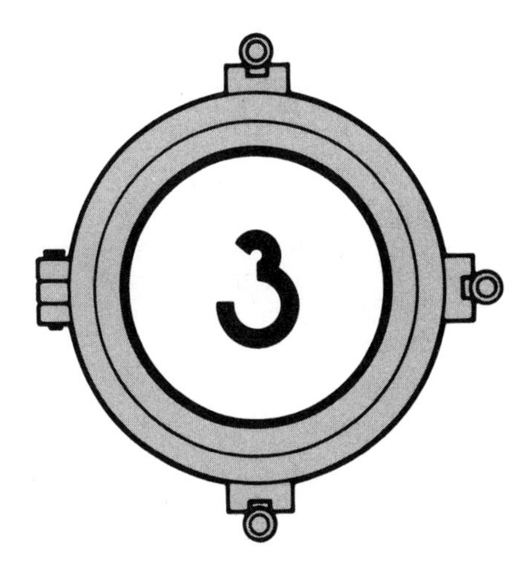

Maritime history includes many accounts of fire aboard ship. In some cases, efficient seamanship and the firefighting efforts of the crew saved the ship, its cargo and everyone aboard. In others, mistakes were made; inadequate firefighting could not prevent the loss of lives and property. This chapter contains several brief episodes of both types.

These case histories make fascinating reading, but they have a very serious purpose: They are presented so that seamen who have not had personal experience with shipboard fires may benefit from the experiences—both good and bad—of those who have. (*See* Chapters 1 and 2 for a discussion of safety measures that could have prevented at least some of the fires that are described here.) The remainder of this book should be read in the light of these accounts. They are, in a sense, demonstrations of correct and incorrect firefighting techniques. As such, they are valuable to seamen who face the possibility of having to fight fire in a very wide range of shipboard situations.

MORRO CASTLE

The *Morro Castle,* 508 feet in length, 70 feet wide, 11,520 gross tons and propelled by turbine electric drive, was considered one of the most attractive Caribbean cruise ships of the Ward Line. Every Saturday, for 173 voyages, she had embarked from New York for a 7-day round-trip cruise to Havana. In addition to passengers and crew, she also transported mail and cargo on these weekly voyages.

The *Morro Castle,* carrying 318 passengers and a crew of 231, representing six different nationalities, sailed from Havana on Wednesday, September 5, 1934. She was due to arrive in New York on Saturday morning and to sail for Havana again that same evening. However, 135 persons were to die as a result of a fire on that return voyage to New York.

On the evening of Friday, September 7, the master became seriously ill while dining in his cabin. The ship's doctor administered a heart stimulant, but to no avail; the master died of heart failure. This misfortune placed the chief officer in command of the vessel and moved other officers upward by one rank.

The Fire

At 0245 on Saturday, September 8, 1934, smoke and possibly flames were detected by several passengers and crewmen coming from the writing room on B (promenade) deck. At approximately the same time the night watchman detected smoke issuing from the cargo ventilating system. He was unfamiliar with the ventilating system but left to find the cause of the smoke, without notifying the bridge. He never reached hold 2 or 3, where he believed the fire to be located; he was stopped at the promenade deck by fire in the writing room and two forward suites.

Stewards were vainly attempting to fight the fire with portable fire extinguishers. Fire hoses were then advanced. However, there was a considerable delay, owing to the master's prior order to remove certain hoses and cap some firemain hydrants. Hoselines had to be brought to the fire scene from two decks below. The master's order had been prompted by a lawsuit brought against the company by a passenger who was injured because other passengers were playing with the fire hose.

The writing room had a locker in which 100 or more blankets were stored. The blankets had been cleaned commercially with a flammable substance, and it was in this locker that the fire actually started. A steward reported that the locker was a mass of flames when he opened its door.

There was a large quantity of highly polished paneling in the writing room, corridors, salon, stairways and other passenger accommodations. The fire spread rapidly along this paneling, into the corridor, to adjoining salons and down the staircase to the deck below.

The vessel had an electric fire sensor system capable of lighting a monitoring panel in the wheelhouse. The system could detect fire in 217 staterooms, officer and crew quarters; however, there were no fire detectors in the lounge ballroom, library, writing room or dining room. At 0256 lights began flashing on the monitoring panel. Every stateroom was either afire or hot enough to transmit an alarm. Smoke conditions were becoming worse, as smoke was being forced throughout the accommodations by the ventilating system which had not been shut down.

Most of the telephones were unserviceable at the time. Thus, the first officer had to run five decks below to the engine room, to order an increase in pressure on the firemain.

At 0257 the acting master gave the signal to stand by the lifeboats.

The ship's Lyle gun and powder were stored directly above the writing room. Shortly before 0300 there was a huge explosion as the Lyle gun and 100 pounds of powder for charges became involved. The explosion blew out many windows in the immediate area, increasing the flow of oxygen to the fire.

The seas were choppy, with winds of approximately 20 knots; yet the acting master continued straight into the wind, thereby driving the fire aft. Finally, at 0300, the acting master called for a left rudder, turning the vessel toward shore.

By 0310 the electric wiring was damaged by fire, and the ship was thrown into darkness. The gyrocompass, electric steering apparatus and standby hydraulic system were all out of action. There was an emergency steering system at the stern that could have been reached through the shaft tunnels. However, the acting master never issued an order to utilize this equipment.

Slow speed was ordered by the acting master, and he began to steer the ship with the engines. He headed toward shore on a zigzag course. The order to stop all engines came at 0321. There was a delay when the order came to drop anchor, because the crewmen at the release levers were unfamiliar with their operation. The anchor was finally dropped by a ship's officer and the signal was then given to abandon ship.

The *Morro Castle's* chief radio operator heard the *Andrea Luckenbach* calling WSC, the radio station at Tuckerton, requesting information about a ship on fire. He then transmitted a CQ (stand by for an important message) without orders from the bridge. The reply from Tuckerton was to stand by for 3 minutes, in compliance with the 3-minute silent period observed each half hour so that emergency messages could be received. Finally, at 0318, the acting master gave the order to transmit an SOS.

Of the 12 lifeboats carried, only 6 were launched. These 6 lifeboats, with a capacity of 408 persons, carried only 85 people, mostly crewmen. The chief engineer left in the first lifeboat with 28 other crewmen and only 3 passengers. The officers in the lifeboats that were lowered made no effort to remain close to the ship, to offer assistance to persons in the water or still aboard. Passengers had to jump or lower themselves into the water by means of ropes.

The *Andrea Luckenbach* was the first vessel to arrive at the scene, and she rescued 62 people in the water. At least three other vessels arrived and assisted in the rescue.

Lack of Fire Protection

The *Morro Castle* was fitted with 42 firemain outlets, thousands of feet of hose and many portable fire extinguishers. She had enough lifeboats, rafts, life buoys and life preservers to accommodate more than three times her maximum capacity of passengers and crew. Her cargo holds had a smoke detection system that was monitored in the wheelhouse. The cargo holds and engine room had fixed fire-extinguishing systems. The vessel had automatic fire doors, installed every 130 feet, in addition to such doors in public areas. One might have believed that she was a very safe ship.

However, most of these safety features were rendered ineffective by alterations to the equipment and/or a lack of proper training. The master had ordered the cargo hold smoke detector system vented outside the wheelhouse because of the offensive odor of some of the cargo. The automatic fire doors had become manual doors when the automatic trip wires were removed.

The crew was not trained in the handling of fire emergencies—either in fighting fire or in evacuating passengers. The capping of many firemain hydrants caused a delay in advancing and charging hoselines and getting water to the fire.

Proper lifeboat drills were not held during the cruise, as they would have disturbed the passengers. Many passengers did not know how to don a life jacket properly. The policy seemed to be that the passengers' enjoyment came first.

Some Conclusions

Among the many lessons to be learned from the *Morro Castle* disaster are are following:

1. The entire crew must be thoroughly trained in order to provide the discipline needed to function promptly, efficiently and effectively both in their normal assignments and in any emergency that may arise.
2. The master must have enough confidence in his officers and crewmen to train them for positions higher in rank than their normal assignments. (When the master died, the chief officer did not have the knowledge required to perform the duties of the master.)
3. Well planned and well conducted training is absolutely necessary, not only in seamanship but also in the handling of emergencies—especially in preventing and combating fire and in abandoning ship.
4. Drills must be held for the benefit of the crew and passengers. Passengers do not need to be frightened, but should be made to realize that an emergency may arise during which they may have to abandon ship.
5. Officers must guide, direct and assist passengers, and lead and supervise crew members. There were complaints from passengers about the crew, but many complimented stewards and bellboys for their handling of the situation without leadership.
6. A sprinkler system, or a continuously monitored smoke and fire detection system, should be installed throughout every passenger vessel. Interior structural members, bulkheads, overheads and decks should be fire resistive, noncombustible or fire retardant.
7. Firestops should be installed in horizontal and vertical voids, wherever smoke, heat and fire can move from one space to another. Ducts should be equipped with dampers that can be controlled both remotely and locally.
8. The alarm must be sounded *without delay* whenever smoke or fire is detected.
9. Fire detection and firefighting systems and equipment should not be altered, removed or changed in any way that reduces their effectiveness.
10. Firefighting plans and drills should include methods for controlling ventilation systems during firefighting operations.
11. Flares and other pyrotechnics should be isolated and stored in a fire resistive space on the highest deck.
12. Every effort should be made to keep passengers and crew informed during an emergency, to guide them and to allay their fears.

NORMANDIE

The *Normandie* was France's entry in the transatlantic crossing competition of the 1930s. She was a quadruple-screw electric steam turbine vessel slightly over 1000 feet in length, 80,000 gross tons and capable of a speed of 30 knots. She mustered a crew of 1300 officers and men and could carry about 2000 passengers. There were 7 decks (A to G) below the main deck and 3 decks (promenade, boat and sun) above the main deck, for a total of 11 decks.

Since shortly before the commencing of hostilities between France and Germany in 1939, she had been tied up at Pier 88, North River, New York. Her French crew maintained the boilers, machinery and other equipment necessary for a ship in idle status. In May 1941, a U.S. Coast Guard detail was placed aboard the vessel "to provide safety and prevent sabotage." Should the United States enter the war, which she did some 7 months later, the *Normandie* would become the USS *Lafayette,* a large, fast, troop carrier. One week after this country entered the war against the Axis powers, the French crew was removed and the Coast Guard took over the ship's maintenance.

Plans were made, scrapped, remade and made again to refit the ship and have her ready to proceed to Boston within two months for further work. Since no drydock was available, a ship repair company was contracted to do the work at Pier 88. Thousands of men climbed aboard, some as the new crew but most as workmen. Tons of equipment and stores were hoisted aboard. Included in these stores was an item that would seal the doom of this fine ship—almost 20 tons of kapok in the form of canvas-covered life preservers. Kapok is a highly flammable, oily fiber that is very apt to produce quick-spreading flash fires. It also has a high combustibility; once ignited, it supports an intense fire that is difficult to extinguish.

When the vessel was in the Port of New York as the French Line's *Normandie,* she had a direct telephone line to the American District Telegraph Company (ADT). The company is a central supervisory agency that receives fire alarm signals from subscribing customers and relays them to the New York City Fire Department. As the USS *Lafayette,* the ship had no fire alarm connection to a shore receiver. The French Line, no doubt feeling that the safety of the ship was no longer their responsibility, had, a month prior to the fire, discontinued the ADT fire alarm service. No government agency made any effort to continue or renew the fire alarm coverage.

At the time of the fire, the vessel's elaborate fire detection system, with 224 fire alarm stations, was not working. Fire guards were employed by the contractor in accordance with the government contract, but they had been given little or no training. As a result, they were practically useless when the fire occurred. Many of the fire extinguishers aboard were empty, while others had instruction labels that were written in French. It would be safe to say that not many of the people aboard the ship that day spoke French. Hose connections aboard ship were in the process of being converted from the French coupling to the American hose thread. There was a U.S. Coast Guard fire brigade aboard but, to accommodate conversion work, they were relocated to a part of the ship remote from the central fire control station. So, while it appeared that fire protection features were maintained, in truth these features were of a cosmetic nature: They looked good on the surface but they only covered up the horrendous vulnerability of the ship.

Every comparable French liner had been destroyed earlier by fire; the *Normandie* was to be no exception. On the afternoon of February 9, 1942, sparks from a burner's blowtorch started a fire that would, within the next 12 hours, leave the ship a helpless wreck. Lying on her side, she denied the use of the pier to other vessels in the busiest wartime port in the United States.

The Fire

Normally a fire occurs when heat is introduced into an area containing sufficient oxygen and fuel. The oxygen (in the air) and the fuel (in this case the highly flammable kapok) were present when the high temperature of the cutting torch was brought into the area. Hot work, that is, welding and cutting with heat, calls for the greatest of surveillance. If heat in the form of an oxyacetylene flame is brought into a space, one of the other alternatives is to remove the fuel. Removing the fuel, tons of kapok, would have been a difficult and time consuming operation, so an attempt was made to separate the kapok from the burner's torch. Portable equipment was used; fire guards would place a 2 × 3 ft asbestos board and a 36-inch semicircular metal shield between the hot work and the kapok.

During the last 20 seconds before completion of the job, the people holding the protective shields in place started to walk away. A flame appeared at the base of the surrounding kapok pile. Workmen attempted to extinguish the fire with their hands. Extinguishers were either unavailable or ineffective, and nearby fire hose contained no water. The untrained fire guards were useless; the one nearby fire bucket was kicked over by a clumsy workman. The alarm had to be sent to the isolated fire brigade by messenger.

About 15 minutes after the start of the fire, the New York City Fire Department was finally alerted via a street fire alarm box. The heavy

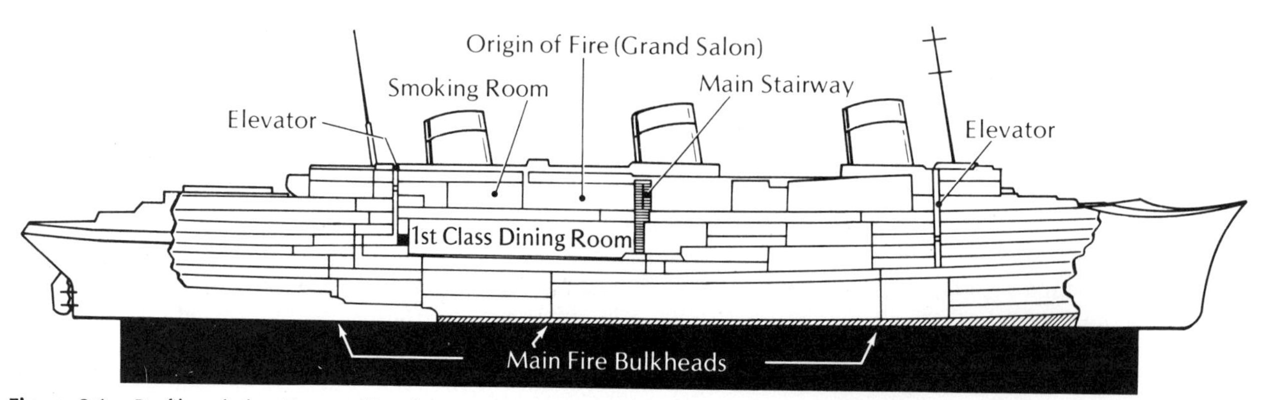

Figure 3.1. Profile of the *Normandie,* showing the grand salon where the fire originated.

smoke had chased the fire watchmen, fire brigade members and others who tried to control the fire away from the grand salon, where the fire had started (Fig. 3.1). The heavy smoke also forced the evacuation of the engine room. The city firefighters ran hoses aboard. (Fire departments prefer to use their own hose from their own pumpers, rather than rely on ship's hose, mains or pumps.) Fireboats directed their deck guns onto the burning ship, and private tow boats added their fire streams. The sheer amount of water had a good and a bad effect. By 0630 the fire was declared under control; however, the ship had a 10° list to port, i.e., away from the pier. The list became progressively worse; at about midnight it was 35°, with water pouring in through open ports and a garbage chute that had been left open on the port side of the hull. The *Normandie* capsized 2 hours and 45 minutes later.

The Causes

What was the cause of this disaster? A quick answer might be "carelessness in handling hot work"; but there was more to it than that. In fact, there was no *single* cause. The destruction of this valuable ship was the result of many faults of omission and commission. Some of the more obvious were the following:

1. *Poor planning.* Within a few short months the ship had been under the jurisdiction of the French Line, the Army, the U.S. Coast Guard and the Navy. Only 2 days before the fire, the Maritime Commission affirmed that the Navy had assumed full responsibility for the ship. A congressional investigative committee made an unsuccessful attempt to find out who (which individual) was actually responsible for the safety of the ship before and during the fire. Responsibility for the ship was respectfully declined by everyone questioned. The committee found that the lines separating the responsibilities of the various levels of command, staff and support agencies and the ship repair people were too hazy to be defined.
2. *Poor use of water.* If care is not used in directing hose streams onto the upper parts of a ship, the water can and will affect the stability of the vessel. In the case of the *Normandie,* many tons of water were thrown onto the fire, 10 decks above the keel. The water could not run off, and it caused the ship to capsize. (*See* Chapter 8 for a discussion of the problem of free surface water.)
3. *Welding and burning.* Hot work—burning and welding—has been the cause of disastrous fires, both ashore and at sea. (*See* Chapter 1 for a discussion of the regulations governing hot work.) In general, repairs involving hot work should be kept to a minimum. The method of burning and the so-called precautions taken in the grand salon of the *Normandie* left much to be desired. The carelessness exhibited there was the immediate cause of the fire. Even so, if a charged hoseline had been available, or if the material in the vicinity had been something other than kapok, there might not have been a disaster. The obvious conclusion is that hot work should be done aboard a vessel only under the personal supervision of a ship's officer, who must ensure that real safety precautions are observed.
4. *Surveillance—fire protection.* Ships require greater surveillance and more fire protection when undergoing repairs and alterations than at any other time.
5. *Previous fire experience.* Those in authority should have been aware that every comparable French liner was lost through fire. In addition, the *Normandie,* like other ocean greyhounds, was a "tender" ship; with its low metacentric height, the shifting of a small amount of weight from one side to the other could cause it to list. When ships of a certain class are subject to the same hazards or have suffered the same unfortunate fate, steps should be taken to counteract the vulnerabilities.

SS LAKONIA

The 20,314-ton *Lakonia* was built at Amsterdam in 1930 for the Dutch Nederland Line and originally named the *Johan van Olbenbarnevelt.* She was rebuilt in 1951 and again in 1959. In 1962 she was purchased by the Greek Line and converted to a first-class cruise liner. At this time she was renamed *Lakonia.*

On the evening of December 19, 1963, the *Lakonia* sailed from Southampton. The crew consisted of men of various nationalities, including Greek, British, Italian, German and Cypriot, which probably created a communication problem. The passengers were mostly British, including a large percentage of older persons looking forward to a holiday cruise.

A boat drill was held on December 20. As was usual on cruise ships during the period, this drill was completed as quickly as possible. Passengers were assembled at their stations, but no boats were lowered; the passengers were not properly instructed in the use of their life jackets.

The Fire

The captain was notified by a crew member at 2250 that smoke was issuing from the main forward staircase and the main ballroom. Passengers stated that approximately 10 minutes earlier they observed the stewards breaking down the door to the hairdressing salon. The stewards attempted to control the fire using portable fire extinguishers, but their efforts were unsuccessful. No fire alarm was sounded before the attempt to extinguish the fire, and this delay consumed precious time. As soon as he was notified of the fire, the master ordered the radio officer to send an SOS giving their position, approximately 180 miles north of Madeira. This order, even though precautionary, was proper.

An order was given for passengers to assemble in the restaurant, which was three decks below the promenade deck and had only one staircase. Fortunately, many passengers refused to obey this order, which would have created a great deal of congestion and possibly panic. Some commotion was caused by passengers looking for relatives and friends, but no real panic.

The master ordered the boats lowered at 2400. The sea was calm, help was on the way and the 24 lifeboats were more than adequate to handle the 1036 persons aboard. However, 95 passengers and 33 crewmen lost their lives in this marine casualty. How and why did it happen?

Probable Causes for the Tragedy

The public address system broke down, causing a lack of communication and creating much confusion. Uniformed deck officers were not present to supervise the loading of the lifeboats. Many seamen were fighting the fire, so the lowering of the boats was left to untrained stewards.

Many boats were not properly stocked with equipment. Their launching gear was in disrepair—jammed and improperly lubricated. Seventeen boats were lowered successfully, but many of these were not fully loaded. Some passengers were reluctant to enter boats after seeing one dump its passengers into the water. The absence of uniformed leaders also influenced passengers' behavior.

Many passengers and crewmen were still aboard when fire disabled the remaining lifeboats. To survive, these people had to go over the side, relying on their life jackets and floating objects for support.

British and American aircraft dropped life rings and rafts to people in the water, greatly reducing the number of casualties. The Argentine liner *Salta,* British freighter *Montcalm,* United States *Rio Grande* and Pakastani *Mahdi* responded to the scene and assisted in rescue work. This, no doubt, was a result of the prompt transmission of the SOS on the orders of the master.

Among the lessons to be learned from this incident are the following:

1. Crew members must receive training in firefighting operations that will not hamper Abandon Ship procedures. Hoselines should be operated from positions between the fire and the lifeboats, to enable passengers to safely reach and use the boats.
2. Lifeboat drills should require passenger participation under the direction of knowledgeable crewmen.
3. Uniformed crew members must be at their assigned positions, in accordance with the station bill, to function as a trained team.
4. During training, emphasis must be placed on the importance of sounding the alarm at the earliest possible moment.
5. Effective communication throughout a vessel is essential, to keep passengers and crew informed and to coordinate crew operations.
6. Passenger and crew accommodations should be free of combustible structural material and furnishings.
7. Smoke detection devices and sprinkler systems provide excellent protection for passenger and crew accommodations.
8. Lifeboats, davits and jackets must be inspected frequently to ensure serviceability during an emergency.

MV RIO JACHAL

The *Rio Jachal* was a cargo and passenger ship of Argentine registry, 527 feet in length, with a 65-foot beam and 18,000-ton displacement. It had completely air-conditioned accommodations for 116 passengers, and four cargo holds with a 4000-ton capacity. It had been placed in service in 1950.

On the morning of September 28, 1962, the *Rio Jachal* left the Todd Shipyard in Brooklyn for Pier 25, North River, to take on 3000 tons of cargo and about 70 passengers for a September 30 sailing. She tied up on the north side of Pier 25. Late that evening, two U.S. Customs port investigators having business on the pier learned that there was a fire on board the *Rio Jachal*. The alarm was relayed to a pier watchman who left the pier and crossed a wide street to send an alarm from a city fire alarm box, bypassing telephones on the vessel and pier. The alarm had been seriously delayed on the vessel and was further delayed by the watchman.

The passenger space aboard the *Rio Jachal* could be described as a combustible, wooden framed, floating hotel within steel bulkheads. These steel bulkheads were covered with wood studding and plywood paneling, which concealed voids varying to 12 inches in depth. The hanging plywood ceilings covered voids up to 24 inches. The passageways were also covered with plywood panels, and the staterooms had wood veneer doors that were not self-closing. The staterooms themselves were separated by combustible partitions of wood studding and plywood. However, these staterooms, in groups of 2, 3 or 4, were located within steel transverse bulkheads. Readily ignitable furnishings were used throughout the staterooms and public spaces.

The Fire

Crew members discovered dense smoke in unoccupied stateroom number 309 on B deck at approximately 2120. The ship was equipped with an automatic fire detection system. The involved stateroom (Fig. 3.2) was equipped with a thermostatic device, actuated at 140°F, that registered an audible and visual alarm on the bridge. Unfortunately, the bridge was unoccupied and locked when the ship was in port, so the system was not monitored.

In attempting to put out the fire, the crew first employed fire extinguishers and then hoselines. However, the fire developed in intensity and volume and forced the crewmen to retreat so rapidly that they were unable to close the watertight doors in the area near the fire site (Fig. 3.2). There were 11 watertight doors on the vessel, 9 of them on B deck. However, the locations of the doors were untenable, owing to the rapid extension of the fire. The crew did manage to close two watertight doors immediately forward of the fire area on B deck, with control wheels located on the deck directly above. This effectively stopped the forward advance of the fire.

The fire extended from stateroom 309 to the passageway and then moved fore and aft. The

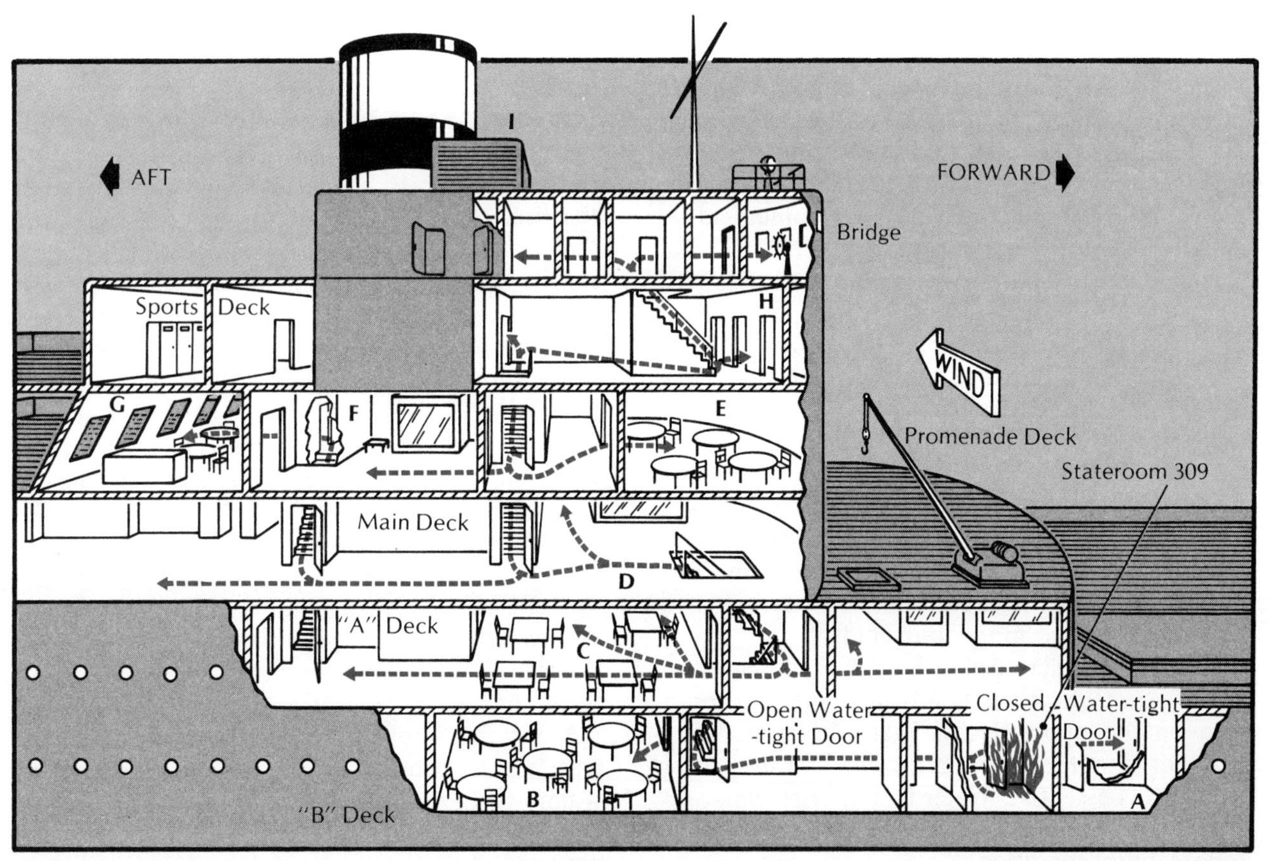

Figure 3.2. Cutaway view of part of the *Rio Jachal*. The arrows show the extension of the fire from stateroom 309 (the site of the fire) to the bridge.

strongest draft was aft. The fire moved through the open watertight door and up the unenclosed semicircular stairway to A deck. From there, it extended from deck to deck, along passageways and stairways, until it reached the bridge.

On the main deck one fire door was closed, but the other three fire doors in the passageways did not close. One switch on the bridge could have closed all these doors simultaneously, but the crew was unable to reach it. In fact, the bridge area was burned out, so that all above-deck power, pump, and fuel-transfer control systems were inoperative. In all, six deck levels were afire through the midship section.

There was no evidence of fire spread other than through stairways and passageways. Even the 29 individual air conditioning systems, with ducts concealed in the dropped ceiling, communicated little or no fire. Spread was horizontal and vertical—essentially a surface fire feeding on the combustible frame interior and furnishings. Concealed spaces were not a major problem.

Firefighting Operations

The first-arriving firefighters promptly stretched hoselines down the pier and then to A deck and the main deck, in an attempt to confine the fire. Other firefighters checked the holds and engine room for fire, removed passengers, and awakened crew members asleep in cabins. Aboard at the time were 12 officers, 40 crew members and 7 passengers who were using the ship as their hotel.

Firefighters were met by the master, officers and crew members, but they encountered language difficulties and had trouble communicating effectively. However, one of the firemen spoke fluent Spanish, and he was immediately enlisted as an interpreter. Because of the length of time taken to transmit the alarm, the fire had a good start and was difficult to control. Not long after arrival, the firefighters had to call for additional help.

A decision was made to leave the ship at the pier, rather than move it. Though the ship was not fully loaded, it could not be considered light. It was, however, in stable condition, with the ballast tanks filled, diesel fuel close to capacity and 935 tons of cargo in the four holds.

Firefighters were ordered to reach the heart of the fire on B deck and to advance lines into position on the upper decks. Many hoselines were brought into the ship, as an aggressive interior attack was essential. Water from the streams was trapped on the upper decks, creating an unfavorable stability condition.

Ventilation of the fire area had to be very closely coordinated with advancement of lines, to ensure that drafts did not draw fire into uninvolved spaces. Each deck had to be vented horizontally; the prevailing wind, blowing across the vessel from starboard to port, was a factor in venting through the exterior windows. The only vertical ventilation was through the stairways, which ended at the bridge.

Three fireboats were moored to the port side of the *Rio Jachal*. After the initial hoselines from the first-arriving firefighting units were in position, all water could have been supplied by the fireboats. Fireboat deck pipes were not generally used. Their heavy streams would have extinguished very little interior fire, but would have added large volumes of water to the decks, worsening the stability problem. They also could have seriously hampered personnel operating in the interior. Deck pipes were used only to prevent extension of the fire to wooden lifeboats, when flames issued from the portholes. Even then, they were used only momentarily.

The Starboard List

During these firefighting operations, the ship's stability had to be monitored continually. Before the fire, the ship was in balance, with the center of gravity and the center of buoyancy on the same vertical plane. As tons of water collected on the starboard side, the center of gravity shifted causing the ship to list toward the pier; when the list reached an angle of about 15°, the master and fire chief became very concerned.

Good progress had been made in controlling the fire. However, the master, the fire chief and their staffs knew that final extinguishment could not be accomplished until the list was corrected, otherwise the continued use of water would cause the ship to capsize and spread fire to the pier. They decided to shut down all water as a first step in correcting the list. For safety, all personnel were ordered off the vessel. Then a few men equipped with breathing apparatus returned to the ship to contain the fire with minimal use of water.

Meanwhile, three crew members and six firefighters, all equipped with breathing apparatus, descended into the engine room to try to correct the list. (The clinometer at the operations platform showed a 15° list to starboard when they reached the smoke-filled engine room.) The list was to be corrected by discharging ballast water—a hazardous operation requiring knowledge and skill. Since the ship was without power, a generator had to be started. The party in the engine room located the compressed-air tank; after about 10 minutes they succeeded in turning over the generator. Immediately the bowels of the ship lit up as power was supplied to the lighting system. Once power was available, the transfer and ballast pumps could be operated, and work could be started on correcting the list.

Five hundred tons of water were discharged from the starboard ballast tanks by one pump operating for 15–20 minutes. Only 2.5 tons of diesel fuel could be transferred from the starboard to the port tanks because the latter were almost filled to capacity. Now the clinometer read 6° to starboard. Measures were then taken to relieve the upper decks of water, e.g., removing stateroom windows and setting up eductors.

As the ship was righted, the previously burned-out portions of the upper decks afforded less fuel for fire extension. All firefighters then returned to their lines. The strategically placed personnel readily extinguished the remaining fire. Within 2 hours after the first alarm was sounded the fire was under control and the *Rio Jachal* was very close to an even keel.

Successful Conclusion

The crew and professional firefighters had worked valiantly and efficiently to confine the fire and thereby save the vessel. Although some areas were severely damaged, the fire was confined to the midship section of the vessel. This successful operation emphasizes several important factors:

1. Early detection of smoke and/or fire is vital for the protection of passengers, crew, vessel and cargo. This is true at sea, at anchorage, moored at a pier or in a shipyard. It is therefore important to monitor all smoke and fire detection devices at all times. At sea, this fire may have been detected earlier and suppressed quickly. The fire detection system would have been monitored, the bridge and engine room fully manned and a trained fire party ready to respond to room 309 on B deck.
2. Remote locations for the control of watertight doors must be readily accessible to crew members.
3. The stability of the vessel is of paramount importance. During a firefighting operation, water must be used wisely because of the adverse effect it may have on stability.
4. When language difficulties are encountered, an interpreter should be found; complete and reliable communication is extremely important.
5. When a vessel in drydock, at anchorage or at a pier experiences a fire, the local fire department will probably assist with, or take over, the firefighting operations. The master should be prepared to provide general arrangement plans and other plans that may be needed, and to assign crew members to guide and assist the professional firefighters.
6. Professional firefighters coming aboard a vessel usually will not use the ship's firemain. They prefer to employ their own equipment, since it is more familiar and they are certain that it is dependable.
7. In passenger ship fires, it is important to locate and control stairways to contain the vertical extension of smoke, heat and fire.
8. Whenever possible, hoselines should be advanced to the fire from below or along the same deck level to avoid the rising heat of the fire. Hoselines should not be stretched down ladders in the vicinity of the fire, because they may be enveloped in heat and smoke.
9. A delay in transmitting the fire alarm gives the fire more time to extend and to increase in severity.

YARMOUTH CASTLE

From the time she was constructed in 1927 for the Eastern Steamship Company, the *Evangeline*, as she was originally named, sailed under several national flags and ownerships. This 379-foot, 5000-gross ton, 2474-net ton vessel also served as a troopship during World War II. In 1965 she was renamed the *Yarmouth Castle* and sailed as a cruise ship for the Chadade Steamship Company of Panama.

Under the command of a 35-year-old master the *Yarmouth Castle* sailed for Nassau on the evening of November 12, 1965, on a regular biweekly cruise. The ship carried 165 crew members and 376 passengers, including 61 members of the North Broward, Florida, Senior Citizens Club. This organization was to lose 22 of its members as a result of the fire that ensued.

The Fire

At approximately 0100 Saturday, November 13, as the *Yarmouth Castle* was in Northwest Providence Channel abeam of Great Stirrup Bay, the odor of smoke was detected in the engine room. The smoke was thought to be coming from the galley bakeshop area via the ventilating system. This area was searched, but no fire was found.

Within minutes passengers and other crew members smelled smoke and began searching for the source. Many believed the fire to be in the men's toilet on the promenade deck, since smoke was issuing from that location. Unknown to the search party, which was increasing in size, the fire was in storage room 610, one deck below on the main deck. There was considerable confusion among the search party, which now included the master and the cruise director.

When the fire was finally located in room 610, members of the crew fought the fire with hand fire extinguishers. When this proved futile, hoselines were advanced, and the engine room was ordered to start the fire pumps. This too proved ineffective, and fire spread into the corridors and toward the stairwells. The inexperienced firefighters were driven back by the intense heat and smoke. At this point the master returned to the bridge.

No alarm had been sounded yet, but some passengers were awakened by the noise, and more became aware of the odor of smoke. The passengers in the ballroom were among the last to learn of the problem as a woman dashed into the room screaming "Fire."

The bridge ordered the engines and ventilating system shut down at 0120. Watertight doors in the engine room were closed. On orders of the master, an SOS was transmitted. Because the radio shack was afire, the message had to be sent by blinker to two ships that were sighted.

The order to abandon ship was given at 0125, however, it could not be transmitted because the wheelhouse was burning and had been abandoned. At this time the fire was very heavy amidships, with flames leaping high into the air. Only 4 of the 14 lifeboats were lowered, and the master was in one of the first boats to leave. He later returned to the *Yarmouth Castle* and explained that he left to get assistance. One of the early boats to leave had mostly crew aboard; only four passengers were lowered with the boat.

At 0155 the U.S. Coast Guard was notified by the Finnish ship SS *Finnpulp* that she had sighted a ship afire. Aircraft were dispatched from Miami for confirmation, observation and rescue if possible. The *Finnpulp* and the Panamanian liner *Bahama Star* both steamed toward the *Yarmouth Castle* to offer assistance. One passenger stated that the *Bahama Star* put 14 boats into the water. The *Finnpulp* put both her boats into the water.

The master returned to his ship at approximately 0300, after most of the passengers and crew had left the vessel. Many had to jump into the water, hoping to be picked up by lifeboats from the two rescue ships.

The *Yarmouth Castle* sank at approximately 0600. The toll in lives was 85 passengers and 2 crewmen. If the *Finnpulp* and *Bahama Star* had not been close by at the time of the disaster, many more lives would have been lost.

Room 610, where the fire originated, was not equipped with a sprinkler system. Even a minimal system would have resulted in early detection and confinement, and possibly extinguishment. This storage room contained mattresses, chairs, paneling and other combustible materials—a relatively high fire loading for a small space.

Lack of Fire Protection

There are many lessons to be learned from a tragedy of this magnitude. Among them are the following:

1. Early detection of fire and the prompt sounding of the alarm are most essential.
2. The crew must be aware of the importance of fire prevention. A crew with the proper attitude toward fire prevention would not have stored so much combustible material in a room without the protection of a sprinkler system.
3. It is the responsibility of the master and other officers to continually train and drill crew members, so that they will function effectively in an emergency. Practical training in the use of portable fire extinguishers, hoselines and breathing apparatus is essential.
4. The orders to transmit an SOS and to abandon ship cannot be delayed until it is no longer possible to do either safely and effectively.
5. A smoke detecting and/or sprinkler system should be required in passenger and crew accommodations.
6. Combustible interior construction and furnishings should be eliminated wherever they can be replaced with fire resistive, noncombustible and fire retardant materials.
7. The master should never have left his ship. His responsibility was to direct and lead his crew in containing the fire and safely evacuating the passengers and themselves.
8. This fire might have been confined and extinguished by the crew if they had been trained to report the fire immediately, order the fire pump started, use portable fire extinguishers skillfully and quickly advance hoselines while wearing self-contained

breathing apparatus. In other words, prompt and efficient firefighting could have been successful in this case.

MV ALVA CAPE AND SS TEXACO MASSACHUSETTS; TUGBOATS ESSO VERMONT AND TEXACO LATIN AMERICAN

The Kill Van Kull is a narrow estuary connecting Newark Bay with the Upper Bay of New York harbor; it lies between the southern tip of Bayonne, New Jersey, and the north shore of Staten Island. On the afternoon of June 16, 1966, visibility in the channel was excellent, the temperature was 85°F, the wind was in a southwesterly direction at 8 knots and the water temperature was approximately 62°F. It was hardly a day on which a nautical catastrophe would be likely to occur. Yet, before the sun set that evening, one of the worst ship disasters ever experienced in New York harbor took place; 33 men lost their lives, and 64 were injured.

A collision between the tankers MV *Alva Cape* and SS Texaco *Massachusetts* resulted in an explosion and fire that involved these two vessels and two tugboats, the Esso *Vermont* and the Texaco *Latin American*. The story will long be remembered, not only by those fortunate enough to survive the holocaust, but also by the men who fought the blaze. Their valiant efforts were successful in subduing the four separate ship fires that raged almost simultaneously.

The Collision

At 1357, the *Massachusetts* left her berth at Texaco's Bayonne Terminal, having just discharged 2,242,800 gallons of gasoline, bound for Port Arthur, Texas. At this time her 27 cargo tanks were tightly closed and empty except for seawater ballast in her nos. 3, 5 and 7 center and no. 5 wing tanks. The tug *Latin American,* alongside on her port bow approximately 150 feet aft, was no longer assisting the tanker to maneuver out into the channel. Aboard the *Massachusetts* were a total of 41 persons—a crew of 39 and 2 pilots.

The *Alva Cape* was heading up the Kill Van Kull with 132,854 barrels (5,579,868 gallons) of naphtha in 21 tanks. The naphtha was to be discharged at Bayway, New Jersey. The majority of the 44 men aboard the *Alva Cape* were orientals from the British Crown Colony of Hong Kong.

By 1407 the *Alva Cape* was just under the Bayonne Bridge, slightly to the right of the middle of the channel (Fig. 3.3). The *Vermont* was moving up fast astern of the *Alva Cape,* along her starboard quarter, in an effort to overtake her. The *Massachusetts* reduced her speed from Slow Ahead to Dead Slow Ahead; 2 minutes later the *Massachusetts* went from Dead Slow Ahead to Full Astern. The two vessels were now approximately 1½ ship lengths apart. Because a collision seemed certain, both vessels dropped anchor. Contact was made at 1412; the prow of the *Massachusetts* sliced the no. 1 starboard wing tank of the *Alva Cape* 10 feet deep and 15 feet below the waterline.

Explosion and Fire

The *Massachusetts,* with her engines full astern, immediately began to back away from the *Alva Cape*. For about 3 minutes there was no evidence of any smoke or fire. However, as naphtha cascaded from the ruptured tank, a mist of vapor could be seen spreading over the water, encircling the two vessels.

Shortly thereafter a tremendous explosion was heard near the *Alva Cape,* and then a second

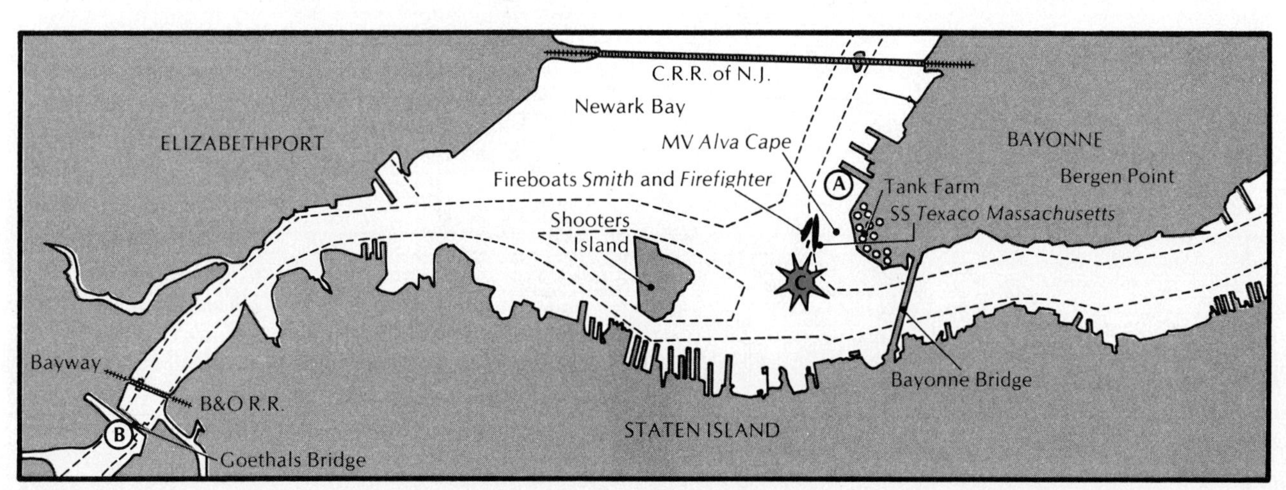

Figure 3.3. The area of the collision between the *Massachusetts* and the *Alva Cape*. The *Massachusetts* had left from point A, and the *Alva Cape* was headed for point B. The ships are shown in their final positions after the collision.

blast near the *Massachusetts*. The water between the two tankers, approximately 450 feet apart at this time, was engulfed in flames. The current quickly carried the flames around the starboard side of the *Massachusetts*. Within a few minutes, a third explosion was heard. Most probably the *Vermont* was the source of ignition for the explosion near the *Alva Cape*, and the explosion near the *Massachusetts* was caused by the *Latin American*.

Via radio, a Moran tugboat captain reported to his dispatcher that two tankers had collided and were afire. The dispatcher immediately telephoned the Marine Division of the New York City Fire Department. A tall dark column of smoke, clearly observed in the vicinity of Bayonne, New Jersey, confirmed the fire. Immediately, orders were transmitted to three fireboats, the *Smith* (Marine Company 8), the *McKean* (Marine Company 1) and the *Firefighter* (Marine Company 9), to respond to the fire.

Conditions Upon Arrival

The first land units to arrive observed two tankers and two tugboats afire beyond their immediate reach. The *Alva Cape* was pivoting on her anchor in a clockwise direction, because of the current and wind. The water surrounding the *Alva Cape* was covered with flaming naphtha; the paint was burning off her hull above the waterline, from bow to stern. Considerable fire was visible on the main deck, in the midship and after superstructures, and from the hole in the hull (Fig. 3.4). The intensity of the fire made survival seem hopeless for those who had not already abandoned ship.

Both vessels drifted toward the tank farm at Bayonne, with the *Vermont* between them. The *Alva Cape* dropped her starboard anchor and the *Massachusetts* dropped her port anchor, which limited their movement. They came to a temporary halt with the *Massachusetts'* starboard stern alongside the port stern of the *Alva Cape*, and here they continued to burn. The *Massachusetts* was burning mainly in and around the stern superstructure, with paint burning on the hull and gasoline vapors aflame at tank vents. It was learned later that because of her deep draft and the mud flats in the immediate area, the *Alva Cape* could not drift closer to Bayonne.

Obviously, the first land-based firefighters to arrive had to wait for the fireboats before starting an attack. The ships were beyond the reach of land apparatus, and the fire was too large for any equipment that could be transported by police launch or tugboat. All available U.S. Coast Guard and police launches, tugboats and private boats were fully occupied with the recovery of persons in the water.

When the fireboats pulled up to the burning tankers, firefighters attempted to determine what specific liquid was burning. The flames coming from the hole in the *Alva Cape* indicated a very volatile substance. Other questions arose. Was there just one kind of liquid, or was this a "drugstore" tanker, carrying a wide variety of liquid cargo? How much did each tanker carry? From the way they rode in the water, it was evident that the *Alva Cape* was carrying near capacity, while the *Massachusetts* was carrying a minimum load.

The scent of naphtha vapors was not discernible until members approached close to the starboard side of, or boarded, the *Alva Cape*. Then they knew what the cargo was! Naphtha has a flash point (closed cup) below −25°F and an initial boiling point of 105°F. It is lighter than water and is not soluble in water. Its flammability limits are approximately 1% and 6%.

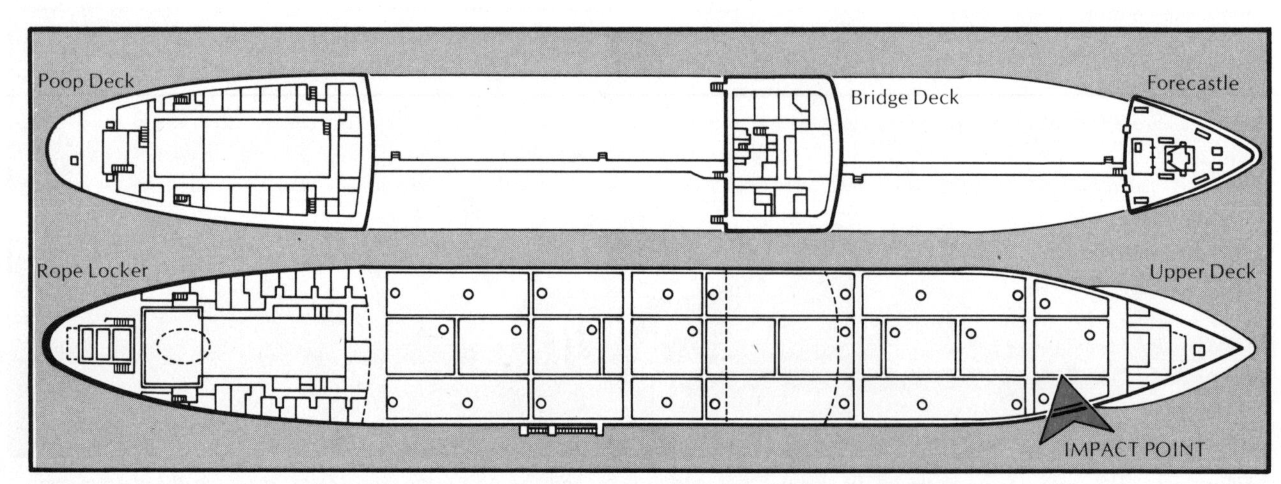

Figure 3.4. Superstructure and tank decks for the *Alva Cape*. Arrow marks the point of impact.

Extinguishment

The fireboat *Smith,* when it arrived at the scene of the collision, proceeded directly to the burning vessels. The *Firefighter* first pulled into a Staten Island pier to take firefighters aboard, and then also went to the fire. The fireboats' first operation was to separate the two tankers, to protect the *Massachusetts* from becoming further involved in fire. The *Smith* was ordered to operate as a wedge between the tankers, using all monitors to extinguish fire and cool down both tankers.

The *Smith* maneuvered between the two tankers, followed by the *Firefighter*. The full water discharge capacities of both boats were needed to protect the firefighters aboard the *Smith* from the intense heat. With the fireboats in position, tugs were able to raise the *Massachusetts*' anchor and tow her away from the *Alva Cape*. The fireboat *Harvey* (Marine Company 2) was dispatched to accompany the *Massachusetts* to anchorage. Enroute, her crew extinguished the remaining fire on the tanker.

When the *Massachusetts* was towed off, all fireboats maneuvered to the ruptured starboard side of the *Alva Cape*. Fireboats amidships and astern concentrated their monitor attacks to make the *Alva Cape* tenable for boarding with handlines. The *Smith* employed a foam attack to control the fire on the water and in the no. 1 starboard wing tank. Once the fire in the gaping hole in the bow was controlled, firefighters from the three fireboats boarded the burning tanker with fog and foam handlines. Aboard the vessel, the firefighters were faced with four fires of major proportion; these were located in

1. The forward storage tanks and the hole in the starboard wing tank
2. The amidship storage tanks
3. The amidship superstructure
4. The stern crew quarters.

Two fire attack teams were formed—one to handle the fire in the forward storage tanks and damaged starboard wing tank, and one to handle the amidship storage tanks. Fog streams were used to cool the deck plates and protect the men while they introduced foam into the ullage openings (inspection holes) and through breaks in the steel combings of the expansion tanks, to control the fires in the storage tanks.

At this time, the naphtha issuing from the hole in the wing tank reignited and burned several of the firefighters. This second fire was more difficult to control than the first but eventually was brought under control by alternately using fog streams and foam lines.

To ensure extinguishment, the heat in the tanks and deck plates had to be dissipated, and the rate of vaporization of the naphtha reduced. The gaskets of the hatch covers on the storage tanks were smoldering and had to be quenched and removed. The covers of the expansion tanks were then opened for better access to the storage tanks; the foam streams were redirected to seal the foam blanket. Opening the tank covers also allowed the heat to dissipate more quickly by convection.

During this period, the naphtha vapors continued to penetrate the foam blanket; when a hatch cover fell and created a mechanical spark, the escaping vapors were reignited. However, the foam blanket restricted the fire to a limited area, and it was quickly controlled. At this point, a holding force could control the tanks, and the main firefighting force could direct their attention to the bridge and amidship quarters.

While the major problem had been solved, the most punishing one remained—the control and search of the after crew quarters. Even though considerable time elapsed before firefighters attempted to enter this area, the heat was unbelievable. The fire in this area had been particularly intense, and the furnishings and contents of the compartments were completely incinerated. Self-contained breathing apparatus and frequent rotation of personnel were required, even in the upper-deck compartments. Progress was very slow, particularly in penetrating the rope locker and other below-deck areas. Through effective leadership, close control, rotation of forces and the determination and courage of the men, control was achieved around midnight.

The tugboats *Latin American* and *Vermont* were towed to Shooters Island and a Staten Island shipyard. They were carefully examined, and all fire was extinguished by land-based firefighters.

The fire was declared under control fairly early. However, it took approximately 12 hours to uncover and extinguish the very last traces of fire and ensure that there was no reignition of vapors, in or out of the tanks.

Stability was never a serious problem, even though the *Alva Cape* developed a list to starboard as the tide receded. Her port side was up against the mud flats, with the starboard side floating. A tugboat was used to keep the *Alva Cape* against the mud flats, to help maintain stability. Eductors were used in the two superstructures, but the water that collected there did not adversely affect stability.

Conclusions

1. If the vessels involved had communicated

with each other via radiotelephone, this tragic accident might have been prevented.

2. Once a fire involving flowing volatile and combustible liquids is extinguished, precautions must be taken against reignition. Reignition can cause serious injuries, and the fire may be more difficult to extinguish the second time. Flotation rings, hose streams, emulsifiers and foam blankets must be employed to prevent reignition.
3. In this kind of situation, where flammable vapors can exist over a large area, the actions of each individual must be closely supervised. A carelessly dropped tool, a nail in a shoe, a short in some emergency electrical equipment or a disturbance in the pool of flammables can cause an explosion and loss of life. Caution and discipline are imperative.
4. Traffic control in the vicinity of such a fire is critical. Collisions, the ignition of escaping vapors and interference with the maneuvering of marine firefighting units must be prevented.
5. In a situation as hazardous and potentially explosive as this one, only minimal manpower should be committed. To minimize the risk to firefighters, the fire problem can be separated into parts and the parts attacked in sequence on a priority basis.
6. In salvage operations involving tankers, flammable products should only be removed by methods that safeguard against the formation of explosive atmospheres.
7. Firefighting can be very punishing physically. Crewman fighting a fire of this caliber should wear full protective clothing and breathing apparatus. They must be rotated to minimize their exposure time.

SS SAN JOSE

The SS *San Jose* was on a voyage from San Francisco to Viet Nam, via Guam. Approximately 95 miles west-northwest of Guam, on November 11, 1967, at 1850, there was a fire in the vessel's boiler room. The vessel subsequently lost all power and had to be abandoned. However, she was reboarded later and towed to Guam. There were no deaths as a result of the incident, and only one crew member was slightly injured.

The vessel was classed as a freighter (reefer), was 6573 gross tons and 3410 net tons, and had a length of 433 feet and a beam of 61.2 feet. Propulsion was by steam, at 12,000 horsepower. She was built in Alabama in 1945 and was owned by the United Fruit Company.

At the time of the casualty, there was a northeasterly wind at 20–25 knots, with a moderate northeasterly sea and swell. The air temperature was 80°F, and the water temperature 82°F. The sky was overcast, with visibility limited only by dusk and darkness. Typhoon Gilda was located approximately 440 miles southwest of Guam, moving westerly at 12 knots.

On July 19, 1967, the ship's firefighting equipment was serviced and found to be in proper working order. Remote shutdowns for the vessel's fuel and ventilation systems were tested on July 25, 1967, and found to be satisfactory. Outlets for the fixed CO_2 system in the boiler room were located both below the floor plates and around the boiler room perimeter, at a height of about 8 feet above the floor plates. The CO_2 manifold was recessed in the boiler room casing in the port passageway of the crew quarters.

At San Francisco, the vessel loaded 3400 tons of refrigerated cargo and 6308 barrels of bunker C fuel oil. The fuel was carried in double-bottom and deep tanks. Its specific gravity was 1.027 and its flash point between 225° to 230°F. The *San Jose* left for Guam on October 27, 1967. During the passage to Guam, oil was consumed from nos. 1, 2 and 3 double-bottom tanks and was transferred daily to the port and starboard settling tanks in the boiler room.

The *San Jose* arrived at Apra Harbor, Guam, on November 10, 1967, at 1300, and discharged 350 tons of cargo. Five tons of ammunition was loaded on deck abreast of each side of the foremast, and 5766 barrels of fuel, commonly referred to as "Navy special," was taken on board via the port fueling station. This fuel has a specific gravity of 0.878, and a flash point of 192°F. The *San Jose* left Guam before 1200 on November 11, 1967.

The port and starboard settling tanks each had a capacity of 615 barrels of oil. The height of the oil in a full tank was 23 feet, 8 inches. Operating procedures aboard the *San Jose* required pumping the oil to a height of 22 feet, or about 570 barrels. Each tank terminated in a raised hatch or expansion trunk with a hinged cover secured by nuts. The pipe nipple at the center of the cover was fitted with a non-self-closing gate valve whose disk was manually operated by a horizontal lever. Next to the trunk and rising from the top of the tank was a 3-inch sounding pipe, about 2½ feet high. The sounding pipe was fitted with a weighted lever-operated device for opening and closing the line.

It was the duty of the second assistant engineer and junior engineer to fuel the ship and transfer fuel oil. They pumped up the settling tanks daily, using electric pumps that were situated against the aft bulkhead of the boiler room. Both had been at sea in the engine departments of merchant ships for many years.

The chief engineer required that the oil level be checked in two ways when the settling tanks were pumped up: *1)* by observing the height of the column of mercury in the pneumercator for each tank, and *2)* by obtaining ullage readings. The pneumercators were located on the side of the starboard settling tank. Above the pneumercators was an oil high-level audible alarm panel. No program was established for periodic testing of the audible alarm, and there was no valid record as to when it was last tested.

The Fire

The first assistant engineer obtained an ullage reading from the port settling tank and instructed the junior engineer to commence pumping. He then performed other work that took him away from the port fueling station, but not out of the machinery spaces. The junior engineer took a pneumercator reading of 11 feet 9 inches, and then started the transfer pump at about 1810. He adjusted the speed of the pump with the rheostat until the pump was running at about one-third capacity. He did not remain at the pump or the pneumercator but attended to other duties in the boiler room. None of these duties took him away from the pneumercator or pump for more than 5 minutes.

The junior engineer, in addition to transferring oil, was engaged in blowing tubes on all three boilers. Since this was an almost fully automated operation, it required very little of his attention. At the time of the casualty he had started blowing tubes on the starboard boiler.

About 2 or 3 minutes before the fire, the fireman/watertender noted that the pneumercator registered 20 feet 6 inches. He gave this information to the junior engineer, who "pumped" the pneumercator and took a reading of 20 feet, 9 inches–21 feet even. The fluctuation was attributed to the rolling of the ship.

The first assistant engineer went down a ladder in the engine room. As he walked toward the port side, the vessel took a moderate roll and he saw oil flowing on the walkway adjacent to the open doorway of the port fueling station. The oil overflowed the walkway coaming, cascaded down into the boiler room and flashed into fire. He quickly shouted a warning to the junior engineer and fireman/watertender on watch below and then ran to advise the chief engineer, who was in the engineer's office. The time was estimated to be 1850.

The fireman/watertender looked to port and saw "balls of fire" falling down around the Bailey board. The fire appeared to be coming from above, but he could not see exactly where. The junior engineer read the pneumercator. He immediately shut down the fuel oil transfer pump and electric ventilation system and closed the valves at the fuel oil manifold. No audible alarm was heard.

The junior engineer and fireman/watertender rapidly unreeled the hose on the semiportable CO_2 system (two 50-pound cylinders) and directed the CO_2 gas at the fire around the Bailey board. They were able to extinguish fire at their level, but it kept reigniting as it was fed by burning oil from above. They continued to fight the fire until the chief engineer ordered them from the boiler room. The boiler fires were secured by the fireman/watertender before he left the boiler room.

Believing that it was impossible to fight the fire using semiportable and portable fire extinguishers, the chief engineer ordered the boiler room evacuated, all doors closed and all ventilation secured. He then activated the remote controls for the fixed CO_2 system in the boiler room, releasing the entire 5600 pounds of CO_2. No attempt was made to fight the fire in the boiler room with water.

The chief engineer arrived on the bridge and reported to the master that the port settling tank had overflowed and the oil was burning in the boiler room. He also informed the master that he had ordered the plant secured and all ventilation to the boiler room closed, and that he had released CO_2 from the fixed system into the boiler room. The master then activated the remote controls that closed all fire doors and stopped the mechanical ventilation to quarters and cargo holds. He sounded the general alarm.

All accessible ventilation and access openings to the boiler room had been closed, but the space could not be totally cut off from outside air. There was a circular opening to the atmosphere between the inner and outer stacks, and no means was available to secure this opening.

Approximately 10 minutes after the fire broke out, the master sent an advisory message to MSTS, Guam, apprising them of the situation and requesting assistance. Two nearby vessels, the USS *Hissem* and the SS *Coeur D'Alene Victory,* were alerted and sent to the *San Jose's* position.

The chief engineer returned to the engine room on several occasions to check on the effectiveness of the CO_2 he had released. He looked into the boiler room through the glass port in the fire door between the lower engine and boiler rooms. He also checked by cracking the watertight door from the machine shop to the boiler room. He was not able to see clearly because of the dense smoke, but he did not observe any actual flames.

The master directed that the ammunition on the fore deck be thrown overboard. He ordered crewmen to fight the fires in the quarters and at the stack with portable extinguishers and fire hoses. The portable extinguishers did little to contain the fire and the fire pump stopped when the ship's generators tripped off because of the falling steam pressure. The water pressure in the firemain and fire hoses soon was reduced to a trickle.

The emergency diesel generator was located on the deck below the wheelhouse, forward of the boiler room. It started automatically and provided power for topside emergency circuits. The power cables that ran from the emergency generator switchboard to the machinery spaces passed through the boiler room. The cables were burned out, so that no power was available in these spaces to operate emergency lighting and a fire pump. The emergency diesel generator continued to run properly under the control of the ship's electrician. However, when the area began to fill with smoke, the master ordered it secured.

Decision to Leave the Ship

A little over an hour after the start of the fire, unable to control it, the master decided to transfer most of his crew to the *Hissem* and *Coeur D'Alene Victory,* which were now standing by. This was accomplished with two lifeboats; 21 men went to the *Coeur D'Alene Victory,* 19 men to the *Hissem* and 13 men remained aboard the *San Jose.*

Various attempts were made by the *Hissem* to send damage control parties and firefighting equipment to the *San Jose.* However, they were all unsuccessful, owing to worsening weather and sea conditions. About midnight, when it appeared that the fire was burning itself out, the master requested that the *Hissem* tow the *San Jose* out of the path of the approaching typhoon Gilda. The *Hissem* began towing the *San Jose* at about 0700 on November 12, 1977. After 2 hours, a broaching sea caused the loss of the towing wires.

The master received instructions from the Commander, Naval Forces, Marianas, and from his employer, United Fruit Company, to abandon the *San Jose.* Typhoon Gilda seemed to be headed straight for the ship and the extremely rough sea made it inadvisable to use either of the two remaining lifeboats to leave the *San Jose.* The master therefore elected to use an inflatable life raft. The passage of the life raft to the *Hissem* was without incident. The *Hissem* then took action to clear the course of typhoon Gilda.

The *San Jose* was subsequently relocated by aircraft. The master and several members of his crew were placed aboard by the USS *Joaquin County* on November 16, 1977. The vessel was riding easily, and no fire was observed. The USS *Cree,* a U.S. Navy salvage tug, arrived on November 17 and towed the *San Jose* to Apra Harbor, Guam.

Discharging of the *San Jose's* refrigerated cargo was begun immediately, so as to salvage as much as possible. The unloading proceeded without incident until no. 2 lower hold was opened. At that time fire, which had been smoldering undetected behind insulation, broke out anew. It was extinguished by flooding the lower hold with water.

Conclusions

1. The fire was caused by an unsafe but common practice. The second assistant engineer on the *San Jose* failed to close the ullage opening in the port settling tank after taking his initial ullage measurement. As a result, the oil in the topped-up tank overflowed through the open ullage pipe. The overflowing oil struck an uninsulated superheated steam line, flange or fitting and quickly flashed into fire.
2. A contributing cause was a malfunction in the operation of the pneumercator for the port settling tank. It failed to indicate the true level of the oil in the tank.
3. A further contributing cause was the failure of the junior engineer to realize fully that the lighter "Navy special" would pump faster than bunker C oil, and to closely observe the pneumercator and transfer pump.
4. The fire in the boiler room continued to burn after the fixed CO_2 system was activated. This was due primarily to the fact that the source of fuel for the fire was approximately 12 feet above the CO_2 discharge outlets in the boiler room. In addition, oxygen was available through the space between the inner and outer stacks.

5. Because of the intensity and location of the fire, water could not be used effectively. The releasing of CO_2 from the fixed system by the chief engineer was warranted and proper under the existing conditions.
6. Except for portable fire extinguishers, the *San Jose* lost its firefighting capability with the loss of the turbine generators and the activation of the fixed CO_2 system.
7. The high-level alarm for the port settling tank did not sound a warning. This could have been due to a malfunction in the sensing element in the tank. If so, the malfunction would not necessarily have been revealed by periodically testing the alarm with the test switch on the panel.
8. Only the approach of typhoon Gilda necessitated the abandoning of the *San Jose* by her crew.
9. All the required firefighting and lifesaving equipment on the *San Jose* operated satisfactorily and as intended.
10. The crew was not properly indoctrinated on the limitations of CO_2 as an extinguishing agent, or the fire extinguishing ability of water streams from the fog nozzles and applicators provided in the engineering spaces.

MV SAN FRANCISCO MARU

Early Saturday afternoon, March 30, 1968, the MV *San Francisco Maru* entered New York harbor and tied up to the Mitsui O.S.K. Line pier in Brooklyn. The vessel, launched in Japan only 9½ months earlier had an overall length of 511 feet, a beam of 71 feet, net tonnage of 5794, gross tonnage of 10,087, diesel engines and one propeller. There were 34 Japanese crewmen manning the vessel.

At 1508, as crewmen were working in hold no. 5 preparing cargo for offloading, they detected smoke in the hold. The alarm was given to alert the master and crew. The ship was equipped with a smoke detecting and CO_2 extinguishing system, with the monitoring cabinet on the bridge. Unfortunately, when the ship was in port and moored to a pier, the bridge was unoccupied.

Extinguishment

The smoke condition was investigated, and heat and heavy smoke were found in the lower hold. Crewmen were ordered out of the hold, hatches were closed by 1520 and a decision was made to call the local fire department and to use the ship's CO_2 extinguishing system. At 1536 the CO_2 extinguishing system was activated, and CO_2 was discharged into lower hold no. 5.

When the professional firefighters came aboard, they requested the stowage plan for holds 4 and 5 and the general arrangement plan for the vessel. These were provided by the master and studied by the fire officers and ship's officers.

At the same time, fire department personnel checked the main deck hatch covers and ventilators, to ensure a tight seal, preventing the entrance of air and the escape of CO_2. The MV *San Francisco Maru* was a new vessel, so the four hydraulically operated hatch covers over hold no. 5 were rather tight. Crewmen had shut down ventilating fans, closed dampers, covered ventilators and dogged hatch covers. Owing to the excellent condition of the hatch covers, it did not appear necessary to fill or cover the joints, but they were examined for escaping smoke or CO_2.

The master and a fire chief inspected the CO_2 room and discussed the initial discharge of CO_2 into lower hold no. 5. The instructions for the CO_2 system were in Japanese; and, although they were set up so that a person unfamiliar with the language could determine how CO_2 was to be applied initially and periodically to the various protected spaces, it was determined that CO_2 was not being applied in accordance with the instructions. Therefore, it was advisable to start anew. The fire department requested that the master order more CO_2. It had to be ordered as soon as possible, so that the supplier would have time to make the delivery before the ship's supply was exhausted.

Crewmen and firefighters worked together to attach four thermometers to bulkheads in hold no. 4, as follows:

1. Lower hold aft
2. Lower 'tween deck aft
3. Upper 'tween deck aft
4. Lower hold forward.

The exterior of the hull on the port and starboard sides was examined and watched for discoloration and blistering. One location at the lower hold level, where heat and discoloration were detected, was marked with white chalk. The forward bulkhead of the engine room, immediately aft of hold no. 5, was also examined and watched for discoloration and blistering, especially at the lower hold level.

At about 1600, a chart (similar to that in Fig. 10.16) was started. The time of day and the temperature reading of each thermometer were recorded, along with the atmospheric temperature and the number of 100-pound CO_2 cylinders discharged into the lower hold.

Examination of the bulkheads and hull plates surrounding hold no. 4 and the temperatures obtained from the thermometers soon confirmed that the seat of the fire was in the lower hold. Moreover, the temperature was not rising. By 2100, thermometers 1, 2 and 3 showed a decline in temperature. This indicated that CO_2 was being discharged into the correct level of the hold; the oxygen content was being reduced below 10% thereby inhibiting combustion.

Examination of the stowage plan revealed that the lower hold contained cardboard, wood, plastics, rubber, fabrics and other combustible items. These were ordinary combustible materials, not extremely hazardous, as was to be expected in a lower hold. Firefighters advanced two hoselines to holds 4 and 5, and both lines were charged to the nozzle. This was a precautionary measure; there was little expectation that they would be used.

Several hours after the master ordered the additional CO_2, the supplier's truck arrived with a large number of 100-pound cylinders. It also contained the equipment and tools needed to discharge CO_2 from the cylinders on the pier directly into the ship's system. No attempt was made to replace the empty cylinders in the CO_2 room; that could be done after the fire was extinguished. Now it was essential to maintain the tight seal on hold no. 5; continue the periodic discharge of CO_2 into the lower hold; keep charting the temperatures obtained from the four thermometers; continue to examine the hull and engine room bulkhead for signs of heating; and, most important, exercise patience.

The temperatures indicated by thermometers 1, 2 and 3 had dropped sharply by 0900 on Sunday, March 31. However, they were still higher than the outside temperature. Therefore, the master and his staff, Mitsui's port captain, and the fire chief and his officers held a conference to plan for the opening of hold no. 5. If it were opened too soon, the fire might rekindle when fresh air reached an area that was not completely extinguished. That would require a repeat of the entire CO_2 extinguishing procedure.

It was decided to open the hold on Monday, April 1, at 0900, after the stevedores had reported for work and firefighting personnel had been changed. When that time arrived, the temperature was about 60°F on all thermometers—a very favorable condition.

Before the hatch was opened, a company of firefighters was sent into the hold. They wore self-contained breathing apparatus, used a lifeline, and carried flashlights, portable hand radios, and an oxygen meter. Their objective was to reach the lower hold if cargo stowage permitted, to determine if there was any fire, smoke or heat in the vicinity of the fire. The fire officer reported by radio and then in person on the main deck that no fire, smoke or heat was found. The oxygen meter registered approximately 6% oxygen—too low for the combustion of ordinary materials.

The forward port section of the hydraulic hatch cover over hold no. 5 was opened at 1011. No forced ventilation was employed. Firefighters wearing breathing apparatus and carrying an oxygen meter entered to the upper 'tween deck, to offload containers that were on the upper 'tween deck hatch cover. When the oxygen content of the atmosphere was 21% the firefighters were able to work without the breathing apparatus. After the hatch cover was clear of cargo, it was time to open that cover.

Firefighting personnel were rotated from the hold to the main deck and vice versa. The firefighters donned their breathing apparatus when the hatch cover was opened. The atmosphere in the lower 'tween space was tested and found to be about 6% oxygen. The procedure employed in the upper 'tween space was repeated in the lower 'tween space. When cargo was removed from the lower 'tween hatch cover, the discoloration and warping of the hatch cover and deck plates readily revealed that the fire had been in the hold below.

Two charged hoselines were lowered into the hold, in case water was needed to extinguish any remaining fire or to reduce heat in the fire area, but not one drop of water was needed. When the lower 'tween deck hatch cover was opened, it exposed a large area of charred cargo but no fire, smoke or heat. It was 1145, and the fire was definitely extinguished.

Extensive investigation and legal action could not determine the cause of the fire. Though several possible causes were discussed, there was not sufficient evidence to support any one of them.

Conclusions

This general cargo hold fire was successfully extinguished with CO_2, without the use of water. The method can be used on other vessels equipped with CO_2 extinguishing systems and manned by well-trained crews. Certain factors, including the following, should be remembered:

1. Crewmen must be thoroughly trained in the use of CO_2 for extinguishing fire, so that they have confidence in the fire protection equipment and their own capabilities.
2. Early detection of smoke and/or fire is vital. Therefore, the automatic smoke detection system must be monitored at all times.
3. Frequent drills must be conducted with the CO_2 extinguishing system, to avoid errors when an actual fire occurs. Errors such as discharging the incorrect amount of CO_2 or discharging it into the wrong space have been made in the past.
4. CO_2 is the safest and most efficient extinguishing agent for hold fires. Water is not as effective as CO_2, may create a stability problem and could seriously damage cargo. CO_2 will not damage cargo.
5. A temperature chart showing the extinguishing process is helpful at the time of the fire, for the record and later for drill purposes.
6. In port the CO_2 supply can be replenished, so more CO_2 can be used in the periodic applications. At sea, the fire can be contained and extinguished with a very tight seal on the involved space, the correct initial application and smaller periodic applications. However, the space may have to remain sealed for a longer period of time.
7. Premature opening of the hold, and a resulting rekindling of the fire, can be disastrous. Crewmen may lose confidence in the CO_2 extinguishing method and resort to water hoselines—a more dangerous and less effective method of extinguishment.
8. Thermometer 4 should have been placed on the forward bulkhead in the engine room at the lower hold level. The temperature of the forward bulkhead of hold no. 4 was not important in this case.
9. Patience is extremely important when CO_2 flooding is used to extinguish a hold fire.

SS AFRICAN STAR

On the morning of March 16, 1968, at about 0340, the dry-cargo vessel SS *African Star* collided in a meeting situation with the tank barge *Intercity no. 11* in the lower Mississippi River, in the vicinity of mile 46 Above Head of Passes (AHP). The *African Star's* bow penetrated the *Intercity no. 11* on the after port side, at an angle of 45°. The motor towing vessel *Midwest Cities* was pushing two tank barges, *Intercity no. 11* and *Intercity no. 14* (the forward barge). The vessels are described in Table 3.1. The two tank barges were identical.

A few minutes before the collision, the *African Star* was making about 16 knots on a 140° true course; the *Midwest Cities* was making 6 knots on a 320° true course with a relative closing speed of 22 knots (Fig. 3.5). Visibility was good and each vessel had been advised of the other vessel's movements on its own radio frequency. Because of the lack of a common radiotelephone frequency, direct communication between the vessels was not possible.

Both vessels were equipped with marine radar units. Both units were in operation prior to and at the time of the casualty, but neither unit was being continuously observed by watch personnel. The pilot of each vessel sighted the navigation

Table 3.1. Descriptions of the Vessels.

	African Star	*Midwest Cities*	*Intercity nos. 11 and 14*
Service	Freight vessel	Tug	Tank barge
Gross tons	7971	165	1319
Net tons	4624	129	1319
Length	468.6 ft	83.2 ft	264 ft
Breadth	69.6 ft	24 ft	50 ft
Depth	29.2 ft	7.2 ft	11.1 ft
Propulsion	Steam	Diesel	None
Horsepower	8500	850	
Owner	Farrell Lines, Inc.	Natural Marine Service Inc.	Intercity Barge Co. Inc.

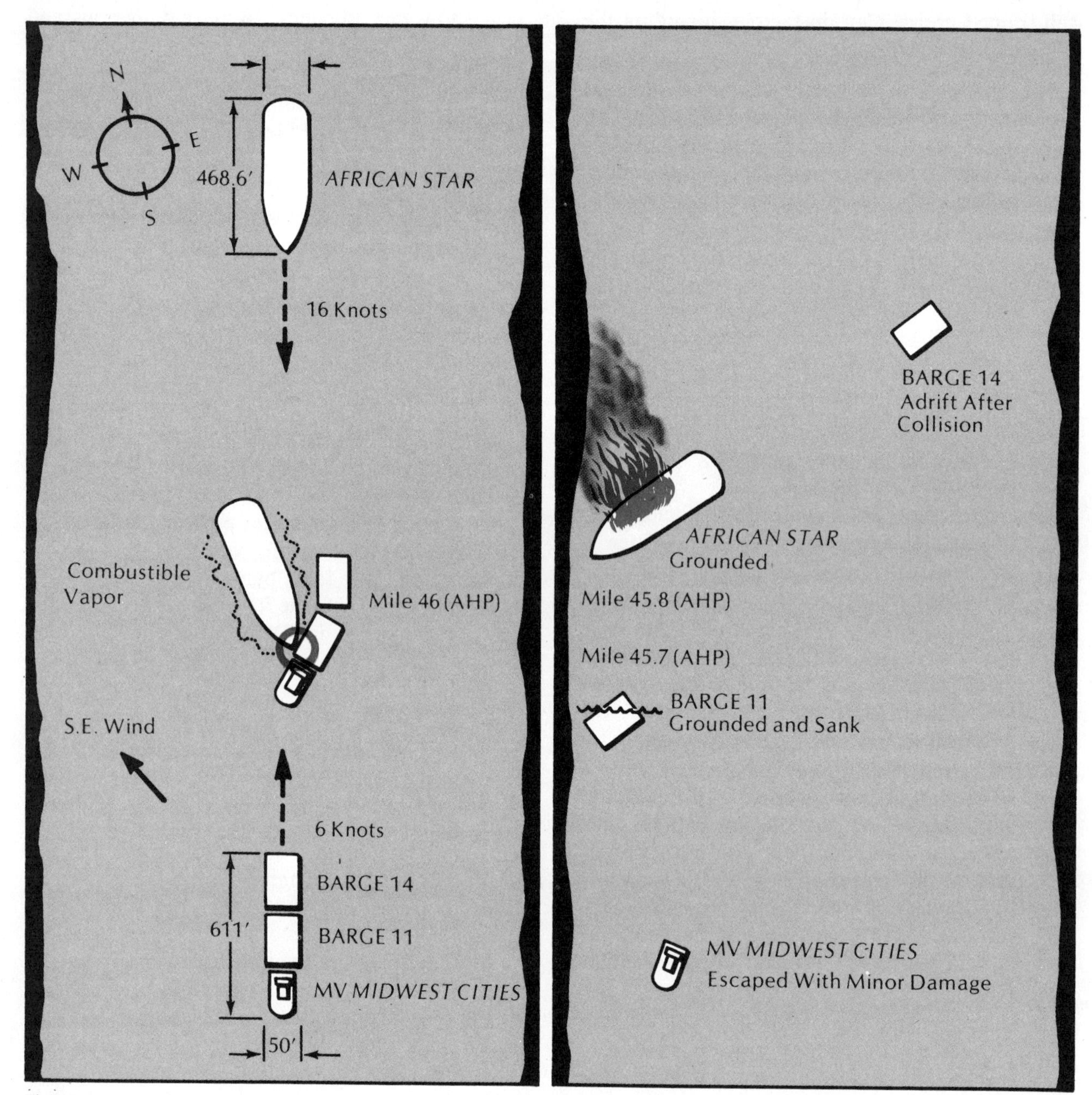

Figure 3.5. A. The situation before and during the collision between the *African Star* and *Intercity no. 11*. **B.** The results of the collision.

lights of the other vessel at about 1½ miles, and later sighted the other vessel on radar.

Witnesses in passing vessels reported that they could easily see the navigation lights on the *Midwest Cities, Intercity no. 14* and *African Star*. The movements of the vessels were not materially affected by wind or current. The steering gear and machinery of both vessels were in good operating order.

The *African Star* had a licensed pilot, but the *Midwest Cities* had an unlicensed pilot; however, both pilots had extensive experience on the Mississippi River. There was a lookout on the bow of the *African Star*, but none on the *Midwest Cities*. The master, third mate and helmsman were also on the bridge of the *African Star*.

The Collision

Different versions of the maneuvers were given by personnel on each of the two vessels.

Midwest Cities Version. The *Midwest Cities* was running parallel to the side of the river, about 250 feet from the east bank. The pilot considered it to be a head-and-head meeting situation, and the pilot sounded the appropriate one-blast whistle signal for a port-to-port passing. The *African Star* responded with one blast. He assumed a safe passage until the *African Star* sounded two blasts when her bow was abeam the lead barge. He saw the *African Star*'s green sidelight and responded with one blast. He then blew four blasts on the whistle, backed full astern from full ahead and put the rudder hard right. However, it was too late to avert a collision between the *African Star* and barge *Intercity no. 11*.

African Star Version. The pilot of the *African Star* stated that his vessel was slightly west of mid-river when he sighted the *Midwest Cities*' two white tow lights and green sidelights on his starboard bow. The tow appeared to be favoring the west bank and running parallel to it. It appeared to him to be a normal starboard-to-starboard meeting situation, not a head-and-head meeting. When the *Midwest Cities* tow was ½ to ¾ mile ahead, he sounded two short blasts on the whistle, but no reply was heard. As the pilot headed for the radar, the third mate called his attention to the tow crossing his starboard bow showing red sidelights. This was about 2 minutes after the two-blast signal was sounded. Hard right rudder, one blast and then emergency full astern were ordered and executed. By this time, the situation was beyond the point of corrective action—a collision was unavoidable. Full astern was in effect a minute before the collision.

In his analysis of the incident, the commandant of the U.S. Coast Guard concluded that the witnesses gave such conflicting testimony that it was impossible to reconstruct the events leading up to the collision.

The Fire

Intercity no. 11 was loaded to a draft of about 9 feet 6 inches, corresponding to approximately 19,000 barrels of crude oil. An analysis of the Louisiana "sweet" crude it carried revealed a 30.6° API a flash point (Pensky Martens) of 80.0°F, and a Reid vapor pressure of 3.2 psia, which categorized the product as a grade C flammable liquid.

When the collision occurred, the general alarm was sounded on the order of the master of the *African Star*. At this time, the oncoming watch personnel were in varying degrees of readiness and, except for those on watch, all crew members and passengers were asleep or resting in their quarters.

In less than a minute, fire broke out and several explosions occurred. The most likely source of ignition was high heat due to metal-to-metal friction or sparks, produced when the barge was sheared by the bow of the *African Star*. Another possible source of ignition was sparks generated by the severing of the electrical cable leading to the navigation lights on *Intercity no. 14*.

When fire broke out on the barge and in the surrounding water, the pilot of the *Midwest Cities* backed full to break the port wire and to clear the intense fire. He estimated it took about a minute to get free; his vessel was backing toward the west bank. *Intercity no. 11* grounded and sank near the west bank at mile 45.7 (AHP). The *Midwest Cities* was downwind of the point of collision and escaped with only minor damage.

The southeasterly wind carried flammable vapors over the *African Star* from bow to stern (because of the vessel's position relative to the wind direction). The flammable vapors ignited, engulfing the vessel in flames. The pilot backed clear and intentionally grounded the vessel on the west bank at mile 45.8 (AHP). The tarpaulins had been ignited, and there were fires in holds 2, 4 and 5. Containers and other deck cargo were burning, as was the paint on the ship. Dense smoke filled the engine room and accommodation spaces.

Firefighting and Rescue

Problems were encountered in lowering the lifeboat and launching the inflatable life raft; the boat cover and man ropes had burned, and the plastic cover of the life raft had ignited. The intense fire, heat and smoke in the quarters gutted the passageways, and a number of passengers and crew members were trapped. Several people tried to escape through portholes when they found that the passageways outside their quarters were impassable. Others were burned when their life preservers and clothing ignited.

For a while the fire and heat on the port side were too intense to endure. There was some minor confusion during the first few minutes after the alarm was sounded. However, this was quickly dispelled under the leadership of the master and his officers. After the *African Star* was grounded, the master went to the cabin deck to see to the safety of the passengers and crew. During this time, he became seriously burned about the feet, face and hands. As a result, he was immobilized and had to be carried back to the bridge by the crew.

At first, burning oil on the water surrounding the vessel prevented personnel from jumping overboard to get away from the burning vessel. The second mate gathered a number of passengers and crew into a small room on the *African Star* for refuge until the fire subsided. He then supervised the extinguishment of small fires in and around no. 1 lifeboat. By this time, the current and the movement of the *African Star* had separated the vessel from the oil burning on the water, the lifeboat was lowered to the edge of the deck and the injured crew members and passengers were assisted into the boat and lowered to the water's edge. Other crewmen and passengers were able to climb or jump into the water and swim ashore.

The second mate observed large fires burning aft on the main deck. He organized a firefighting team that advanced hoselines to the area. They were successful in confining the deck fires and cooling the flammable-liquid cargo.

An oiler in the engine room was forced to leave because of difficulty breathing in the smoke. However, the chief engineer, third assistant engineer, and fireman/watertender continued to maintain the engine room plant in full operation. Power was maintained to keep the vessel aground, the lights on and the fire and bilge pumps in operation.

Rescue operations had commenced swiftly following the *Midwest Cities* request for immediate assistance via the marine operator in New Orleans. Badly burned victims were quickly evacuated by U.S. Coast Guard helicopters. This operation is credited with saving the lives of a number of people injured on the *African Star*. The *Midwest Cities,* a New Orleans fireboat and a local ferry with fire apparatus on board assisted Coast Guard boats in fighting the fire.

Firefighting was complicated by inaccessibility to the cargo manifest of hazardous materials located in the chief mate's room. In addition, a number of deck fire hoses had been burned. The combustibles on deck and in the holds continued to burn after the vapor and oil spray fires had subsided.

Firefighting by the *African Star* crew controlled the fire until the U.S. Coast Guard vessels and other help arrived. The fire in hold no. 5 was contained by use of the ship's CO_2 extinguishing system.

At about 0530, the fires on board the *African Star* had been brought under control, and the *Midwest Cities* departed to retrieve *Intercity no. 14,* adrift in the river. *Intercity no. 14* was undamaged.

Consequences

The many fatalities and injuries sustained on board the *African Star* were due to the rapid spread of fire, the heat and smoke in living spaces and the burning oil on the water surrounding the vessel, which kept most personnel from immediately jumping overboard. Of a total of 11 passengers and 52 crewmen on the freighter, 2 passengers were killed and 9 were injured; 15 crew members were killed, 4 were missing and presumed dead, 31 were injured, and 2 escaped injury. Many more lives would have been lost, but for the gallant efforts and bravery of *African Star* crewmen and others involved in the rescue and firefighting operations.

A collision and fire of this magnitude must point up both weaknesses (i.e., areas where seamen can learn from the mistakes of others) and strengths (i.e., examples of leadership, teamwork and heroism). Some of the more important lessons to be learned include the following:

1. Whistle signals are not of themselves a reliable means of communicating a vessel's passing or turning intentions. Bridge-to-bridge radiotelephone communication on a single frequency would probably have prevented this tragedy. It is now required by law.
2. Uncertainties and difficulties are experienced in applying the inland rules of the road to arrange a safe passing. Passing requires the use of visual and verbal communication in both directions, plus good judgment.
3. A properly equipped vessel can withstand a serious collision and fire. A disciplined and well-trained crew can keep the vessel afloat, maintain control of the wheelhouse and engine room and successfully combat the fire.
4. Leadership, courage and discipline are essential traits for officers and crewmen in the merchant marine. The value of these traits becomes most evident in an emergency situation such as a serious fire.

SS HANSEATIC

At approximately 0730 on September 7, 1966, while the German passenger vessel SS *Hanseatic* was docked at Pier 84, North River, New York, a fire started in the diesel generator room on B deck level. Fuel from a leaking line on No. 4 diesel generator ignited and burned. The fire then extended up an intake ventilator through seven decks and was finally brought under control by the New York City Fire Department at approximately 1430 the same day. There were no reportable personal injuries, but as a result of the fire, the vessel suffered damage estimated at one million dollars.

The SS *Hanseatic* was a passenger vessel 674 feet in length, 48.6 feet in depth and 30,300 gross tons. The vessel, a twin-screw turbine owned by the Hamburg-Atlantic Lines, was built in 1930 at Glasgow, Scotland. The weather was clear and fair and was not a factor in the casualty. The SS *Hanseatic* underwent alterations and renovations in the passenger areas upon her transfer to the German flag in 1958. At the time the fire occurred, the vessel was maintained in class with the American Bureau of Shipping and German Lloyds. Her last quarterly inspection completed on August 11, 1966, at New York, indicated that the vessel was in compliance with the safety certificate required by the International Convention for Safety of Life at Sea (SOLAS 60).

The vessel was constructed with 11 watertight bulkheads and 6 main vertical fire zones. Her machinery spaces included boiler room no. 1, boiler room no. 2, the diesel generator room and the main engine (turbine) room. There were four generators in the diesel generator room, located forward of the turbine room. The vessel also was equipped with two steam-driven turbogenerators, one on either side of the turbine room. The two turbogenerators were capable of carrying the ship's electrical load; however, it was normal operating practice to split the load between the two turbogenerators and two diesel generators.

The four diesel generators were supplied with air by mechanical means. Vent ducts starting on either side of the after funnel passed straight down through all the decks and terminated in branches over each diesel engine. The blower motors were located on the navigating deck. Each duct was fitted with balancing dampers where it branched over the diesel engines and with a positive closing damper at the navigating deck level. The ducts measured approximately 4 × 4½ ft in cross section. From the A deck to a location between the upper and promenade decks, they passed up the outboard after corners of an escape trunk, measuring 10 × 20 ft, located abaft the bulkhead at frame 100. At the upper deck level the escape trunk passed forward of frame 100 into the fiddley, but the vent ducts continued straight up to the navigating deck. From A deck to the upper deck, the escape trunk was insulated with sheet asbestos except on its forward bulkhead, which was common to the boiler casing. The insulated areas included the inboard and forward sides of the vent ducts. The outboard and after sides of the vent ducts faced the passenger areas and were not insulated.

The Fire

The SS *Hanseatic* had arrived in New York at 1200 on September 6, 1966. She was scheduled to sail at 1130 on September 7, with embarkation of passengers to commence at 0800. The watchstanding fourth engineer was in charge of the watch in the turbine and diesel generator rooms. At 0730 he heard a knock in the no. 4 diesel generator, located on the starboard side of the diesel generator room. At this time the no. 1 and no. 4 diesel generators were on the line with the turbogenerators. He told the two engineer cadets then on watch to shut down the no. 4 diesel generator. He called the switchboard room and told the watch electrician to take no. 4 diesel generator off the line. He then proceeded immediately to the diesel generator room and saw that, although the cadet had attempted to shut down the diesel engine, it continued to run. Diesel fuel was spurting from the low-pressure gravity-feed fuel line above no. 5 cylinder onto the no. 4 diesel generator and igniting. The fourth engineer sounded the engineer's alarm, and all three men attempted to extinguish the flames with the portable dry chemical and foam fire extinguishers immediately available in the area.

The engineer's alarm sounded in the engineer's office and quarters located on the sun deck. The staff chief engineer proceeded below and observed the fire from the turbine room doors aft of the diesel generator room. He then went forward through the diesel generator room to the door to boiler room no. 2. From this position, he and other crew members attempted to extinguish the fire with semiportable foam extinguishers obtained from the two boiler rooms. When he started using this portable equipment, the fire was concentrated at the upper level; the foam was directed onto the flames from the lower level, through the gratings at the engine head level. However, the fire continued to spread. The staff chief engineer then proceeded to A deck. With the chief engineer, he began to shut off the diesel fuel and bunker oil tanks at the remote stations located in the port and starboard passageways.

At approximately 0745, while the staff chief engineer was in the process of shutting off the fuel tanks, all power was suddenly lost on the vessel. The staff chief engineer proceeded to the emergency generator room located aft on the main deck, leaving the chief engineer to finish closing the remote shutoffs. Although the emergency generators were in operation within 5 minutes, the emergency lighting was not energized in the machinery spaces or any other areas forward of the diesel generator room. Sometime during the morning, the vessel's boilers were secured because of heavy smoke entering the ventilation system.

The fire was reported to the bridge by telephone approximately 10 minutes after it started. The quartermaster on watch called the first officer, who in turn notified the staff captain and the master. The first officer and the staff captain proceeded to the scene of the fire by different routes. The first officer started into boiler room no. 2 and was proceeding toward the door of the diesel generator room when the power failure occurred. He retraced his steps and met the staff captain in the A deck port passage, outside the switchboard room. The staff captain ordered the first officer to sound the fire alarm and notify the shoreside fire department. Following the sounding of the alarm on the vessel's whistle, the crew reported to their fire stations, securing ventilation and closing fire doors. The crew did not attempt to charge the ship's firemain by means of the emergency bilge and fire pump, because the fire department preferred to use its own hose equipment. However, the emergency bilge and fire pump energized by the emergency diesel generator were used later to pump out the machinery space bilges.

Fire Department Operations

The New York City Fire Department received the first alarm at 0746; fire apparatus and firefighters arrived on the pier at about 0750. By this time the heat was increasing in the vent ducts and in the escape trunk. The paint on the two uninsulated vent duct faces in the passenger spaces on all decks began to smolder. Fire department personnel were led to the scene of the fire and were assisted by crewmen who were familiar with the arrangement of the vessel. As soon as the nature of the fire was known to the fire department, it was evident that quantities of foam would be required. They immediately arranged to bring the necessary equipment to the scene.

Sometime during the first hour, after additional firefighting help responded, they became aware that the diesel generator room was equipped with a fixed CO_2 extinguishing system. However, that system had not yet been utilized. All doors to the diesel generator room were then closed, after personnel were evacuated. At approximately 0940, the CO_2 system was activated. It failed to extinguish the fire.

Up to this time, the primary fire was contained within the machinery spaces by the steel bulkheads of the escape trunk and the vent ducts. However, secondary fires had started in the passenger and service areas on all decks, by direct conduction of heat through the steel bulkheads. Smoke was generated on all decks and became so dense that breathing apparatus had to be used. As noted earlier, the vent ducts at frame 100 passed through all the decks; they were not insulated on the two sides that faced the passenger spaces. The joiner construction consisted of wood furring bolted directly to the bare steel of the bulkhead. Plywood and pressed-wood paneling was attached to the furring to provide the interior decor. Similarly a plywood false ceiling was attached to wood furring suspended below the steel overhead. There was direct communication between the concealed spaces behind the bulkhead paneling and the hanging ceiling.

The furring strips began to smolder where they butted against the uninsulated faces of the vent ducts. On R deck, in way of the vent duct at frame 97, the crackling of fire was heard behind the plywood panels. The paneling in this area was removed from the bulkhead and the overhead. Water was directed on the burning wood to extinguish the fire. This was the most serious outbreak of fire reported outside the machinery space. By this time conditions on all other decks were generally the same: heavy smoke, smoldering furring strips and blistering paint. On each deck a number of firefighters were removing the combustible ceiling and linings.

Although a sprinkler system was installed throughout the accommodation and service areas, it was activated only in the starboard passageway on A deck, frames 90 to 100. The sprinkler discharge ceased when the gravity storage tank was exhausted and the automatic pump did not start because of the power failure.

The fire department had three foam lines in service in an attempt to smother the fire in the diesel generator room. With sufficient personnel on all decks to cool the hot bulkheads, fire department personnel in the turbine room advanced their foam lines closer to the seat of the fire. By 1430 the fire was extinguished, although the residual heat on all decks and in the diesel generator room remained intense.

The use of water was generally limited to cooling hot bulkheads and extinguishing small blazes and smoldering woodwork. As a consequence, firefighting water did not rise above the door sills in the upper decks in the fire area, or above the floor plates in the machinery spaces. During the course of the fire there was no appreciable change in the vessel's draft or trim, and the list never exceeded $1°–1\frac{1}{2}°$.

Postfire Analysis

The vessel's fixed CO_2 system consisted of a central supply of about 3300 pounds of CO_2 gas and the piping connecting this supply to the forward holds, no. 1 boiler room, no. 2 boiler room, the diesel generator room and the after holds. Valves located in three manifolds on A deck controlled the distribution of all or part of the supply to any of these spaces. The CO_2 room was located on the sun deck. Two manually operated release mechanisms were provided, either of which could activate the system. One was located on a bulkhead directly outside the CO_2 room; the other was located at one of the distribution manifolds. The system was not used in the initial firefighting effort because it was believed to be a bilge flooding system, and the seat of the fire was several feet above the bilge. Later the cylinders were triggered by pulling the release cable within the CO_2 room instead of at one of the release stations. Subsequent examination of the system indicated that all 50 cylinders had not been discharged. It was also determined that about half the available CO_2 would have been sufficient for total flooding of the diesel generator room.

Following the casualty, examination of the electrical circuits revealed the following: The main feed cable from the starboard turbogenerator passed straight up to A deck and then forward into the switchboard room, located immediately over the diesel generator room. The main feed cable from the port turbogenerator passed across the turbine room to the vessel's centerline. It then passed forward to a point over the forward end of no. 3 diesel generator, and then vertically through the deck into the switchboard room. The fire had originated approximately 10 feet away from the point where the cables passed through the deck. Apparently, the heat had melted these cables within 15 minutes after the fire started. This caused a dead short in the port turbogenerator. The loss of this generator, and of no. 1 and no. 4 diesel generators after their fuel was shut off, overloaded the starboard generator, tripping it and causing the power failure. Further investigation revealed that some of the circuits from the emergency generator passed through the diesel generator room enroute to the main switchboard; these were also destroyed.

All mechanical and electrical equipment in the diesel generator room was exposed to extreme heat and flame and suffered considerable damage. In addition, the steel bulkhead between boiler room no. 2 and the escape trunk was buckled at the A deck and restaurant deck levels. Fire damage in the accommodation and service areas on all decks was limited to an area within a few feet of the ventilation ducts to the diesel generator room and consisted primarily of charred and burned furring strips and plywood. Firefighting efforts caused additional damage to the ceilings and linings on all decks, broken glass in windows and port lights, and general damage due to smoke and water.

Some conclusions may be drawn from firefighters' and crew members' experience with this fire:

1. The fire originated in the vicinity of the no. 5 cylinder-head pump on the no. 4 diesel generator when a low-pressure feed pipe failed. Diesel oil was sprayed directly onto the engine head and exhaust manifold.
2. Due to subsequent fire damage, the cause of the failure of the fuel line could not be determined. However, the probable cause was one or both of the following:
   a. A malfunction in no. 4 diesel generator had created excessive vibration, which caused the fuel line to fail.
   b. A rubber fuel line in the diesel generator room failed owning to deterioration and/or embrittlement.
3. The flames and heat of the initial fire cracked the gauge glasses on the diesel oil tanks located immediately forward and above the diesel generators. This leaking oil fed the fire with approximately 5½ tons of liquid fuel, causing the fire to spread throughout the bilges of the diesel generator room and to continue to burn until this fuel supply was consumed.
4. The primary fire passed vertically through the ship from the bilges to the navigation deck. However, it was contained within the steel bulkheads forming the boundaries of the vent ducts and the escape trunk extending above the diesel generator room.
5. Secondary fires, smoke and fumes were generated on all decks by direct conduction of the heat of the primary fire through the steel bulkheads of the vent ducts to the wood furring and linings which were fitted to these bulkheads.
6. The horizontal spread of the secondary fires was restricted to the immediate vicinity of the vent ducts on each deck by the following:
   a. The prompt efforts of firefighters in uncovering and combating the secondary fires before they were able to extend.

b. The action of firefighters in cooling down the hot steel bulkheads and removing all combustibles attached directly to these bulkheads.
c. The action of firefighters in directing the opening of the vent dampers and other closures on the navigation deck, directly above the primary fire, to vent the heat and smoke of the primary fire.
d. The fire resistive insulation fitted in certain areas of the escape trunk, which effectively prevented the conduction of sufficient heat to cause fire or smoldering in the combustible materials attached to the steel bulkheads in these locations.

7. The primary fire produced sufficient heat to melt a section of the main power cables where they passed through the diesel generator room. This disconnected all the generators from the main switchboard except for the starboard turbogenerator. The latter became overloaded and tripped out, resulting in a total power failure approximately 15 minutes after the fire started.
8. The power failure deenergized two of the vessel's three fire pumps. This resulted in a complete loss of pressure in the firemain throughout the vessel. The vessel's emergency bilge and fire pump was energized from the emergency generator. Because the fire department supplied its own firefighting water, the pump was not utilized.
9. The sprinkler system functioned initially, both by discharging in the involved area and by indicating the existence of the fire. However, due to the power failure, the sprinkler system did not continue to operate as it should. Because of the circumstances, this did not contribute to the severity of the casualty because areas of fire extension were adequately protected by hoselines.
10. The initial firefighting efforts of the crew were ineffective for the following reasons:
    a. The flammable liquid fire was not sufficiently confined to be brought under control with portable dry chemical extinguishers.
    b. The two portable foam extinguishers that were used could not distribute a cohesive blanket over the fire area because the foam stream had to be directed onto the fire from underneath and through a deck grating.
    c. The vessel's CO_2 system, which had sufficient capacity to totally flood the diesel generator room, was not used when the fire was first discovered.
11. Subsequent efforts of the vessel's crew in assisting the fire department were orderly, efficient and well directed. Their performance contributed materially to the successful extinguishment of the fire.
12. It was the opinion of the Marine Board of Investigation, U.S. Coast Guard, that the vessel's crew could not have successfully combated the fire had it occurred while the vessel was at sea. The intense heat of the primary fire, the effect of the power failure on the vessel's firefighting capability, and the combustible interior paneling in the accommodation and service areas would have made extinguishment extremely difficult.
13. The delay of 10 minutes in reporting the fire to the master and the delay in notifying the local fire department were evidence of a weakness in the firefighting training of the crew.
14. If the crew had been alerted earlier, simultaneous firefighting operations could have been instituted. While the fire was being attacked with portable fire extinguishers, other crewmen could have run out hoselines with fog nozzles and/or applicators. Other members of the fire party could have prepared to activate the CO_2 system and shut down ventilation promptly in the event the direct attack was not successful.

Comparative Study

A detailed comparison of the structural and equipment standards that were applicable to the SS *Hanseatic* with those applicable to large ocean-going passenger vessels of the United States was conducted by the Technical Division of the Office of Merchant Marine Safety. This comparison was limited to the locations involved in or affected by the fire. It was undertaken to find ways in which United States flag vessels could be improved. The most critical items were felt to be the following:

1. Materials within accommodation and service spaces
2. Ventilation ducts

3. Automatic sprinkler systems
4. CO_2 extinguishing systems
5. Tubing used in fuel lines
6. Gauge glasses on diesel oil tanks
7. Routing of main turbogenerator cables
8. Emergency power and lighting systems.

The following are some of the results of the comparison.

1. The *Hanseatic* was constructed and renovated in accordance with method II fire protection as described in the International Convention for Safety of Life at Sea (SOLAS 1948), which permitted extensive use of combustible materials. United States vessels are, and have been since 1936, constructed essentially in accordance with method I fire protection. Method I protection severely limits the use of combustible materials and requires internal divisional bulkheading capable of preventing the passage of flame for extended periods.
2. The primary fire in the *Hanseatic* passed vertically from the diesel generator room via the ventilation ducts. Apparently there were no automatic fusible-link dampers in these ducts, nor was insulation fitted to the ducts where they faced passenger spaces. A U.S. vessel of the same vintage and history would probably have ducts from machinery spaces insulated with fire resistive insulation, and automatic fire dampers would likely be fitted.
3. The *Hanseatic* was equipped with an automatic sprinkler system for the protection of passenger accommodation and service spaces in accordance with construction standards under method II of the 1948 SOLAS convention. A similar U.S. vessel would have employed method I standards, relying on containment by incombustible fire barriers; an automatic sprinkler system would not have been fitted.
4. The quantity of CO_2 protecting the auxiliary machinery space was sufficient to totally flood the space. However, to be effective, this system had to be activated as soon as the fire was discovered. The delay of more than 2 hours in actuating the system was critical and rendered the system ineffective. It is doubtful that systems presently installed on U.S. vessels would be capable of extinguishing a fire of such magnitude after a similar delay.
5. The tubing used in the fuel line that failed was evidently a short length of rubber. Rubber tubing is not permitted on U.S. vessels for this type of service. Where short lengths of flexible nonmetallic hose are permitted, they must be wire reinforced and have a fire resistive cover.
6. There was no testimony as to whether or not the gauge glasses on the diesel oil tanks were constructed of heat resistant glass or were equipped with automatic closure devices. In light of what was observed after the fire, it would appear that they had neither. Both heat resistant materials and automatic closure devices, to protect against spillage if the gauge glass ruptures, are required on U.S. vessels.
7. The failure of normal power was evidently due to the routing of the main turbogenerator cables through the forward diesel generator room, where the fire originated, and then vertically through the deck to the switchboard room. The SOLAS and U.S. Coast Guard regulations governing the relative positions of main generators, cable runs and switchboards are minimal. Arrangements similar to those on the *Hanseatic* may possibly be found in U.S. passenger vessels.
8. After the power failure, which occurred within 15 minutes of the start of the fire, the emergency generators came into operation satisfactorily. However, emergency power and lighting did not come on in the machinery spaces or other areas forward of the diesel generator room at frame 100. Cables running forward from the emergency switchboard must have been routed through the diesel generator room to the boiler room so that they were destroyed by the fire. Moreover, it was necessary to energize the emergency lighting and power system manually. Manual systems are permissible on older U.S. passenger vessels. However, those contracted for since November 19, 1952, are equipped with a self-contained power source with automatic transfer equipment, or diesel generators with automatic starting and transfer equipment.

BIBLIOGRAPHY

SS *San Jose:* Report, Officer in Charge, USCG, Marine Inspection, San Francisco, Calif. 94126, 28 January 1968.

SS *Hanseatic:* Report, Commander, 3rd Coast Guard District, U.S. Custom House, New York, N.Y. 10004, 22 September 1966.

MV *Rio Jachal:* WNYF Magazine, Fire Department City of New York, 1st Issue 1963. Fire Report, NYFD.

MV *Alva Cape* and SS Texaco *Massachusetts:* WNYF Magazine, Fire Department, City of New York, Fire Report, NYFD.

SS *African Star:* Report, Officer in Charge USCG, New Orleans, La.

Barnaby KC: *Morro Castle:* Some Ship Disasters and their Causes.
Great Ship Disasters, by A. A. Hoehling.
Fire Aboard, by Frank Rushbrook.

Barnaby KC: SS *Lakonia:* Some Ship Disasters and their Causes.
Great Ship Disasters by A. A. Hoehling.

SS *Yarmouth Castle:* Report, Commander, 7th Coast Guard District, Miami, Florida, 23 February 1966.
Great Ship Disasters by A. A. Hoehling.

MV *San Francisco Maru:* Fire Reports, Fire Department City of New York, March 30, 1968 and April 1, 1968.

SS *Normandie:* Investigation Committee Report, US Congress.
Some Ship Disasters and their Causes by K.C. Barnaby.
Great Ship Disasters by A. A. Hoehling.
Fire Aboard, Frank Rushbrook.
Other Valuable Sources of Information:

Proceedings of the Marine Safety Council (U.S. Coast Guard)

Marine Casualty Reports (U.S. Coast Guard)

National Transportation Safety Board Reports

Firefighting Part II

If the first part of this book makes only one point, it is this: The surest way to protect a ship and its crew from fire is to prevent shipboard fires. At sea or in port, ship fire of any size will result in damage to the ship, its cargo or both. If the fire has gained headway and is difficult to control, it may also cause injuries or deaths. An uncontrolled fire may mean loss of the vessel and a life-or-death situation for the crew and passengers. In port, such a fire could spread to land installations.

In Part I, Chapters 1 and 2 discuss this first line of defense against fires—prevention—and Chapter 3 shows that shipboard fires do occur. Fires start small but grow quickly. Damage and the danger of injuries and deaths can be minimized by early detection, control and extinguishment, all of which will be discussed in Part II.

To fight fire effectively, it is important to know the enemy. Chapter 4, the first chapter in this part, is a discussion of what fire is and how it destroys. Chapter 5 covers the four classes, or types, of fires. Fires are classified according to the properties of the materials involved and, thus, according to the most effective means of control and extinguishment.

The earlier a fire is discovered, the less chance it has to spread, and the sooner the crew can begin to fight it. Several different types of fire detection systems are used aboard ships. Chapter 6 covers these systems, from patrols to sophisticated automatic alarms.

The next four chapters cover firefighting equipment and techniques once fire is discovered. Extinguishing agents are covered in Chapter 7; portable and semiportable equipment in Chapter 8; fixed, or built-in, equipment in Chapter 9; and crew firefighting operations in Chapter 10. The last two chapters deal with specialized firefighting problems. Chapter 11 covers tugboats and towboats, and Chapter 12 covers offshore installations.

The information contained in Part II represents the work and the experience of many people, including shipbuilders, equipment manufacturers, seamen, engineers, professional firefighters and scientists. Yet the information alone is almost useless in the event of a shipboard fire. It must be combined with a knowledge of the ship's construction features, firefighting equipment and cargo if it is to be used effectively. In other words, the success of a firefighting operation—and perhaps survival—will depend on how well the crew has been trained, and how well they know and maintain their vessel and its firefighting systems.

Fire

Once a fire starts, it will continue to burn as long as there is something to burn. But what causes a fire to start, and how does it burn? Why are some substances more or less flammable than others? Those questions are answered in this chapter. In addition, we look at how fires spread and how they can be kept from spreading.

CHEMISTRY OF FIRE

Oxidation is a chemical process in which a substance combines with oxygen. During this process, energy is given off, usually in the form of heat. The rusting of iron and the rotting of wood are common examples of *slow* oxidation. Fire, or combustion, is *rapid* oxidation; the burning substance combines with oxygen at a very high rate. Energy is given off in the form of heat and light. Because this energy production is so rapid, we can feel the heat and see the light as flames.

The Start of a Fire

All matter exists in one of three states—solid, liquid or gas (vapor). The atoms or molecules of a solid are packed closely together, and those of a liquid are packed loosely. The molecules of a vapor are not packed together at all; they are free to move about. In order for a substance to oxidize, its molecules must be pretty well surrounded by oxygen molecules. The molecules of solids and liquids are too tightly packed to be surrounded. Thus, *only vapors can burn.*

However, when a solid or liquid is heated, its molecules move about rapidly. If enough heat is applied, some molecules break away from the surface to form a vapor just above the surface. This vapor can now mix with oxygen. If there is enough heat to raise the vapor to its ignition temperature, and if there is enough oxygen present, the vapor will oxidize rapidly—it will start to burn.

Burning

What we call burning is the rapid oxidation of millions of vapor molecules. The molecules oxidize by breaking apart into individual atoms and recombining with oxygen into new molecules. It is during the breaking–recombining process that energy is released as heat and light.

The heat that is released is *radiant* heat, which is pure energy. It is the same sort of energy that the sun radiates and that we feel as heat. It radiates, or travels, in all directions. Thus, part of it moves back to the seat of the fire, to the "burning" solid or liquid (the fuel).

The heat that radiates back to the fuel is called *radiation feedback* (Fig. 4.1). Part of this heat

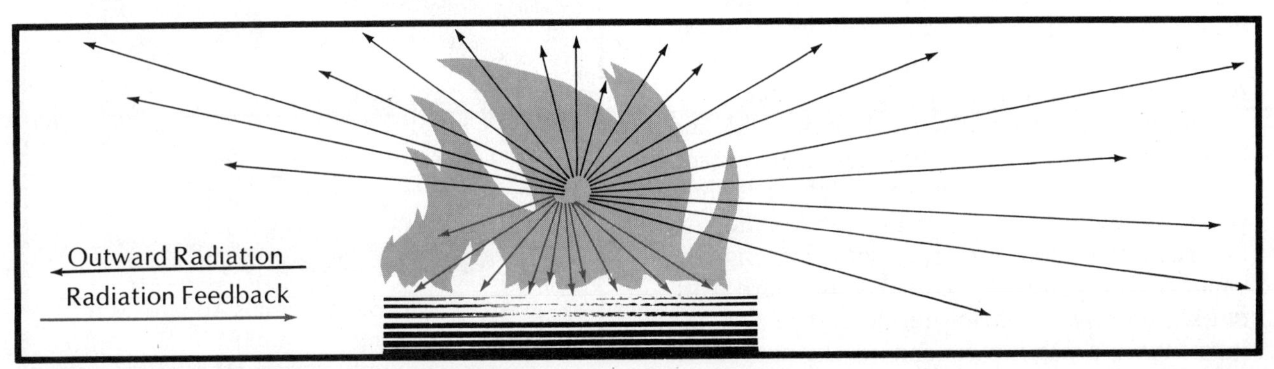

Figure 4.1. Radiation feedback is heat that travels back to the fuel from the flames. It releases vapor from the fuel, then ignites it.

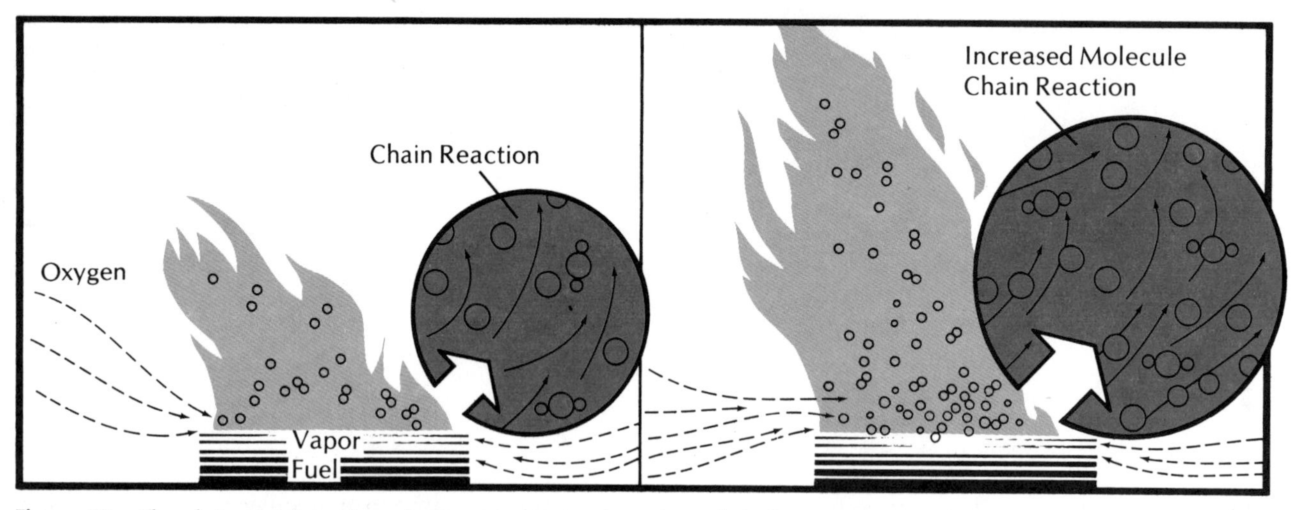

Figure 4.2. The chain reaction of combustion. **A.** Vapor from heated fuel rises, mixes with air and burns. It produces enough heat to release more vapor and to draw in air to burn that vapor. **B.** As more vapor burns, flame production increases. More heat is produced, more vapor released, more air drawn into the flames and more vapor burns. The chain reaction keeps increasing the size of the fire.

releases more vapor, and part of it raises the vapor to the ignition temperature. At the same time, air is drawn into the area where the flames and vapor meet. The result is that the newly formed vapor begins to burn. The flames increase.

The Chain Reaction

This is the start of a *chain reaction:* The burning vapor produces heat which releases and ignites more vapor. The additional vapor burns, producing more heat, which releases and ignites still more vapor. This produces still more heat, vapor and combustion. And so on (Fig. 4.2). As long as there is plenty of fuel available, the fire continues to grow, and more flame is produced.

After a time, the amount of vapor released from the fuel reaches a maximum rate and begins to level off producing a steady rate of burning. This usually continues until most of the fuel has been consumed. Then there is less vapor to oxidize, and less heat is produced. Now the process begins to break down. Still less vapor is released, there is less heat and flame, and the fire begins to die out. A solid fuel may leave an ash residue and continue to smolder for some time. A liquid fuel usually burns up completely.

Although we have discussed only solid and liquid fuels, there are, of course, flammable gases. Gases burn more intensely than solids or liquids, because they are already in the vapor state. All the radiation feedback goes into igniting the vapor, so it is more fully ignited. Gases burn without smoldering or leaving residues. The size and intensity of a gas fire depend on the amount of fuel available—usually as a flow from a gas pipe or bottle.

THE FIRE TRIANGLE

From the preceding section, it is obvious that three things are required for combustion: *fuel* (to vaporize and burn), *oxygen* (to combine with fuel vapor), and *heat* (to raise the temperature of the fuel vapor to its ignition temperature). The *fire triangle* (Fig. 4.3) illustrates these requirements. It also illustrates two facts of importance in preventing and extinguishing fires:

1. If any side of the fire triangle is missing, a fire cannot start.
2. If any side of the fire triangle is removed, the fire will go out (Fig. 4.4).

Solid Fuels

The most obvious solid fuels are wood, paper and cloth. These are found aboard ship as cordage,

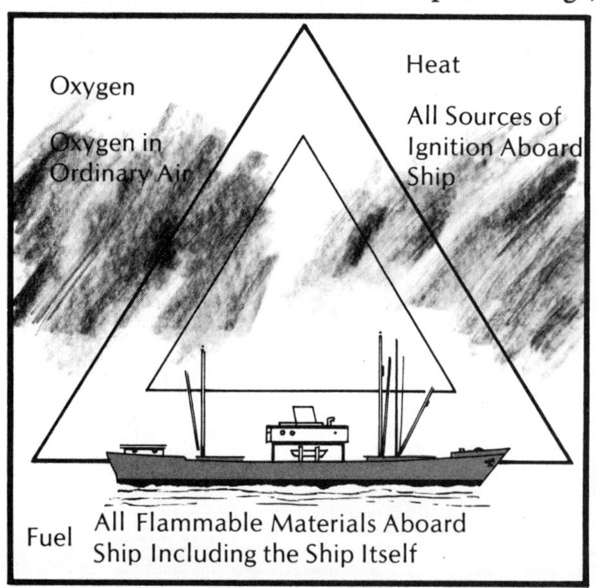

Figure 4.3. The fire triangle: fuel, oxygen and heat are necessary for combustion.

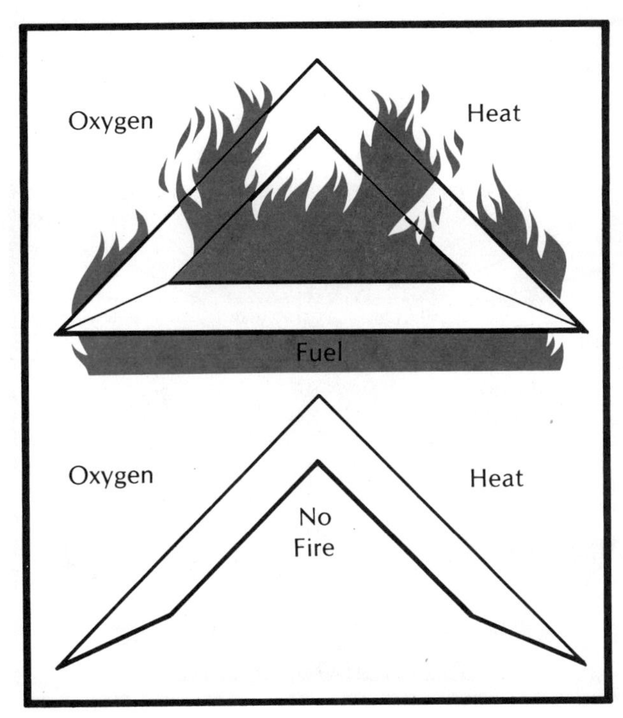

Figure 4.4. Fire cannot occur (or exist) if any part of the fire triangle is missing or has been removed.

canvas, dunnage, furniture, plywood, wiping rags and mattresses. The paint on bulkheads is also a solid fuel. Vessels may carry a wide variety of solid fuels as cargo, from baled materials to goods in cartons, and loose materials such as grain. Metals such as magnesium, sodium and titanium are also solid fuels that may be carried as cargo.

Pyrolysis. Before solid fuel will burn, it must be changed to the vapor state. In a fire situation, this change usually results from the initial application of heat. The process is known as *pyrolysis,* which is generally defined as "chemical decomposition by the action of heat." In this case, the decomposition causes a change from the solid state to the vapor state (Fig. 4.5). If the vapor mixes sufficiently with air and is heated to a high enough temperature (by a flame, spark, hot motor, etc.), combustion results.

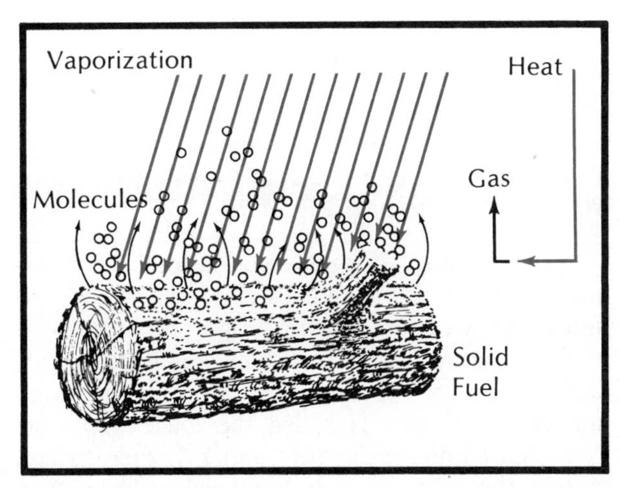

Figure 4.5. Pyrolysis: The conversion of solid fuel to flammable vapor by heat.

Burning Rate. The burning rate of a solid fuel depends on its configuration. Solid fuels in the form of dust or shavings will burn faster than bulky materials (that is, small wood chips will burn faster than a solid wooden beam). Finely divided fuels have a much larger surface area exposed to the heat. Therefore, heat is absorbed much faster, and vaporization is more rapid. More vapor is available for ignition, so it burns with great intensity and the fuel is quickly consumed. On the other hand, a bulky fuel will burn longer than a finely divided fuel.

Dust clouds are made up of very small particles. When a cloud of flammable dust (such as grain dust) is mixed well with air and ignited, it burns extremely quickly, often with explosive force. Such explosions have occurred on ships during the loading and discharging of grains and other finely divided materials.

Ignition Temperature. The *ignition temperature* of a substance (solid, liquid or gas) is the lowest temperature at which sustained combustion will occur without the application of a spark or flame. Ignition temperatures vary among substances. For a given substance, the ignition temperature also varies with bulk, surface area and other factors. The ignition temperatures of common combustible materials lie between 149°C (300°F) and 538°C (1000°F).

Liquid Fuels

The flammable liquids most commonly found aboard ship are bunker fuel, lubricating oil, diesel oil, kerosene, oil-base paints and their solvents. Cargos may include flammable liquids and liquefied flammable gases.

Vaporization. Flammable liquids release vapor in much the same way as solid fuels. The rate of vapor release is greater for liquids than solids, since liquids have less closely packed molecules. In addition, liquids can release vapor over a wide temperature range. Gasoline starts to give off vapor at −43°C (−45°F). This makes gasoline a continuous fire hazard; it produces flammable vapor at normal temperatures (Fig. 4.6A). Heating increases the rate of vapor release.

Heavier flammable liquids such as bunker oil and lubricating oil must be heated to release sufficient vapor for combustion. Lubricating oils can ignite at 204°C (400°F). A fire reaches this temperature rapidly, so that oils directly exposed to a

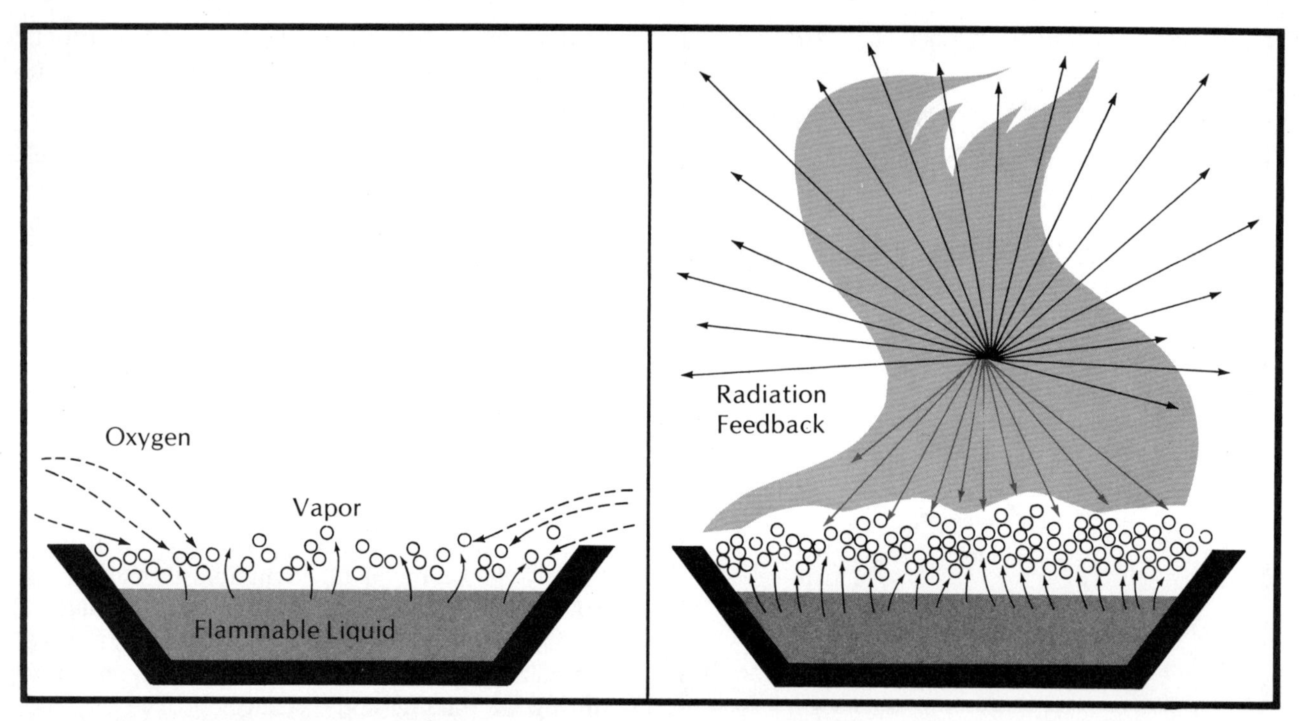

Figure 4.6. Vaporization of a flammable liquid. A. Many flammable liquids produce vapor without being heated. When such a liquid is stored in an open container, it can easily be ignited. B. Once the vapor–air mixture is ignited, radiation feedback causes a massive release of fuel vapor.

fire will soon become involved. Once a light or heavy flammable liquid is burning, radiation feedback and the chain reaction quickly increase flame production (Fig. 4.6B).

The vapor produced by a flammable liquid is heavier than air. This makes the vapor very dangerous, because it will seek low places, dissipate slowly, and travel to a distant source of ignition. For example, vapor escaping from a container can travel along a deck and down deck openings until it contacts a source of ignition (such as a spark from an electric motor). If the vapor is properly mixed with air, it will ignite and carry fire back to the leaky container. The result can be a severe explosion and fire.

Burning. Pound for pound, flammable liquids produce about 2.5 times more heat than wood. This heat is liberated 3 to 10 times faster from liquids than from wood. These ratios illustrate quite clearly why flammable liquid vapor burns with such intensity. When flammable liquids spill, they expose a very large surface area, release a great amount of vapor and thus produce great amounts of heat when ignited. This is one reason why large open tank fires and liquid-spill fires burn so violently.

Flash Point. The *flash point* of a liquid fuel is the temperature at which it gives off sufficient vapor to form an ignitable mixture near its surface. An ignitable mixture is a mixture of vapor and air that is capable of being ignited by an ignition source, but usually is not sufficient to sustain combustion.

Sustained combustion takes place at a slightly higher temperature, referred to as the *fire point* of the liquid. The flash points and fire points (temperatures) of liquids are determined in controlled tests.

Gaseous Fuels

There are both natural and manufactured flammable gases. Those that may be found on board a vessel include acetylene, propane and butanes.

Burning. Gaseous fuels are already in the required vapor state. Only the proper intermix with oxygen and sufficient heat are needed for ignition. Gases, like flammable liquids, always produce a visible flame; they do not smolder. Radiation feedback is not necessary to vaporize the gas; however, some radiation feedback is still essential to the burning process, to provide continuous reignition of the gas (Fig. 4.7).

Explosive Range (Flammable Range). A flammable gas or the flammable vapor of a liquid must mix with air in the proper proportion to make an ignitable mixture. The smallest percentage of a gas (or vapor) that will make an ignitable air–vapor mixture is called the *lower explosive limit* (LEL) of the gas (or vapor). If there is less gas in the mixture, it is too lean to burn. The

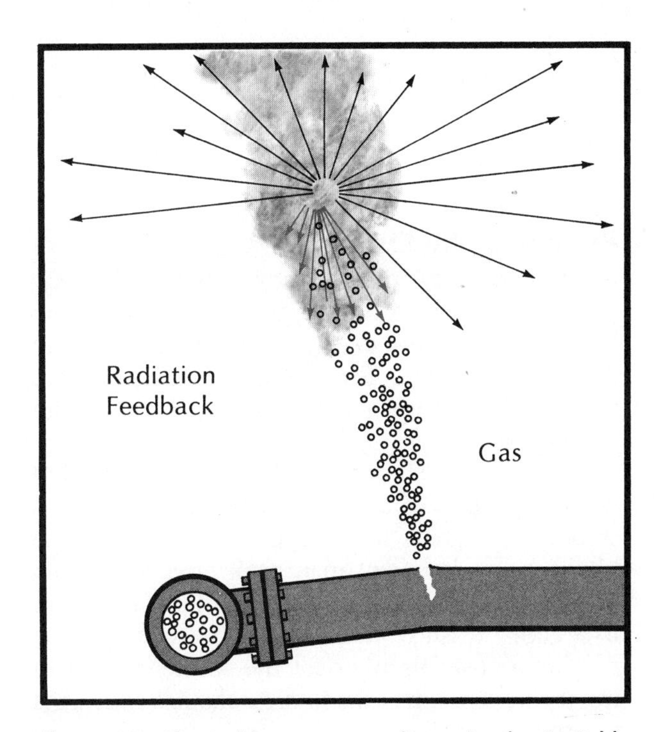

Figure 4.7 Flammable gases are always in the ignitable state. Radiation feedback sustains the combustion.

greatest percentage of a gas (or vapor) in an ignitable air–vapor mixture is called its *upper explosive limit* (UEL). If a mixture contains more gas than the UEL, it is too rich to burn. The range of percentages between the lower and upper explosive limits is called the *explosive range* of the gas or vapor.

Table 4.1 gives the LEL and UEL for a number of substances. It shows, for example, that a mixture of from 1.4% to 7.6% gasoline vapor and from 98.6% to 92.4% air will ignite. However, a mixture of 9% gasoline vapor and 91% air will not ignite, because it is too rich (above the UEL). Thus, a large volume of air must intermix with a small amount of gasoline vapor to form an ignitable mixture.

Table 4.1. Typical upper and lower flammable limits.*

Product	*Lower explosive limit (LEL)*	*Upper explosive limit (UEL)*
Gasoline	1.4	7.6
Kerosene	0.7	6.0
Propane	2.1	9.5
Hydrogen	4.0	74.2
Methane	5.0	15.0
Ethylene Oxide	2.0	100.0
Ammonia	15.5	27.0
Naphtha	0.9	6.7
Butane	1.8	8.4
Benzene	1.4	8.0

* Percent by volume in air.

A mixture of a gas or vapor in air that is below the LEL may burn under some special circumstances. This fact is the basis of certain devices that utilize a Wheatstone bridge to detect the presence of potentially hazardous concentrations of hazardous or explosive gases. Such devices as the combustible-gas indicator (Chapter 16) make it unnecessary to memorize the explosive ranges of fuels. It is much more important to realize that certain ranges of vapor–air mixtures can be ignited, and to use caution when working with these fuels.

The explosive ranges of specific types of fuels are published in the NFPA *Fire Protection Handbook* and the US Coast Guard *Chemical Data Guide for Bulk Shipment by Water,* CG388.

Oxygen

The oxygen side of the fire triangle refers to the oxygen content of the surrounding air. Ordinarily, a minimum concentration of 16% oxygen in the air is needed to support flaming combustion. However, smoldering combustion can take place in about 3% oxygen. Air normally contains about 21% oxygen, 78% nitrogen and 1% other gases, principally argon.

Heat

Heat is the third side of the fire triangle. When sufficient heat, fuel and oxygen are available, the triangle is complete and fire can exist. Heat of ignition initiates the chemical reaction that is called combustion. It can come from the flame of a match, sparks caused by ferrous metals striking together, heat generated by friction, lightning, an oxyacetylene torch cutting or welding metal, an electric short circuit, an electric arc between conductors, or the overheating of an electric conductor or motor. Sufficient heat may also be produced internally, within the fuel, by a chemical reaction (*see* Spontaneous Ignition, Chapter 1).

THE FIRE TETRAHEDRON

The fire triangle (Fig. 4.3) is a simple means of illustrating the three requirements for the existence of fire. However, it does not explain the *nature* of fire. In particular, it does not include the chain reaction that results from chemical reactions among the fuel, oxygen and heat.

The fire tetrahedron (Fig. 4.8) is a better representation of the combustion process. (A tetrahedron is a solid figure with four triangular faces. It is useful for illustrating and remembering the combustion process because it has room for the chain reaction and because each face touches

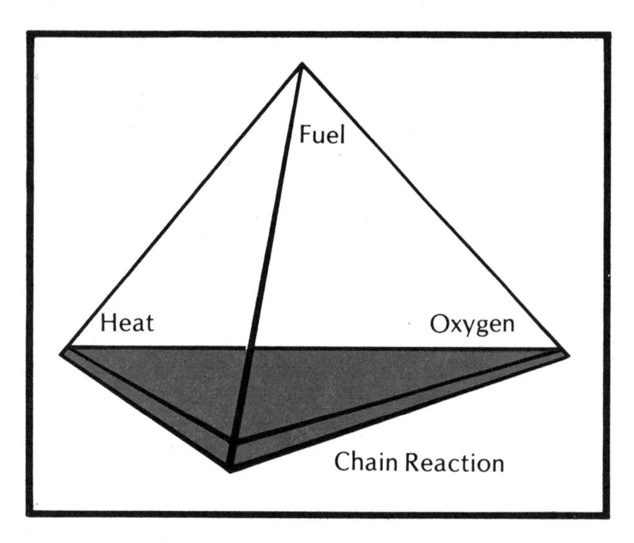

Figure 4.8. The fire tetrahedron.

the other three faces.) The basic difference between the fire triangle and the fire tetrahedron is this: The tetrahedron illustrates how flaming combustion is supported and sustained through the chain reaction. In a sense, the chain reaction face keeps the other three faces from falling apart. This is an important point, because the extinguishing agents used in many modern portable fire extinguishers, automatic extinguishing systems and explosion suppression systems directly attack and break down the chain reaction sequence.

EXTINGUISHMENT VIA THE FIRE TETRAHEDRON

A fire can be extinguished by destroying the fire tetrahedron. If the fuel, oxygen or heat is removed, the fire will die out. If the chain reaction is broken, the resulting reduction in vapor and heat production will extinguish the fire. (However, additional cooling with water may be necessary where smoldering or reflash is a possibility.)

Removing the Fuel

One way to remove the fuel from a fire is to physically drag it away. In most instances, this is an impractical firefighting technique. However, it is often possible to move nearby fuels away from the immediate vicinity of a fire, so that the fire does not extend to these fuels.

Sometimes the supply of liquid or gaseous fuel can be cut off from a fire. When a fire is being fed by a leaky gasoline or diesel line, it can be extinguished by closing the proper valve. If a pump is supplying liquid fuel to a fire in the engine room, the pump can be shut down to remove the fuel source and thereby extinguish the fire. Fire in a defective fuel-oil burner can be controlled and extinguished by closing the supply valve. Fire involving acetylene or propane can often be extinguished by shutting the valve on the cylinder.

Removing the Oxygen

A fire can be extinguished by removing its oxygen or by reducing the oxygen level in the air to below 16%. Many extinguishing agents (carbon dioxide and foam, for example) extinguish fire with a smothering action that deprives the fire of oxygen.

This extinguishment method is difficult (but not impossible) to use in an open area. Gaseous smothering agents like carbon dioxide would be blown away from an open deck area, especially if the ship is under way. On the other hand, fire in a galley trash container can be snuffed out by placing a cover tightly over the container, blocking the flow of air to the fire. As the fire consumes the oxygen in the container, it becomes starved for oxygen and is extinguished.

Tank vessels that carry petroleum products are protected by foam systems with monitor nozzles on deck. When used quickly and efficiently, the foam is capable of extinguishing a sizable deck fire.

To extinguish a fire in an enclosed space such as a compartment, engine room or cargo hold, the space can be flooded with carbon dioxide. When the carbon dioxide enters the space and mixes with the atmosphere, the percentage of oxygen in the atmosphere is reduced below 16%, and extinguishment results. This method is used to combat fires in cargo holds. For the technique to be successful, the hold must be completely sealed to keep fresh air out. (For further discussion of this method of extinguishment, *see* Chapter 10.)

Oxidizing Substances. An oxidizing substance is a material that releases oxygen when it is heated or, in some instances, when it comes in contact with water. Substances of this nature include the hypochlorites, chlorates, perchlorates, nitrates, chromates, oxides and peroxides. All contain oxygen atoms that are loosely bonded into their molecular structure. That is, they carry their own supply of oxygen, enough to support combustion. This oxygen is released when the substances break down, as in a fire. For this reason, burning oxidizers cannot be extinguished by removing their oxygen. Instead, large amounts of water, limited by ship stability safety needs, are used to accomplish extinguishment. Oxidizers are hazardous materials and, as such, are regulated by the U.S. Coast Guard.

Removing the Heat

The most commonly used method of extinguishing fire is to remove the heat. The base of the fire is attacked with water to destroy the ability of the fire to sustain itself. Water is a very effective heat absorber. When properly applied, it absorbs heat from the fuel and absorbs much of the radiation feedback. As a result, the chain reaction is indirectly attacked both on the fuel surface and at the flames. The production of vapor and radiant heat is reduced. Continued application will control and extinguish the fire.

When fire is attacked with a hoseline, water in the proper form must be directed onto the main body of fire to achieve the quickest heat reduction. Water spray can be a highly efficient extinguishing agent. For complete extinguishment, water must be applied to the seat of the fire. (*See* Chapter 10 for a discussion of the uses of water as an extinguishing agent.)

Breaking the Chain Reaction

Once the chain reaction sequence is broken, a fire can be extinguished rapidly. The extinguishing agents commonly used to attack the chain reaction and inhibit combustion are dry chemicals and Halons. These agents directly attack the molecular structure of compounds formed during the chain reaction sequence. The breakdown of these compounds adversely affects the flame-producing capability of the fire. The attack is extremely rapid; in some automatic systems the fire is extinguished in 50 to 60 milliseconds. Because of their ultrafast action, Halons and dry chemicals are used in automatic explosion suppression systems.

It should be borne in mind that these agents do not cool a smoldering fire or a liquid whose container has been heated above the liquid's ignition temperature. In these cases, the extinguishing agent must be maintained on the fire until the fuel has cooled down naturally. Otherwise, a cooling medium such as water must be used to cool the smoldering embers or the sides of the container. In most situations, it is best to use water for cooling along with extinguishing agents that attack the chain reaction only.

FIRE SPREAD

If a fire is attacked early and efficiently, it can easily be confined to the area in which it started. If it is allowed to burn unchecked, it can generate great amounts of heat that will travel away from the fire area, igniting additional fires wherever fuel and oxygen are available. And both are in plentiful supply throughout most ships. Steel bulkheads and decks and other fire barriers can stop or delay the passage of heat to some extent, but not completely. As the original fuel source is consumed, the heat, and thus the fire, will extend to new fuel sources.

Heat from a fire is transferred by one or more of three methods: conduction, radiation and convection.

Conduction

Conduction is the transfer of heat through a solid body. For example, on a hot stove, heat is conducted through the pot to its contents. Wood is ordinarily a poor conductor of heat, but metals are good conductors. Since most ships are constructed of metal, heat transfer by conduction is a potential hazard. Fire can move from one hold to another, one deck to another, and one compartment to another via heat conduction (Fig. 4.9).

In many cases the skillful application of water, particularly in the form of a spray, will retard or halt the transmission of heat by conduction. The water cools the affected structural members, bulk-

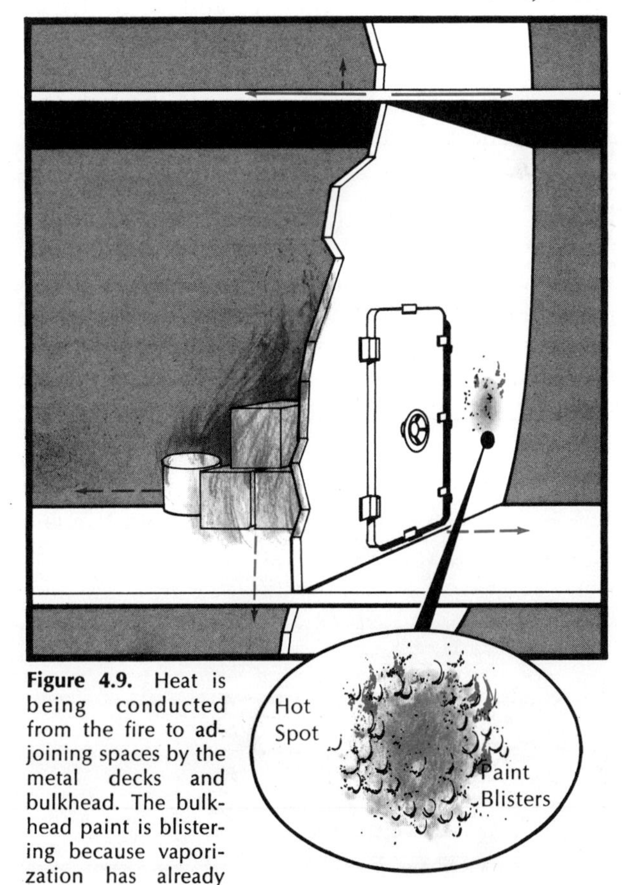

Figure 4.9. Heat is being conducted from the fire to adjoining spaces by the metal decks and bulkhead. The bulkhead paint is blistering because vaporization has already begun.

heads and decks. A water spray pattern absorbs heat more efficiently than a solid stream, because the smaller water droplets present more surface to the heat source. At the same time, less water is used, so there is less of a water runoff problem and less danger of affecting the stability of the vessel.

Radiation

Heat radiation is the transfer of heat from a source across an intervening space; no material substance is involved. The heat travels outward from the fire in the same manner as light, that is, in straight lines. When it contacts a body, it is absorbed, reflected or transmitted. Absorbed heat increases the temperature of the absorbing body. For example, radiant heat that is absorbed by an overhead will increase the temperature of that overhead, perhaps enough to ignite its paint.

Heat radiates in all directions unless it is blocked. Radiant heat extends fire by heating combustible substances in its path, causing them to produce vapor, and then igniting the vapor (Fig. 4.10).

Within a ship, radiant heat will raise the temperature of combustible materials near the fire or, depending on the ship's design, at quite some distance from the fire. Intense radiated heat can make an approach to the fire extremely difficult. For this reason, protective clothing must be worn by firefighters, and the heat reduced through the use of a heat shield such as water spray or dry chemical.

Convection

Convection is the transfer of heat through the motion of heated matter, i.e., through the motion of smoke, hot air, heated gases produced by the fire, and flying embers.

When it is confined (as within a ship), convected heat moves in predictable patterns. The fire produces lighter-than-air gases that rise toward high parts of the ship. Heated air, which is lighter than cool air, also rises, as does the smoke produced by combustion. As these heated combustion products rise, cool air takes their place; the cool air is heated in turn and then also rises to the highest point it can reach (Fig. 4.11). As the hot air and gases rise from the fire, they begin to cool; as they do, they drop down to be reheated and rise again. This is the *convection cycle.*

Heat originating at a fire on a lower deck will travel horizontally along passageways, and then upward via ladder and hatch openings. It will ignite flammable materials in its path. To prevent fire spread, the heat, smoke and gases should be released into the atmosphere. However, the structural design of a ship makes it next to impossible to rapidly cut openings through decks, bulkheads or the ship's hull for ventilation. Thus, it is imperative that the fire be confined to the smallest possible area. For this purpose, doors and hatchways should be kept closed when they are not in use. If a fire is discovered, attempts should be made to close off all openings to the fire area until firefighting personnel and equipment can be brought into position to fight the fire.

THE HAZARDOUS PRODUCTS OF COMBUSTION

Fire produces flames, heat, gases and smoke. Each of these combustion products can cause serious injuries or death.

Flames

Direct contact with flames can result in totally or partially disabling skin burns and serious dam-

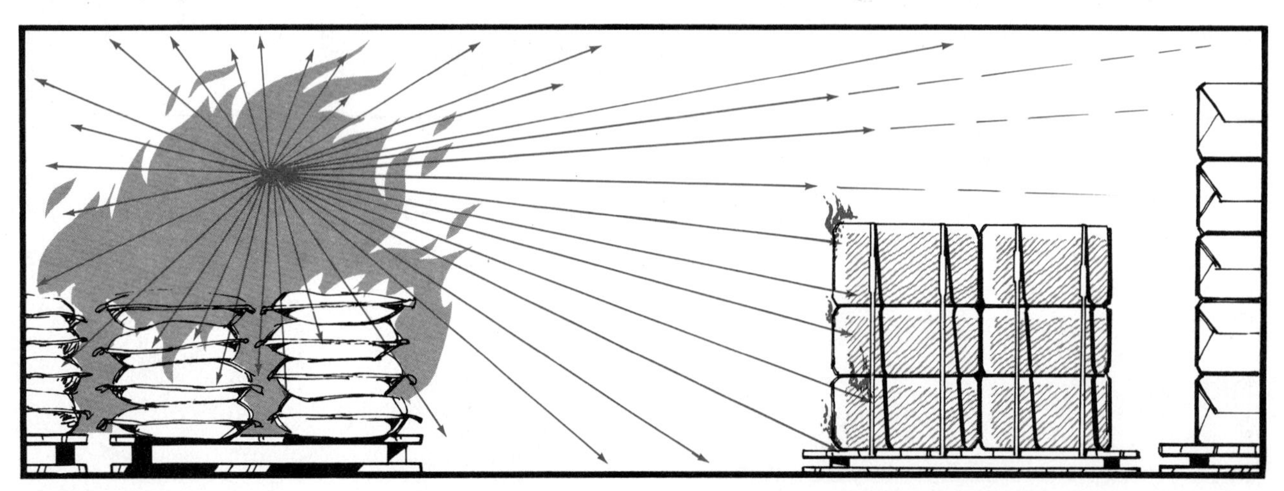

Figure 4.10. Radiated heat travels in straight lines to combustible materials, igniting them and spreading the fire.

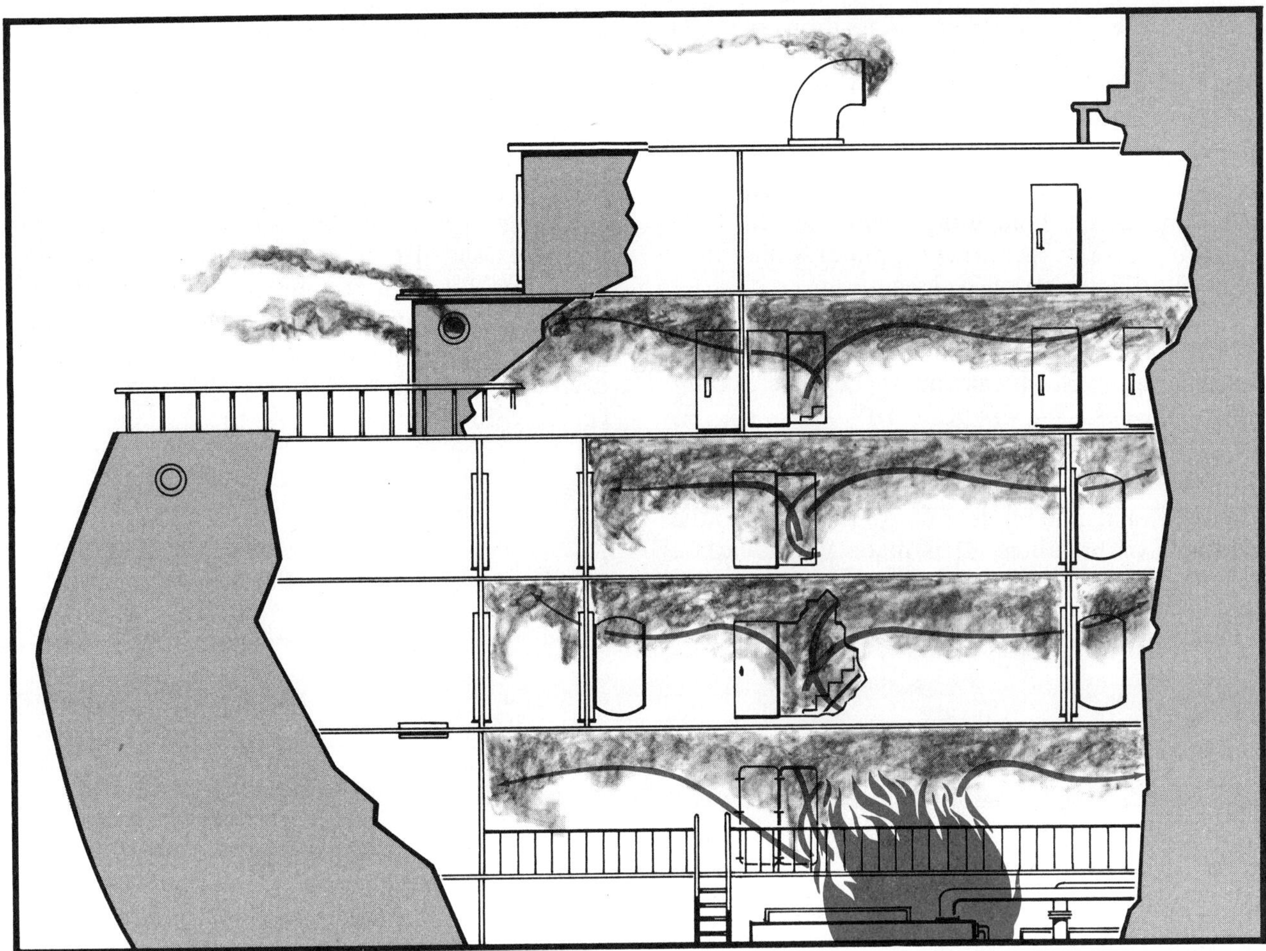

Figure 4.11. Convection carries heated air, gases and smoke upward through the ship. When vertical passage is blocked, they move horizontally.

age to the respiratory tract. To prevent skin burns during a fire attack, crewmen should maintain a safe distance from the fire unless they are properly protected and equipped for the attack. Protective clothing (Chapter 15) should be worn when combating a serious fire.

Respiratory tract damage can be prevented by wearing breathing apparatus. However, firefighting personnel must remember that breathing apparatus does not protect the body from the extreme heat of a fire.

Heat

Fire generates temperatures in excess of 93°C (200°F) very rapidly, and the temperature can build up to over 427°C (800°F) in an enclosed area. Temperatures above 50°C (122°F) are hazardous to humans, even if they are wearing protective clothing and breathing apparatus. The dangerous effects of heat range from minor injury to death. Direct exposure to heated air may cause dehydration, heat exhaustion, burns and blockage of the respiratory tract by fluids. Heat also causes an increased heart rate. A firefighter exposed to excessive heat over an extended period of time could develop hyperthermia, a dangerously high fever that can damage nerve centers. (*See* Chapter 14 for a detailed discussion of burns and methods of treatment.)

Gases

The particular gases produced by a fire depend mainly on the fuel. The most common hazardous gases are carbon dioxide (CO_2), the product of complete combustion, and carbon monoxide (CO), the product of incomplete combustion.

Carbon monoxide is the more dangerous of the two. When air mixed with carbon monoxide is inhaled, the blood absorbs the CO before it will absorb oxygen. The result is an oxygen deficiency in the brain and body. Exposure to a 1.3% concentration of CO will cause unconsciousness in two or three breaths, and death in a few minutes.

Carbon dioxide works on the respiratory system. Above normal CO_2 concentrations in the air reduce the amount of oxygen that is absorbed in the lungs. The body responds with rapid and deep breathing—a signal that the respiratory system is not receiving sufficient oxygen.

When the oxygen content of air drops from its normal level of 21% to about 15%, human muscular control is reduced. At 10% to 14% oxygen in air, judgment is impaired and fatigue sets in. Unconsciousness usually results from oxygen concentrations below 10%. During periods of exertion, such as firefighting operations, the body requires more oxygen; these symptoms may then appear at higher oxygen percentages.

Several other gases generated by a fire are of equal concern to firefighters. Therefore, anyone entering a fire must wear an appropriate breathing apparatus. (*See* Chapter 15.)

Smoke

Smoke is a visible product of fire that adds to the problem of breathing. It is made up of carbon and other unburned substances in the form of suspended particles. It also carries the vapors of water, acids and other chemicals, which can be poisonous or irritating when inhaled.

Smoke greatly reduces visibility in and above the fire area. It irritates the eyes, nose, throat and lungs. Either breathing a low concentration for an extended period of time or a heavy concentration for a short time can cause great discomfort to a firefighter. Firefighters who do not wear breathing apparatus in the fire area will eventually have to retreat to fresh air or be overcome by smoke.

BIBLIOGRAPHY

Fire Chief's Handbook

CG-329 Fire Fighting for Tank Vessels

NFPA Handbook 14th Ed.

Engine Company Fireground Operations
Richman, R. J. Brady Co.

Basic Fireman's Training Course, Md. Fire
& Rescue Inst. Univ. of Md. College Park, Md. 1969

5

Classification of Fires

Title 46 CFR requires the master of a vessel to post a station bill outlining the duties and duty station of each crew member during various emergencies. The master also is required to conduct drills and give instructions to ensure that all hands are familiar with their emergency duties. One of these emergencies is fire aboard ship.

To extinguish a fire successfully, it is necessary to use the most suitable type of extinguishing agent—one that will accomplish the task in the least amount of time, cause the least damage and result in the least danger to crew members. The job of selecting the proper extinguishing agent has been made easier by the classification of fires into four types, or classes, lettered A through D. Within each class are all fires involving materials with similar burning properties and requiring similar extinguishing agents. Thus, knowledge of these classes is essential to efficient firefighting operations, as well as familiarity with the burning characteristics of materials that may be found aboard ship.

NFPA CLASSES OF FIRE

The fire classification scheme used by the U.S. Coast Guard was originally devised by the National Fire Protection Association (NFPA). In this scheme, fires are classed according to the *fuel* and the *most effective extinguishing agents,* as follows:

- *Class A fires:* Fires involving common (ash-producing) combustible materials, which can be extinguished by the use of water or water solutions. Materials in this category include wood and wood-based materials, cloth, paper, rubber and certain plastics (Fig. 5.1).

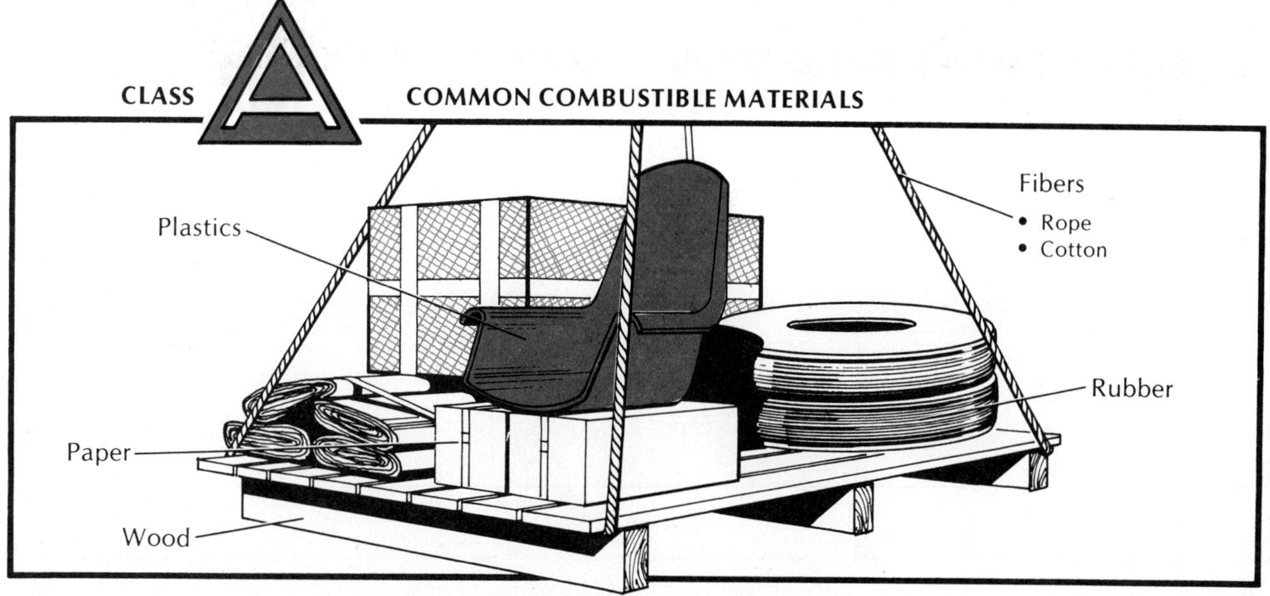

Figure 5.1. Class A fires are those involving common combustible materials.

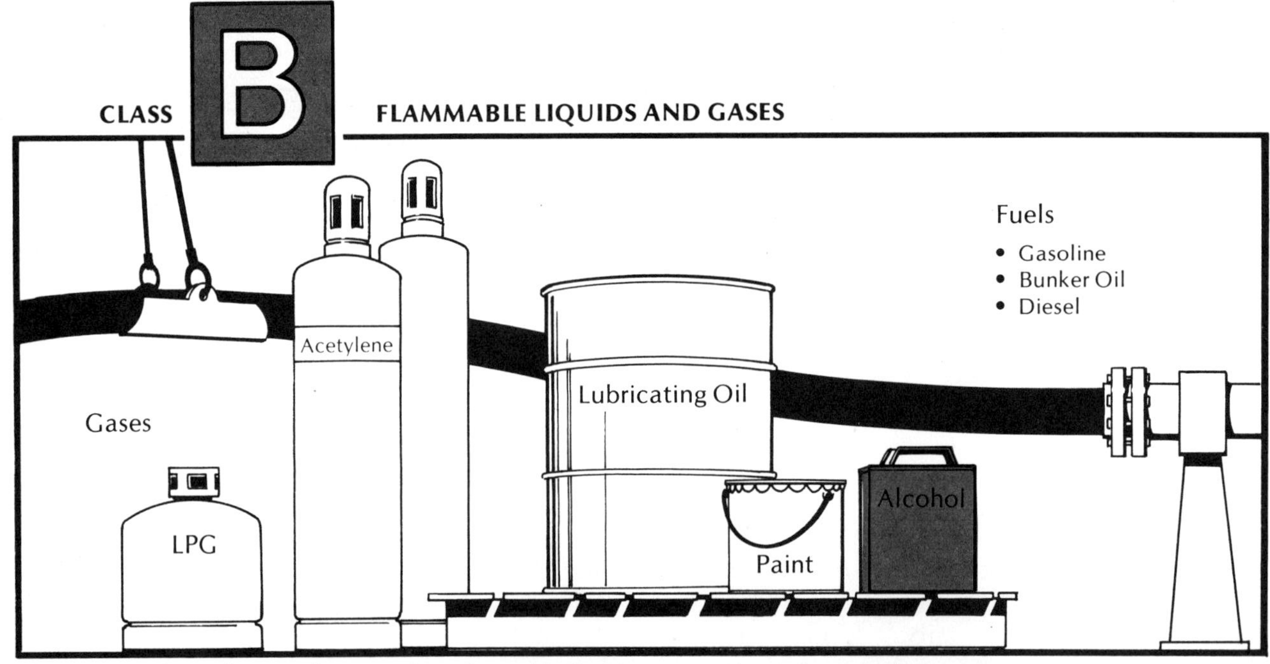

Figure 5.2. Class B fires are those involving flammable liquids, gases and petroleum products.

- *Class B fires:* Fires involving flammable or combustible liquids, flammable gases, greases and similar products (Fig. 5.2). Extinguishment is accomplished by cutting off the supply of oxygen to the fire or by preventing flammable vapors from being given off.
- *Class C fires:* Fires involving energized electrical equipment, conductors or appliances (Fig. 5.3). Nonconducting extinguishing agents must be used for the protection of crew members.
- *Class D fires:* Fires involving combustible metals, e.g., sodium, potassium, magnesium, titanium and aluminum (Fig. 5.4). Extinguishment is effected through the use of heat-absorbing extinguishing agents such as certain dry powders that do not react with the burning metals.

The main objective of this classification scheme is to aid crew members in selecting the appropriate extinguishing agent. However, it is not enough to know that water is best for putting out a class A fire because it cools, or that dry chemical works well in knocking down the flames of a burning liquid. The extinguishing agent must be applied properly, and sound firefighting techniques must be used.

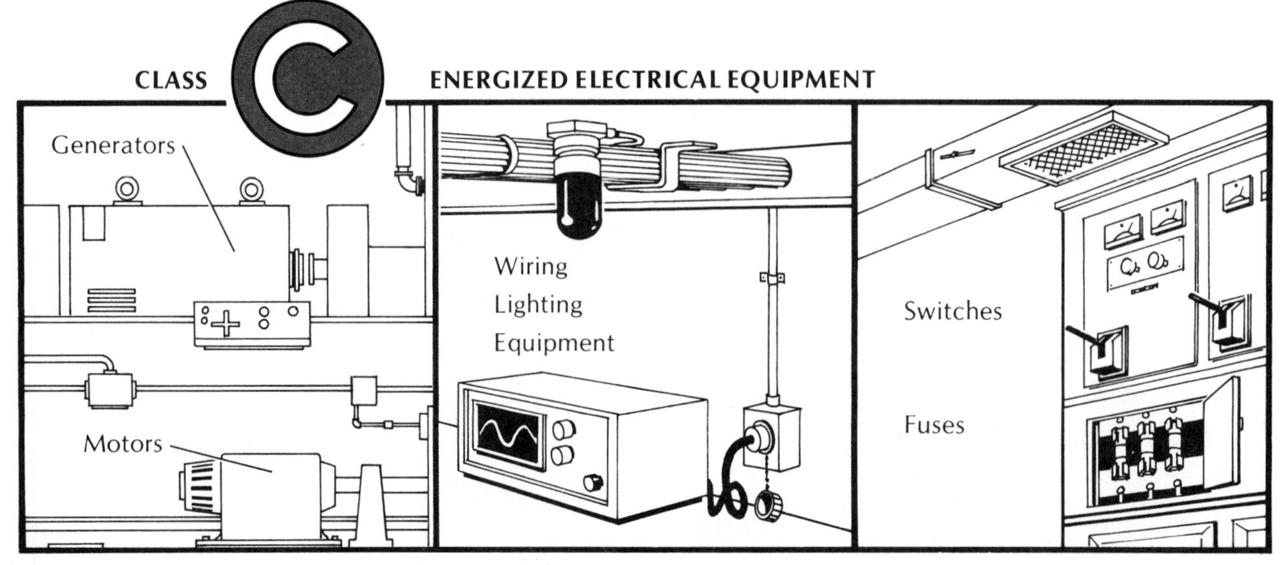

Figure 5.3. Class C fires are those involving energized electrical equipment and wiring.

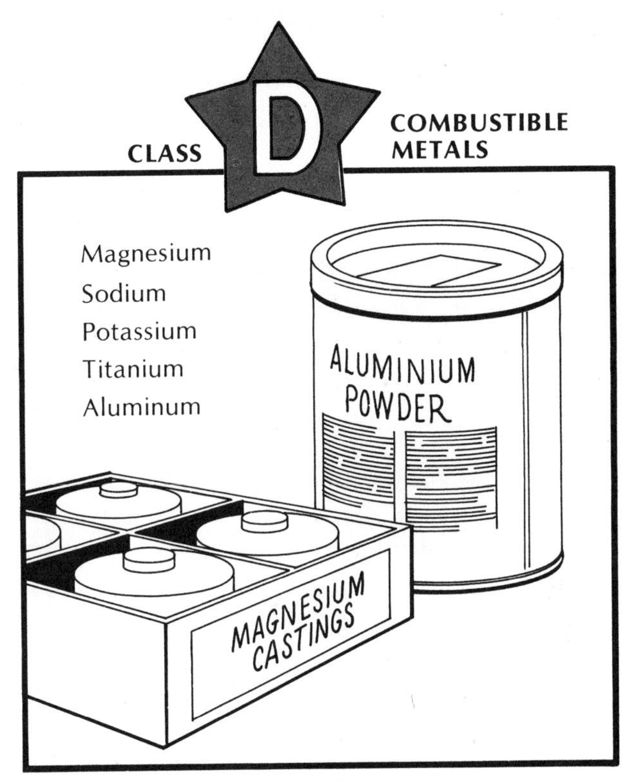

Figure 5.4. Class D fires are those involving combustible metals.

In the remainder of this chapter, the fuels within each fire class are discussed in some detail. Further information regarding extinguishing agents and firefighting techniques may be found in Chapters 7–10.

CLASS A FIRES INVOLVING MATERIALS COMMONLY FOUND ABOARD SHIP

The materials whose involvement leads to class A fires may be placed in three broad groups: *1)* wood and wood-based materials, *2)* textiles and fibers, and *3)* plastics and rubber. We shall discuss each of these groups of fuels individually.

Wood and Wood-based Materials

Wood is very often involved in fire, mainly because of its many uses. Marine uses include decking and the interior finish of bulkheads (on small boats only), dunnage and staging, among many others. Wood-based materials are those that contain processed wood or wood fibers. They include some types of insulation, ceiling tiles, plywood and paneling, paper, cardboard and pressboard.

Figure 5.5. Flashover. **A.** Radiant heat or heat conducted through the bulkhead causes the wood paneling to produce combustible vapor. **B.** Once the vapor is properly mixed with air, any ignition source will ignite the entire vapor–air mixture.

The properties of wood and wood-based materials depend on the particular type involved. For example, seasoned, air-dried maple (a hardwood) produces greater heat on burning than does pine (a softwood) that has been seasoned and dried similarly. However, all these materials are combustible; they will char, smolder, ignite and burn under certain heat conditions. Normally self-ignition does not occur. A source of ignition such as a spark, open flame, contact with a hot surface or exposure to heat radiation is usually necessary. However, wood can be pyrolyzed to charcoal, which has a lower ignition temperature.

Wood is composed mainly of carbon, hydrogen and oxygen, with smaller amounts of nitrogen and other elements. In the dry state, most of its weight consists of cellulose. Some other ingredients found in dry wood are sugars, resins, gums, esters of alcohol and mineral matter (from which ash is formed when wood burns).

Burning Characteristics. The ignition temperature of wood depends on many factors, such as size, shape, moisture content, and type. Generally, the ignition temperature of wood is about 204°C (400°F). However, it is believed that 100°C (212°F) is the maximum temperature to which wood can be subjected over a long period of time without self-ignition taking place.

The rate of combustion of wood and wood-based materials depends heavily on the physical form of the material, the amount of air available, the moisture content and other such factors. However, the wood must be vaporized by heat before complete combustion can proceed.

A slowly developing fire or a source of radiant heat may gradually transmit enough energy to begin the pyrolysis of wood products at bulkhead or overhead surfaces. The combustible vapor that is released will mix with the surrounding air. When this mixture is within the flammable range, any source of ignition may ignite the entire mass almost instantly. This condition is called *flashover* (Fig. 5.5). Crewmen must guard against flashover while fighting fires involving such combustible solids as wood-paneled walls and furniture in the confined spaces of older ships. In modern ships, noncombustible materials are used in cabins, passageways and other confined spaces.

Flames move slowly across most combustible solids. Before the flame can spread, flammable vapor must be released by the solid fuel. Then the vapor must be mixed in the proper proportion with air.

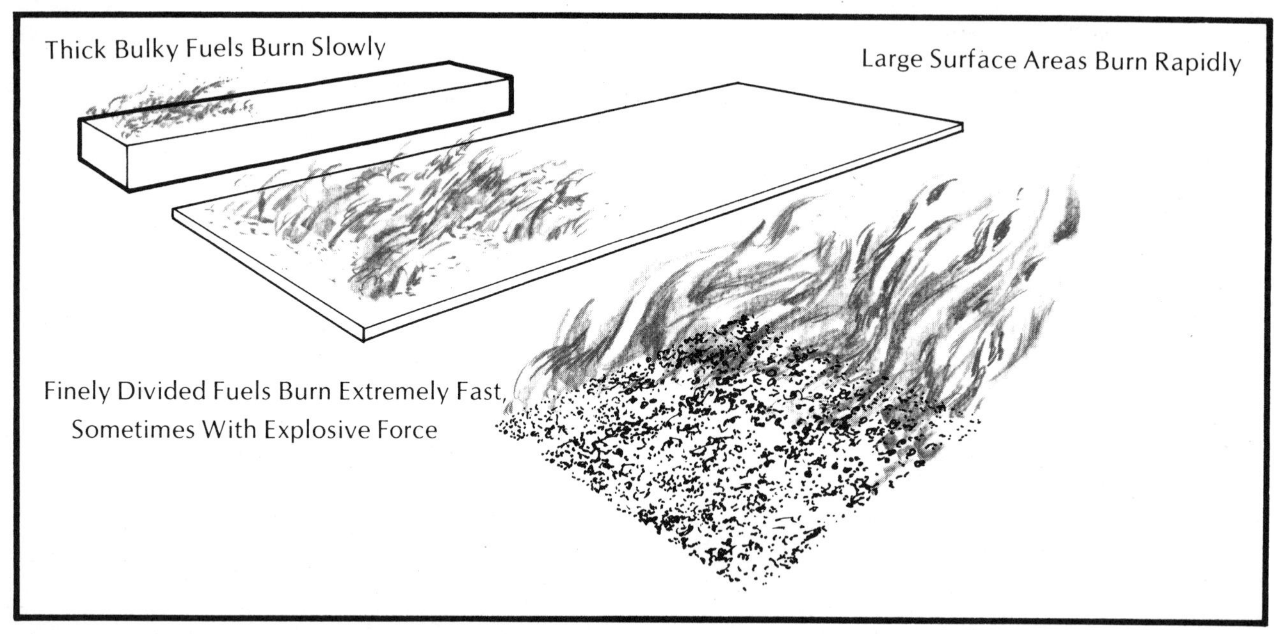

Figure 5.6. The burning rate increases with the surface area of the fuel.

Bulky solids with a small surface area (for example, a heavy wood beam) burn more slowly than thinner solids with a large surface area (for example, a sheet of plywood). Solids in chip, shaving or dust form (wood, metal shavings, sawdust, grains and pulverized coal) burn most rapidly, since the surfaces of the individual particles add up to a very large total area (Fig. 5.6). In general, the thicker the fuel is, the more time it requires to release vapor into the air. Therefore it will burn longer. The larger the surface area, the more rapidly the fuel burns: The larger surface allows combustible vapor to be released at a greater rate and to mix more quickly with air. (This is also true of flammable liquids. A shallow liquid spill with a large area will burn off more rapidly than the same volume of liquid in a deep tank with a small surface area.)

Products of Combustion. Burning wood and wood-based materials produce water vapor, heat, carbon dioxide and carbon monoxide, as discussed in Chapter 4. The reduced oxygen levels and the carbon monoxide present the primary hazard to crew members. In addition, wood and wood-based materials produce a wide range of aldehydes, acids and other gases when they burn. By themselves or in combination with the water vapor, these substances can cause severe irritation at least. Because of the toxicity of most of these gases, the use of breathing apparatus should be mandatory in and near the fire area (Fig. 5.7).

Burns can be caused by direct contact with flames, or by heat radiated from the fire. Flames are rarely separated from the burning material by any appreciable distance. However, in certain types of smoldering fires, heat, smoke and gas can

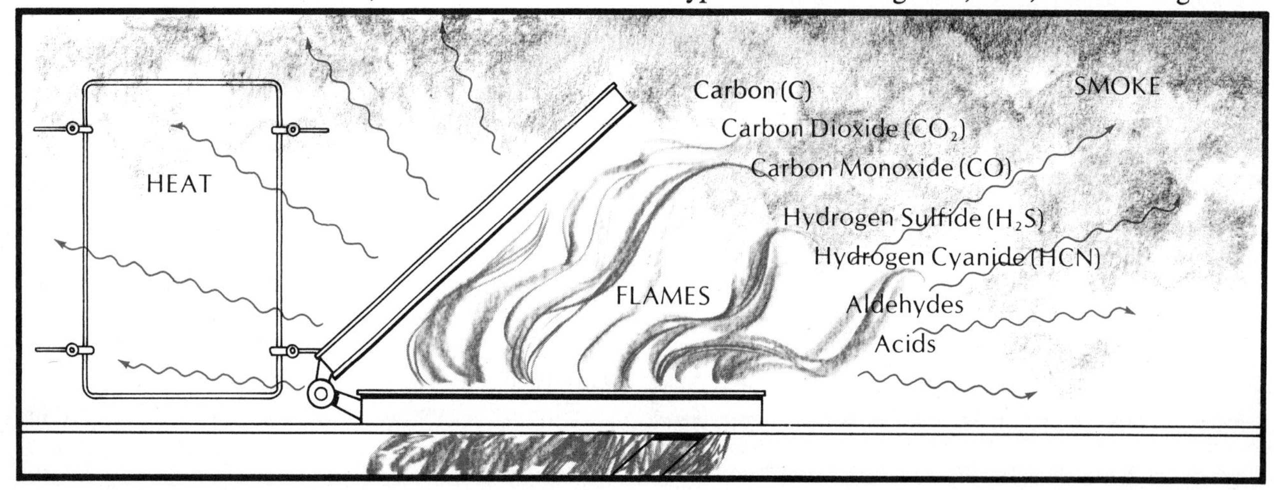

Figure 5.7. The products of combustion subject firefighters to burns, oxygen depletion, heat exhaustion and dehydration, respiratory tract irritation, and poisoning.

develop without visible flames. Air currents can carry them far in advance of the fire.

As is true of many organic substances, wood and related materials can produce large quantities of smoke in the beginning stages of fire. In some very special circumstances, materials can burn without producing visible combustion products; however, smoke generally accompanies fire and, like flame, is visible evidence of fire.

Smoke frequently provides the early warning of fire. At the same time, its blinding and irritating effects can contribute to panic.

Textiles and Fibers

Textiles in the form of clothing, furniture, carpets, canvas, burlap, ropes and bedding are used extensively in the marine environment; others are carried as cargo. Almost all textile fibers are combustible. These two facts explain the frequency of textile-related fires and the many deaths and injuries that result.

Natural-Fiber Textiles. *Vegetable fibers* consist largely of cellulose. They include cotton, jute, hemp, flax and sisal. Cotton and the other plant fibers are combustible (the ignition temperature of cotton fiber is 400°C (752°F)). Burning vegetable fibers produce heat and smoke, carbon dioxide, carbon monoxide and water. They do not melt. The ease of ignition, rate of flame spread, and amount of heat produced depend on the construction and finish of the textile and on the design of the finished product.

Animal fibers such as wool and silk are solid and are chemically different from cotton. They do not burn as freely, and they tend to smolder. For example, wool is basically protein. It is more difficult to ignite than cotton (the ignition temperature of wool fiber is 600°C (1112°F)), burns more slowly, and is easier to extinguish.

Synthetic Textiles. Synthetic textiles are fabrics woven wholly or mainly of synthetic fibers. Such fibers include rayon, acetate, nylon, polyester, acrylic and plastic wrap. The fire hazards involved with synthetic textiles are sometimes difficult to evaluate, owing to the tendency of some of them to shrink, melt or drip when heated. Rayon and acetate resemble plant fibers chemically, whereas most other synthetic fibers do not. Most are combustible to varying degrees but differ in ignition temperature, burning rate and other combustion features.

Burning Characteristics. Many variables affect the way in which a textile burns. The most important are the chemical composition of the textile fiber, the finish on the fabric, the fabric weight, the tightness of weave and any flame retardant treatment.

Vegetable fibers ignite easily and burn readily, giving off large amounts of heavy smoke. Partially burned vegetable fibers may present a fire risk, even after they have been extinguished. Half-burned fibers should always be removed from the fire area to a location where reignition of the material would not create an additional problem. Most baled vegetable fibers absorb water readily. The bales will swell and increase in weight when large quantities of water are used to extinguish fires in which they are involved.

Wool is difficult to ignite; it tends to smolder and char rather than to burn freely unless it is subjected to considerable external heat. It will, however, contribute toward a fierce fire. Wool can absorb a large amount of water—a fact that must be considered during prolonged firefighting operations.

Silk is the least dangerous fiber. It is difficult to ignite, and it burns sluggishly. Combustion usually must be supported by an external source of heat. Once set on fire, silk retains heat longer than any other fiber. In addition, it can absorb a great amount of water. Spontaneous ignition is possible in wet silk. There may be no external evidence that a bale of silk has ignited, until the fire burns through to the outside.

The burning characteristics of synthetic fibers vary according to the materials used in their manufacture. The characteristics of some of the more common synthetics are given in Table 5.1.

Table 5.1. Burning Characteristics of Some Common Synthetic Fibers.

Synthetic	*Burning characteristics*
Acetate	Burns and melts ahead of the flame; ignition about the same as cotton.
Acrylic	Burns and melts; ignition temperature 560°C (1040°F); softens at 235–330°C (455–626°F).
Nylon	Supports combustion with difficulty; melts and drips; melting point, 160–260°C (320–500°F); ignition temperature 425°C (797°F) and above.
Polyester	Burns readily; ignition temperature, 450–485°C (842–905°F); softens at 256–292°C (493–558°F) and drips.
Plastic wrap	Does not support combustion. Melts.
Viscose	Burns about the same as cotton.

These characteristics are based on small-scale tests and may be misleading. Some synthetic fabrics appear to be flame retardant when tested with a small flame source, such as a match. However, when the same fabrics are subjected to a larger flame or a full-scale test, they may burst into flames and burn completely while generating quantities of black smoke.

Products of Combustion. As noted above and in Chapter 4, all burning materials produce hot gases (called *fire gases*), flame, heat and smoke, resulting in decreased oxygen levels. The predominant fire gases are carbon monoxide, carbon dioxide and water vapor. Burning vegetable fibers such as cotton, jute, flax, hemp and sisal give off large amounts of dense smoke. Jute smoke is particularly acrid.

Burning wool gives off dense, grayish-brown smoke. Another product of the combustion of wool is hydrogen cyanide, a highly toxic gas. Charring wool forms a sticky, black, tarlike substance.

Burning silk produces a large amount of spongy charcoal mixed with ash, which will continue to glow or burn only in a strong draft. It emits quantities of thin gray smoke, somewhat acrid in character. Silk may produce hydrogen cyanide gas under certain burning conditions.

Plastics and Rubber

A wide variety of organic substances are used in manufacturing plastics. These include phenol, cresol, benzene, methyl alcohol, ammonia, formaldehyde, urea and acetylene. The cellulose-based plastics are largely composed of cotton products; however, wood flour, wood pulp, paper and cloth also play a large part in the manufacture of many types of plastic.

Natural rubber is obtained from rubber latex, which is the juice of the rubber tree. It is combined with such substances as carbon black, oils and sulphur to make commercial rubber. *Synthetic rubbers* are similar to natural rubber in certain characteristics. Acrylic, butadiene and neoprene rubbers are some of the synthetic types.

Burning Characteristics. The burning characteristics of plastics vary widely. They depend to a significant extent on the form of the product—solid sections, films and sheets, foams, molded shapes, synthetic fibers, pellets or powders. The fire behavior of plastic materials also depends on their shape, their end use, the manner in which they are exposed to ignition and their chemical makeup. All the major plastic materials are combustible, and, in a major fire, all contribute fuel.

Plastics may be divided roughly into three groups as regards burning rates:

1. Materials that either will not burn at all or will cease to burn if the source of ignition is removed. This group includes asbestos-filled phenolics, some polyvinyl chlorides, nylon and the fluorocarbons.
2. Materials that are combustible, burn relatively slowly, but may or may not cease to burn when the source of ignition is removed. These plastics include the wood-filled formaldehydes (urea or phenol) and some vinyl derivatives.
3. Materials that burn without difficulty and can continue to burn after the source of ignition is removed. Included in this group are polystyrene, the acrylics, some cellulose acetates and polyethylene.

In a class of its own is the oldest well-known form of plastic, celluloid, or cellulose nitrate plastic. It is the most dangerous of the plastics. Celluloid decomposes at temperatures of 121°C (250°F) and above with great rapidity, and without the addition of oxygen from the air. Flammable vapor is produced by the decomposition. If this vapor is allowed to accumulate and is then ignited, it can explode violently. It will burn vigorously and is difficult to extinguish.

The caloric value of rubber is roughly twice that of other common combustible materials. For example, rubber has a heating value of 17.9×10^6 kilojoules (17,000 BTU/lb); pine wood, a value of 8.6×10^6 (8200 BTU/lb). Most types of rubber soften and run when burning and may thus contribute to rapid fire spread. Natural rubber decomposes slowly when first heated. At about 232°C (450°F) or above, it begins to decompose rapidly, giving off gaseous products that may result in an explosion. The ignition temperature of these gases is approximately 260°C (500°F).

Synthetic rubbers behave similarly, though the temperature at which decomposition becomes rapid may be somewhat higher. This temperature ranges upward from 349°C (660°F) for most synthetics, depending on the ingredients. Latex is a water-based emulsion and so does not present fire hazard.

Products of Combustion. Burning plastic and rubber produce the fire gases, heat, flame and smoke described in Chapter 4. These materials may also contain chemicals that yield additional combustion products of a toxic or lethal nature.

The type and amount of smoke generated by a burning plastic material depend on the nature of the plastic, the additives present, whether the fire is flaming or smoldering and what ventilation is available. Most plastics decompose when heated, yielding dense to very dense smoke. Ventilation tends to clear the smoke, but usually not enough for good visibility. Those plastics that burn cleanly yield less dense smoke under conditions of heat and flame. When exposed to flaming or nonflaming heat, urethane foams generally yield dense smoke; in almost all cases, visibility is lost in a fraction of a minute.

Hydrogen chloride is a product of combustion of chlorine-containing plastics such as polyvinyl chloride, a plastic used for insulating most electrical wiring. Hydrogen chloride is a deadly gas, but it has a pungent and irritating odor. No one would be likely to inhale it voluntarily.

Burning rubber produces dense, black, oily smoke that has some toxic qualities. Two of the noxious gases produced in the combustion of rubber are hydrogen sulfide and sulfur dioxide. Both are dangerous and can be lethal under certain conditions.

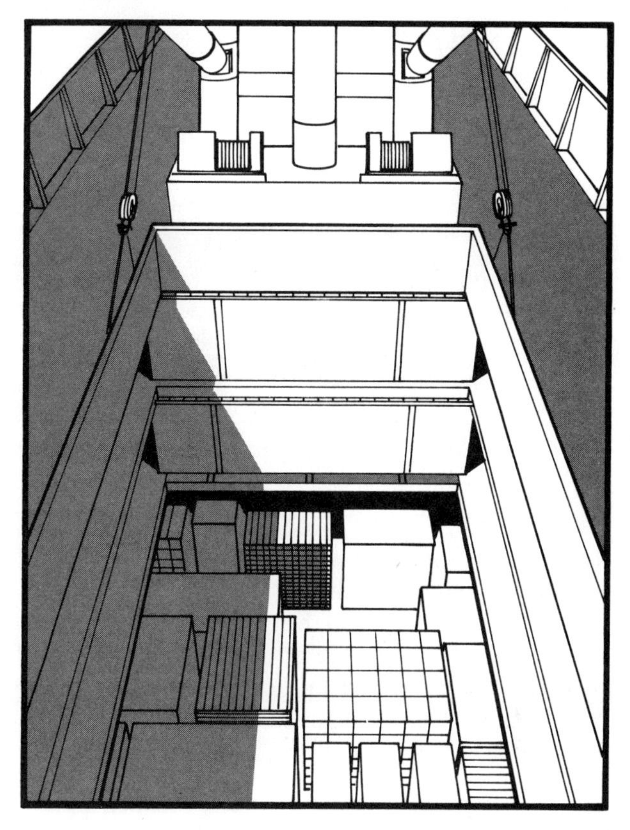

Figure 5.8. Cargo holds contain a wealth of fuel for all classes of fires.

Usual Locations of Class A Materials Aboard Ship

Although vessels are constructed of metal and may appear incombustible, there are many flammable products aboard. As noted in Chapter 1, practically every type of material (class A and otherwise) is carried as cargo. It may be located in the cargo holds or on deck, stowed in containers or in bulk stowage (Figs. 5.8 and 5.9). In addition, class A materials are used for many purposes throughout the ship.

According to Coast Guard regulations, bulkheads, linings and overheads within a room, excluding corridors and hidden spaces, may have a combustible veneer not exceeding 0.18 cm (2/28 in.) in thickness. These veneers are usually constructed of some type of plastic material with a backing of wood-based materials. In addition, the furnishings found in passenger, crew and officer accommodations are usually made of class A materials. Lounges and recreation rooms may contain couches, chairs, tables, bars, television sets, books and other items constructed wholly or partly of class A materials.

Combustible veneers, trim, or decoration may not be used in corridors or hidden spaces, according to Coast Guard regulations. However, deck coverings not exceeding 0.95 cm (3/8 in.) are not restricted. It is probable that tiles used for deck covering would be affected by fire. Title 46 CFR 164.006, Subchapter Q, contains Coast Guard requirements for deck coverings for U.S. merchant vessels.

Other areas in which class A materials may be located include the following:

- The bridge contains wooden desks, charts, almanacs and other such combustibles.
- Wood in many forms may be found in the carpenter shop.
- Various types of cordage are stowed in the boatswain's locker (Fig. 5.10).
- The emergency locker on the bridge wing contains rockets and/or explosives for the line throwing gun.
- The undersides of metal cargo containers are usually constructed of wood or wood-based materials.
- Lumber for dunnage, staging and other uses may be stored below decks.
- Large numbers of filled laundry bags are sometimes left in passageways, awaiting movement to and from the laundry room.
- Rubber and plastics are used extensively for the insulation on electrical wiring.

Extinguishment of Class A Fires

It is a fortunate coincidence that the materials most often involved in fire, class A materials, may best be extinguished by the most available

Figure 5.9. Containers also may be filled with a variety of fuels.

extinguishing agent, water. U.S. Coast Guard regulations specify a system whereby water will be available for firefighting purposes by requiring that a firemain system be installed on self-propelled vessels. Firemain systems and other fixed systems are described in Chapter 9. The use of water and other extinguishing agents in fighting class A fires is discussed in Chapters 7, 8 and 10.

CLASS B FIRES INVOLVING MATERIALS COMMONLY FOUND ABOARD SHIP

The materials whose involvement leads to class B fires may be grouped as flammable and combustible liquids, paints and varnishes, and flammable gases. Again we shall discuss each group individually.

Flammable and Combustible Liquids

Flammable liquids as defined by Title 46 CFR (30.10–22) are those that give off flammable vapors at or below 26.7°C (80°F) and having a Reid vapor pressure not exceeding 40 pounds per square inch absolute (psia) at 37.8°C (100°F). There are three grades of flammable liquids; their definitions are given in Chapter 1. Examples of common flammable liquids are ethyl ether, gasoline, acetone and alcohol. All flash at or below 26.7°C (80°F).

Combustible liquids are those with a flash point above 26.7°C (80°F). There are two grades of combustible liquids; their definitions are given in Chapter 1. The heavier petroleum products, such as kerosene, diesel oil and fuel oil, are considered to be combustible liquids; their flash points range from 26.7°C (80°F) to 65.5°C (150°F). Some other combustible liquids are acids, vegetable oils and lubricating oils, all of which have flash points above 65.5°C (150°F).

Burning Characteristics. As noted in Chapter 4, it is the vapor of a flammable or combustible liquid, rather than the liquid itself, that burns or explodes when mixed with air and ignited. These liquids will vaporize when exposed to air—and at an increased rate when heated. They should be stored in the proper type of closed containers to minimize the fire hazard; even in use they should be exposed to air as little as possible.

Flammable vapor explosions most frequently

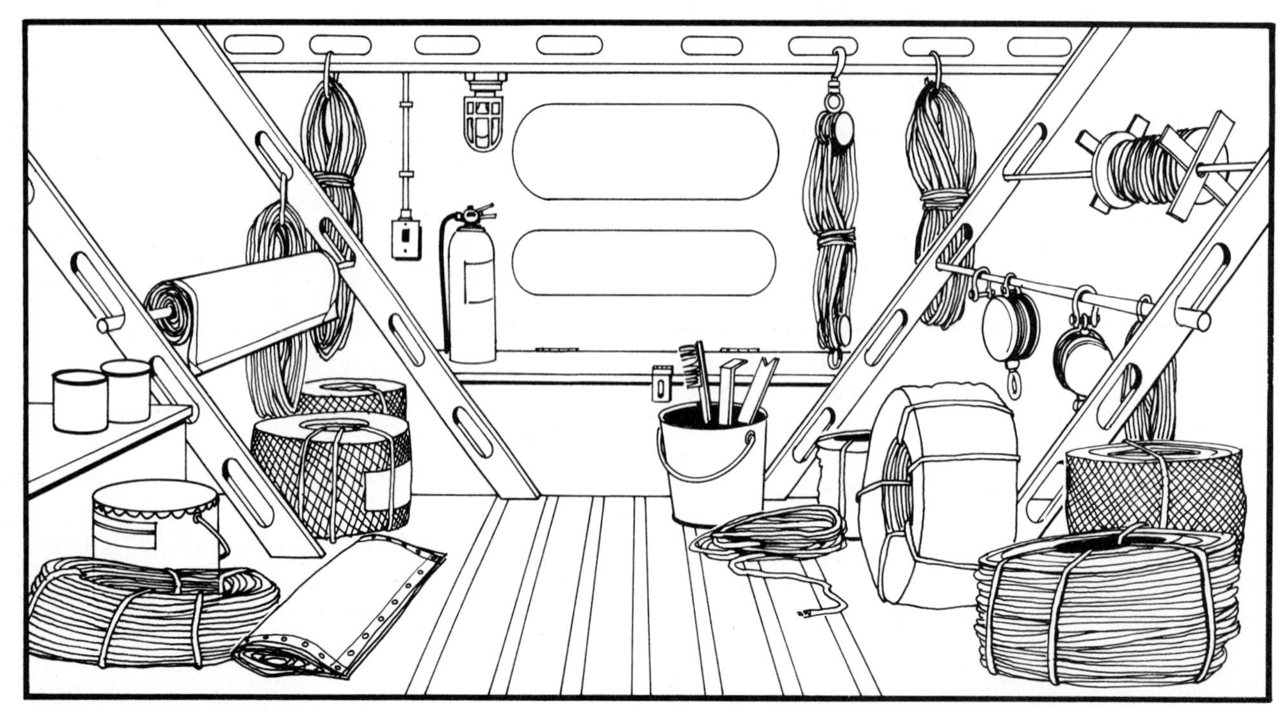

Figure 5.10. The boatswain's locker contains rope and many other class A materials.

occur within a confined space such as a container, tank, room or structure. The violence of a flammable vapor explosion depends upon

- The concentration and nature of the vapor
- The quantity of vapor–air mixture present
- The type of enclosure in which the mixture is confined.

The flash point is the commonly accepted and most important factor determining the relative hazard of a flammable or combustible liquid. However, it is not the only factor involved. The ignition temperature, flammable range, rate of evaporation, reactivity when contaminated or exposed to heat, density and rate of diffusion of the vapor also determine how dangerous the liquid is. However, once a flammable or combustible liquid has been burning for a short time, these factors have little effect on its burning characteristics.

The burning rates of flammable liquids vary somewhat, as do their rates of flame travel. The burning rate of gasoline is 15.2–30.5 cm (6–12 in.) of depth per hour; for kerosene, the rate is 12.7–20.3 cm (5–8 in.) of depth per hour. For example, a pool of gasoline 1.27 cm (½ in.) deep could be expected to burn itself out in 2.5 to 5 minutes.

Products of Combustion. In addition to the usual combustion products, there are some that are peculiar to flammable and combustible liquids. Liquid hydrocarbons normally burn with an orange flame and give off dense clouds of black smoke. Alcohols normally burn with a clean blue flame and very little smoke. Certain terpenes and ethers burn with considerable ebullition (boiling) of the liquid surface and are difficult to extinguish. Acrolein (acrylic aldehyde) is a highly irritating and toxic gas produced during the combustion of petroleum products, fats, oils and many other common materials.

Usual Locations Aboard Ship. Flammable and combustible liquids of all types are carried as cargo by tank vessels. In addition to this bulk stowage, these liquids are transported in portable tanks that, according to U.S. Coast Guard regulations, can be "barrels, drums or other packages" having a maximum capacity of 416 liters (110 U.S. gallons). Flammable and combustible liquids in smaller packages may be found in holds and in large shipping containers.

Large quantities of combustible liquids, in the form of fuel and diesel oil, are also stowed aboard ship, for use in propelling and generating electricity. The hazards involved in stowing and transferring these fuels are covered in Chapter 1. Figures 5.11–5.13 show typical stowage locations.

Fuel and diesel oil are most hazardous when they have been heated prior to feeding into the burners. Cracks in the piping will then allow the oil to leak out, exposing it to ignition sources. If the resultant spill is large, an extensive, hot fire will result.

Other locations where combustible liquids may be found include the galley (hot cooking oils) and the various shops and spaces where lubricating oils are used and stored. Fuel and diesel oil may also be found as residues and films on and under oil burners and equipment in the engine room.

Extinguishment. U.S. Coast Guard regulations provide for the installation of firemains and fixed fire-extinguishing systems using foam, carbon

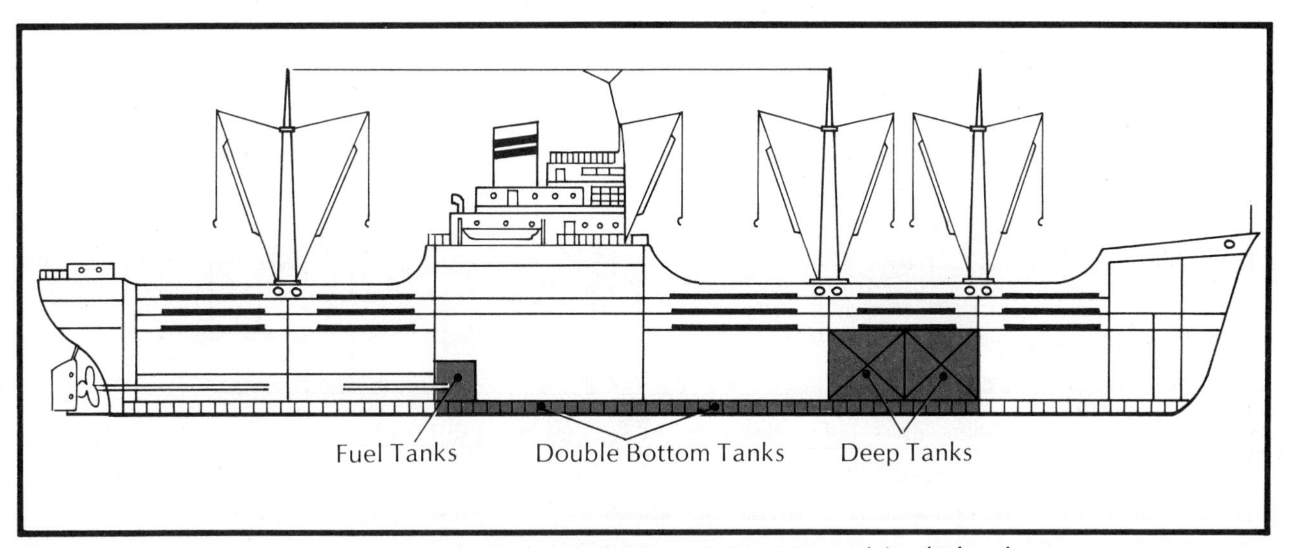

Figure 5.11. Location stowage of combustible liquid aboard break bulk cargo vessel for shipboard use.

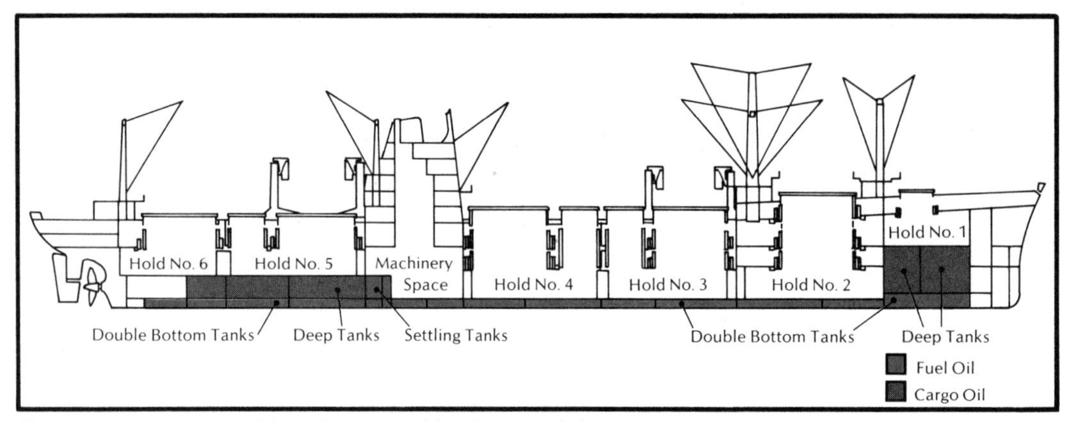

Figure 5.12. Location of liquid cargo and bunker tanks aboard dry cargo vessel.

dioxide, steam and water spray in appropriate locations. In addition, foam, dry chemical, carbon dioxide and water extinguishers of various size and portability must be placed in specified areas throughout the ship. These appliances are discussed in Chapters 8 and 9.

The source of the flammable or combustible liquid involved in fire should be cut off as soon as possible (and if possible). This will halt the supply of fuel feeding the fire, and allow firefighters to employ the following general methods of extinguishment:

- *Cooling*. Using water from the firemain system, in spray or solid-stream form, to cool tanks and exposed areas.
- *Smothering*. Using foam to blanket the liquid and thus shut off the supply of oxygen to the fire; discharging steam or carbon dioxide into burning areas; eliminating oxygen by sealing off the ventilation to the fire.
- *Inhibiting flame propagation*. Applying dry chemicals above the burning surfaces.

It is difficult to establish rigid procedures for extinguishing specific types of fires, since no two fires are alike. However, the following general guidelines apply for fires involving flammable and combustible liquids:

- *Minor spills:* Use dry chemical or foam extinguishers, or water fog.
- *Large spills:* Use large dry chemical extinguishers, backed up by foam or fog lines. Use water streams to protect objects that are exposed to fire.
- *Spills on water:* If contained, use foam to smother fire. Otherwise use large-volume fog stream.
- *Sighting or ullage ports:* Apply foam, dry chemical or high- or low-velocity water fog horizontally across the opening until it can be closed.
- *Cargo tanks:* Use the deck foam system and/or carbon dioxide or steam smothering system, if so equipped. Water fog may be used for heavy oils.
- *Ship's galley:* Use carbon dioxide or dry chemical extinguishers.
- *Oil-burning equipment:* Use foam or water fog.

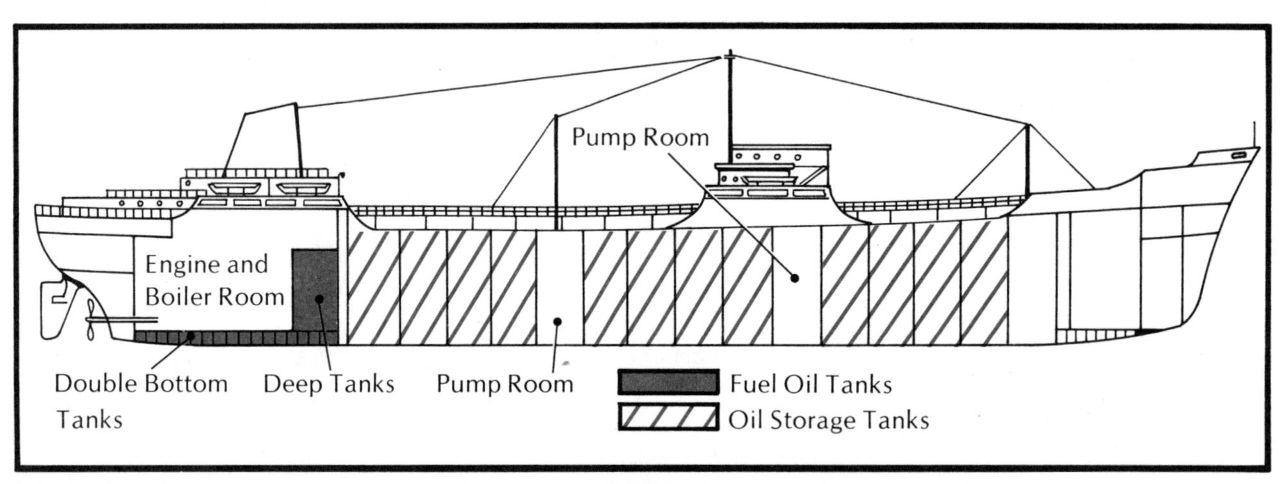

Figure 5.13. Location of liquid cargo and bunker tanks aboard tanker vessel.

Additional information is given in Chapters 7–10.

Paints and Varnishes

Most paints, varnishes, lacquers and enamels, except those with a water base, present a high fire risk in storage or in use. The oils in oil-base paints are not themselves very flammable (linseed oil, for example, has a flash point of over 204°C (400°F)). However, the solvents commonly used in these paints are flammable and may have flash points as low as 32°C (90°F). The same is true for enamels and oil varnishes. And, normally, all the other ingredients of most paints and varnishes are combustible.

Most paints and varnishes are still combustible after they have dried, though their flammability is much reduced when the solvent has evaporated. In practice, the flammability of dry paint depends on the flammability of its base. Thus, the usual oil-base paints are not very flammable.

Burning Characteristics and Products of Combustion. Liquid paint burns fiercely and gives off much heavy black smoke. It also, obviously, can flow, so that a paint fire resembles an oil fire in many ways. Because of the dense smoke and the toxic fumes given off by liquid paint and varnish, breathing apparatus should always be used by crewmen fighting a paint fire in an enclosed area.

Explosions are another hazard of liquid paint fires. Since paint is normally stored in tightly sealed cans or drums (of up to 150–190 liters (40–50 gal.) capacity), fire in any paint storage area may easily heat up the drums and burst them. The contents are likely to ignite instantly and with explosive force on exposure to air.

Usual Locations Aboard Ship. Paints, varnishes, enamels, lacquers and their solvents are stored in paint lockers. These are usually located either fore or aft, in compartments below the main deck. U.S. Coast Guard regulations require that paint lockers be constructed of steel or wholly lined with metal. Such spaces must also be serviced by a fixed carbon dioxide extinguishing system or other approved system. (As indicated previously, paint is still combustible after drying, and it will burn in a fire.)

Extinguishment. Because liquid paints contain low-flash-point solvents, water is not a suitable extinguishing agent. Foam must be used if any substantial quantity of paint is involved. Surrounding materials may have to be cooled with water. On small quantities of paint or varnish, a carbon dioxide or dry chemical extinguisher may be used in place of a foam extinguisher. Water is the proper extinguishing agent for dry paint (Fig. 5.14).

Flammable Gases

In the gaseous state, the molecules of a substance are not held together, but are free to move about. As a result, a gas has no shape of its own, but rather takes the shape of its container. Most solids and liquids can be vaporized (become gases) when their temperature is increased sufficiently. However, we shall use the term *gas* to mean a substance that is in the gaseous state at so-called normal temperature and pressure (NTP) conditions. These are approximately 21°C (70°F) and 101.4 kilopascals (14.7 psia).

Any gas that will burn in the normal concentrations of oxygen in air is a *flammable* gas. As with other gases or vapors, a flammable gas will burn only when its concentration in air is within its combustible range and the mixture is heated to its ignition temperature.

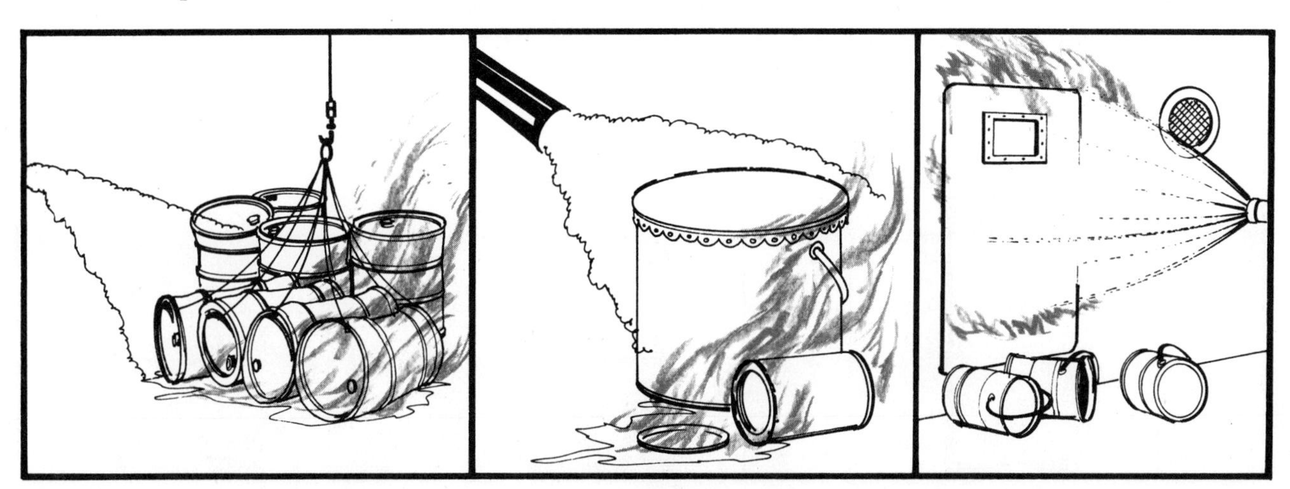

Figure 5.14. Extinguishment of paint and varnish fires. **A.** Apply foam to large fires. **B.** Carbon dioxide or dry chemical may be used on small fires. **C.** Dry paint is a class A material; water is the primary extinguishing agent.

Flammable gases are usually stored and transported aboard vessels (Fig. 5.15) in one of three ways:

- *Compressed.* A compressed gas is one that, at normal temperatures, is entirely in the gaseous state under pressure in its container.
- *Liquefied.* A liquefied flammable gas is one that, at 100°F has a Reid vapor pressure of at least 40 psia. At normal temperatures it is partly in the liquid state and partly in the gaseous state under pressure in its container.
- *Cryogenic.* A cryogenic gas is one that is liquefied in its container at a temperature far below normal temperatures, and at low to moderate pressures.

Basic Hazards. The hazards presented by a gas that is confined in its container are different from those presented when the gas escapes from its container. We shall discuss them separately, even though these hazards may be present simultaneously, in a single incident.

Hazards of confinement. When a confined gas is heated, its pressure increases. If enough heat is applied, the pressure can increase sufficiently to cause a gas leak or a container failure. In addition, contact with flames can reduce the strength of the container material, possibly resulting in container failure.

To prevent explosions of compressed gases, pressure relief valves and fusible plugs are installed in tanks and cylinders. When gas expands in its container, it forces the relief valve open allowing gas to flow out of the container, thereby reducing the internal pressure. A spring loaded device closes the valve when the pressure is reduced to a safe level. A plug of fusible metal that will melt at a fixed temperature is also used. The plug seals an opening in the body of the container, usually near the top. Heat from a fire, threatening the tank or cylinder, causes the metal plug to melt allowing the gas to escape through the opening. Explosive pressure within the tank is prevented. However, the opening cannot be closed, therefore, the gas will continue to escape until the container is empty.

Explosions can occur when these safety devices are not installed or should they fail to operate. Another cause of explosion is a very rapid buildup of pressure in a container. The pressure cannot be relieved through the safety valve opening fast enough to prevent pressure buildup of explosive force. Tanks and cylinders are also subject to explosion when flame impinges on their surface causing the metal to lose its strength. Flame impingement above the liquid level is more dangerous than impingement on the container surface area that is in contact with the liquid. Heat from flames above the liquid line is absorbed by the metal itself; below the liquid line most of the heat is absorbed by the liquid. This is not to be construed as a safe condition because absorption of heat by the liquid also causes a dangerous, although less rapid, pressure increase. Spraying the surface of the container with water can help keep the pressure from building up to explosive force. Cooling with water is not a guarantee an explosion can be averted, especially when flame impingement is occurring.

Container failures. Compressed or liquefied gas represents a great deal of energy held in check by its container. When the container fails, this energy is released—often very rapidly and vio-

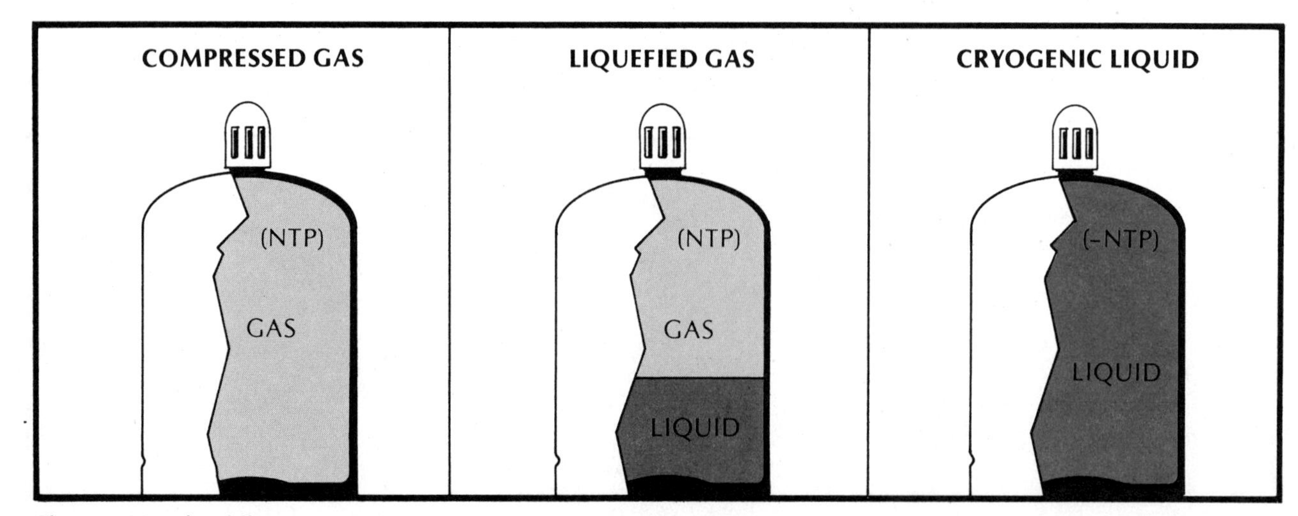

Figure 5.15. The differences among compressed, liquefied and cryogenic gases.

lently. The gas escapes and the container or container pieces are thrown about.

Failures of liquefied flammable gas containers from fire exposure are not rare. This type of failure is called boiling liquid–expanding vapor explosion, or BLEVE (pronounced "blevey"). In most BLEVEs, the container fails at the top, where it is in contact with gas (see Fig. 5.16). The metal stretches, thins out, and tears lengthwise, until it finally gives way.

The magnitude of the explosion (BLEVE) depends mainly on how much liquid vaporizes when the container fails and on the weight of the container pieces. Most BLEVEs occur when containers are from slightly less than half full to about three-fourths full of *liquid.* A small, uninsulated container can experience a BLEVE in a very few minutes, and a very large container in a few hours, in the absence of water cooling.

Uninsulated liquefied gas containers that are exposed to fire can be protected from BLEVEs by applying water. A film of water should be maintained on the upper portion of the container, the portion that is in the internal contact with vapor.

Hazards of gases released from confinement. The hazards of a gas that has been released from its container depend on the properties of the gas and where it is released. All gases except oxygen and air are hazardous if they displace breathing air. Odorless and colorless gases such as nitrogen and helium are particularly hazardous, as they give no warning of their presence.

Toxic, or poisonous, gases are obviously hazardous to life. When released in the vicinity of a fire, they will prevent access by firefighters or force firefighters to use breathing apparatus.

Oxygen and other oxidizing gases are nonflammable. However, these gases can cause combustible substances to ignite at lower than usual temperatures.

Contact with liquefied gas can cause frostbite, which can be severe if the exposure is prolonged. In addition, many structural materials can become brittle and fail when exposed to low temperatures. Carbon steel and plastics are affected in this way.

Released flammable gases present the danger of explosion or fire or both. A released flammable gas will explode when enough gas has collected and mixed with air in a confined space before it is ignited. It will burn without exploding if a sufficient quantity of gas–air mixture has not accumulated—either because it ignited too quickly or because it is not confined and can dissipate. Thus, when a flammable gas escapes into open deck positions, the result is usually fire. However, if a massive release occurs, the surrounding air or the ship's superstructure can confine the gas sufficiently to cause an explosion. This type of explosion is known as an *open air explosion* or *space explosion.* Liquefied noncryogenic gases, hydrogen and ethylene are subject to these open air explosions.

Properties of Some Common Gases. The important properties of a number of flammable gases are discussed in the following pages. These properties lead to varying degrees and combinations of hazards when the gases are confined or released.

Acetylene. Acetylene is composed of carbon and hydrogen. It is used primarily in chemical processing and as a fuel for oxyacetylene cutting and welding equipment. It is nontoxic and has been used as an anesthetic. Pure acetylene is odorless, but the acetylene in general use has an odor due to minor impurities mixed in with the gas.

Acetylene is shipped and stored mainly in cylinders. For safety acetylene cylinders are filled with a porous packing material usually diatomaceous earth containing very small pores or cellular spaces. In addition, the packing material is saturated with acetone, a flammable liquid in which acetylene dissolves easily. Thus acetylene cylinders contain much less of the gas than they appear to hold. A number of safety fuse plugs are installed in the top and bottom of the cylinder. The plugs release the gas to the atmosphere in case of a dangerously high temperature or pressure within the container.

Acetylene is subject to explosion and fire when released from its container. It is easier to ignite than most flammable gases, and it burns more rapidly. This increases the severity of explosions and the difficulty of venting to prevent explosion. Acetylene is only slightly lighter than air, which means it will mix well with air upon leaving its container.

Anhydrous ammonia. Anhydrous ammonia is composed of nitrogen and hydrogen. It is used primarily as fertilizer, as a refrigerant, and as a source of hydrogen for the special atmospheres needed to heat-treat metals. It is a relatively toxic gas, but its sharp odor and irritating properties serve as warnings. However, large clouds of anhydrous ammonia, produced by large liquid leaks, have trapped and killed people before they could evacuate the area.

Anhydrous ammonia is shipped in cylinders,

cargo trucks, railroad tank cars and barges. It is stored in cylinders, tanks, and in cryogenic form in insulated tanks. BLEVEs of uninsulated anhydrous ammonia containers are rare, mainly because of the limited flammability of the gas. Where BLEVEs have occurred, they have resulted from exposure to fires involving other combustibles.

Anhydrous ammonia is subject to explosion and fire (and presents a toxicity hazard) when released from its container. However, its high LEL and low heat of combustion tend to minimize these hazards. In unusually tight locations such as refrigerated process or storage areas, the release of the liquid or a large quantity of gas can result in an explosion.

Ethylene. Ethylene is composed of carbon and hydrogen. It is used principally in chemical processing, for example the manufacture of polyethylene plastic; smaller amounts are used to ripen fruit. It has a wide flammable range and burns quickly. While nontoxic, ethylene is an anesthetic and asphyxiant.

Ethylene is shipped as a compressed gas in cylinders and as a cryogenic gas in insulated cargo trucks and railroad tank cars. Most ethylene cylinders are protected against overpressure by frangible (bursting) discs. (Medical cylinders may have fusible plugs or combination safety devices.) Tanks are protected by safety relief valves. Cylinders are subject to failure from fire exposure but not BLEVEs, as they do not contain liquid.

Ethylene is subject to explosion and fire when released from its container. Its wide flammable range and high burning rate accentuate these hazards. In a number of cases involving rather large outdoor releases, open air explosions have occurred.

Liquefied natural gas (LNG). LNG is a mixture of materials, all composed of carbon and hydrogen. The principal component is methane, with smaller amounts of ethane, propane and butane. LNG is nontoxic but is an asphyxiant. It is used as a fuel.

LNG is shipped as a cryogenic gas in insulated cargo trucks by Department of Transportation (DOT) permit, and in tank vessels under U.S. Coast Guard authorization. It is stored in insulated tanks, protected against overpressure by safety relief valves.

LNG is subject to explosion and fire when released from its container into an enclosed space, such as inside a hatch. Test data and experience indicate that escaping LNG is not subject to open air explosions.

Liquefied petroleum gas (LPG). LPG is a mixture of materials, all composed of carbon and hydrogen. Commercial LPG is mostly either propane or normal butane, or a mixture of these with small amounts of other gases. It is nontoxic but is an asphyxiant. It is used principally as a fuel and, in domestic and recreational applications, sometimes known as "bottled gas."

LPG is shipped as a liquefied gas in uninsulated cylinders and tanks and in cargo trucks, railroad tank cars and vessels. It is also shipped in cryogenic form in insulated marine vessels. It is stored in cylinders and insulated tanks. LPG containers are generally protected against overpressure by safety relief valves. Some cylinders are protected by fusible plugs and, occasionally, by a combination of these (Fig. 5.16). Most containers are subject to BLEVEs.

LPG is subject to explosion and fire when released from its container. As most LPG is used indoors, explosions are more frequent than fires. The explosion hazard is accentuated by the fact that 3.8 liters (1 gal) of liquid propane or butane produces 74.7–83.8 cubic meters (245–275 cubic ft) of gas. Large releases of liquid-phase LPG outdoors have led to open air explosions.

Usual Locations Aboard Ship. Liquefied flammable gases such as LPG and LNG are transported in bulk on tankers. Flammable gases in cylinders may be carried only on deck on cargo vessels. Additionally, such flammable gases as acetylene will be found stored in cylinders for use on board.

The Department of Transportation regulates the shipment of hazardous materials on cargo vessels, and flammable gases are in this category. According to Coast Guard regulations, flammable gas cylinders may be stowed on deck or under deck (meaning in a compartment or hold), depending on how hazardous the gas is. Acetylene, for example, can only be stowed on deck, and it must be shaded from radiant heat. Anhydrous ammonia, on the other hand, is classified as a nonflammable gas and may be stowed on deck or under deck. Ethylene and LPG are flammable and can also be stowed on deck or under deck. LNG, however, may be shipped only after a thorough case review and authorization by the Department of Transportation.

Extinguishment. Flammable gas fires can be extinguished with dry chemicals. Carbon dioxide

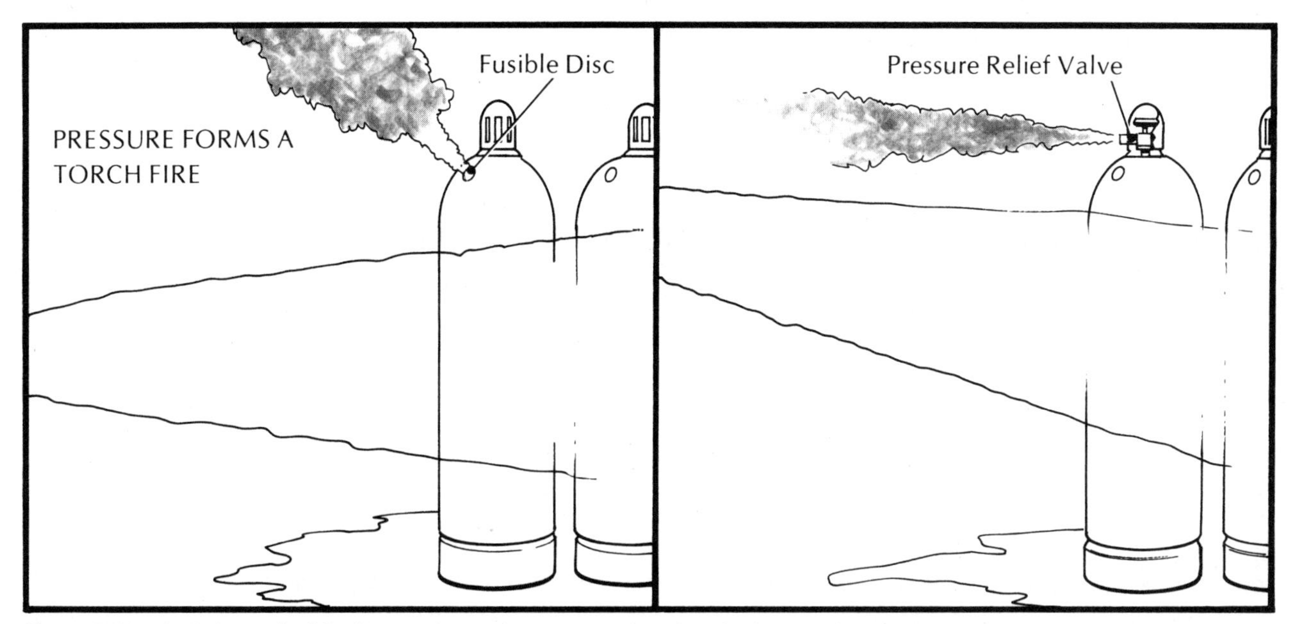

Figure 5.16. **A.** When a fusible disc or plug melts, it cannot be closed. The entire contents of the container will escape and will burn if ignited. Water should be applied to cool the container and prevent explosion. **B.** When a relief valve opens, the container should be cooled with water. This will reduce the pressure within the container, allowing the relief valve to close automatically.

and vaporizing liquids may extinguish certain gas fires. However, these fires present a severe radiant heat hazard to firefighting forces. Additionally, there is the danger of the gas continuing to escape after the fire is extinguished, thus creating another fire and explosion problem. Dry chemical and water spray offer good heat shields from the radiant heat of gas fires while CO_2 and vaporizing liquid do not.

The standard procedure for control is to allow the gas to burn until the flow can be shut off at the source. Extinguishment should not be attempted unless such extinguishment leads to shutting off the fuel flow. Until the flow of gas supplying the fire has been stopped, firefighting efforts should be directed toward protecting exposures. (*Exposures* are combustible materials that may be ignited by flames or radiated heat from the fire. Water in the form of straight streams and fog patterns is usually used to protect exposures.) When the gas is no longer escaping from its container, the gas flames should go out. However, where the fire was extinguished before shutting off the gas flow, firefighters must be careful to prevent the ignition of gas that is being released.

Fires involving liquefied flammable gases (such as LPG and LNG) can be controlled, and often extinguished, by maintaining a thick blanket of foam over the surface area of the spilled fuel.

CLASS C FIRES INVOLVING ELECTRICAL EQUIPMENT ABOARD SHIP

Electrical equipment involved in fire, or in the vicinity of a fire, may cause electric shock or burn to firefighters. In this section we discuss some electrical installations found aboard ship, their hazards and the extinguishment of fires involving electrical equipment.

Types of Equipment

Generators. Generators are machines that produce electrical power. These machines are usually driven by machines which utilize steam as produced in an oil-fired boiler or internal combustion engine burning a fuel in its cylinders. The electrical wiring in the generator is insulated with a combustible material. Any fire involving the generator or its prime mover will involve a high risk of electrical shock to the firefighter.

Panelboards. A panelboard has fuses and automatic devices for the control and protection of lighting and power circuits. The switches, fuses, circuit breakers and terminals within a panelboard all have electrical contacts. These contacts may develop considerable heat, causing dangerously high temperatures and unnecessary operation of overcurrent devices, unless they are maintained in good condition. Overcurrent devices are provided for the protection of conductors and electrical equipment. They open a circuit if the current in that circuit produces an excessively high temperature.

Switches. Switches are required for the control of lights and appliances and for disconnecting motors and their controllers. They are also used to isolate high voltage circuit breakers for maintenance operations. Switches may be of either the air-break or the oil-break type. In the oil-break type, the device that interrupts the circuit is immersed in oil.

The chief hazard is the arcing produced when the switch is opened. In this regard, oil-break switches are the more hazardous of the two types. The hazard increases when a switch is operated much beyond its rated capacity, when its oil is in poor condition or when the oil level is low. Then the arc may vaporize the remaining oil, rupture the case and cause a fire. However, if properly used and maintained, these switches present no hazard.

Electric Motors. Many fires are caused by electric motors. Sparks or arcs, from short circuiting motor windings or improperly operating brushes, may ignite the motor insulation or nearby combustible material. Other causes of fires in motors include overheating of bearings due to poor lubrication and grimy insulation on conductors preventing the normal dissipation of heat.

Electrical Faults that May Cause Fires

Short Circuits. If the insulation separating two electrical conductors breaks down, a *short circuit* occurs. Instead of following its normal path, the current flows from one conductor to the other. Because the electrical resistance is low, a heavy current flows and causes intense local heating. The conductors become overloaded electrically, and they may become dangerously overheated unless the circuit is broken. If the fuse or circuit breaker fails to operate, or is unduly delayed, fire can result and spread to nearby combustible material.

Overloading of Conductors. When too large an electrical load is placed on a circuit, an excessive amount of current flows and the wiring overheats. The temperature may become high enough to ignite the insulation. The fuses and circuit breakers that are installed in electric circuits will prevent this condition. However, if these safety devices are not maintained properly, their failure may result in a fire.

Arcing. An arc is pure electricity jumping across a gap in a circuit. The gap may be caused intentionally (as by opening a switch) or accidentally (as when a contact at a terminal becomes loose). In either case, there is intense heating at the arc. The electrical strength of the arc and amount of heat produced depend on the current and voltage carried by the circuit. The temperature may easily be high enough to ignite any combustible material near the arc, including insulation. The arc may also fuse the metal of the conductor. Then, hot sparks and hot metal may be thrown about, and set fire to other combustibles.

Hazards of Electrical Fires

Electric Shock. Electric shock may result from contact with live electrical circuits. It is not necessary to touch one of the conductors of a circuit to receive a shock; any conducting material that is electrified through contact with a live circuit will suffice. Thus, firefighters are endangered in two ways: First, they may touch a live conductor or some other electrified object while groping about in the dark or in smoke. Second, a stream of water or foam can conduct electricity to firefighters from live electrical equipment. Moreover, when firefighters are standing in water, both the chances of electric shock and the severity of shocks are greatly increased.

Burns. Many of the injuries suffered during electrical fires are due to burns alone. Burns may result from direct contact with hot conductors or equipment, or from sparks thrown off by these devices. Electric arcs can also cause burns. Even persons at a distance from the arc may receive eye burns.

Toxic Fumes from Burning Insulation. The insulation on electrical conductors is usually made of rubber or plastic. The toxic fumes given off by burning plastics and rubber have been discussed previously. One plastic deserves special attention because of its widespread use as electrical insulation and its toxic combustion products. This is polyvinyl chloride, also known as PVC. This plastic releases hydrogen chloride, which attacks the lungs with serious consequences. It is also believed that PVC contributes to the severity and hazards of fires.

Usual Locations of Electrical Equipment Aboard Ship

Electric power is essential to the operation of a modern vessel. The equipment that generates, controls and delivers this power is found throughout the ship. Some of this equipment, such as lighting devices, switches and wiring, is common and easily recognized. The locations of some of the less familiar and more hazardous electrical equipment are covered here.

Engine Room. The source of the ship's electric power is its generators. Two generators are located in the engine room. One is always in use, and the other is available in case the first is shut down. The generators supply power to the main electrical panelboard, which is in the same area as the generators in the engine room. The main panelboard houses the generator control panel and the distribution panels. If fire breaks out in the vicinity of the generator switches or the main panelboard, the ship's engineer can stop the generator by mechanical means. This will deenergize the panelboard and switches. Also nearby is the engine room console, which contains controls for the fire pumps, ventilating fans, engineer's signal alarm panel, temperature scanner system and other engine room equipment.

Emergency Generator Room. An emergency generator and switchboard are available for use on most ships in case the main generator fails. It will provide power for emergency lighting and equipment only. They are located in the emergency generator room, which is always at some distance from the engine room. In case of fire this generator shuts down automatically when carbon dioxide from the total flooding extinguishment system is released into the room.

Passageways. Electrical control lockers are situated at the ends of some passageways. (Controls shall be outside the space protected.) Electrical distribution panels for the ventilation system and for boat and ladder winches are placed in these lockers. Lighting panelboards are located along passageway bulkheads. Much of the ship's electrical wiring is placed in the passageway overheads. Access panels are provided in these overheads to allow work on the wiring; these panels can be removed to check the area for fire extension.

Other Locations. The bridge contains much electrical equipment, including the radar apparatus, bridge console, smoke detector indicating panel and lighting panelboards. Below decks, in the bow and stern, are electrical control panels for the capstan and winch motors. A power panelboard in the machine shop controls the electric-arc welding machine, buffer and grinder, drill press and lathe. There is still much more electrical equipment located throughout every vessel. The important point is that the hazards of live electrical equipment must be considered whenever a shipboard fire is being fought.

Extinguishment of Class C Fires

"When any type of electrical equipment is involved with fire, its circuit should be deenergized. However, whether the circuit is deenergized or not, the fire must be extinguished using a nonconducting agent, such as dry chemical, CO_2 or Halon. Firefighters should always consider an electrical circuit to be energized. The use of water in any form is not permitted. Firefighters must wear appropriate breathing devices when entering spaces where electrical equipment has been burning since toxic gases are given off by burning electrical insulation.

Crew members must remember two things when combating electrical fires: First, all electrical equipment in the fire area must be treated as "live" until it is known that the deenergizing process has been completed. Second, breathing apparatus must be worn because of the toxic gases given off by burning insulation and metal wiring.

CLASS D FIRES INVOLVING METALS FOUND ABOARD SHIP

Metals are commonly considered to be nonflammable. However, they can contribute to fires and fire hazards in a number of ways. Sparks from the ferrous metals, iron and steel, can ignite nearby combustible materials. Finely divided metals are easily ignited at high temperatures. A number of metals, especially in finely divided form, are subject to self-heating under certain conditions; this process has caused fires. Alkali metals such as sodium, potassium and lithium react violently with water, liberating hydrogen; sufficient heat is generated in the process to ignite the hydrogen. Most metals in powder form can be ignited as a dust cloud; violent explosions have resulted. In addition to all this, metals can injure firefighters through burning, structural collapse and toxic fumes.

Many metals, such as cadmium, give off noxious gases when subjected to the high temperatures of a fire. Some metallic vapors are more toxic than others; however, breathing apparatus should be used whenever fires involving metals are fought.

Hazards and Characteristics of Some Specific Metals

Aluminum. Aluminum is a light metal with good electrical conductivity. In its usual forms it does not present a problem in most fires. However, its melting point of 660°C (1220°F) is low enough to cause the collapse of unprotected aluminum structural members. Aluminum chips and shavings have been involved in fire, and aluminum dust is a severe explosion hazard. Aluminum does not ignite spontaneously and is not considered to be toxic.

Iron and Steel. Iron and steel are not considered combustible. In the larger forms, such as structural steel, they do not burn in ordinary fires. However, fine steel wool or dust may be ignited, and iron dust is a fire and explosion hazard when exposed to heat or flame. Iron melts at 1535°C (2795°F), and ordinary structural steel at 1430°C (2606°F).

Magnesium. Magnesium is a brilliant white metal that is soft, ductile and malleable. It is used as a base metal in light alloys for strength and toughness. Its melting point is 648.8°C (1200°F). Dust or flakes of magnesium are easily ignited, but in solid form it must be heated above its melting point before it will burn. It then burns fiercely with a brilliant white light. When heated, it reacts violently with water and all moisture.

Titanium. Titanium is a strong white metal, lighter than steel, that melts at 2000°C (3632°F). It is mixed with steel in alloys to give high working temperatures. It is easily ignited in smaller forms (titanium dust is very explosive), though larger pieces offer little fire hazard. Titanium is not considered toxic.

Usual Locations of Class D Materials Aboard Ship

The metal principally used in the construction of vessels is steel. However, aluminum, its alloys and other lighter metals are used to build the superstructures of some ships. The advantage of aluminum lies in the reduction of weight. A disadvantage, from the firefighting viewpoint, is the comparatively low melting point of aluminum as compared to that of steel.

In addition to the material used for the ship itself, metals are carried in most forms as cargo. Generally, there are no stowage restrictions regarding metals in solid form. On the other hand, the metallic powders of titanium, aluminum and magnesium must be kept in dry, segregated areas. The same rules apply to the metals potassium and sodium.

It should be noted here that the large containers used for shipping cargo are usually made of aluminum. The metal shells of these containers have melted and split under fire conditions, exposing their contents to the fire.

Extinguishment of Class D Fires

Fires involving most metals present an extinguishment problem to firefighters. Frequently there is a violent reaction with water, which may result in the spreading of the fire and/or explosion. If only a small amount of metal is involved and the fire is confined, it may be advisable to allow the fire to burn itself out. Exposures should, of course, be protected with water or another suitable extinguishing agent.

Some synthetic liquids have been employed in extinguishing metal fires, but these are not usually found aboard ship. The ABC or multipurpose dry chemical extinguisher has been used with some success on fires involving metals. Such extinguishers may be available to shipboard firefighters.

Sand, graphite, various other powder extinguishing agents and salts of different types have been applied to metallic fires with varying success. No one method of extinguishment has proven effective for all fires involving metals.

BIBLIOGRAPHY

Accident Prevention Manual for Industrial Operations, 6th Ed. Chicago, National Safety Council, 1974.

Coast Guard Rules and Regulations for Cargo and Miscellaneous Vessels. Department of Transportation, Washington, D.C., 1973.

Coast Guard Rules and Regulations for Tank Vessels, 1973.

Coast Guard Rules and Regulations, Subchapter J, Electrical Engineering, 1977.

Eyres, D. J.: Ship Construction. London, England, Heinemann, 1974.

Fire Protection Handbook, 14th Ed. Boston, National Fire Protection Association, 1976.

Guides for Fighting Fires in and Around Petroleum Storage Tanks. Washington, D. C., American Petroleum Institute, 1974.

Haessler, Walter M.: The Extinguishment of Fire, Boston, Mass., National Fire Protection Association, 1974.

Hazardous Materials Regulations. Department of Transportation, Materials Transportation Bureau, Federal Register, 1976.

Ifshin, Sidney, Deputy Chief, New York Fire Department: Symposium: Products of Combustion of (Plastics) Building Materials. Lancaster, Pa., Armstrong Cork Company Research and Development Center, 1973.

International Oil Tanker and Terminal Safety Guide, 2nd Ed. Oil Companies International Marine Forum. New York, Halsted Press, John Wiley and Sons, 1974. (17) pp. 177-178.

Manual of Firemanship, Part 6-C. London, England, Her Majesty's Stationery Office, 1964.

Rushbrook, Frank: Fire Aboard. New York, Simmons—Boardman, 1961.

Fire Detection Systems

A fire detector is a device that gives a warning when fire occurs in the area protected by the device. The fire detection system, including one or more detectors, relays the alarm to those endangered by the fire and/or those responsible for firefighting operations. Ashore, a fire detector sounds an alarm so that occupants can leave a burning building promptly, and the fire department can be summoned. The detection system can also activate fire extinguishing equipment. At sea, however, there are no fire escapes, and no professional fire department to call. A shipboard fire detection system alerts the ship's crew, who must cope with the emergency using the resources they have on board.

Early discovery of fire is essential. The fire must be confined, controlled and extinguished in its early stages, before it gets out of control and endangers the ship and the lives of those on board. A well designed fire detection system, properly installed and maintained, and understood by those who must interpret its signals, will give early warning of a fire in the area it protects and its location.

Fire detection systems on board a ship are so arranged that in case of a fire, both a visible and audible alarm is received in the pilothouse or fire control station (normally the bridge) and for vessels of over 150 feet in length there should be an audible alarm in the engine room. The receiving equipment (or consoles) indicates both the occurrence of a fire and its location aboard the ship. Consoles are located on the bridge and in the CO_2 room. The CO_2 room is the space that contains the fire extinguishing mechanisms. Only a bell is required in the engine room to alert the engineer to an emergency outside the machinery space.

Upon hearing a fire alarm, the watch officer on the bridge sounds the general alarm to call the crew to their fire and emergency stations as listed on the station bill. However, in all cases the master must be alerted immediately and the cause of the alarm must be investigated. If the alarm was for an actual fire, action should be taken to confine, control and extinguish it. The crew must respond as per the station bill, under the direction of the master. If it was a false alarm, its cause should be investigated and corrected, if possible. In either event the fire detection system should be checked and the system put back in service after the proper action is taken. Losses have occurred when a system was not reactivated after an alarm, and hence did not send a signal when a subsequent real fire or reflash occurred.

The types of fire protective systems approved for use aboard ship include the following:

1. Automatic fire detection systems
2. Manual fire alarm systems
3. Smoke detection systems
4. Watchmen's supervisory system
5. Combinations of the above.

Coast Guard regulations (title 46 CFR) require that certain types of detecting equipment be used in specified spaces aboard certain ships. The USCG permits other types of systems where the equivalent protection is demonstrated. They may also allow the installation of systems that are not actually required by law or regulation. Approved types of fire protective systems are carried in the Coast Guard *Equipment Lists* (CG190). If any doubt exists as to whether any item of fire protection equipment may be installed or carried aboard ship, inquiries can be made at a Coast Guard marine inspection or marine safety office. These offices should be consulted for clarification or permission to install unlisted equipment.

AUTOMATIC FIRE DETECTION SYSTEMS

Automatic fire detection systems consist of normal and emergency power supplies, a fire detection control unit, fire detectors and vibrating bells.

Normal Power Supply

The normal power may be supplied either by a separate branch circuit from the ship's main switchboard or by storage batteries. When the power is supplied by storage batteries, they must be used *only* for the fire alarm and fire detection systems. The storage batteries must be in pairs, with one of each pair in use, and the other being charged. Otherwise, single batteries, connected to a charging panel, may be used.

Emergency Power Supply

Emergency power may be supplied by a separate branch circuit taken from the temporary emergency lighting and power system switchboard or by storage batteries. If duplicate storage batteries supply the normal power, the battery being charged may serve as the emergency power source.

Fire Detection Control Unit

The fire detection control unit consists of a drip-proof enclosed panel containing the fire alarm signaling, trouble-alarm and power-failure alarm devices. These devices must register both a visual and an audible signal. The visible signals are lights:

- A *red* light indicates fire or smoke.
- A *blue* light indicates trouble in the system.
- A *white* light indicates that the power is on in the system.

The control unit also contains a power supply transfer switch to engage the emergency power supply if the normal power supply fails. Over-current protection devices are incorporated into the system to prevent damage in the event of an electrical malfunction. If battery charging equipment is employed, it may be located in the control unit.

Fire Detectors

Fire detectors sense (and initiate a signal in response to) heat, smoke, flame or some other indication of fire. Not all types of detectors are used aboard ship—some are not practicable, and some are not necessary. The types of detectors that are in common use aboard vessels are discussed in the next few sections.

Vibrating Bells

Vibrating bells are, like the red lights on the control unit, fire alarm signals. The operation of any automatic fire detection system (or manual fire alarm in a manual fire alarm system) must automatically cause the sounding of

1. A vibrating-type fire bell with a gong diameter not smaller than 15.24 cm (6 in.) on the control panel
2. A vibrating-type fire bell with a gong diameter not smaller than 20.32 cm (8 in.) located in the engine room.

These signals must be sounded in addition to the red light on the control panel and an indication of the fire detection zone from which the signal originated.

Light and Bell Signals

When fire is detected, the alarm lights stay on and the bells keep ringing until a resetting device is operated manually. A shutoff device may be used to silence the bells. However shutting off the bells will not extinguish the alarm lights. The alarm lights can be shut off only with the manual resetting device. Like the modern fire alarm boxes on land, shipboard fire alarms are noninterfering; any number of alarms can be received simultaneously. An alarm that is being received on one circuit will not prevent an alarm from being received on another circuit.

Power Failure

A power failure in the system is announced by the ringing of a bell, which is reserved for this purpose in the control panel. The emergency power source provides the power to actuate the bell. The power failure bell can be shut off by switching its signal to a visible lighted indicator.

An open circuit in the wiring from the control unit to the detectors or in the wiring from the normal source of power is indicated in two ways: A blue signal light comes on in the control unit and the trouble bell rings. In some cases, an open circuit may result in a fire alarm. Such a false alarm can be received when there is a break in a circuit of a system that uses closed circuit series-connected detectors.

HEAT-ACTUATED FIRE DETECTORS

As their name implies, heat-actuated fire detectors sense (and are activated by) the heat of a fire. The main classes of heat-actuated devices are fixed-temperature detectors and rate-of-rise detectors. Some devices are combinations of both.

Fixed-Temperature Detectors

A fixed-temperature detector initiates a fire alarm when the temperature of the device reaches a preset value. Note that the device operates only when the detector itself, not the surrounding air, reaches the preset temperature. The difference between these two temperatures, that of the surrounding air and that necessary to actuate the detector, is called the *thermal lag*. It results because heat must be transferred from the surrounding air to the detector, to bring the detector up to its operating temperature. This heat transfer takes time; it is never so perfect that the air and the detector are at the same temperature. Thus, when a fixed-temperature detector is actuated, the surrounding air is always hotter than the detector. The thermal lag, or delay, is proportional to the speed at which the temperature is rising in the area.

Temperature Classifications. Title 46 CFR 161.002-11(c) classifies fixed-temperature detectors according to use as follows:

1. *Ordinary degree,* for use where the normal temperature at the device does not exceed 38°C (100°F).
2. *Intermediate degree,* for use where the normal temperature at the devices exceeds 38°C (100°F) but not 66°C (150°F).
3. *Hard degree.* For use where the normal temperature at the device exceeds 66°C (150°F) but not 107°C (225°F).

Note that the temperatures listed are not the temperatures that will actuate the detectors, but rather the expected normal temperatures of the area in which the detectors are placed.

These fixed-temperature detectors should be actuated within the temperature limits given in Table 6.1.

By comparing their normal and actuating temperatures, it can be seen that fixed-temperature detectors are designed to operate when there is a substantial increase of temperature over the normal temperature in the protected area. This is exactly what takes place when a fire occurs.

Types of Fixed-Temperature Detectors. Fixed-temperature detectors differ in their design and how they function. More specifically, they differ in how their sensing elements detect and react to heat. The common types are bi-metallic, electric (resistance- or cable-type), fusible metal and liquid expansion. Some authorities may include the bi-metallic with the electric type, while others may consider fusible metal and liquid expansion detectors as a single type.

Bi-metallic Strip Detector. In a bi-metallic strip heat detector, the sensing element is made up of two strips of different metals, welded together.

Table 6.1. Limits of Rated Temperature of Operation, °C (°F).

Rating	Maximum	Minimum
Ordinary	74°C (165°F)	57°C (135°F)
Intermediate	107°C (225°F)	79°C (175°F)
Hard	149°C (300°F)	121°C (250°F)

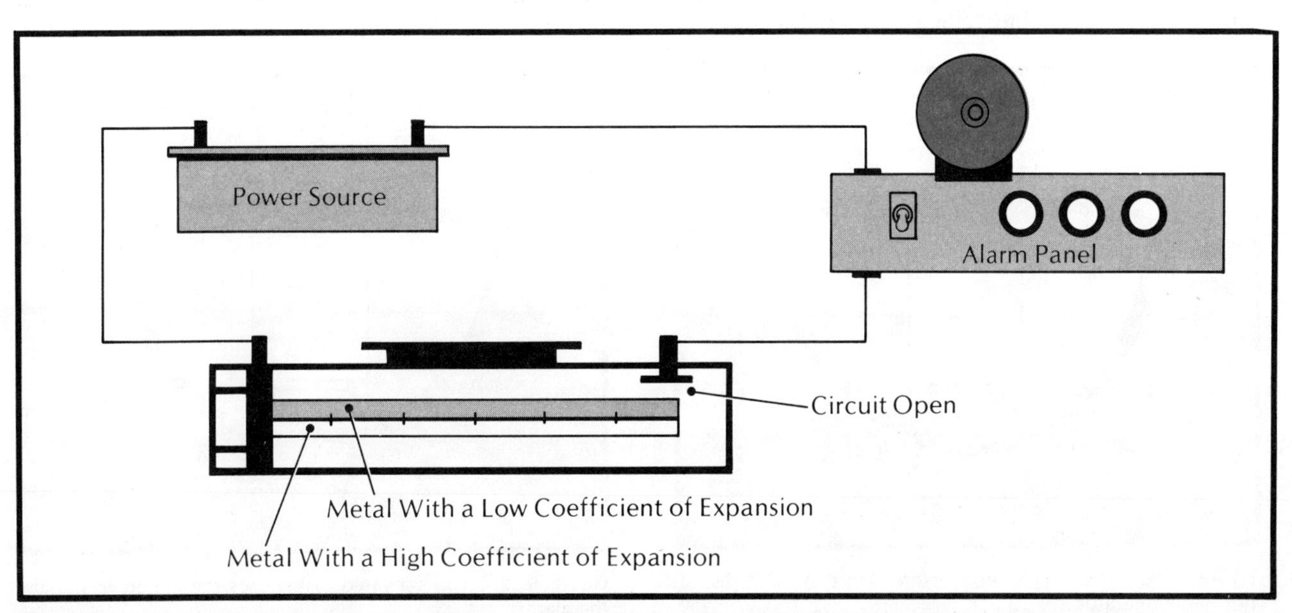

Figure 6.1A. The bi-metallic strip heat detector. At normal temperatures, the strip is straight.

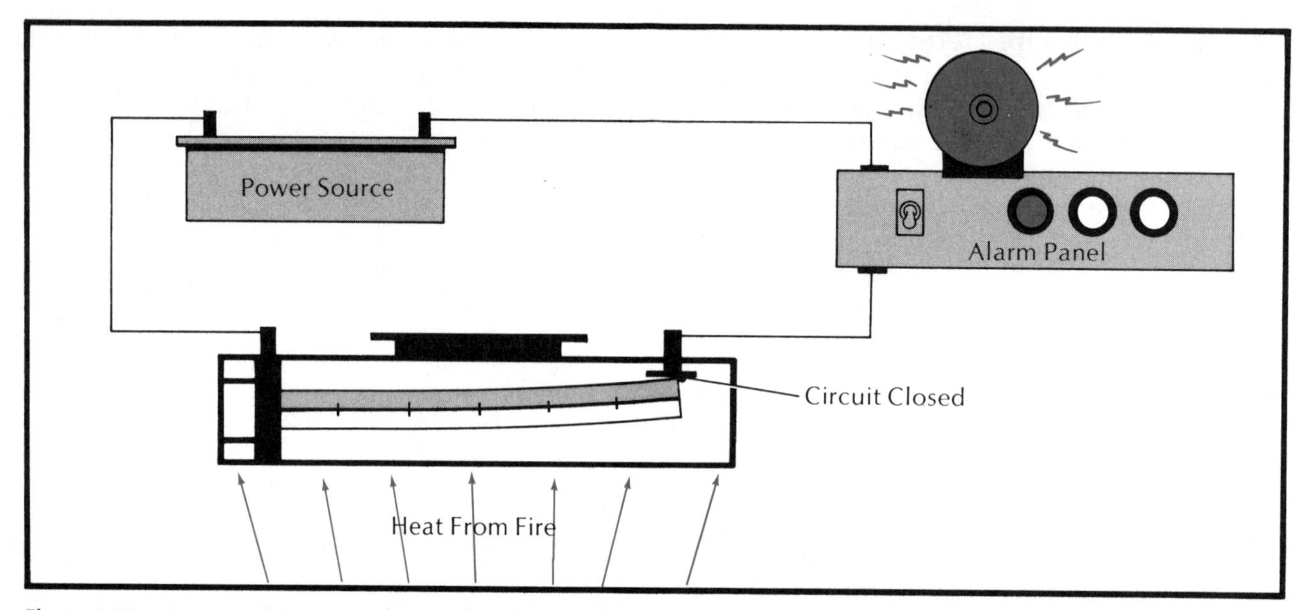

Figure 6.1B. As temperature rises, the strip bends upward, because the lower metal expands more than the upper metal. At temperatures within its activating range, the strip bends enough to touch the contact and complete the alarm circuit.

The two metals have differing coefficients of expansion—one expands faster than the other when heated.

At normal temperatures, the strip is straight (Fig. 6.1A). When the temperature increases, the bi-metallic strip bends because one metal expands faster than the other. This bending causes the metal strip to touch a contact point, closing an electric circuit and transmitting an alarm (Fig. 6.1B.)

An advantage of the bi-metallic strip is that it returns to its original shape after the heat is removed. If it is not destroyed by fire, it can remain in place and be used again. A disadvantage of this type element is that it is prone to false alarms. When the strip is close to the contact, but the temperature is less than the minimum fire alarm rating, ship vibrations or a physical shock to the detector housing could cause the circuit to close and there is no permanent indication of which detector operated.

Snap-Action Bi-metallic Disk. Like the bi-metallic strip, the snap-action disk changes its shape when it is heated sufficiently. However, it does so with a surer, more positive movement. Instead of slowly approaching the electrical contact, the disk snaps against the contact at its activating temperature.

In Figure 6.2A, the disk is at its normal temperature and is concave upward. The circuit is open. In Figure 6.2B, the temperature has increased to the detector's actuating temperature. The disk has snapped so it is convex upward, closing the contacts and completing the alarm

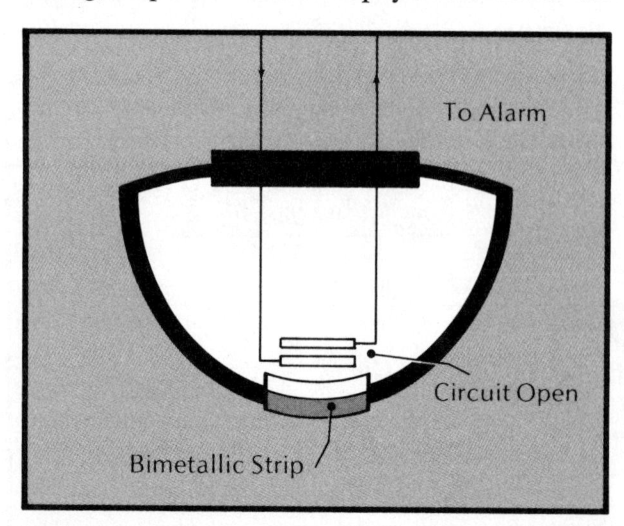

Figure 6.2A. The snap action bi-metallic disk. The disk at normal temperatures.

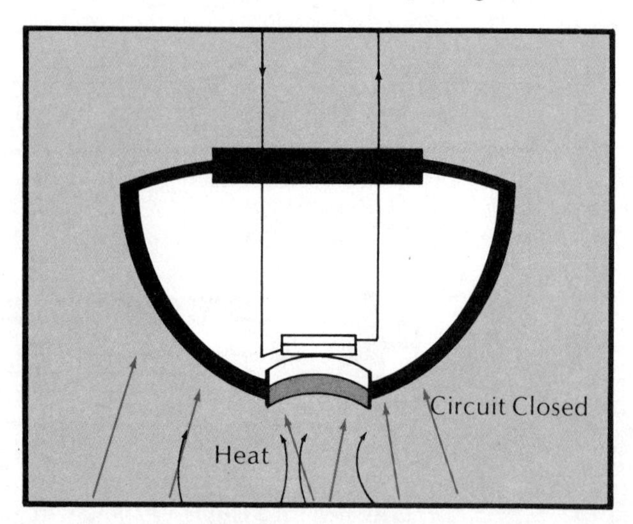

Figure 6.2B. The activated disk closes the circuit to initiate the alarm.

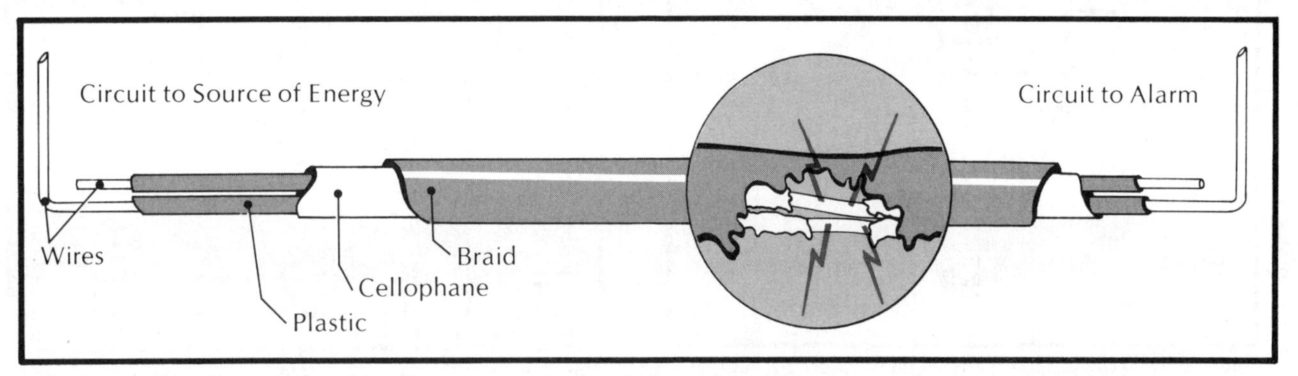

Figure 6.3. The thermostatic cable. The top wire is electrically energized; the bottom wire is not. At the activating temperature, the insulating material melts, allowing the two wires to touch and complete the alarm circuit.

circuit. The snap-action disk returns to its original shape when the temperature is reduced.

Bi-metallic strips and disks are *spot* detectors, in that each device senses the temperature at a single location. A number of detectors housed in small cases are placed in the protected area and wired to the control unit.

Thermostatic Cable. The thermostatic cable is an electrical heat-actuated detector. It consists of two wires, enclosed in a protective cover, with an insulating material between them. One wire is electrically energized; the other is not. At a preset temperature the insulating material melts; the energized wire then contacts the second wire, completing an electrical circuit (Fig. 6.3).

The thermostatic cable is a *line*-type detector. A single length of cable, enough to protect an entire area, is strung through that area. The full length of the cable acts as the detecting element. This type of detector must be replaced if it is actuated since the actuated portion of the cable is destroyed in the process.

Metallic Cable. Another electrical line-type detector is composed of a metallic cable enclosing a nickel wire. The cable and wire are separated by a heat-sensitive salt. When the temperature increases to the detector's actuating range, the resistance of the salt decreases enough to allow a current to flow from the outer cable to the nickel wire. The current in the nickel wire initiates an alarm at the control unit. If it is not damaged by fire, this detector automatically readjusts itself when the temperature decreases to the normal range.

Fusible Metal. A fusible metal is one that melts at some preset temperature. In a fire detector, a fusible metal part is used to hold back a movable switch contact (Fig. 6.4A). When the fusible part melts, the contact moves to close the circuit and sound the alarm (Fig. 6.4B).

Fusible metals are also used in sprinkler heads. When the metal melts, the water is released and an alarm is activated. Fusible metal devices must be replaced when the fire detection system is put back in service.

Liquid Expansion. Liquid expansion devices are similar in operation to fusible metal devices. They are used to restrain something—the water in a sprinkler head or a movable contact in an electric switch. A frangible (breakable) glass bulb is partly filled with a liquid. An air bubble is left above the liquid (Fig. 6.5A). As the temperature rises, the liquid expands. If the temperature continues to rise, the liquid expands further. At a preset temperature, the bulb bursts, allowing whatever action it had been holding back (Fig.

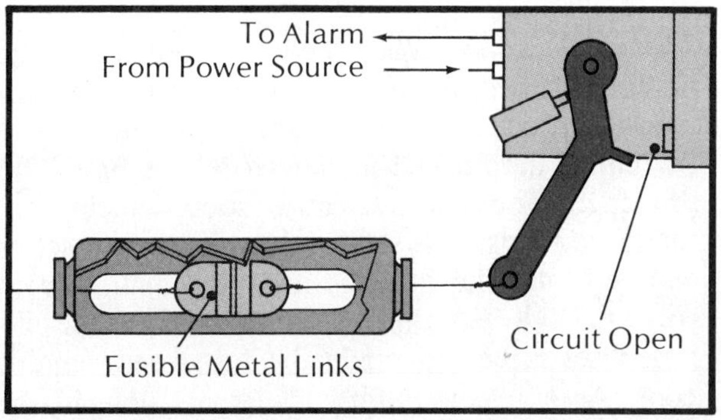

Figure 6.4A. The fusible metal link. At normal temperatures, the link keeps the spring-loaded movable contact from moving to the right.

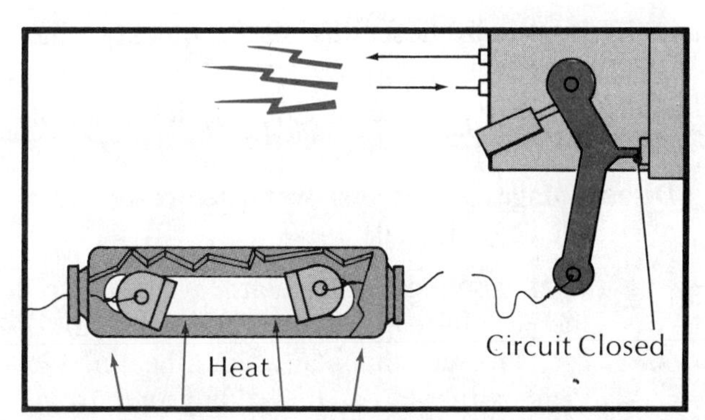

Figure 6.4B. When the link melts, the movable contact is free to complete the alarm circuit.

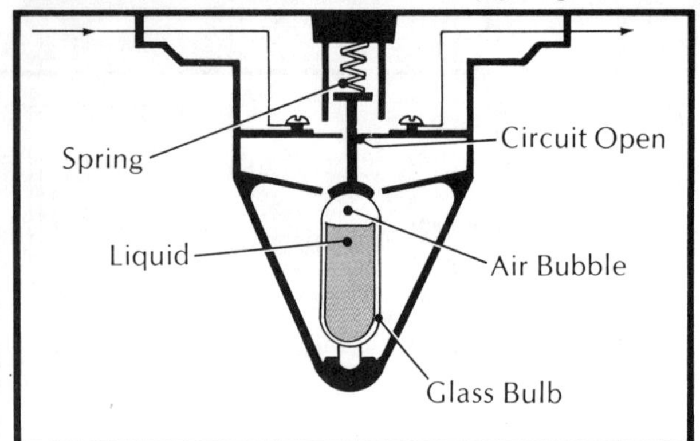

Figure 6.5A. The liquid expansion bulb. At normal temperatures, the bulb holds back the spring-loaded plunger.

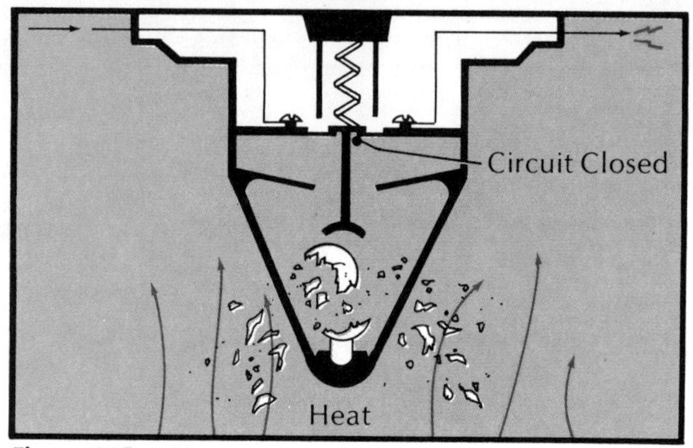

Figure 6.5B. At a preset temperature, liquid in the bulb has expanded enough to break the bulb. The plunger drops to complete the alarm circuit.

6.5B). The bulb must be replaced when the system is put back in service.

Rate-of-Rise Detectors

Rate-of-rise detectors sense temperature changes rather than the temperature itself. They are actuated when the temperature increases faster than a preset value. For example, suppose a detector is set for a rate of increase of 8.3°C (15°F) per minute. If the temperature of the detector were to rise from 38°C to 46°C (100°F to 115°F) in 1 minute, the alarm would be sounded. However, if the temperature rose from 41°C to 46°C (105°F to 115°F) in a minute, this detector would not be actuated. The temperature rate of rise that actuates this type of detector depends on its design, which in turn depends on where it is being used. For instance, the pneumatic detectors approved for passenger and cargo vessels are set to actuate when the temperature rises at approximately 22°C (40°F) per minute at the center of the circuit.

Advantages. Among the advantages of the rate-of-rise detector are the following:

1. Slow rises in temperature will not activate the device.
2. It can be used in low-temperature areas (refrigerated spaces) as well as in high-temperature areas (boiler rooms).
3. It usually responds more quickly than fixed-temperature devices.
4. Unless destroyed by fire, it quickly adjusts for reuse.

Disadvantages. The disadvantages of the rate-of-rise detector include these:

1. It may sound a false alarm when a rapid increase in temperature is not the result of fire. This may happen when a heating element is turned on, or welding or burning operations in the immediate area cause a rapid rise in temperature.
2. It may not be activated by a smoldering fire that increases the air temperature slowly, such as in baled cotton or other tightly packed cargo.

Types of Rate-of-Rise Detectors. Two types of rate-of-rise detectors, pneumatic and thermoelectric, are in common use.

Pneumatic. The pneumatic-type detector operates on the principle that an increase in temperature causes an increase in the pressure of a confined gas. There are two forms of pneumatic detectors, line and spot. In the line type, a small diameter copper tube is strung high in the compartment to be protected. An increase in the temperature of the tube raises the pressure of air within the tube. A small vent allows some of this air to escape (Fig. 6.6), reducing the pressure in the tube. But if the temperature of the device rises at or faster than a preset rate, the pressure builds up faster than the vent can reduce it. This stretches a diaphragm that closes a pair of contacts to trigger the alarm.

The spot detector is usually employed in small spaces or rooms. Increased air pressure in the spot detector may be conveyed by a tube to a remote control point. Otherwise, the pressure actuates a switch close by or in the detector, which in turn sends an electric signal to the control point.

Thermoelectric. When heat is applied to the junction of two dissimilar metals, the rise in temperature produces a small but measurable electric current. Thermoelectric detectors are based on this fact. The thermoelectric spot detector actually contains two sets of junctions; one set is exposed, and the other is insulated against heat (Fig. 6.7). When the temperature rises, the exposed set is heated while the other set remains cool. As a result, different currents flow in the two sets of junctions. The difference between the currents is monitored. If it increases at a preset rate or above it, an alarm is actuated.

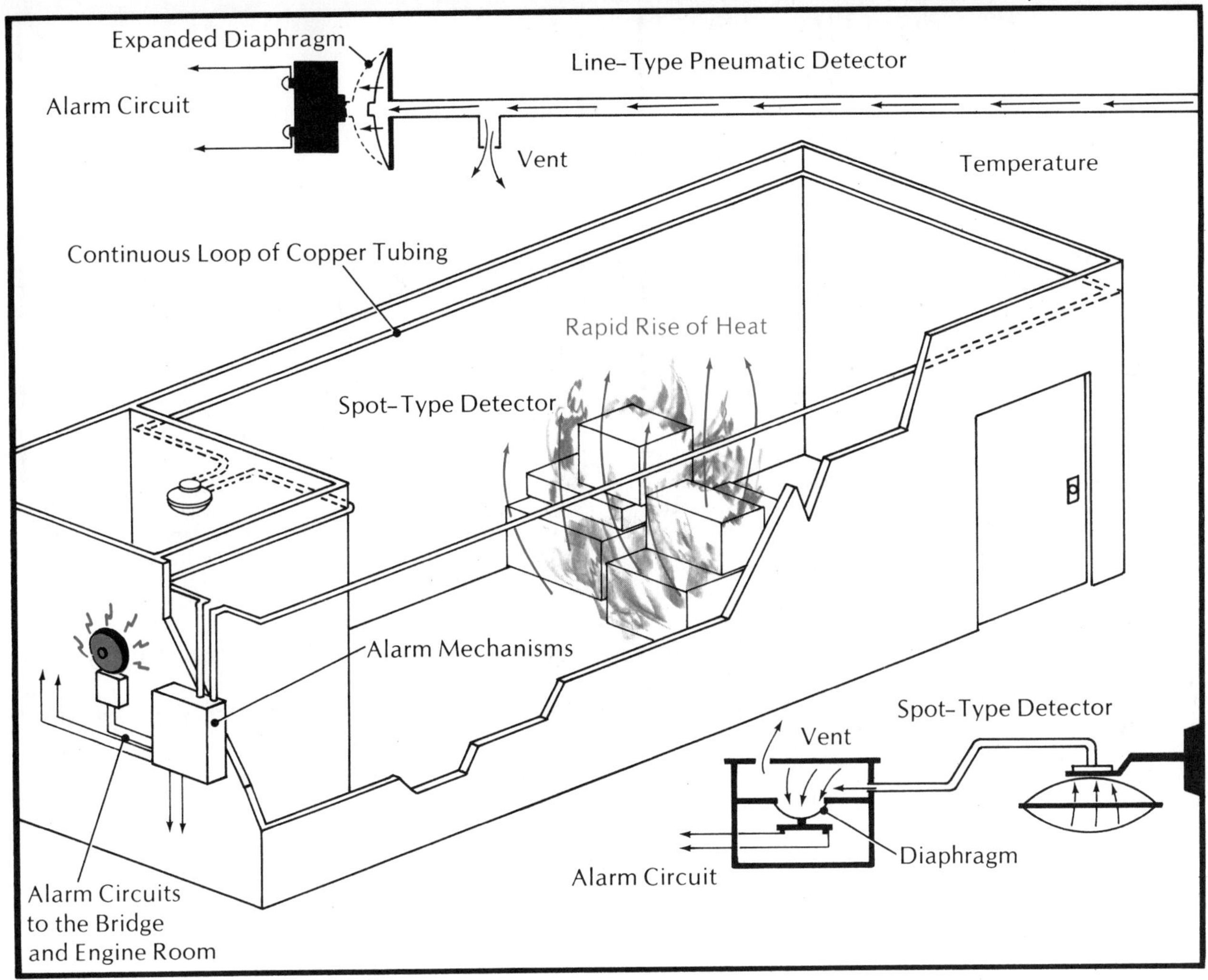

Figure 6.6. Pneumatic rate-of-rise detectors. Heat expands the air inside the tube or bulb, increasing its pressure. If the expansion is slow, the vent releases enough of the pressure to keep the detector from being actuated. If the expansion is fast, pressure builds up enough to stretch the diaphragm and complete the alarm circuit.

Line-type thermoelectric detectors are also available. Two pairs of wires are enclosed in a sheath (to protect them from physical damage). One of each pair has a high coefficient of heat resistance, and the other has a low coefficient of heat resistance. Two wires with the same coefficient of heat resistance (one from each pair) are insulated against heat. The other two wires are open to temperature changes in the protected space. The wires are connected to a device that measures the resistance of the wires. An increase in temperature in the protected space shows up as an unbalance in the resistance of the wires. A high enough rate of unbalancing causes the alarm to be activated.

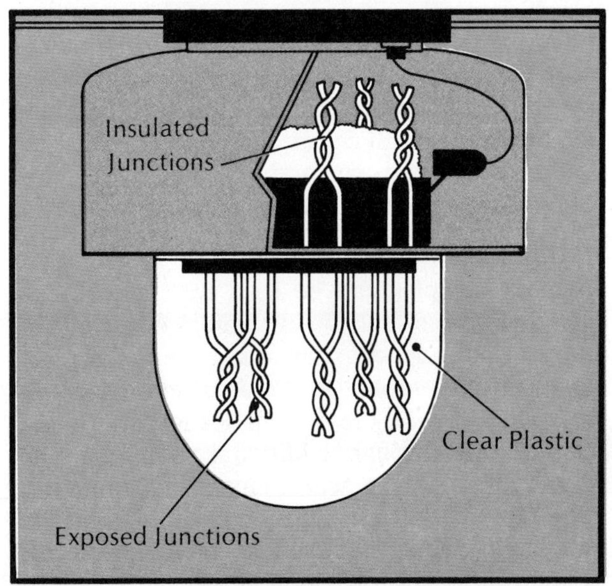

Figure 6.7. Thermoelectric spot rate-of-rise detector, with its two sets of junctions of dissimilar metal wires.

The Combined Fixed-Temperature and Rate-of-Rise Detector

The combined-type detector contains both a fixed-temperature device and a rate-of-rise device. It is activated when the temperature rises at, or faster than a preset rate. However, if the temperature rises slowly but continuously, the rate-

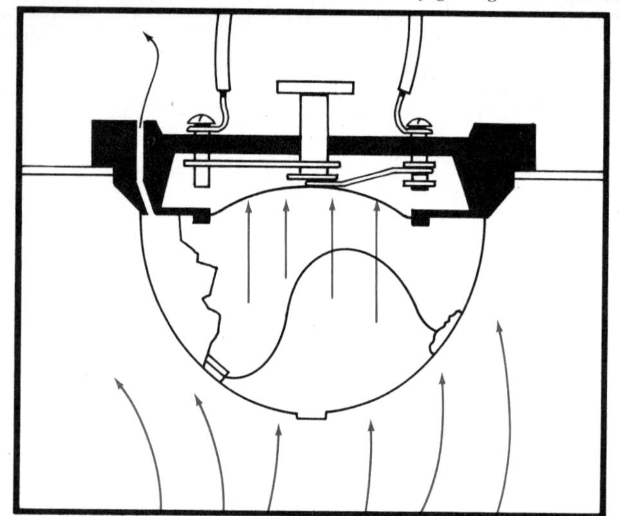

Figure 6.8A. Combined heat-actuated detector. The rate-of-rise device is a diaphragm that is stretched upward to close the contacts, when heat increases the pressure in the shell.

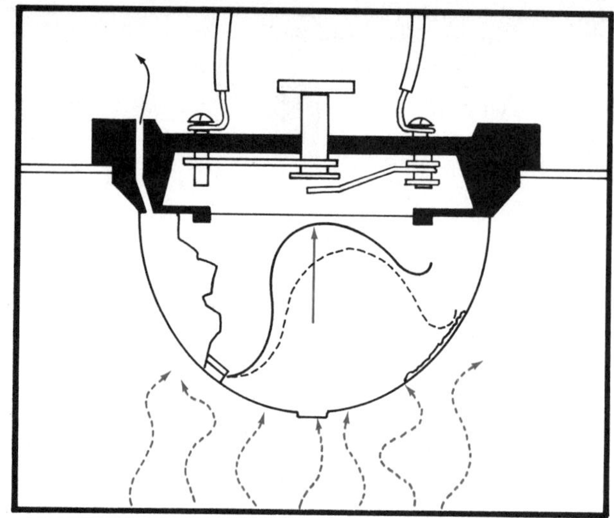

Figure 6.8B. The fixed-temperature device is a spring that is released when a fusible metal melts. The spring pushes up on the diaphragm and the contacts to transmit the alarm.

of-rise device may not be activated. Then the fixed-temperature device will eventually initiate an alarm.

A combination spot-type detector is shown in Figure 6.8. Heat absorbed on the shell raises the temperature of the enclosed air. The air expands, as in a pneumatic detector. If the temperature increases at a high enough rate, the diaphragm stretches up. It pushes the contacts together to close the alarm circuit (Fig. 6.8A). The fixed-temperature feature comes into play as follows: One end of a piece of spring metal is permanently affixed to the shell. The other end is held against the opposite side of the shell by a fusible-metal seal. When the activating temperature is reached, the fusible metal melts and releases its end of the strip (Fig. 6.8B). The strip presses the diaphragm up, which again closes the contacts to complete the alarm circuit.

The main advantage of the combined detector is the added protection: The fixed-temperature device responds to a slowly building fire that may not activate the rate-of-rise device. In addition, one combined detector could protect a space that might otherwise require both the fixed-temperature and rate-of-rise types.

The rate-of-rise device in the detector resets itself, but the fixed-temperature, fusible metal part does not. Thus, the only disadvantage is that the entire device must be replaced if the fixed-temperature part is activated. Some combination detectors utilize a bi-metallic strip as the fixed-temperature device, so that replacement is not necessary. However, these detectors are subject to false alarms, as noted above.

Automatic Sprinkler Systems

Automatic sprinkler systems are considered to be both fire detection and fire extinguishing systems because they fulfill both functions. The system piping is usually charged with water to the sprinkler heads. The water is held back by a fixed-temperature seal in each head. The seal is either a piece of fusible metal or a liquid-expansion bulb. Either one will allow water to flow through the sprinkler head when the temperature reaches a preset value.

Aboard ship, automatic sprinkler systems are arranged so that the release of water from a sprinkler head automatically activates visible and audible alarms in the pilot house or fire control station. On vessels over 45.7 m (150 ft) in length, there must also be an audible alarm in the engine room.

SMOKE DETECTION SYSTEMS

A smoke detection system is a complete fire detection system. Aboard ship, smoke detection systems consist generally of a means for continuously exhausting air samples from the protected spaces; a means of testing the air for contamination by smoke of all colors and particle sizes, and a visual (or visual and audible) means for indicating the presence of smoke.

Smoke Sampler

A smoke sampler can be used with any smoke detection device that draws samples of air out of the protected space. This sampled air usually moves through tubing to the detection device. A tee connection in the system leads part of the sampled air, through additional tubing, to the wheelhouse. At the wheelhouse, this tubing is normally uncovered. Thus, any smoke in the sampled air would be noted by the watch officer, as well as by the detection device. The wheelhouse tubing has a cap. It can be placed over the tubing to keep heavy smoke from a fire out of the wheelhouse.

Combinations of detectors and detection systems are frequently used. The most common are photoelectric smoke detectors combined with a smoke sampler.

Types of Smoke Detectors

The smoke detector is the device that tests the air samples for smoke. The available types include photoelectric, ionization, smoke sampler, resistance bridge, and cloud chamber detectors. Of these, some lend themselves to shipboard use, while others are more suitable to large buildings on land.

Photoelectric. Photoelectric smoke detectors are employed on ships and in land installations. In the *beam*-type photoelectric smoke detector, a light beam is usually reflected across the protected space. In some cases, air from the protected space is blown into a sampling chamber, and the light beam is reflected across the chamber. The beam of light shines on a photoelectric receiving surface. The receiving surface does not activate the alarm as long as it senses the light beam. However, when smoke particles are present in the air, they obscure the path of the light beam. This reduces the amount of light falling on the receiving surface, which then activates the alarm (Fig. 6.9).

The *refraction*-type smoke detector contains a light source and a photoelectric receiving element that is *not* in the path of the light beam (Fig. 6.10). If the air is clear, no light falls on the receiving surface; this is the normal condition. However, if particles of smoke pass in front of the light beam, they refract (deflect) light onto the receiving element. When the receiving element senses the light, it actuates the alarm.

Ionization. Ionization smoke detectors are now approved for shipboard use. In operation, sampled air passes through the detector. As it does, a small amount of radioactive material at the inlet of the detector ionizes (adds or removes electrons from) the air. This causes a small electric current. Smoke in the air interferes with the flow of ionized particles and the current is decreased; an alarm is triggered by this decrease in current. The minute amount of radioactive material used in the detector is not considered a health hazard.

Resistance Bridge. Resistance-bridge smoke detectors are activated by an increase in smoke particles or in moisture. (Water vapor is given off during the early stages of a fire.) These detectors are more applicable to land installations than to ships.

Cloud Chamber. The use of a cloud chamber (sometimes called the Wilson cloud chamber) as a smoke detector is relatively new. This detector tests sampled air. If smoke particles are present, moisture causes them to form a cloud that is denser than normal air. A photoelectric device scans the sampled air. It sets off an alarm when the air is denser than some preset value.

Federal Specifications

Title 46 CFR 161.002–15 specifies that the type of smoke detection system be one of the following:

1. Visual detection, where the presence of smoke is detected visually and by sense of smell

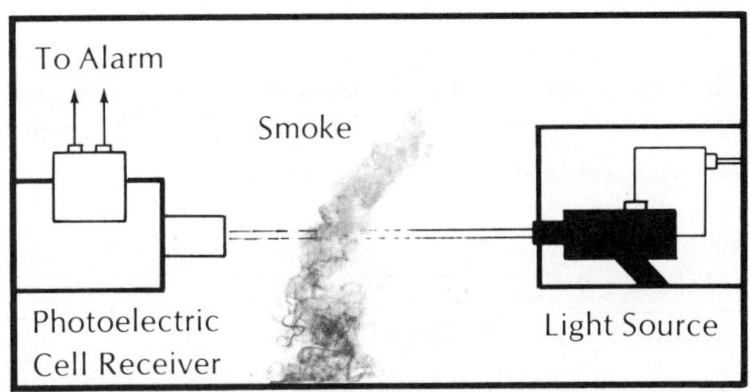

Figure 6.9. Beam-type photoelectric smoke detector. The receiving surface activates the alarm only when it senses a decrease in the intensity of the light beam.

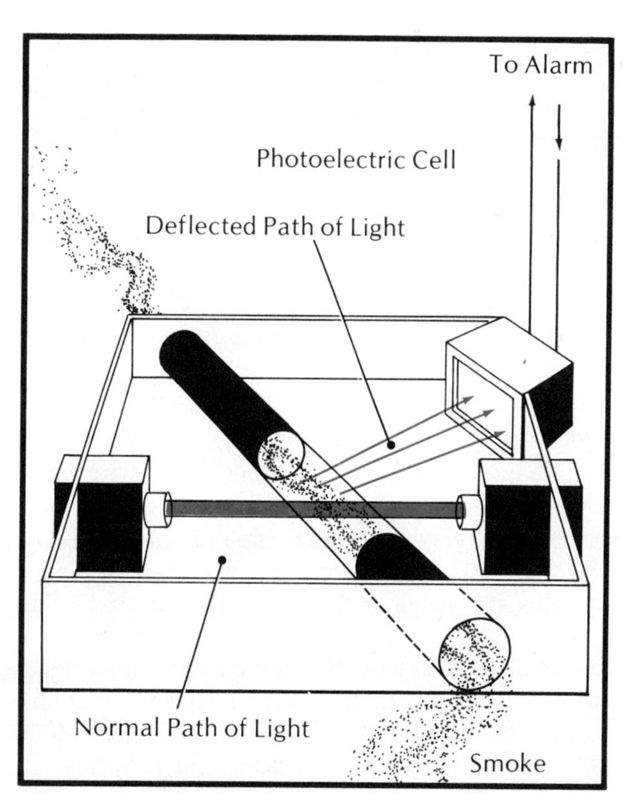

Figure 6.10. Refraction-type photoelectric smoke detector. Smoke in the air causes light to fall on the photoelectric sensor, which then activates the alarm.

2. Audible detection, where the presence of smoke is detected visually and audibly and by sense of smell
3. Visual or aural type combined with a carbon dioxide extinguishing system
4. Other types that may be developed and approved

FLAME DETECTORS

Flame detectors are designed to recognize certain characteristics of flames—the light intensity, the flicker (pulsation) frequency or the radiant energy. While flame detectors are used in shore installations such as warehouses, piers and aircraft hangers, they are unlikely to be found aboard ships for a number of reasons. First, a flame must be directly in front of the detector to be recognized. If the flame is off to the side or obscured by smoke, the detector will not activate. Second, some flame detectors transmit a false alarm when subjected to radiant energy from a source other than a fire. Some activate when they sense flickering light reflections (for example, light reflected off the water surface) or arcs from welding operations. Third, some flame detectors respond to the flickering of flames. Electric lamp bulbs aboard vibrating ships could imitate this flickering closely enough to cause a false alarm.

MANUAL FIRE ALARM SYSTEMS

Manual fire alarm systems consist of normal and emergency power supplies, a fire control unit to receive the alarm and the necessary fire alarm boxes. The fire control unit is similar to the automatic fire detection control unit; it must contain means for receiving alarm signals and translating these signals into audible and visible alarms. It must also have provision for registering trouble signals. And, as with automatic systems, vibrating bells are required for engine room notification.

Combined Manual and Automatic Systems

Where both manual and automatic alarm systems are installed on a ship, the U.S. Coast Guard may approve a single console capable of receiving the signals of both systems. In fact, manual alarm systems are usually combined with automatic detection systems. If the automatic system fails, a crewman who discovers a fire can promptly send an alarm via the manual alarm system. In addition, the manual system is important even when the automatic system is functioning properly. If a manual alarm is received on the bridge shortly after an automatic alarm, the watch officer can be fairly certain that there is an actual fire and not a false alarm.

Automatic extinguishing systems (*see* Chapter 9) have provision for manual operation. With one exception, automatic extinguishing systems transmit an automatic alarm when they are operated manually. The one exception is the automatic sprinkler system. An automatic sprinkler system cannot operate until the heat of a fire activates a sprinkler head. The opening of the sprinkler head releases the water and, at the same time, activates the alarm.

Manual fire alarm stations may be superimposed on and connected as an integral part of the wiring of an automatic fire detection system. An electrical system using manually operated fire alarm boxes may also be employed. Here again, the U.S. Coast Guard could approve other arrangements that may be developed.

Alarm Boxes

There must be at least one manual fire alarm box in each fire zone on the vessel. Framed charts or diagrams in the wheelhouse and fire control station, adjacent to the fire alarm receiving equipment, should indicate the locations of the fire zones in which the alarm boxes are installed.

Manual fire alarm boxes are usually located in main passageways, stairway enclosures, public spaces and similar areas. They should be readily available and easily seen in case of need. Manual alarm boxes must be placed so that any person evacuating a fire area will pass one on the way out.

All new alarm boxes must be clearly marked: IN CASE OF FIRE BREAK GLASS. Older alarm boxes not so marked must be identified with the same instruction printed on an adjacent bulkhead in 1.27-cm (½-in.) letters. Every alarm box must be numbered to agree with the number of the fire zone in which it is located. The box must be painted red, with the operating instructions printed in a contrasting color.

Newer boxes are equipped with an operating lever. When the lever is pulled, the glass is broken and the alarm box mechanism transmits the alarm. Older boxes may not have a lever; instead, they may have a small hammer, attached with a chain, to be used in breaking the glass. Once the glass is broken, the lever must be operated to sound the alarm.

SUPERVISED PATROLS AND WATCHMEN'S SYSTEMS

The purpose of a supervised patrol is basically the same as that of a system—to guard against fire and sound an alarm if fire is discovered. The difference between the two systems is the way in which the vigilance is maintained. The supervised patrol system is similar to a system of police officers walking their beats. Each ship's patrolman follows a prescribed route, designed to ensure that he visits each station on his round. The watchman, on the other hand, is more of a fixed sentry. He is stationed in a specific area and remains in that area.

Supervised Patrols

Supervised patrol systems are required on passenger vessels whenever passengers are on board. Cargo vessels are not required to have supervised patrol systems. However, if they are installed, they must meet the requirements set down for passenger vessels. On these vessels, supervised patrols must be maintained between the hours of 10 PM and 6 AM. Patrolmen must cover all parts of the vessel accessible to passengers and crew, except occupied sleeping accommodations and machinery and similar spaces where a regular watch is maintained.

To verify that patrolmen make their appointed rounds, recording apparatus is installed in the zones that must be visited. Generally, the apparatus can be either of two types:

1. A mechanical system consisting of a portable spring-motor-driven recording clock and key stations located along each patrol.
2. An electrical system consisting of a recorder located at a central station and key stations along each patrol route.

In the mechanical system, there is a key at each station along the patrolman's route. On reaching a key station, the patrolman inserts the key into his portable clock. This causes the time and the station to be recorded on a tape within the sealed clock. The entries on these tapes should be examined regularly, to ensure that the prescribed visits were made at the proper times. The portable clocks have antitampering devices that automatically register any unauthorized opening (Fig. 6.11).

In the electrical system, each patrolman carries a key. When he reaches a key station, he inserts his key into the station mechanism. The placing of the key in the mechanism registers on a recorder at the central station. When a signal is not received from a key station within a reasonable time, there may be a problem. The patrolman may not be making his rounds, or he may have become ill or had an accident. An immediate investigation should be made.

Figure 6.11. A patrolman's portable clock for recording the stations visited and the times of the visits.

When a ship is not equipped with an electrical recording apparatus, patrolman must report to the bridge at least once an hour. However, where there are two or more patrol routes, one patrolman may contact the others and make a joint report.

Watchmen's Systems

Watchmen are used on vessels that are not required to have supervised patrols. At night, a suitable number of watchmen must be stationed in the passenger accommodation areas on each deck. The watchmen are under the direct control of the master or watch officer and must report to that officer at fixed intervals not exceeding one hour.

Duties of Patrolmen and Watchmen

Patrolmen and watchmen should be given specific instructions concerning their duties. They must be made aware that their primary duty is to transmit an alarm on discovering fire, or even on seeing or smelling smoke. Their first action should be to use the nearest manual fire alarm box. Valuable time may be lost if a patrolman or watchman suspects that an alarm is unnecessary and instead goes to the bridge to report his findings.

After transmitting the alarm, the patrolman or watchman should take such action as is necessary: awaken passengers and crew in the area, use a fire extinguisher or simply report to the officer in charge of the emergency squad.

The patrolman, watchman or other crew member who first discovers a fire is very important to the investigation of the cause of the fire. He should be encouraged to write down what he knows about the fire as soon as possible while the facts are fresh in his mind. Important items include:

- Time of discovery
- Exact location where smoke or fire was seen
- What doors were open or closed
- Who, if anyone, was in the area prior to the discovery
- The condition of any fire extinguisher he used
- Any other condition or circumstance that might have a bearing on the fire.

It might be well to take statements from all witnesses on a recording device.

The watchman's uniform is conspicuously different from the clothing of other crewmen so that it can readily be identified. A rating badge marked "Watchman" should be worn on the left sleeve; the front of the hat should be marked similarly. Patrolmen may either wear a distinctive uniform or be identified by a distinctive badge. Both patrolmen and watchmen must carry flashlights; notebooks could also be of use.

EXAMPLES OF DETECTION SYSTEMS USED ABOARD SHIP

Air Sampling Smoke Detection System

An automatic air-sampling apparatus designed to sense the presence of smoke in protected cargo

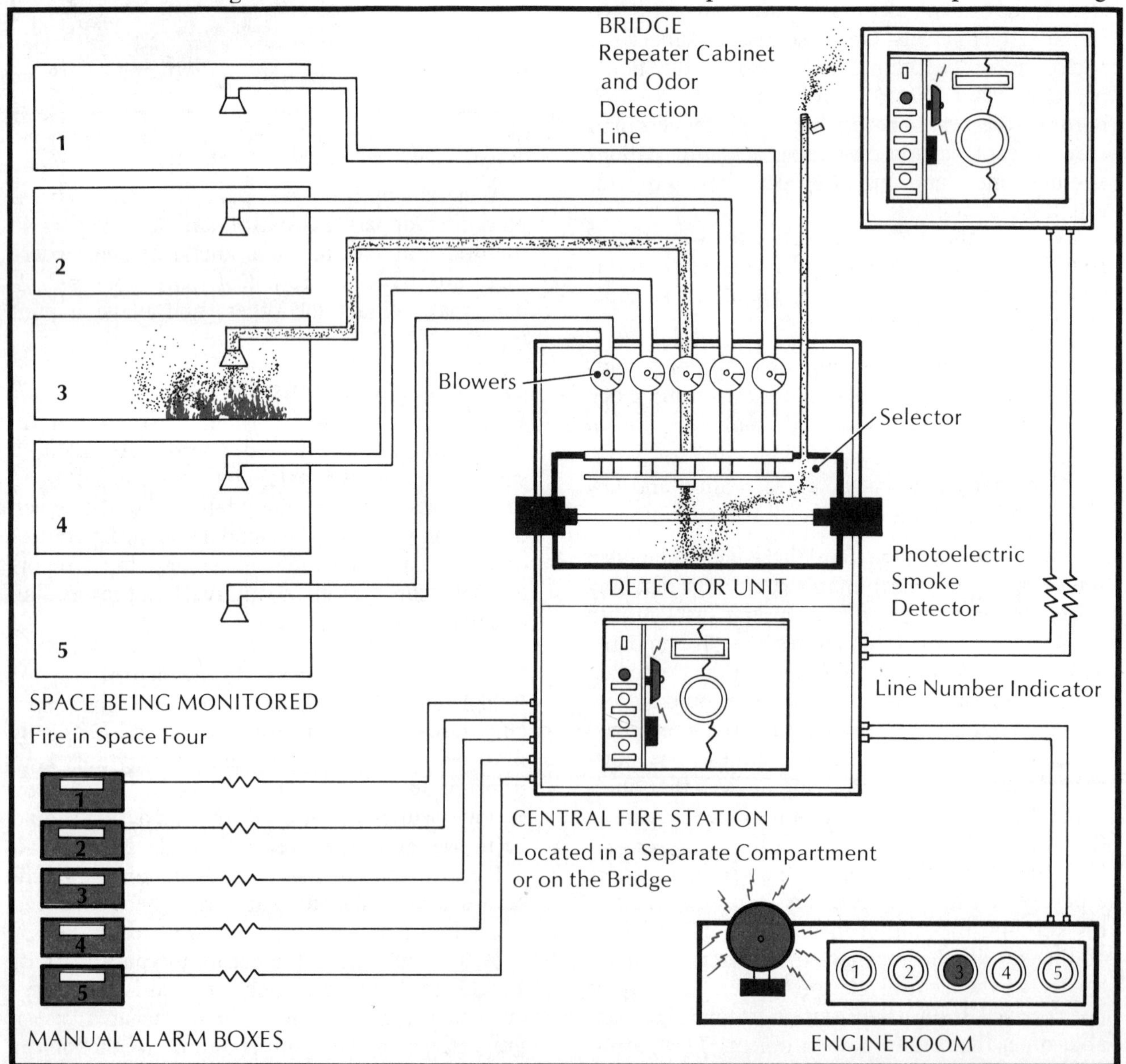

Figure 6.12. Simplified schematic drawing of the automatic smoke-sampling system. The apparatus has detected smoke in an air sample from space 3. That number is indicated on the main cabinet and the repeater cabinet. The alarm is sounded at those cabinets and in the engine room.

areas is installed on most vessels. It is equipped with a photoelectric smoke detector and three visual smoke detectors. Air samples are continually drawn from each protected space and conveyed to the main cabinet through an individual 1.9-cm (¾-in.) pipe. Since each protected space has its own sampling pipe, positive identification of the location generating smoke is ensured.

A photoelectric smoke detector examines the air sample from each pipe individually and in sequence, for a period of 5 seconds, by means of an automatic selector valve. When smoke is sensed by the photoelectric detector, both visible (red light) and audible (gong) alarms are activated, and a vibrating bell is sounded in the engine room. The selector valve stops and locks in on the code number identifying the space from which the smoke is coming (Fig. 6.12).

Air samples from each pipe may be examined individually, at any time, by observation of the visible smoke detectors in the main cabinet. The three visual detectors allow simultaneous observation of air drawn from three different spaces. Smoke that might be too weak to be noticed by a single detector becomes conspicious when contrasted with smoke-free air from adjacent detectors. A translucent cylinder mounted on the selector valve rotor carries a series of numbers. When the visual detector switch is depressed, these numbers are illuminated by a lamp and made visible by three mirrors mounted inside the number cylinder. These numbers visually align below the smoke detector viewing field to designate the space that is delivering air to the corresponding detector.

The main cabinet, in which the receiving apparatus is housed, is usually remote from the wheelhouse. In most cases, it is located in the CO_2 room. A repeater cabinet (Fig. 6.13), wired from the main cabinet, is located in the wheelhouse. The mechanisms in this cabinet repeat the signals received in the main cabinet, activate the alarms signals and lock the selector on the same zone number registering on the main cabinet. Close to both the main cabinet and the repeater cabinet are framed charts under glass or a transparent laminate covering. They show the space associated with each code number. If a fire alarm signal is sounded in the wheelhouse, the watch officer notes the code number held stationary on the selector. By checking this code number on the chart, he determines the fire zone or space from which the smoke is coming.

To ensure that a fire does exist and its location has been accurately determined, the detecting system may be reactivated by pushing the reset button. This causes the smoke detector to repeat the complete cycle of air sampling. If the original information was correct, the alarm should again sound, and the same code number should be indicated on the selector.

The ability to visually examine air samples from various spaces is especially useful during a fire. For example, Figure 6.12 shows fire in hold no. 3. The detection system has indicated that there is fire in that zone. By monitoring air samples on the visual detectors, it could be determined whether there was smoke in hold no. 4 or no. 2. Smoke in either of these two holds would indicate that the fire was extending or there was

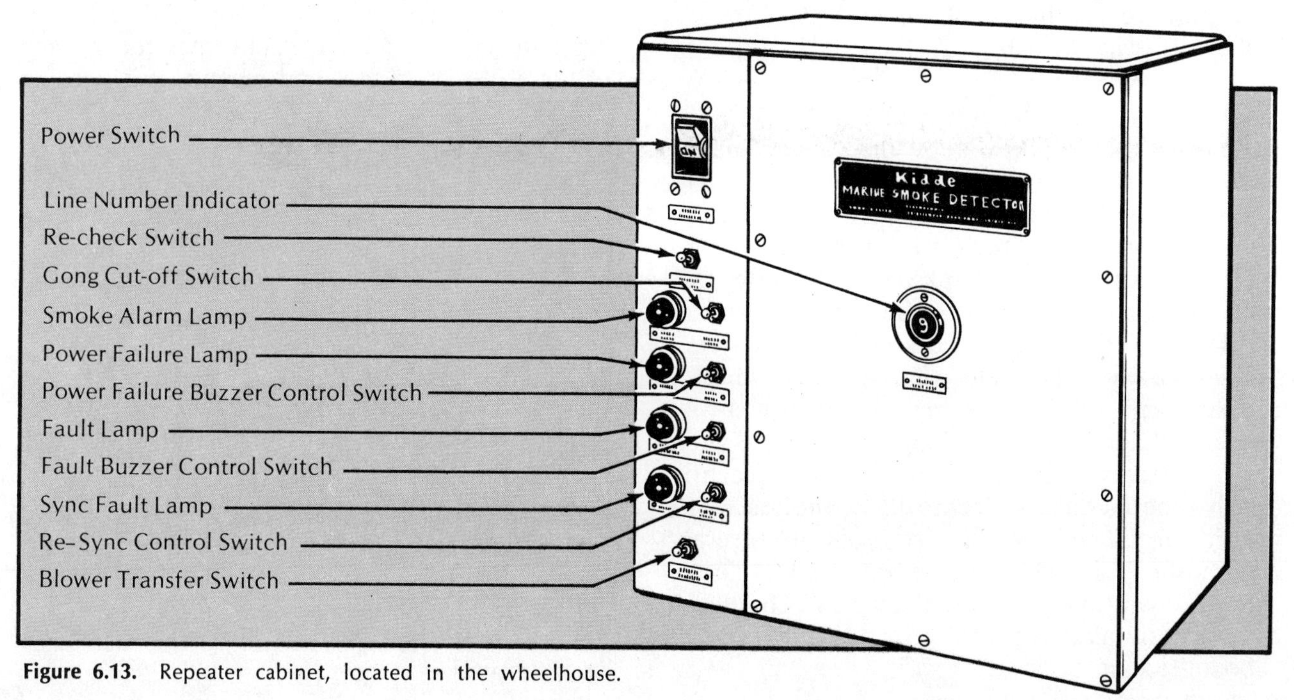

Figure 6.13. Repeater cabinet, located in the wheelhouse.

seepage of smoke into an adjoining space. However, the absence of smoke in either adjoining hold would be a fairly good indication that the fire was contained in hold no. 3.

The equipment described is available in two sizes: one will monitor 32 zones, and the other 48 zones. Each zone has at least one sampling pipe, although large zones may have more. In some installations, a pipe is run to the wheelhouse (or equivalent) to discharge a portion of the mixed air drawn from all protected spaces, permitting smoke detection by smell. In addition to the alarm signals at the main and repeater cabinets, some installations have additional signals, such as gongs, howlers or lights, located in particular areas.

A carbon dioxide extinguishing system (*see* Chapter 9) may be coupled into the smoke detection system. In this arrangement, the pipes that transmit air samples to the main cabinet are also used to carry CO_2 to the protected spaces. Carbon dioxide gas cylinders are stowed in the CO_2 room and are connected to distribution lines leading to the protected spaces. When a fire is detected, its location determined, and CO_2 is to be used to control the fire, a three-way valve located at the distribution manifold is operated. This closes the line to the main cabinet while connecting the burning space with the CO_2 gas supply lines. A chart in the CO_2 room indicates the number of CO_2 cylinders to be discharged initially into the space to create an inert atmosphere. The chart also indicates the number of additional cylinders to be periodically discharged into the space to maintain this inert atmosphere. (The use of CO_2 in fighting cargo hold and engine room fires is detailed in chapter 10).

Supervised Fire Alarm System

Another system approved by the Coast Guard is designed to receive alarms of fire from manual fire alarm boxes and from automatic heat detectors. The system is a two-wire supervised system that operates from two banks of 24-V batteries. One set of batteries supplies power to the system while the other is being charged. The system includes a centrally located (usually in the wheelhouse) control unit consisting of a main control panel and zone modules, along with audible and visible signals and control and test equipment. Each zone module is electrically connected to a particular fire zone. Within each zone are several thermostatic fire detecting sensors and at least one manually operated fire alarm box. The thermostats and alarm boxes are strategically placed about the zone.

The zone modules are mounted in the control unit, beneath the main panel. One control unit can accommodate up to 40 zone modules. The components of each module are mounted on a bracket and consist of two lights, two lever switches, associated relays and a terminal bar. One light indicates red for fire alarms; the other indicates blue for trouble in its circuit. Below these lights are corresponding lever switches, with TEST and RESET positions. Each module is identified with its fire zone number. A chart showing the locations of all fire zones is kept near the control unit, readily visible, for quick reference.

The main control panel (Fig. 6.14) is mounted above the zone modules. The panel consists of five lever switches, four indicator lights, a voltmeter and a rotary switch, with associated relay coils and terminal bars. Two of the lever switches are used to perform ground tests. A third switch is used to test and reset the engine room gong circuit. The remaining two switches are cutout switches. One is used to silence the trouble buzzer and transfer to light indication. The other performs the same funcion for the power failure bell. Two of the four indicator lights are associated with the cutout switches. Another light is for the trouble bell in the engine room gong circuit. The fourth light, designated "fire alarm off," is a warning light that indicates that the audible signal is not operative and the system is in a silent alarm condition. When the cover door of the unit is opened, a switch disconnects the fire alarm bell and energizes the warning light. The unit must never be left unattended in this condition.

The voltmeter has a range of 30–0–30 v. It has two functions: First, it indicates the voltage of the batteries in use, and second, it is used to

Figure 6.14. Main control panel of the supervised fire alarm system. The 40 individual zone modules are located below this panel.

perform ground tests of the positive and negative lines of each zone. The rotary switch is a two-position switch that is used to transfer the batteries from "in service" to "on charge," and vice versa. This also allows the voltage of each battery to be checked.

The fire detectors used in the system are fixed-temperature detectors. The alarm boxes are the standard shipboard manual fire alarm boxes described earlier in this chapter.

TESTING FIRE DETECTION EQUIPMENT

At each annual inspection, all fire detection (and extinguishing) systems, piping controls, valves and alarms must be checked to ensure that they are in operating condition. Smoke detection systems must be checked by introducing smoke into the accumulators. Fire detection and manual alarm systems must be checked by means of test stations or by actuating detectors or pull boxes. Sprinkler systems must be checked by means of test stations or by opening heads.

In addition to the annual or biannual inspections required for the issuance of a certificate of inspection, fire detection systems must be tested at regular intervals. For instance, it is the duty of the master to see that the smoke detection system is checked at least once in each 3 months. Smoke inlets in cargo holds must be examined to determine if the inlets are obstructed by corrosion, paint, dust or other foreign matter. Smoke tests must be made in all holds; the system must be found operable or made operable. The date of the test and the conditions found must be entered in the log. Title 46 CFR 111.05–10 requires that fire detecting thermostats be tested at regular intervals. The intervals are not spelled out, but the regulations specify that 25% of the thermostats (heat sensitive detectors) be tested annually. The regulations further suggest how a thermostat may be tested:

> A portable handlight with an open end sheet metal shield (such as a No. 3 fruit can) replacing the usual guard and globe would serve as a source of heat to operate the thermostat without damage to paint work or the thermostat itself. Any thermostat requiring a time to operate materially different from the average when covered with the heating device should be suspected of being defective and forwarded to Coast Guard Headquarters for further testing.

The operating instructions issued by manufacturers of detection systems usually contain instructions for the periodic testing (monthly and weekly) of some components. Records of all tests must be maintained—if not in the log then in a record book kept in the vicinity of the main cabinet. Title 46 CFR, part 76, Fire Protection Equipment, requires that an officer of the ship make the inspection and initial the entry in the record book.

GAS DETECTION SYSTEMS

Two types of systems are used on ships to warn of dangerous concentrations of combustible gas. These are the catalytic detection system and the infrared gas monitor. While they are not fire detection systems as such, they do detect the presence of situations that could lead to explosion and fire. The catalytic type requires an air-enriched atmosphere; the infrared monitor will operate within any atmosphere. Both systems can be manufactured and installed to monitor combustible gases at a single location or at several locations.

Catalytic Combustible-Gas Detection System

The catalytic system is designed to continuously sample the atmosphere of the protected space and to detect the presence of flammable gases or vapors up to the lower explosive limit (LEL). The components of the system are one or more detector heads, a control-indicating unit and alarms.

How the System Works. The detector head contains two electrically heated elements; each element forms one half of a balanced electrical circuit (a Wheatstone bridge). When a combustible-gas mixture is drawn across the circuit, it burns. This changes the resistance of the circuit and, therefore, its electrical output. The output is transmitted to the control-indicating unit, where it is calibrated and displayed on a meter. A reading on the meter between 0 and 100% shows how closely the atmosphere being monitored approaches the minimum concentration required for a flammable mixture. When dangerous gas concentrations accumulate, warning lights, bells or horns are activated (Fig. 6.15).

To operate properly, catalytic detectors require air that contains enough oxygen to support combustion. For this reason, the catalytic system is not used to detect the presence of combustible gases in inert atmospheres or steam-saturated spaces. The system is usually used in dead air, void spaces, bilges and pump rooms on tank vessels carrying LNG or other combustible gases.

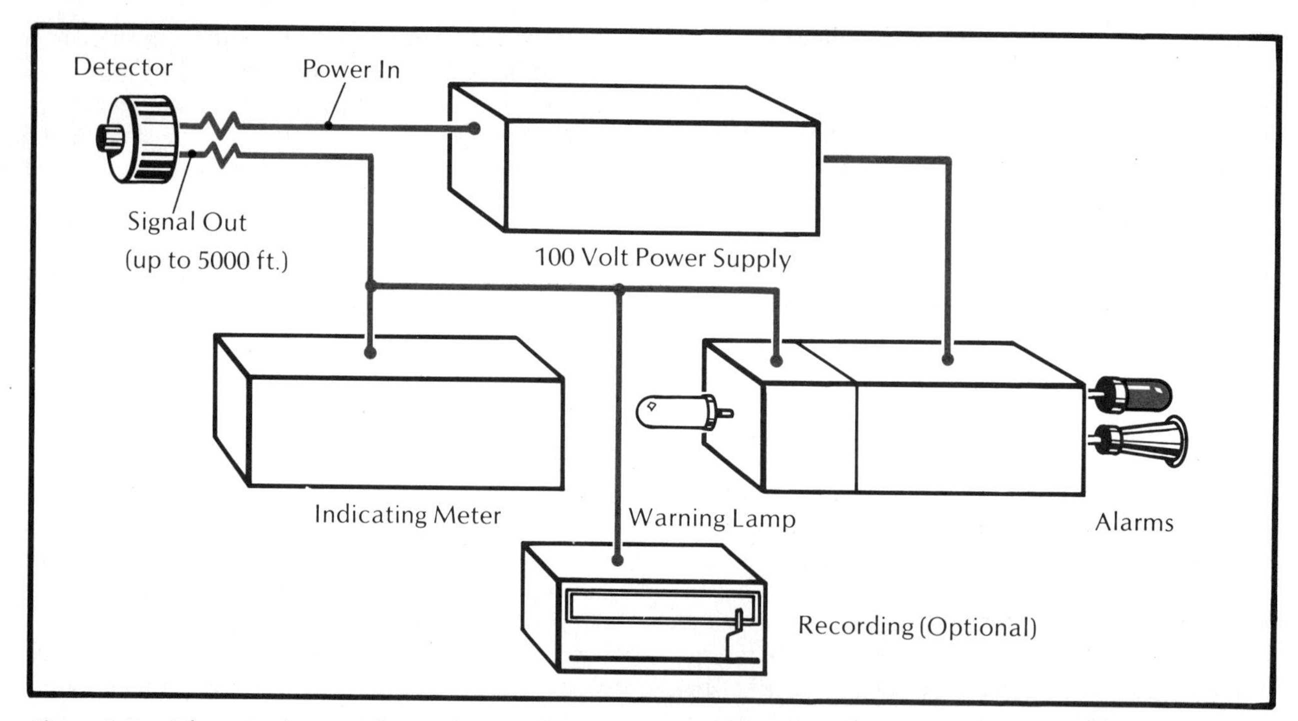

Figure 6.15. Schematic drawing of a catalytic combustible gas detection system.

Infrared Combustible-Gas Leak Detector and Air-Monitoring System

The infrared system automatically detects either toxic noncombustible gases or combustible gases. It may be used to monitor either inert or air-filled atmospheres. For instance, the system can be used to detect methane gas in an inert nitrogen atmosphere, such as that used to protect LNG tanks. As another example, it could be used to monitor the level of carbon monoxide in the air-filled hold of a roll on–roll off vessel. While the system is adaptable to many gases, it can only be set up to detect one particular gas at any given time. Moreover, it will not detect gases, such as nitrogen and oxygen, that do not absorb infrared radiation.

A sample of the atmosphere of each protected space is pumped to a central station through its own sampling line. At the central station, a nondispersive infrared gas analyzer screens the sample for abnormal gas concentrations. If it detects a dangerous level of gas, the system sounds an audible alarm and lights an indicator that shows which protected space is involved.

How the System Works. Figure 6.16 is a block diagram of the infrared monitoring system. There is one sample line, or stream, for each point to be monitored. The stream is a tube through which a sample is drawn from the protected space. The filter on the end of each stream keeps dust from entering the system; it should be replaced periodically.

The stream-selector manifold contains a set of electrically operated valves that are used to connect individual streams to the sample pump. The valves are controlled by the stream selector. The stream selector operates the valves so that the streams are connected to the sample pump in turn, at fixed intervals. After the last stream has been sampled, the cycle begins again with the first stream. The stream selector also identifies the particular stream that is being sampled at any time, through the alarm and indicator unit.

The bypass pump maintains samples in all the streams, up to the stream-selector manifold valves. This reduces the time needed to draw each sample up to the infrared gas analyzer. It thus reduces the length of the complete sampling cycle. The sample pump sends the selected sample to the gas-selector manifold. Normally, the sample passes immediately to the gas analyzer. (The zero gas and span gas shown in Fig. 6.16 are used to calibrate the analyzer periodically. They do not enter into the sampling cycle.)

In the infrared gas analyzer, an infrared beam is passed through the sample. The amount of infrared radiation absorbed by the sample indicates the concentration of the toxic or combustible gas of interest. It is measured and shown on a meter on the analyzer (Fig. 6.17). The information is also transmitted to the alarm and indicator unit.

The alarm and indicator unit visually displays the results of each sample analysis. It also sounds an audible alarm and flashes an indicator light if the concentration of gas in a sample is above

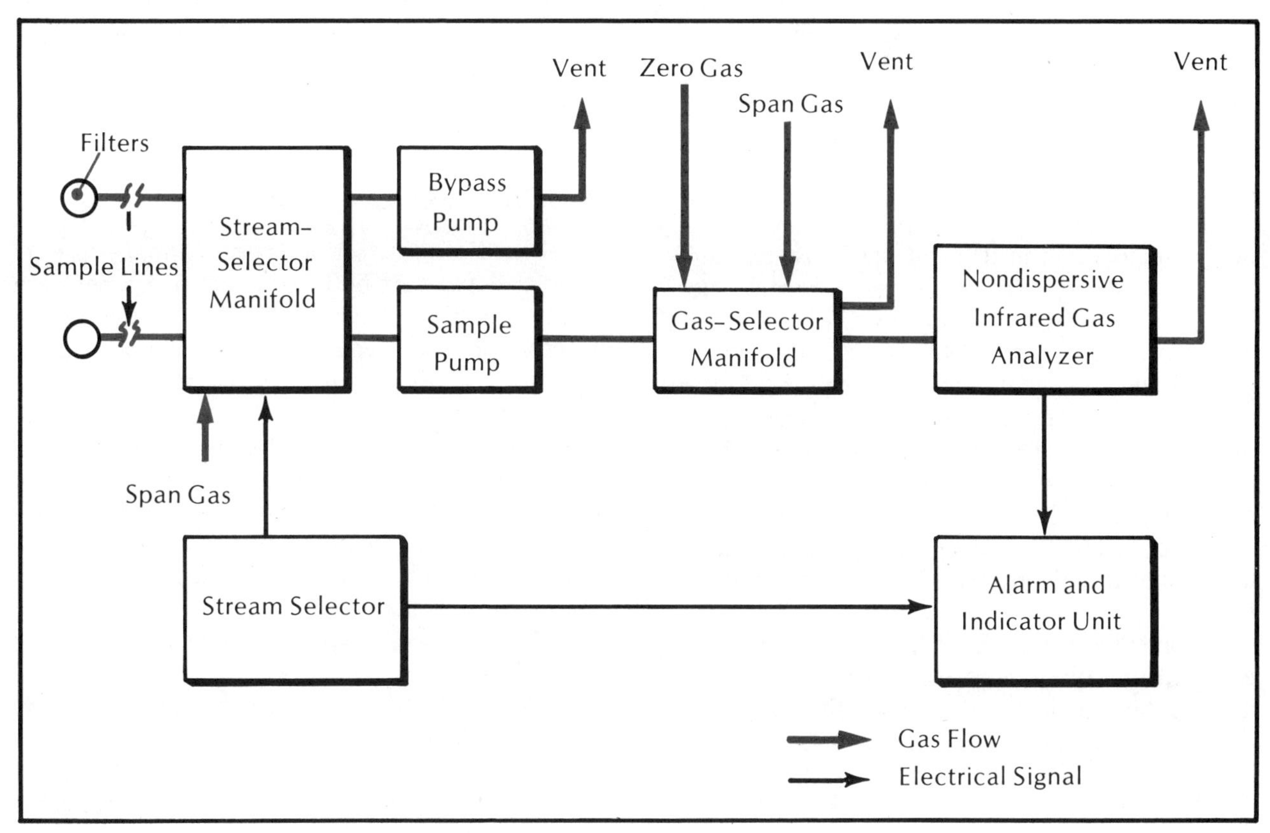

Figure 6.16. The main components of an infrared shipboard gas-monitoring system. (Courtesy Beckman Instruments, Inc., Mountainside, N.J.)

a preset value; the light indicates which stream the sample was drawn from. The watch officer may push a button to silence the audible alarm. Once he notes which indicator is flashing, he may push a second button to stop the flashing (the indicator light simply remains lit).

The watch officer then must take the necessary corrective action. This may include evacuating the area, shutting off valves in pipe lines running through the area, turning on exhaust fans and shutting down electrical equipment. If the problem is corrected, the indicator light automatically shuts off the next time the atmosphere of the involved space is sampled. If the problem remains, the light remains on but the audible alarm does not sound.

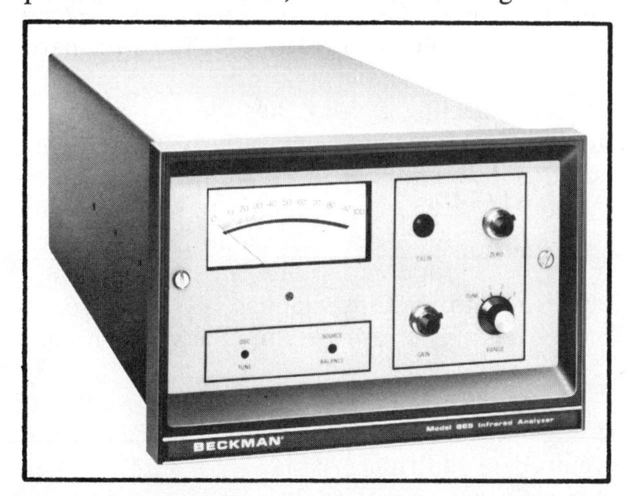

Figure 6.17. Nondispersible infrared gas analyzer. The meter indicates the concentration of gas in each sample as it is being analyzed.

The alarm and indicator light can be set to respond to varying concentrations of the gas of interest. Usually, for safety, they are set for some fraction of the lower explosive level. The system also checks itself once during each cycle, as follows: It analyzes a sample of gas whose concentration is high enough to cause an alarm. However, in this case, the system sounds the alarm only if the high concentration goes undetected. That is, if the system is working correctly and detects the high concentration, no alarm is sounded. The stream selector simply selects the next sample.

Maintenance. As with all safety devices, maintenance is important to the proper operation of the infrared analyzer. Proper maintenance includes

- Periodic calibration as detailed in the manufacturer's operating manual
- Replacement of calibration-gas cylinders when their pressure falls below the minimum
- Periodic checks of equipment operation, including light bulbs
- Lubrication and cleaning of pumps

- Periodic replacement of stream filters
- General cleaning of cabinets and other equipment and the areas occupied by this equipment.

PYROMETERS

A pyrometer is an instrument for measuring temperatures too great for an ordinary thermometer. It is used to find the temperature of a fire. An important use of pyrometers is in checking the progress of a fire that cannot be seen, e.g., a fire that has been confined in a closed compartment or hold. By taking readings at the same location at various times, one can tell if the fire is gaining or lessening in intensity. By moving the pyrometer to different locations along a bulkhead or deck, one may determine if the fire is extending laterally.

Pyrometers are attached to, or embedded in, either of two types of bases. The usual type base may be placed on the deck over the fire space. The magnetic type can be "slapped" onto the outside of a bulkhead of a burning space. A chain should be attached to the base of the pyrometer. It can be used to pull the instrument across a deck that is too hot for personnel. It is also useful in lowering the pyrometer into a hot area.

A pyrometer can be useful in evaluating the success or lack of success when flooding a burning compartment with carbon dioxide. It must be remembered that great patience is needed to successfully extinguish cargo hold fires with carbon dioxide. One cannot "take a peek" to see how things are going. Opening up would significantly dilute the extinguishing gas within the cargo compartment, thereby destroying its effectiveness. Using a pyrometer and checking the variations in temperature should give meaningful information. A rising temperature after carbon dioxide has been introduced would indicate two possibilities: *1)* the amount of carbon dioxide introduced is insufficient and more is required, or *2)* the carbon dioxide is not reaching the fire (directed to the wrong fire zone, a control valve is closed or malfunction of the system). A steady lowering of the temperature would indicate that the carbon dioxide has either extinguished the fire or has it under control. However, though a steady lowering of the temperature is observed or even if the temperature reading is down to 66°C (150°F) or less, these encouraging readings should not be interpreted as a signal to open the compartment for examination. At the risk of being repetitive, it is stated again that great patience is needed with carbon dioxide. There should be no need to open a cargo hatch until port is reached. After all, the damage to the cargo has already been done by the fire; carbon dioxide can cause no further damage.

A COMMENT ON SHIP SAFETY

Shipboard fire detection systems make use of people, devices or both to detect a fire before it does damage to people, the ship or both. The International Convention of Seafaring Nations (London, 1960) recognized the validity of this basic principle and embodied it in their regulations. The participating nations agreed to promote laws and rules to ensure that, from the point of view of safety to life, a ship is fit for the service for which it was intended. But long before such a convention, the regulatory bodies of the United States enacted laws and regulations for safety at sea; for instance, a sprinkler law for passenger ships was passed as early as 1936. Some ship owners had already voluntarily installed fire safety equipment and procedures. Today, ships flying the flag of the United States are the safest in the world. To keep them the safest, the applicable parts of the Code of Federal Regulations are constantly being revised by the Coast Guard. The object is to stay abreast of current needs, based upon experience and the expertise, inventiveness and productiveness of engineers and manufacturers in the fire protection field. Current regulations require that ships be equipped with certain devices or personnel procedures (patrols and watchmen) as a minimum standard for safety. Wise ship owners and masters insist upon greater protection than the minimum standards require. They want the best equipment that can be provided; crews that have been instructed in fire prevention, fire protection and firefighting; frequent and meaningful fire drills; and present equipment tested and maintained to the highest obtainable standard of perfection.

The detecting systems and devices described in this chapter are those currently available. Some, such as the smoke detector and the electric (thermostat) detector, are particularly applicable for shipboard use. Others, such as the flame detector, resistance bridge and cloud chamber, are more applicable to land installations. The infrared gas monitor is a very new system; others have been in use for some time.

The Coast Guard leaves open the possibility of using fire detection equipment that may be developed in the future. However before any new type of fire detection equipment may be installed aboard ship, it must be subjected to thorough and exhaustive examination. If it is found to comply with government specifications, it may then be approved by the Commandant, U.S. Coast Guard, for use aboard vessels.

Regardless of the type or age of a system installed aboard a vessel, all ship's personnel must be familiar with its operation and maintenance. Instruction and maintenance manuals must be available and kept near the signal receiving equipment. If a manual is lost, or an additional copy is desired, the manufacturer will gladly supply one. When requesting a manual from a manufacturer include all information, such as the name, type and model number of the equipment. This information is stamped or printed on the receiving cabinet. The addresses of equipment manufacturers may be found in the U.S. Coast Guard Equipment Lists (CG190), obtainable at any Coast Guard inspection office.

BIBLIOGRAPHY

Bryan JL: Fire Suppression and Detecting Systems. Beverly Hills, Glencoe Press, 1974

Detex Corporation: Detex Newman Watchclock System. New York, 1977

Haessler WM: Systems Approach Vital to Design of Early Alert Detection Installation. In Fire Engineering. New York, January, 1975

Henschel Corporation: Henschel Instruction Book #574. Amesbury, Mass, 1976

Johnson JE: Concepts of Fire Detection. Cedar Knolls, NJ, Pyrotronics, 1970

W. Kidde and Co.: Operation and Maintenance of the Kidde Marine Smoke Detector. Belleville, NJ

Lein H: Automatic Fire Detecting Devices and Their Operating Principles. *In* Fire Engineering. New York, June, 1975

McKinnon CP: Fire Protection Handbook. 14th ed. Boston, NFPA, 1976

National Fire Protection Association: Automatic Fire Detectors. NFPA Standard No. 72E. Boston, 1974

Osbourne AA, Neild AB: Modern Marine Engineer's Manual. Cambridge, Md, Cornell Maritime Press, 1965

Underwriters Laboratories, Inc: Standards for Smoke Detectors. Combustion Products Type for Fire Protective Signaling Systems. UL 167, Melville, NY, 1974

Instruction Manuals, Norris Industries. Marine Smoke Detector, Newark, N.J.

Beckman Instruments Inc. Operation and Maintenance Manuals for Gas Detection Systems. Mountainside, N.J.

Mine Safety Appliances. Operation and Maintenance Manuals for Gas Detection Systems. Pittsburgh, Pa.

Extinguishing Agents

An extinguishing agent is a substance that will put out a fire. Every extinguishing agent operates by attacking one or more sides of the fire tetrahedron (Fig. 4.8). The specific actions involved are the following (Fig. 7.1):

- *Cooling:* to reduce the temperature of the fuel below its ignition temperature. This is a direct attack on the heat side of the fire tetrahedron.
- *Smothering:* to separate the fuel from the oxygen. This can be considered as an attack on the edge of the fire tetrahedron where the fuel and oxygen sides meet.
- *Oxygen dilution:* to reduce the amount of available oxygen below that needed to sustain combustion. This is an attack on the oxygen side of the tetrahedron.
- *Chain breaking:* to disrupt the chemical process that sustains the fire (the chain reaction side of the tetrahedron).

Eight extinguishing agents are in common use. Each is applied to the fire as a liquid, gas or solid, depending on its extinguishing action and physical properties (Fig. 7.2). Some may be used on several types of fires, whereas others are more limited in use. We shall discuss the agents listed in Figure 7.2 (and some others) in the remainder of this chapter, after a brief discussion of the classes of fires that may be encountered aboard ship.

CLASSES (AND COMBINATIONS) OF FIRES

Fires are grouped into four classes labeled A through D, according to their fuels (*see* Chapter 5). However, some fuels are found in combinations, and electrical fires always involve some solid fuel. Thus, for firefighting purposes, there are actually six possible combinations of fire classes:

1. Class A fires (common flammable solid fuel)
2. Class B fires (flammable liquid or gaseous fuel)
3. Combined class A and B fires (solid fuel combined with liquid or gaseous fuel)
4. Combined class A and C fires (solid fuel combined with electrical equipment)
5. Combined class B and C fires (liquid or gaseous fuel combined with electrical equipment)
6. Class D fires (combustible-metal fuel).

This list includes every known type of fire. Note that the environment of a fire, i.e., *where* it occurs, does not affect its classification. For example, class B fires are class B fires whether they occur in an engine room or on a pier. The choice of extinguishing agent depends on the class of fire, the hazards involved and the agents available (Fig. 7.3).

Class A Fires

Fires involving common combustible solids such as wood, paper, cloth and plastics are most effectively extinguished by water, a cooling agent. Foam and dry chemical may also be used; they act mainly as smothering agents.

Class B Fires

For fires involving oils, greases, gases and other substances that give off large amounts of flammable vapors, a smothering agent is most effective. Water fog, dry chemical, foam and carbon dioxide (CO_2) may be used. However, if the fire

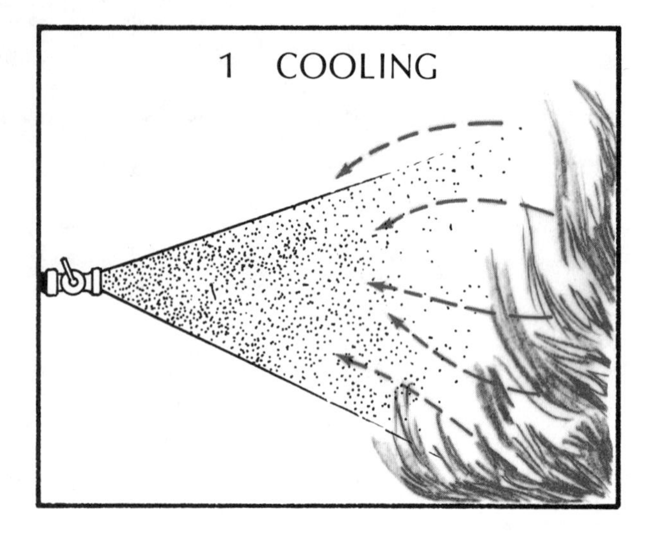

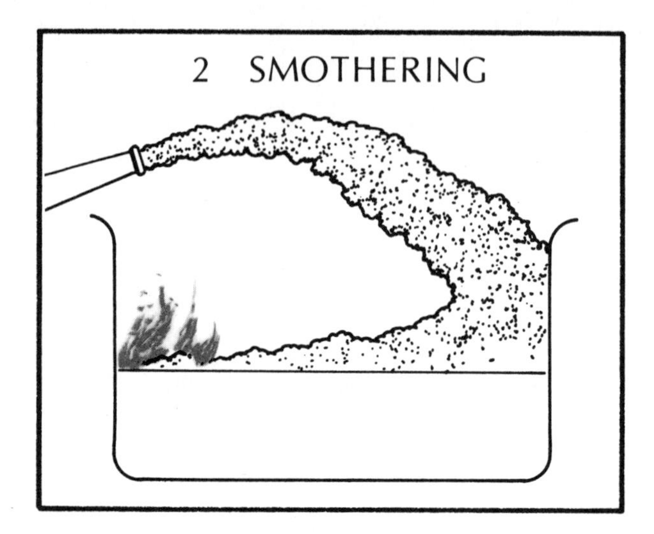

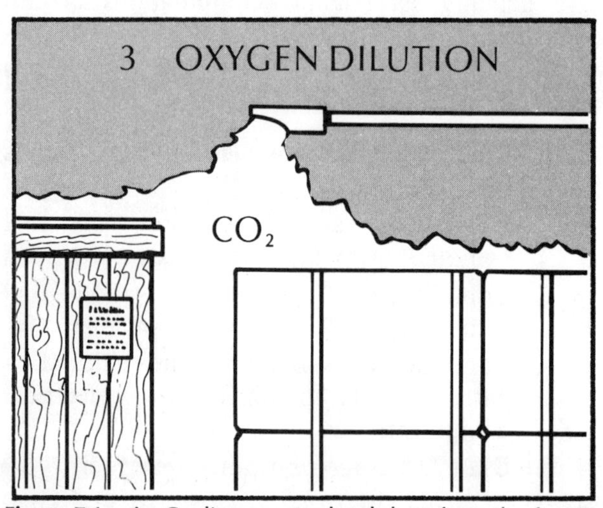

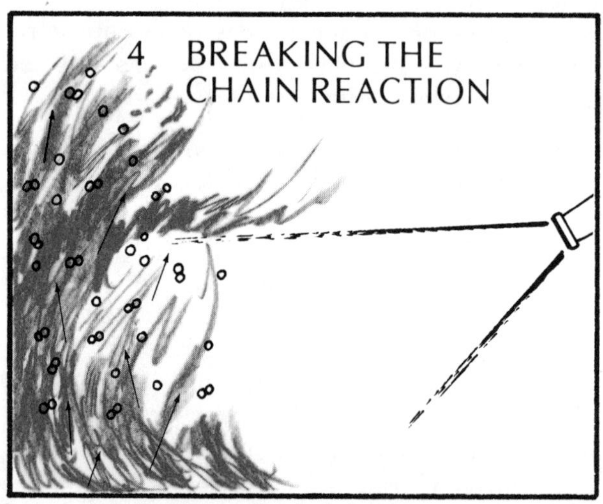

Figure 7.1. A. Cooling agents absorb heat from the fire. **B.** Smothering agents separate the fuel from its oxygen supply. **C.** Oxygen diluting agents reduce the amount of oxygen available. **D.** Chain breakers attack the chemical process that keep the fire going.

is being supplied with fuel by an open valve or a broken pipe, a valve on the supply side should be shut down. This may extinguish the fire or, at least, make extinguishment less difficult and allow the use of much less extinguishing agent.

In a gas fire, it is imperative to shut down the control valve before you extinguish the fire. If the fire were extinguished without shutting down the valve, flammable gas would continue to escape. The potential for an explosion, more dangerous than the fire, would then exist. It might be necessary to extinguish a gas fire before shutting down the fuel supply in order to save a life or to reach the control valve; however, these are the only exceptions.

Combined Class A and B Fires

Water spray and foam may be used to smother fires involving both solid fuels and flammable liquids or gases. These agents also have some

LIQUIDS

1. WATER SPRAY
2. FOAM

GASES

3. CARBON DIOXIDE (CO^2)
4. HALON 1211, 1301

SOLIDS (dry chemical)

5. MONOAMMONIUM PHOSPHATE
6. BICARBONATE
7. POTASSIUM BICARBONATE
8. POTASSIUM CHLORIDE

Figure 7.2. The eight common extinguishing agents.

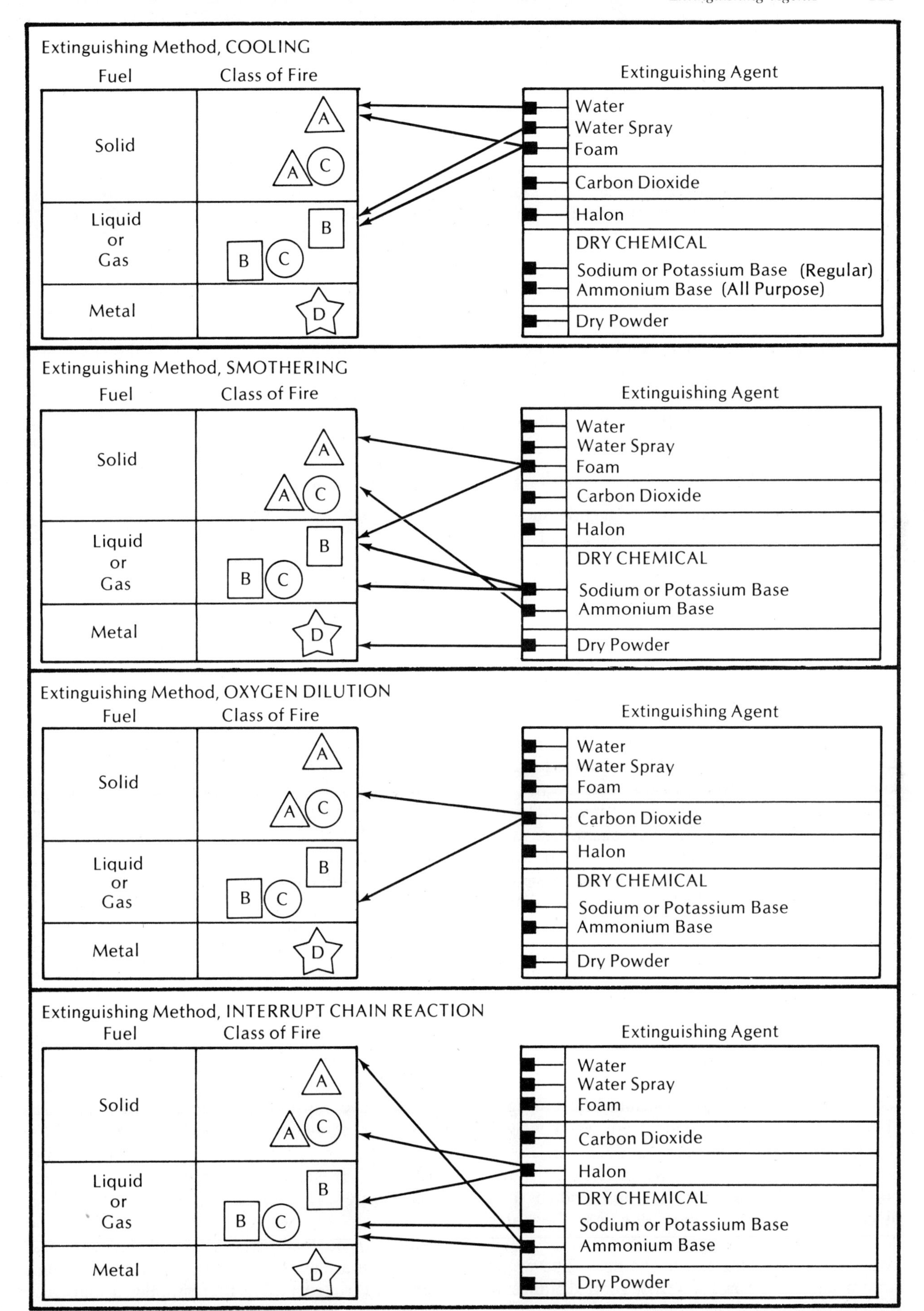

Figure 7.3. The actions of extinguishing agents on the different classes of fires.

cooling effect on the fire. Carbon dioxide has also been used to extinguish such fires in closed spaces.

Combined Class A and C Fires

Because energized electrical equipment is involved in these fires, a nonconducting extinguishing agent must be used. Carbon dioxide, Halon and dry chemical are the most efficient agents. Carbon dioxide dilutes the oxygen supply, while the others are chain breaking agents.

Combined Class B and C Fires

Here again, a nonconducting agent is required. Fires involving flammable liquids or gases and electrical equipment may be extinguished with Halon or dry chemical acting as a chain breaker. They may also, in closed spaces, be extinguished with CO_2.

Class D Fires

These fires involve combustible metals such as potassium, sodium and their alloys and magnesium, zinc, zirconium, titanium and powdered aluminum. They burn on the metal surface at a very high temperature and often with a brilliant flame. Water should not be used on class D fires, as it may add to the intensity or cause the molten metal to spatter. This, in turn, can extend the fire and inflict painful and serious burns on those in the vicinity.

Fires in combustible metals are generally smothered and controlled with specialized agents known as *dry powders*. Dry powders are *not* the same as dry chemicals, although many people use the terms interchangeably. The agents are used on entirely different types of fires: Dry powders are used only to extinguish combustible-metal fires. Dry chemicals may be used on other fires, but not on class D fires.

WATER

Water is a liquid between the temperatures of 0°C and 100°C (32°F and 212°F); at 100°C (212°F) it boils and turns to steam. Water weighs about 1 kg/liter (8.5 lb/gal); fresh water weighs slightly less, and seawater slightly more. Being fluid and relatively heavy, water is easily transported through firemains and hoses when it is placed under pressure. The velocity of the water is increased by forcing it through a restricted nozzle at the working end of the hose. The water stream can be thrown a fairly good distance if sufficient pressure is available.

Extinguishing Capabilities of Water

Water is primarily a cooling agent. It absorbs heat and cools burning materials more effectively than any other of the commonly used extinguishing agents. It is most effective when it absorbs enough heat to raise its temperature to 100°C (212°F). At that temperature water absorbs still more heat, turns to steam, and moves the absorbed heat away from the burning material. This quickly reduces the temperature of the burning material below its ignition temperature, and the fire goes out.

Water has an important secondary effect: When it turns to steam, it converts from the liquid state to the gaseous (vapor) state, and in so doing, it expands about 1700 times in volume. This great cloud of steam surrounds the fire, displacing the air that supplies oxygen for the combustion process. Thus, water provides a smothering action as well as cooling.

Seawater is just as effective in fighting fires as fresh water. In fact, hard water, soft water, seawater and distilled water are all equally effective against class A fires.

Moving Water to the Fire

At sea the supply of water is limitless; however, moving the water is another matter. The amount of water that can be moved to a shipboard fire depends on the number of pumps carried and their capacities. If the total pump capacity is 946 liters/min (250 gal/min), then that is the maximum water flow rate that can be delivered through the ship's firefighting water system. This is one reason for using firefighting water judiciously. But even when water is available in huge quantities, it still must be used economically and wisely. If it isn't, its weight can affect the equilibrium of the ship. This is especially true if large amounts of water are introduced into, and remain at, a high point in the ship: The weight of the water raises the center of gravity of the vessel, impairing its stability (Fig. 7.4). In many cases the vessel will list or even capsize. Water that is not confined but can run to lower portions of the ship may affect the buoyancy of the ship. Ships have capsized and sunk because excessive amounts of water were used during firefighting efforts. Every 1 m^3 (35 ft^3) or about 946 liters (250 gal) of water adds another tonne to be reckoned with.

Aboard ship, water is moved to the fire in two ways: *1)* via the firemain system, through hoselines that are manipulated by the ship's personnel, and *2)* through piping that supplies a manual or automatic sprinkler or spray system. Both are

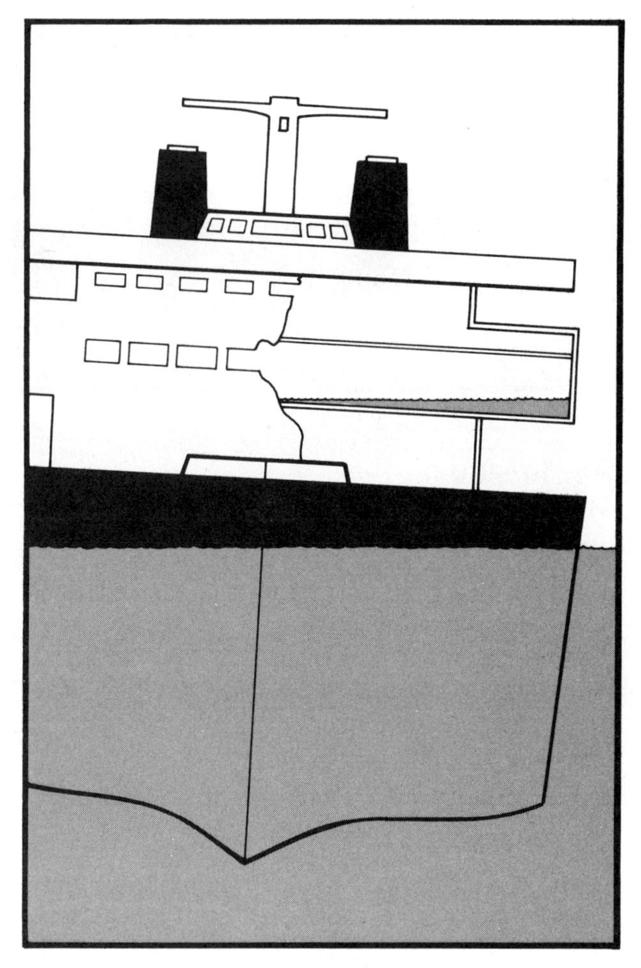

Figure 7.4. Water confined high on the ship has a detrimental effect on the ship's stability.

reliable methods for bringing water to bear on a fire, provided the pumps, piping and all components of the system are maintained. These systems are covered in Chapters 9 and 10.

Automatic fire suppression systems are important to the safety of every vessel. Crewmen should understand how they operate and know how to maintain them. However, the mobility of a hoseline is an equally important asset in most firefighting operations. The hose and nozzle complete the job of moving water to the fire in the proper form. Moreover, hoseline operations represent a much greater involvement of crew members in combating the fire.

This human involvement—and the possibility of human error—make drills of paramount importance. Through periodic drilling, crewmen should become proficient in the use and maintenance of water-moving equipment. U.S. Coast Guard regulations require that each fire station be equipped with a single length of hose, with the nozzle attached. The hose must be situated at its proper location and maintained in good working order. More than one small fire has become a major fire owing to poor maintenance practices.

Straight Streams

The *straight stream,* sometimes called the *solid stream,* is the oldest and most commonly used form of water for firefighting. The straight stream is formed by a nozzle that is specially designed for that purpose. The end from which the water is thrown is tapered to less than one half the diameter of the entry or hose end (Fig. 7.5). The tapering increases both the velocity of the water at the discharge end and the reach.

Efficiency of Straight Streams. The distance that a straight stream travels before breaking up or dropping is called its *reach.* Reach is important when it is difficult to approach close to a fire. Actually, despite its name, a straight stream is not really straight. Like any projectile, it has two forces acting upon it. The velocity imparted by the nozzle gives it reach, either horizontally or at an upward angle, depending on how the nozzleman aims the nozzle. The other force, gravity, tends to pull the stream down, so the reach ends where the stream encounters the deck. On a vessel, the nozzle pressure is usually below 690 kilopascals (100 psi). The maximum horizontal reach is then attained with the nozzle held at an upward angle of 35°–40° from the deck. The maximum vertical reach is attained at an angle of 75°.

Probably less than 10% of the water from a straight stream actually absorbs heat from the fire. This is because only a small portion of the water surface actually comes in contact with the fire—and only water that contacts the fire absorbs heat. The rest runs off, sometimes over the side; but more often the runoff becomes free surface water and a problem for the ship.

Using Straight Streams. A straight stream should be directed into the seat of the fire. This is important: For maximum cooling the water

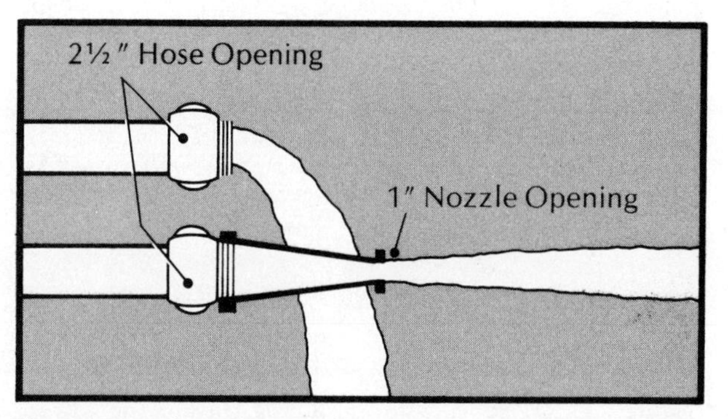

Figure 7.5. The taper greatly increases the velocity of the water coming from the nozzle.

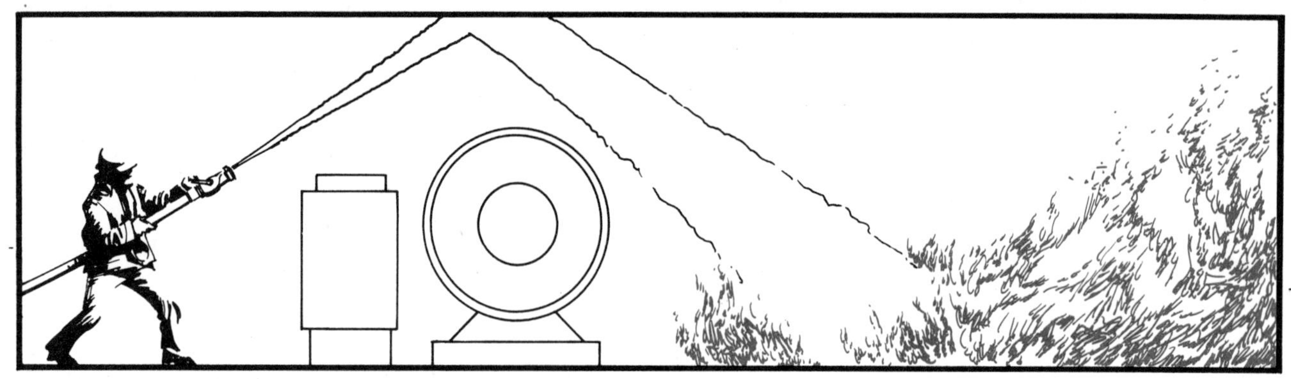

Figure 7.6. A straight stream can be bounced off the overhead to hit a fire located behind an obstruction.

must contact the material that is actually burning. A solid stream that is aimed at the flames is ineffective. In fact, the main use of solid streams is to break up the burning material and penetrate to the seat of a class A fire.

It is often difficult to hit the seat of a fire, even with the reach of a solid stream. Aboard ship, bulkheads with small openings can keep firefighters from getting into proper position to aim the stream into the fire. If the stream is used before the nozzle is properly positioned, the water may hit a bulkhead and cascade onto the deck without reaching the fire. The nozzleman must not open the nozzle until he is sure it is positioned so that the stream will reach into the fire.

In some instances, there may be an obstruction between the fire and the nozzleman. Then the stream can be bounced off a bulkhead or the overhead to get around the obstacle (Fig. 7.6). This method can also be used to break a solid stream into a spray-type stream, which will absorb more heat. It is useful in cooling an extremely hot passageway that is keeping firefighters from advancing toward the fire. (A combination fog–solid nozzle could be opened to the fog position to achieve the same results.)

Fog Streams

The fog (or spray) nozzle breaks the water stream into small droplets. These droplets have a much larger total surface area than a solid stream (Fig. 7.7). Thus, a given volume of water in fog form will absorb much more heat than the same volume of water in a straight stream.

The greater heat absorption of fog streams is important where the use of water should be limited. Less water need be applied to remove the same amount of heat from a fire. In addition, more of the fog stream turns to steam when it hits the fire. Consequently, there is less runoff, less free surface water and less of a stability problem for the ship. Figure 7.8 compares straight and fog streams as extinguishing methods.

High-Velocity Fog Streams. The high-velocity fog stream can be used effectively to reduce heat in compartments, cabins and cargo spaces. In spaces where there is an overhead, the nozzle should be directed upward at an angle of 20–30° from the plane of the deck. This directs the fog toward the overhead, where the most heat is concentrated (Fig. 7.9). The foglike spray quickly

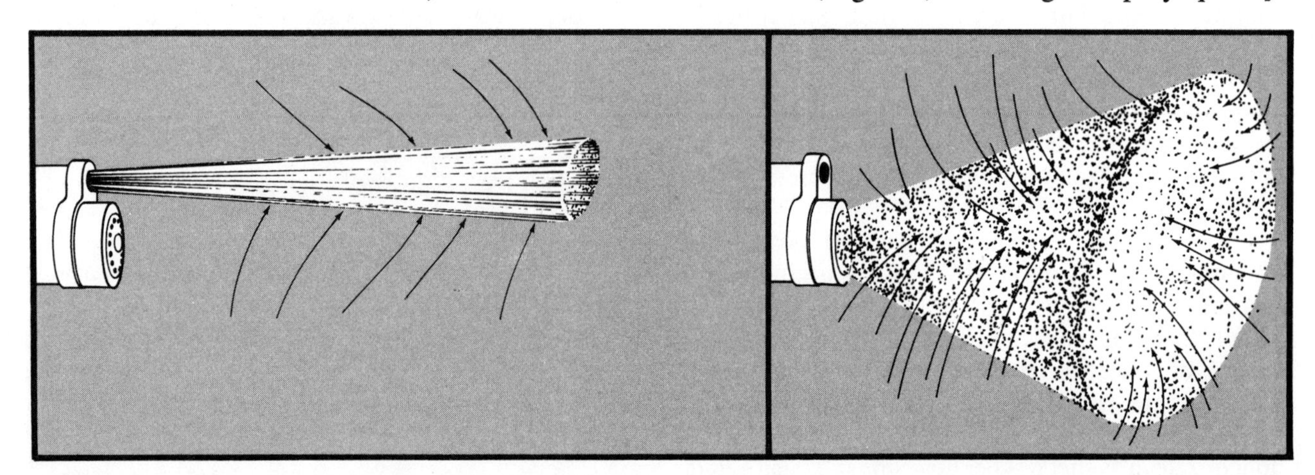

Figure 7.7. **A.** Straight stream. **B.** Fog stream. The fog stream droplets present a greater water surface area to the fire and can absorb more heat.

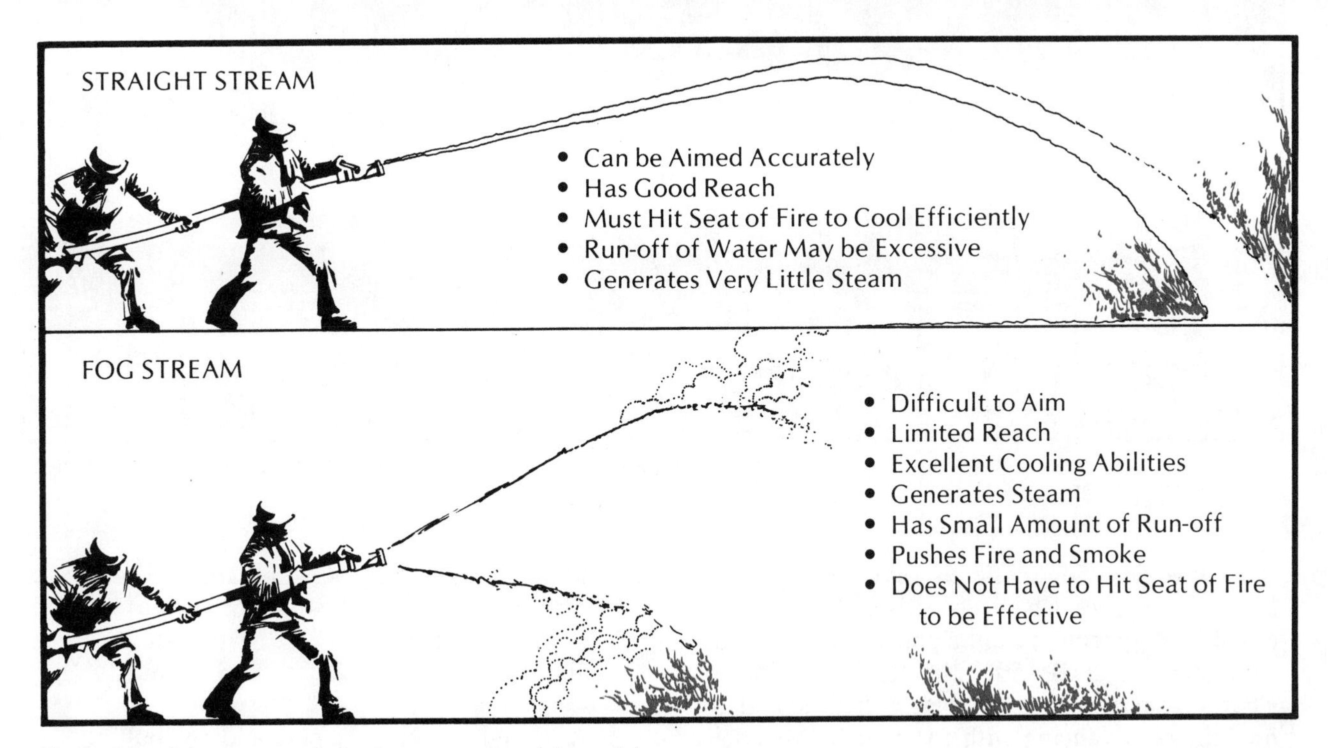

Figure 7.8. Advantages and disadvantages of straight and fog streams.

absorbs heat, allowing firefighters to enter or advance to the fire.

The high-velocity fog stream can also be used to move air in passageways and to drive heat and smoke away from advancing firefighters (Fig. 7.10). This operation can be used to facilitate the rescue of persons who are trapped in staterooms, cabins or other spaces. If at all possible, the far side of the passageway should be opened and kept clear of people. However, if there is no opening in a passageway other than the one from which the nozzle is being advanced, the heat and smoke have no place to go and may burst through or around the fog stream (blow back) and endanger those advancing the nozzle (Fig. 7.11). Therefore, in such a passageway, short bursts of fog should be aimed at the overhead to knock down the flame while minimizing the chance of blowback, or it may be better to use a solid stream.

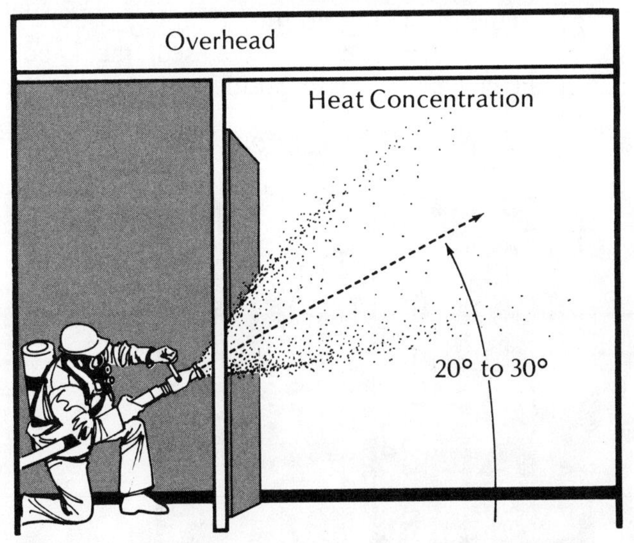

Figure 7.9. The fog nozzle should be directed upward at an angle of 20°–30° to hit heat concentrations at the overhead.

Low-Velocity Fog Streams. Low-velocity fog is obtained by using an applicator along with a combination nozzle. Applicators are tubes, or pipes, that are angled at 60° or 90° at the water outlet end. They are stowed for use with the low-velocity head already in place on the pipe. Some heads are shaped somewhat like a pineapple, with tiny holes angled to cause minute streams to bounce off one another and create a mist. Some heads resemble a cage with a fluted arrow inside. The point of the arrow faces the opening in the applicator tubing. Water strikes the fluted arrow and then bounces in all directions, creating a fine mist.

For 3.8 cm (1½-in.) nozzles, 1.2 m (4-ft) 60°-angle and 3 m (10-ft) 90°-angle applicators are approved for shipboard use. For 6.4 cm (2½-in.) nozzles, 3.7 m (12-ft) 90°-angle applicators are approved (Fig. 7.12). Other lengths with different angles are sometimes found. The 1.2 m (4-ft) applicator is intended for the 3.8 cm (1½-in.) combination nozzles fitted in propulsion machinery spaces containing oil-fired boilers, internal combustion machinery or fuel units.

Figure 7.10. A high velocity fog stream can be used to drive heat and smoke ahead of firefighters when there is an outlet for these combustion products.

Low-velocity fog is effective in combating class B fires in spaces where entry is difficult or impossible. Applicators can be poked into areas that cannot be reached with other types of nozzles. They are also used to provide a heat shield for firefighters advancing with foam or high-velocity fog. Low-velocity fog can be used to extinguish small tank fires, especially where the mist from the applicator can cover the entire surface of the tank. However, other extinguishing agents, such as foam and carbon dioxide, are usually more effective.

Limitations of Fog Streams. Fog streams do not have the accuracy or reach of straight streams. Improperly used, they can cause injury to personnel, as in a blowback situation. While they can be effectively used on the surface of a deep-seated fire, they are not as effective as solid streams in soaking through and reaching the heart of the fire.

Combination Nozzle Operation

The combination nozzle will produce a straight stream or high-velocity fog, depending on the position of its handle. Combination nozzles are available for use with 3.8- and 6.4-cm (1½- and 2½-in.) hose. Reducers can be used to attach a 3.8-cm (1½-in.) nozzle to a 6.4-cm (2½-in.) hose.

A straight stream is obtained by pulling the nozzle handle all the way back toward the operator (Fig. 7.13A).

A fog stream is obtained by pulling the handle back halfway (Fig. 7.13B). In other words, the handle is perpendicular to the plane of the nozzle.

The nozzle is shut down, from any opened position, by pushing the handle forward as far as it will go (Fig. 7.13C).

The low-velocity fog applicator must be attached with the nozzle shut down. First, the high-velocity "button" or tip is removed. Then the straight end of the applicator is snapped into the fog outlet and locked with a quarter turn. A low-velocity fog stream is obtained with the nozzle handle in the fog position (halfway back).

When any nozzle is to be used, the handle should be in the closed position until the water

Figure 7.11. If there is no outlet for combustion products that are being pushed ahead, they may blow back and engulf advancing firefighters.

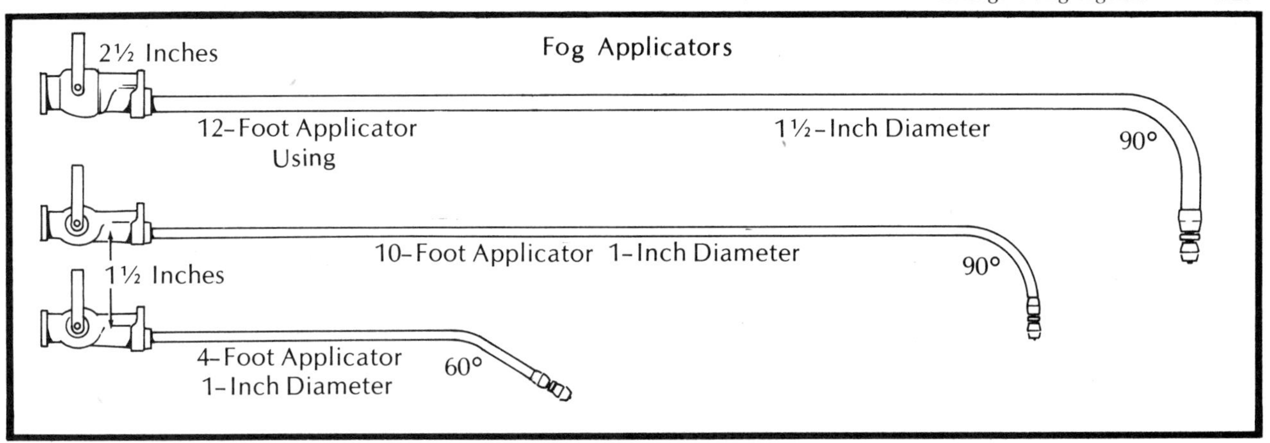

Figure 7.12. Low-velocity fog applicators approved for shipboard use.

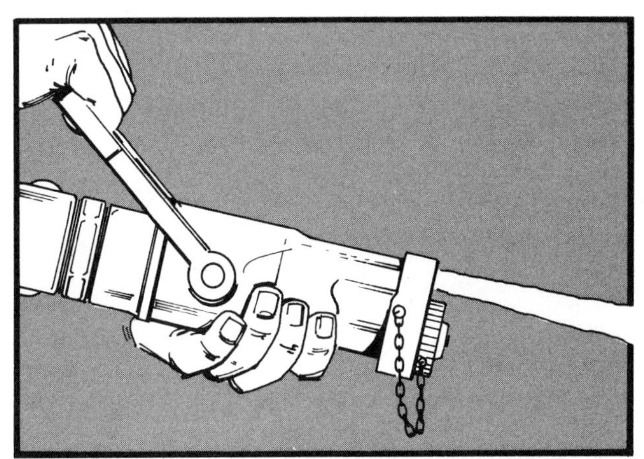

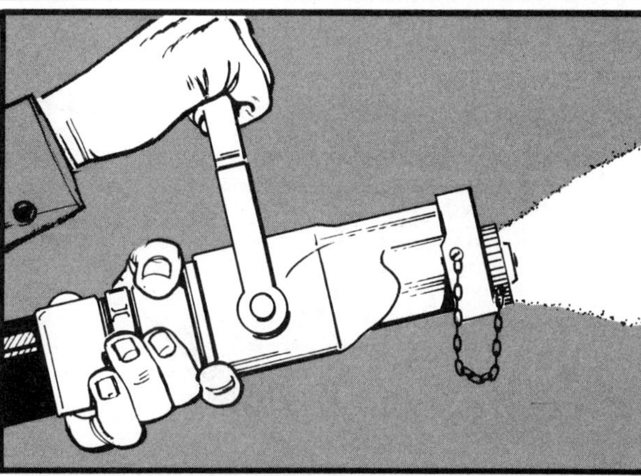

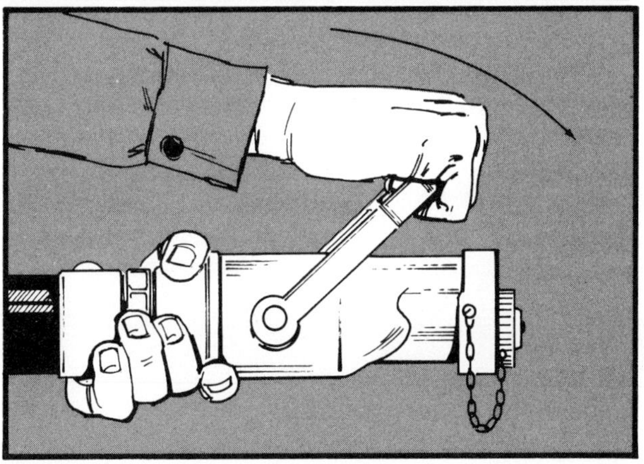

Figure 7.13. **A.** Nozzle open for straight stream. **B.** Nozzle open for high-velocity fog (or low-velocity fog if applicator is attached). **C.** Nozzle shut down.

reaches the nozzle. The hose will bulge out, and the nozzleman will feel the weight of the water. Before pushing the handle to an open position, he should let the entrained air out of the nozzle. This is done by turning a bit sideways with the nozzle and slowly opening it until a spatter of water comes out. Now the nozzle is directed at the target. The backup man closes up to the nozzleman and takes some of the weight of the hose and the back pressure from the nozzle. The nozzle is opened to the desired position, and the fire is attacked.

Straight and fog streams can be very effective against class A fires in the hands of skilled operators. Fog streams can also be used effectively against class B fires. However, it is important that crewman have actual experience in directing these streams during drills. Applicators should also be broken out at drills so crewmen can get the feel of these devices.

Other Types of Water for Firefighting

Wet Water. Wet water is water that has been treated with a chemical agent to lower its surface tension. The treated water penetrates porous materials, such as baled cotton and rolls of fabric, more easily than plain water. Thus it can sink in and extinguish fires that have extended into the interior of the bale or roll.

Thick Water. Thick water is water that has been treated to decrease its ability to flow. It forms a thick wall that clings to burning material and remains in place longer than plain water. However, it does not penetrate as easily as wet or untreated water. Thick water is slippery and makes walking on wet decks difficult.

Rapid or Slippery Water. Rapid water is water to which small quantities of polyethylene oxide have been added. This chemical reduces the viscosity of the water and the friction loss in hoselines. The result is an increase in the reach of the stream.

FOAM

Foam is a blanket of bubbles that extinguishes fire mainly by smothering. The bubbles are formed by mixing water and a foam-making agent (*foam concentrate*). The result is called a *foam solution*. The various foam solutions are lighter than the lightest of flammable oils. Consequently, when applied to burning oils, they float on the surface of the oil (Fig. 7.14).

Foam concentrates are produced in two strengths, 3% and 6%. These percentages do not have the usual meaning. They are the percentages of the concentrate to be used in making the foam solution. Thus, if 3% concentrate is used, 3 parts of concentrate must be mixed with 97 parts of water to make 100 parts of foam solution. If 6% concentrate is used, 6 parts of concentrate must be mixed with 94 parts of water. The 3% foam solution is just as effective as the 6% solution. The difference is in shipping and storing the products. Five containers of 3% concentrate make as much foam as 10 similar containers of 6% concentrate.

Extinguishing Effects of Foam

Firefighting foam is used to form a blanket on the surface of flaming liquids, including oils. The blanket prevents flammable vapors from leaving the surface and prevents oxygen from reaching the fuel. Fire cannot exist when the fuel and oxygen are separated. The water in the foam also has a cooling effect, which gives foam its class A extinguishing capability.

The ideal foam solution should flow freely enough to cover a surface rapidly, yet stick together enough to provide and maintain a vapor-tight blanket. The solution must retain enough water to provide a long-lasting seal. Rapid loss of water would cause the foam to dry out and break down (wither) from the high temperatures associated with fire. The foam should be light enough to float on flammable liquids, yet heavy enough to resist winds.

The quality of a foam is generally defined in terms of its 25% drainage time, its expansion ratio and its ability to withstand heat (burnback resistance). These qualities are influenced by

- The chemical nature of the foam concentrate
- The temperature and pressure of the water
- The efficiency of the foam-making device.

Foams that lose their water rapidly are the most fluid. They flow around obstructions freely and spread quickly. Such foams would be of use in engine room or machinery space fires; they would be able to flow under and around machinery, floorplates and other obstructions.

There are two basic types of foam, chemical and mechanical.

Chemical Foam

Chemical foam is formed by mixing an alkali (usually sodium bicarbonate) with an acid (usually aluminum sulfate) in water. When chemical foam was first introduced, these substances were stored in separate containers; they are now combined in a sealed, airtight container. A stabilizer is added to make the foam tenacious and long-lived.

When these chemicals react, they form a foam or froth of bubbles filled with carbon dioxide gas. The carbon dioxide in the bubbles has little or no extinguishing value. Its only purpose is to inflate the bubbles. From 7 to 16 volumes of foam are produced for each volume of water.

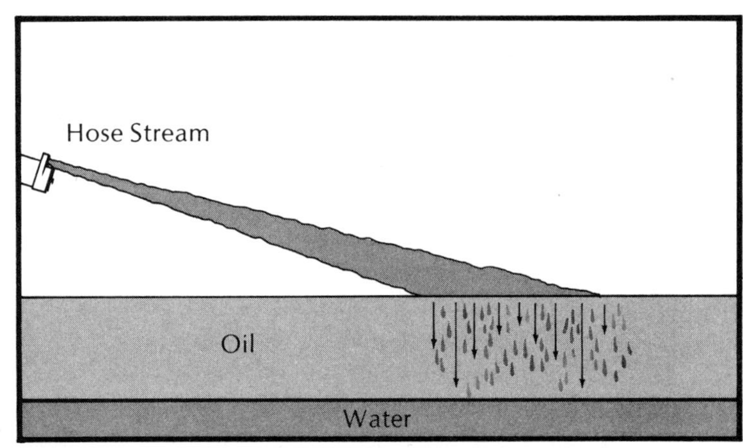

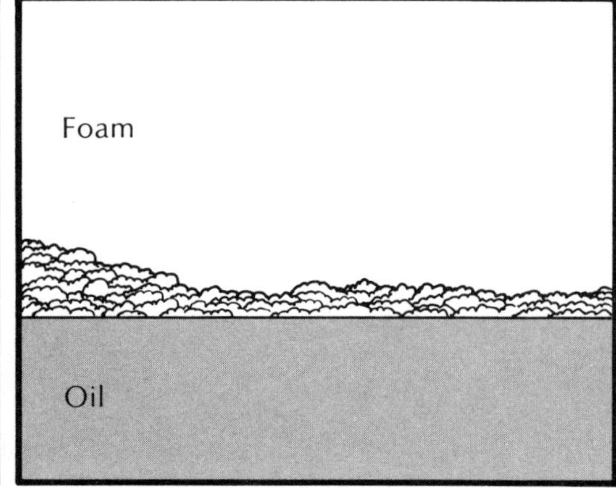

Figure 7.14. **A.** Water is heavier than oil and sinks below its surface. **B.** Foam is lighter than oil and floats on its surface.

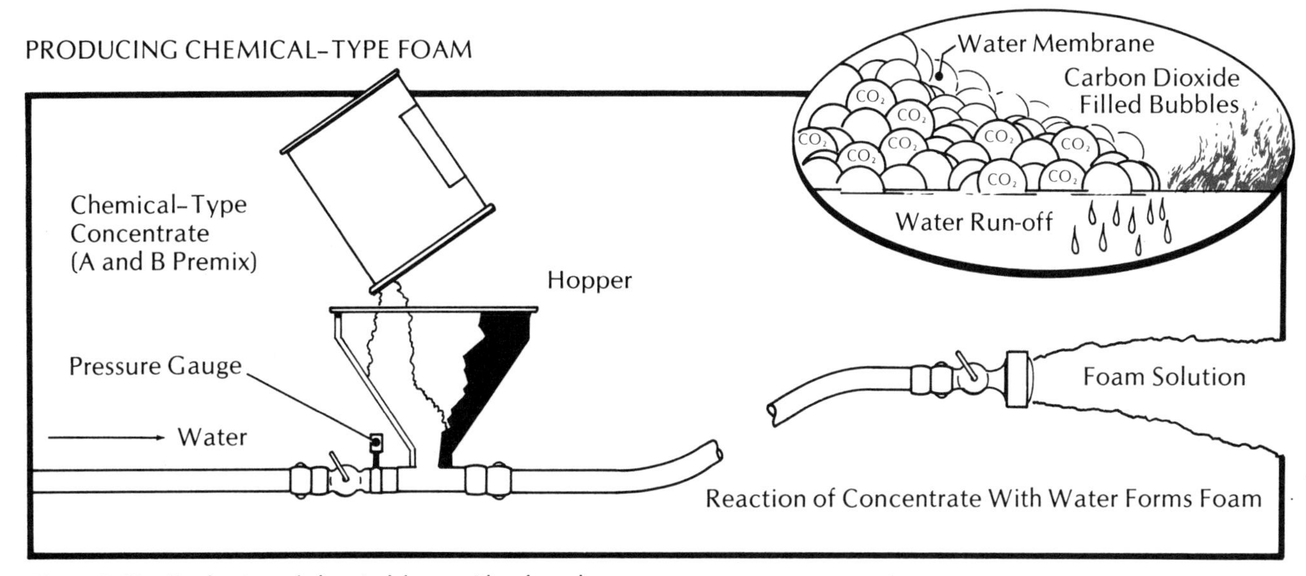

Figure 7.15. Production of chemical foam with a foam hopper.

The premixed foam powder may be stored in cans and introduced into the water during firefighting operations. For this, a device called a foam hopper (Fig. 7.15) is used. Or, the two chemicals may be premixed with water to form an aluminum sulfate solution and a sodium bicarbonate solution. The solutions are then stored in separate tanks until the foam is needed. At that time, the solutions are mixed to form the foam.

Many chemical foam systems are still in use, both aboard ship and in shore installations. However, they are being phased out in favor of the newer mechanical foam or, as it is sometimes called, air foam.

Mechanical (Air) Foam

Mechanical foam is produced by mixing a foam concentrate with water to produce a foam solution. The bubbles are formed by the turbulent mixing of air and the foam solution (Fig. 7.16). As the name air foam implies, the bubbles are filled with air. Aside from the workmanship and efficiency of the equipment, the degree of mixing determines the quality of the foam. The design of the equipment determines the quantity of foam produced.

There are several types of mechanical foams. They are similar in nature, but each has its own special firefighting capabilities. They are produced from proteins, detergents (which are synthetics) and surfactants. The surfactants are a large group of compounds that include detergents, wetting agents and liquid soaps. Surfactants are used to produce aqueous film-forming foam, commonly referred to as AFFF.

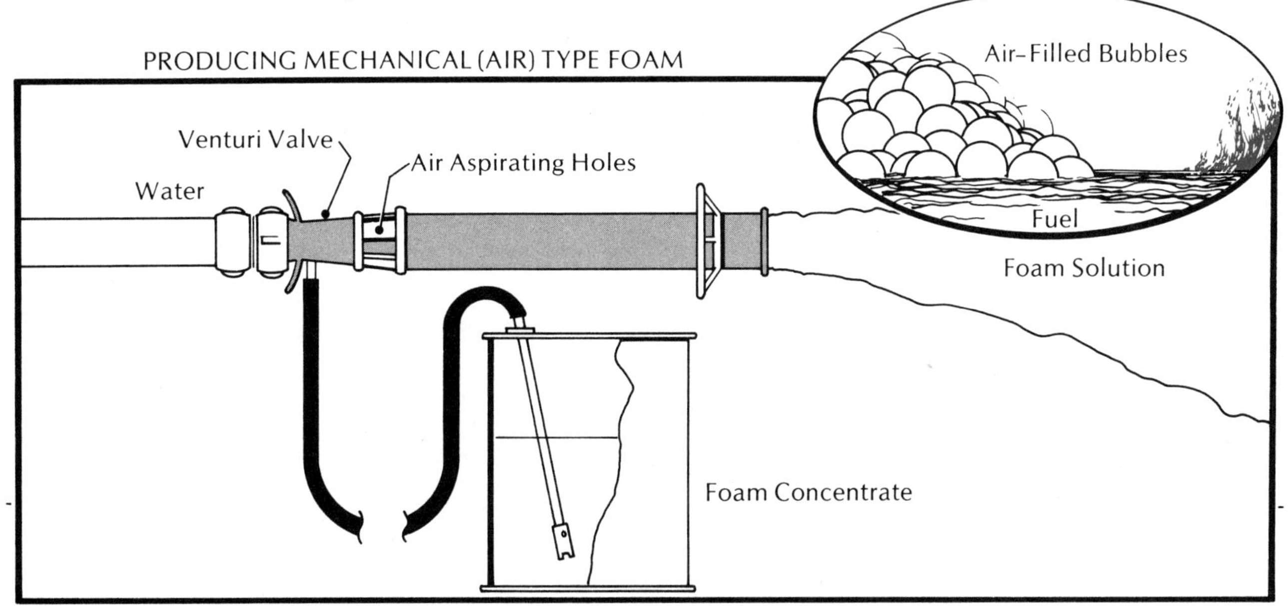

Figure 7.16. Production of mechanical (air) foam by mixing foam concentrate with water and air.

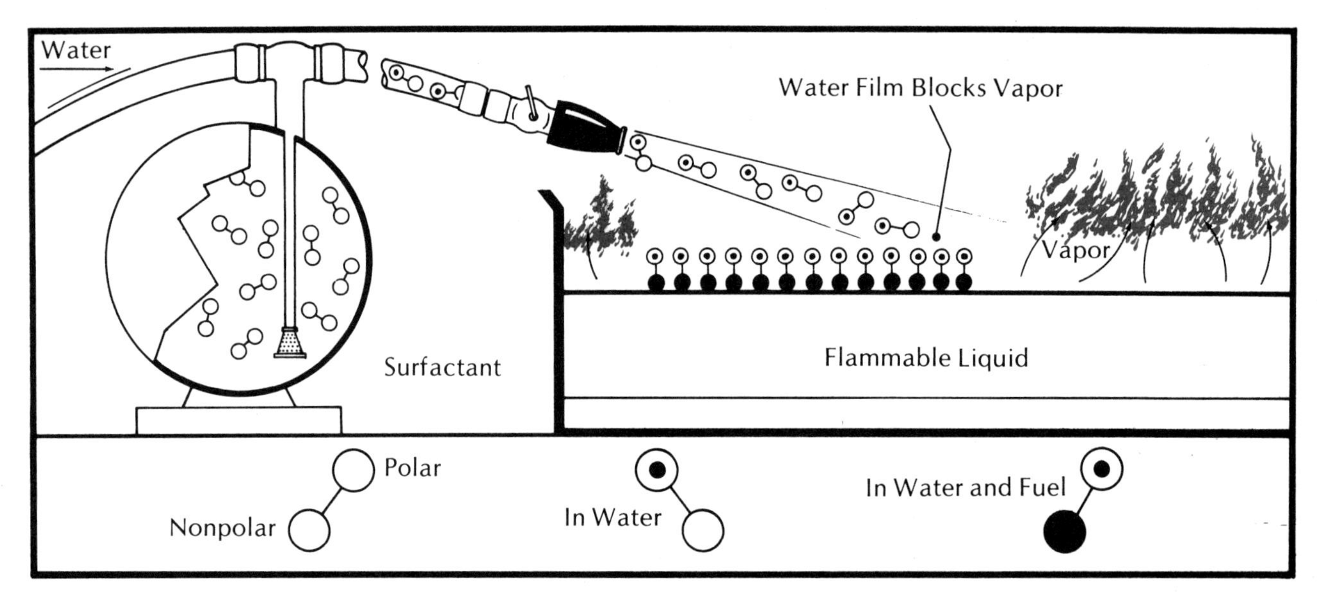

Figure 7.17. The AFFF surfactant molecule holds water at one end and flammable liquid fuel at the other end. It thus forms a thin layer of water on top of the burning fuel.

Protein Foams. The usual protein foams are produced from protein-rich animal and vegetable waste that is hydrolyzed (subjected to a chemical reaction with water that produces a weak acid). Mineral salts are added to increase their resistance to withering, making the foams resistant to burnback. The foam concentrate can produce foam in all types of water, except water that is contaminated with oil. When antifreeze is added, foam can be produced in subfreezing temperatures down to −23.3°C (−10°F).

Protein foam is the oldest type of foam and has been used since its development during World War II. The concentrate is available in 3% and 6% concentrations. Protein foam is not compatible with dry chemical extinguishing agents.

Fluoroprotein is a foam similar to hydrolyzed protein foam, with a fluorinated compound bonded to the protein. This foam can be injected below the liquid surface in a tank. It also works very well with dry chemical agents. Fluoroprotein is available in both 3% and 6% concentrations; with antifreeze, it produces foam in subfreezing temperatures.

Alcohol Foams. Alcohol-resistant protein foam is similar to standard protein foam. However, it is blended with an insoluble soap, to permit its use on water-soluble organic flammable liquids, such as alcohol, ketones, ethers and aldehydes. These water-soluble liquids will break down ordinary protein foam. Tankers that carry such liquids may be equipped with alcohol foam. The application rate depends on the vessel design, products carried and foam system used. Instructions for using the system are posted in each vessel.

Synthetic Foam. Synthetic detergent-based foam is made up of alkyl sulfonates. This form has less burnback resistance than protein formulas, but may be used with all dry chemicals. It foams more readily than the proteins and requires less water. This is important where the water supply is limited.

Aqueous Film-Forming Foam (AFFF). This foam was developed by the U.S. Naval Research Laboratory to be used in a twinned system: A flammable liquid fire would be quickly knocked down with a dry chemical; then AFFF would be applied to prevent reignition. However, the AFFF proved more effective than expected, and it is now used without the dry chemical. AFFF controls the vaporization of flammable liquids by means of a water film that forms as the foam is applied. Like other foams, it cools and blankets. This double action gives a highly efficient, quick-acting foam cover for combustible-liquid spills.

The foam is made from surfactants, through a fairly complex chemical process. The result is an extinguishing agent that is highly effective when used according to the manufacturer's directions.

One end of a surfactant molecule is polar (water soluble), whereas the other end is nonpolar (oil soluble but not water soluble). (This is what gives detergents the ability to clean away grease and oil, which do not dissolve in water.) In use, the surfactant is mixed with water before it reaches the nozzle (Fig. 7.17). As the surfactant mixes with the water, the polar end dissolves; the nonpolar end remains intact.

When the surfactant reaches the surface of the flammable liquid, the nonpolar end dissolves in the fuel. The polar end drags water along with it. Thus, a thin film of water floats on top of the water-insoluble flammable liquid (such as gasoline, kerosene or jet fuel). It remains on the surface even though it is heavier than the burning fuel; the surface tension holding the nonpolar end is greater than the force of gravity. The film is very thin, less than 0.003 cm (0.001 in.) thick. The remainder of the water sinks below the surface of the fuel, to the bottom of the container (Fig. 7.17).

Because AFFF works through surface tension, it spreads the water thinly, but over a larger surface area than untreated water could cover. The thin water film, spread across the flammable liquid, keeps the flammable vapors beneath its surface. When vapor cannot reach the flames, flame production ceases.

The water film can be broken if it is agitated. It may also be broken by the roll and pitch of a ship that is under way, especially in heavy weather.

AFFF is similar in some respects to wet water. It has a low viscosity and spreads quickly over the burning material. Water draining from this type of foam has a low surface tension, so AFFF can be used on mixed class A and B fires. The draining water penetrates and cools the class A material, while the film blankets the class B material.

AFFF can be produced from fresh water or seawater. As noted above, AFFF can be used with, before or after dry chemicals. AFFF concentrates should not be mixed with the concentrates of other foams, although in foam form they may be applied to the same fire successfully.

Low-Temperature Foams. Most foam concentrates can be purchased with additives that protect them against temperatures as low as −6.7°C (20°F) during storage and use. However, the water that is mixed with the concentrate must be above 0°C (32°F) or it will freeze. But as long as the water is above freezing and is running, an effective foam can be produced with either fresh water or seawater.

Foam Supplies

Enough foam concentrate should be available to produce *foam solution* at the rate of 6.5 liters/min for each square meter (1.6 gal/min for each 10 ft^2) of area protected, for at least 3 minutes for spaces other than tanks and at least 5 minutes for tanks. The same discharge rate applies to fixed foam extinguishing systems in tank vessels; there should be enough concentrate on hand to produce foam at this rate for at least 3 minutes. Deck foam systems on tankers carrying the usual petroleum products should produce foam solution at the rate of at least 6.5 liters/min for each square meter (1.6 gal/min for each 10 ft^2) of cargo area, or 9.7 liters/min for each square meter (2.4 gal/min for each 10 ft^2) of the horizontal sectional area of the single tank having the largest area, whichever is greater. Enough concentrate must be available to supply deck foam systems on tankers at this rate for at least 15 minutes.

There need not be enough foam concentrate aboard to supply the maximum amount required by all systems. Instead, the total available amount need only be sufficient to supply the space requiring the greatest amount.

Before foam is used, it is necessary to ensure that there is enough to do the job. There must be a sufficient amount to cover the entire surface of the fuel and to completely extinguish the fire. If not enough foam is available, there is no sense in using foam at all. Half measures do not work with foam. Incomplete coverage allows the fire to burn around the foam and destroy it.

High-Expansion Foams

High-expansion foams are those that expand in ratios of over 100:1 when mixed with air. Most systems produce expansion ratios of from 400:1 to 1000:1. Unlike conventional foam, which provides a blanket a few inches thick over the burning surface, high-expansion foam is truly three dimensional; it is measured in length, width, height and cubic feet.

High-expansion foam is designed for fires in confined spaces. Heavier than air but lighter than water or oil, it will flow down openings and fill compartments, spaces and crevices, replacing the air in these spaces. In this manner it deprives the fire of oxygen. Because of its water content, it absorbs heat from the fire and cools the burning material. When the high-expansion foam has absorbed sufficient heat to turn its water content to steam at 100°C (212°F), it has absorbed as much heat as possible; then the steam continues to replace oxygen and thus combat the fire.

Uses of High-Expansion Foam. High-expansion foam is effective on both class A and class B fires. On class A fires its effectiveness stems from its cooling capability, on class B fires from its smothering action. Class A fires are controlled when the foam covers the burning material. For complete extinguishment, the foam cover must be continuously replenished, to replace water that has been spent in absorbing the heat of the fire. It is the water content of the foam that is important here.

Cooling is also involved in class B fires involving high flash point oils and liquids, such as lubricating oils and cooking oils. Here, the cooling reduces the surface temperature of the liquid below the temperature at which flammable vapors are given off. Class B fires involving low flash point liquids such as gasoline and naphtha are extinguished by high-expansion foam in the same manner as by conventional foam. The fire is deprived of oxygen (smothered); the flammable vapors are prevented from joining with oxygen in the air.

Automatic High-Expansion Foam Systems. Automatic high-expansion foam systems automatically flood the protected space with foam. Such systems are available but are not yet used on vessels. An automatic system requires a fire detector such as those discussed in Chapter 6. It must be wired to sense the fire, sound an alarm and actuate a mechanism to start generating and sending high-expansion foam into the protected space. In addition, the detector must actuate an *evacuation alarm* to warn people who may be in the space that it is about to be flooded. Automatic foam systems generate foam very quickly. A person who does not leave the area immediately could be inundated or cut off from escape.

While the foam itself is not considered toxic, it blocks vision, impairs hearing and makes breathing difficult. It is thus dangerous for anyone to be within the foam; the only valid reason for entering or remaining in a foam buildup is to rescue someone who might be trapped. When it is absolutely necessary for someone to enter the foam to save a life, a lifeline must be attached to the rescuer. A canister-type breathing mask should not be used, because the foam will mix with the chemicals in the mask and suffocate the wearer. Either a fresh-air hose mask, a demand-type compressed-air or OBA self-contained unit may be used.

A coarse water fog stream can be used to cut a path through high-expansion foam. However, it is difficult to cut a path if the foam is higher than about 1.8 m (6 ft), as the foam tends to slide down into the path. All this leads to one valid conclusion: Everyone must get out of the area as soon as the evacuation alarm sounds.

Flooding a Compartment with a Portable Foam Generator. With the approval of the Coast Guard, portable high-expansion foam generators may be used for firefighting aboard ship. To flood a compartment, a hose is run out on the deck above, the generator is attached to the hose, and the foam concentrate is connected to the generator pickup tube. All personnel must leave the space to be flooded, if the fire has not already driven them out.

A hole, through which the foam is to be applied, is cut in the deck (Fig. 7.18). A charged hoseline must be available at this time. (Whenever an opening is made into a fire area, an additional hoseline, charged with water, must be available in case the fire pushes through the opening. The charged line is then used to protect the opening.) Before the high-expansion foam is directed into the opening, another opening must be made. The second opening, some distance away but still over the same space, allows the escape of steam that is generated when the high-expansion foam hits the fire (Fig. 7.18). If the

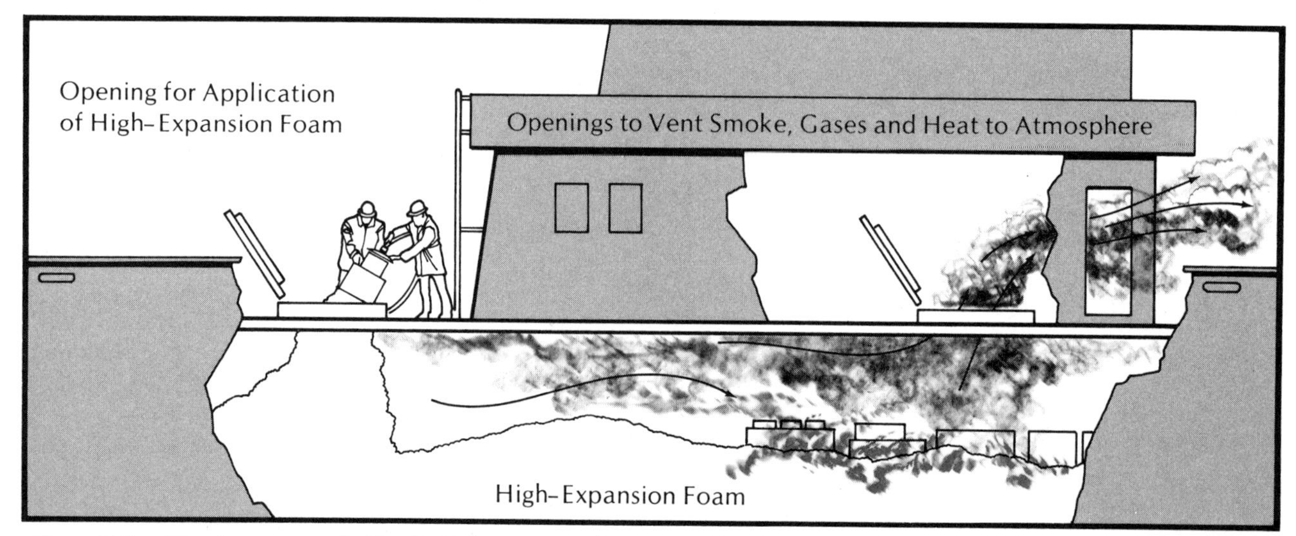

Figure 7.18. Flooding a cargo hold with high-expansion foam. Here, hatches are being used to apply the foam (*left*) and to vent the products of combustion including the hot steam that is produced (*right*).

second opening (the vent) is made on an open deck, the steam and heat will dissipate into the open air. If it is impossible to place the vent in the open, then it must be made in a passageway that leads to the open air. Once firefighters are sure that the passageway is open, no one should remain in the path of the escaping steam. When the foam is applied, it will generate steam that is hot enough to scald.

Steam leaving the vent hole is a good indication that the foam is reaching its target and doing its work. If no steam is seen within a few minutes, either the foam is not reaching its objective or the vent hole is improperly placed.

Production of High-Expansion Foam. An aspirating nozzle is used to produce high-expansion foam. In the nozzle, a solution of foam concentrate and water is sprayed over a meshed screen. Air is drawn into the nozzle and through the screen at high velocity. The air mixes with the solution at the screen. Bubbles are formed at the screen, and high-expansion foam leaves the nozzle at the far side of the screen (Fig. 7.19). If air that has been heated or contaminated by the fire is used to produce foam, the result is a poor grade of foam. Also, soot can clog the openings in the screen and affect the quantity and quality of the foam. The air should be as clean and fresh as possible.

Limitations on the Use of Foam

Foams are effective extinguishing agents when used properly. However, they do have some limitations, including the following:

1. Because they are aqueous (water) solutions, foams are electrically conductive and *should not be used on live electrical equipment.*
2. Like water, foams *should not be used on combustible-metal fires.*
3. Many foams must not be used with dry chemical extinguishing agents. AFFF is an exception to this rule and may be used in a joint attack with dry chemical.
4. Foams are not suitable for fires involving gases and cryogenic (extremely low temperature) liquids. However, high-expansion foam is used on cryogenic liquid spills to rapidly warm the vapors to minimize the hazards of such spills.
5. If foam is placed on burning liquids (like asphalts) whose temperatures exceed 100°C (212°F), the water content of the foam may cause frothing, spattering or slopover. Slopover is different from boilover, although the terms are frequently confused. *Boilover* occurs when the heat from a fire in a tank travels down to the bottom of the tank and causes water that is already there to boil and push part of the tank's contents over the side. Certain oils with a high water content, such as crude oil, have a notorious reputation for boilover. *Slopover* occurs when foam, introduced into a tank of hot oil (surface temperature over 100°C (212°F)) sheds its water content due to the high heat. The water forms an emulsion of steam, air and the foam itself. The forming of the emulsion is accompanied by a corresponding increase in volume. Since tanks are three dimensional, the only place for the emulsion to go is over the sides of open tanks or into the vents of enclosed tanks.
6. Sufficient foam must be on hand to ensure that the entire surface of the burning material can be covered. In addition, there must be enough foam to replace foam that is burned off and to seal breaks in the foam surface.

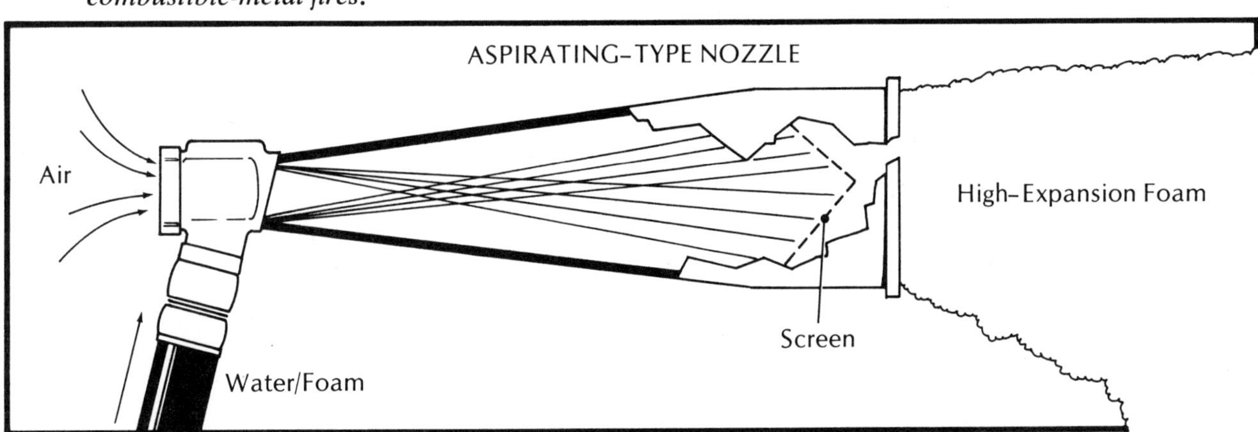

Figure 7.19. Production of high-expansion foam. High-velocity air strikes the water–foam concentrate solution at the screen, producing the foam.

Advantages of Foam

In spite of its limitations, foam is quite effective in combating class A and class B fires.

1. Foam is a very effective smothering agent, and it provides cooling as a secondary effect.
2. Foam sets up a vapor barrier that prevents flammable vapors from rising. The surface of an exposed tank can be covered with foam to protect it from a fire in a neighboring tank.
3. Foam is of some use on class A fires because of its water content. AFFF is especially effective, as are certain types of wet-water foam. Wet-water foam is made from detergents; its water content quickly runs out and seeps into the burning material. It is not usually found aboard vessels; a more likely use is in protecting bulk storage in piers or warehouses.
4. Foam is effective in blanketing oil spills. However, if the oil is running, an attempt should be made to shut down a valve if such action would stop the flow. If that is impossible, the flow should be dammed. Foam should be applied on the upstream side of the dam (to extinguish the fire) and on the downstream side (to place a protective cover over any oil that has seeped through).
5. Foam is the most effective extinguishing agent for fires involving large tanks of flammable liquids.
6. Foam can be made with fresh water or seawater, and hard or soft water.
7. Foam does not break down readily; it extinguishes fire progressively when applied at an adequate rate.
8. Foam stays in place, covers and absorbs heat from materials that could cause reignition.
9. Foam uses water economically and does not tax the ship's fire pumps.
10. Foam concentrates are not heavy, and foam systems do not take up much space.

CARBON DIOXIDE

Carbon dioxide (CO_2) extinguishing systems have, for a long time, been approved for ship installation as well as for industrial occupancies ashore. Aboard ship, carbon dioxide has been approved for cargo and tank compartments, spaces containing internal combustion or gas-turbine main propulsion machinery and other spaces.

Properties of Carbon Dioxide

Carbon dioxide is normally a gas, but it may be liquefied or solidified under pressure. At −43°C (−110°F), carbon dioxide exists as a solid, called "dry ice." The critical temperature of carbon dioxide is 31°C (87.8°F). Above that temperature, it is always a gas, regardless of pressure. Carbon dioxide does not support combustion in ordinary materials. However, there are some exceptions, as when CO_2 reacts with magnesium and other metals.

Carbon dioxide is about 1.5 times heavier than air. This adds to its suitability as an extinguishing agent, because CO_2 tends to fall through air and blanket a fire. Its weight makes it less prone to dissipate quickly. In addition, carbon dioxide is not an electrical conductor; it is approved for extinguishing fires in energized electrical equipment.

Extinguishing Properties of Carbon Dioxide

Carbon dioxide extinguishes fire mainly by smothering. It dilutes the air surrounding the fire until the oxygen content is too low to support combustion. For this reason it is effective on class B fires, where the main consideration is to keep the flammable vapors separated from oxygen in the air. CO_2 has a very limited cooling effect. It can be used on class A fires in confined spaces, where the atmosphere may be diluted sufficiently to stop combustion. However, CO_2 extinguishment takes time. The concentration of carbon dioxide must be maintained until all the fire is out. Constraint and patience are needed.

Carbon dioxide is sometimes used to protect areas containing valuable articles. Unlike water and some other agents, carbon dioxide dissipates without leaving a residue. As mentioned above, it does not conduct electricity and can be used on live electrical equipment. However, fire parties must maintain a reasonable distance when using a portable CO_2 extinguisher or hoseline from a semiportable system on high voltage gear.

Uses of Carbon Dioxide

Carbon dioxide is used primarily for class B and C fires. It may also be used to knock down a class A fire. It is particularly effective on fires involving

1. Flammable oils and greases
2. Electrical and electronic equipment, such as motors, generators and navigational devices
3. Hazardous and semihazardous solid materials, such as some plastics, except those

that contain their own oxygen (like nitrocellulose)

4. Machinery spaces, engine rooms and paint and tool lockers
5. Cargo spaces where total flooding with carbon dioxide may be accomplished
6. Galleys and other cooking areas, such as diet kitchens
7. Compartments containing high value cargo, such as works of art, delicate machinery and other material that would be ruined or damaged by water or water-based extinguishing agents
8. Spaces where after-fire cleanup would be a problem.

Limitations on the Use of Carbon Dioxide

Effectiveness. CO_2 is not effective on substances that contain their own oxygen (oxidizing agents). It is not effective on combustible metals such as sodium, potassium, magnesium and zirconium. In fact, when CO_2 is used on burning magnesium, it reacts with the magnesium to form carbon, oxygen and magnesium oxide. The fire is intensified by the addition of oxygen and carbon, a fuel.

Outside Use. To be fully effective, the gas must be confined. For this reason, CO_2 is not as effective outside as it is in a confined space. This does not mean that it cannot be used outside. Portable CO_2 extinguishers and hoselines have extinguished many fires in the open. An outside fire should be attacked from the windward side; the CO_2 should be directed low with a sweeping motion for a spill fire, or down at the center of a confined fire. The effective range for a portable CO_2 fire extinguisher is about 1.5 m (5 ft).

Possibility of Reignition. Compared with water carbon dioxide has a very limited cooling capacity. It may not cool the fuel below its ignition temperature, and it is more likely than other extinguishing agents to allow reflash. (Its main extinguishing action, as noted above, is oxygen dilution.) When portable CO_2 extinguishers or hoselines from semiportable extinguishers are used, additional backup water hoselines should be brought to the scene. In case of live electrical equipment, an additional nonconducting agent must be brought to the scene.

When a space is flooded with CO_2 the concentration must be kept up to a certain level. After the initial application of a set number of CO_2 cylinders, additional cylinders must be discharged into the space periodically. These backup applications maintain the concentration of CO_2 for periods varying from hours to days. CO_2 works well in confined spaces, but it works slowly; patience is the watchword.

If a flooded space is opened before the fire is completely extinguished, air entering the space may cause reignition. Carbon dioxide cannot be purchased at sea. Reignition requires a second attack, at a time when less CO_2 is available.

Hazards. Although carbon dioxide is not poisonous to the human system, it is suffocating in the concentration necessary for extinguishment. A person exposed to this concentration would suffer dizziness and unconsciousness. Unless removed quickly to fresh air, the victim could die.

Carbon Dioxide Systems

Carbon dioxide extinguishing systems aboard vessels are usually not automatic. However automatic systems may be installed in certain ships and towing vessels with Coast Guard approval. In the manual system, a fire detector senses fire and actuates an alarm. The engine room is alerted, and the bridge and CO_2 room are notified as to the location of the fire (*see* Chapter 6). After it is verified that a fire actually exists, the amount of carbon dioxide required for the involved space is released from the CO_2 room.

Coast Guard regulations require that an evacuation alarm be sounded when CO_2 is introduced into a space that is normally accessible to persons on board, other than paint and lamp lockers and similar small spaces. However on systems installed since July 1, 1957, an alarm is required only if *delayed discharge* is used. Delayed discharge is required where large amounts of CO_2 are released into large spaces. Delayed discharge may also be required for smaller spaces from which there are no horizontal escape routes.

The alarm sounds during a 20-second delay period prior to the discharge of carbon dioxide into the space. It uses no source of power other than the carbon dioxide itself. Every carbon dioxide alarm must be conspicuously identified with the warning "WHEN THE ALARM SOUNDS VACATE AT ONCE. CARBON DIOXIDE IS BEING RELEASED."

Portable and semiportable CO_2 extinguishers may be located in certain spaces. Small systems, consisting of one to four CO_2 cylinders, a hose and a nozzle, are often provided to protect against specific hazards. Those who work in the areas protected by these appliances should be familiar with their operation.

DRY CHEMICAL

Dry chemical extinguishing agents are chemicals in powder form. Again we note that they should not be confused with dry powders, which are intended only for combustible-metal fires.

Types of Chemical Extinguishing Agents

At the present time, five different types of dry chemical extinguishing agents are in use. Like other extinguishing agents, dry chemical may be installed in a fixed system or in portable and semiportable extinguishers. Such systems may be installed aboard ship with the approval of the Coast Guard commandant.

Sodium Bicarbonate. Sodium bicarbonate is the original dry chemical extinguishing agent. It is generally referred to as *regular dry chemical* and is widely used because it is the most economical dry chemical agent. It is particularly effective on animal fats and vegetable oils because it chemically changes these substances into nonflammable soaps. Thus, sodium bicarbonate is used extensively for galley range, hood and duct fires. There is one possible problem with sodium bicarbonate: Fire has been known to flash back over the surface of an oil fire when this agent is used.

Potassium Bicarbonate. This dry chemical was originally developed to be used with AFFF in a twinned system. However it is commonly used alone. It has been found to be most effective on liquid fuel fires in driving flames back and has a good reputation for eliminating flashback. It is more expensive than sodium bicarbonate.

Potassium Chloride. Potassium chloride was developed as a dry chemical that would be compatible with protein-type foams. Its extinguishing properties are about equal to those of potassium bicarbonate. One drawback is its tendency to cause corrosion after it has extinguished a fire.

Urea Potassium Bicarbonate. This is a British development, of which the NFPA says, "Urea potassium bicarbonate exhibits the greatest effectiveness of all the dry chemicals tested." It is not widely used because it is expensive.

Monoammonium Phosphate (ABC, Multipurpose). Monoammonium phosphate is called a multipurpose dry chemical because it can be effective on class A, B and C fires. Ammonium salts interrupt the chain reaction of flaming combustion. The phosphate changes into metaphosphoric acid, a glassy fusible material, at fire temperatures. The acid covers solid surfaces with a fire retardant coating. Therefore, this agent can be used on fires involving ordinary combustible materials such as wood and paper, as well as on fires involving flammable oils, gases and electrical equipment. However, it may only control, but not fully extinguish, a deep-seated fire. Complete extinguishment may require the use of a hoseline. In fact, it is always prudent to run out a hoseline as a backup when any dry chemical extinguisher is used.

Extinguishing Effects of Dry Chemical

Dry chemical agents extinguish fire by cooling, smothering, shielding of radiant heat and, to the greatest extent by breaking the combustion chain.

Cooling. No dry chemical exhibits any great capacity for cooling. However, a small amount of cooling takes place simply because the dry chemical is at a lower temperature than the burning material. Heat is transferred from the hotter fuel to the cooler dry chemical when the latter is introduced into the fire. (Heat is always transferred from a hotter body to a cooler body. The greater the surface area and the temperature difference, the greater the heat transfer.)

Smothering. When dry chemical reacts with the heat and burning material, some carbon dioxide and water vapor are produced. These dilute the fuel vapors and the air surrounding the fire. The result is a limited smothering effect.

Shielding of Radiant Heat. Dry chemical produces an opaque cloud in the combustion area. This cloud reduces the amount of heat that is radiated back to the heart of the fire, i.e., the opaque cloud absorbs some of the radiation feedback that is required to sustain the fire (*see* Chapter 4). Less fuel vapor is produced, and the fire becomes less intense.

Chain breaking. Chain reactions are necessary for continued combustion (*see* Chapter 4). In these chain reactions, fuel and oxygen molecules are broken down by heat; they recombine into new molecules, giving off additional heat. This additional heat breaks down more molecules, which then recombine and give off still more heat. The fire thus builds, or at least sustains itself, through reactions that liberate enough heat to set off other reactions.

Dry chemical (and other agents such as the halogens) attacks this chain of reactions. It is believed that it does so by reducing the ability of molecular fragments to recombine. It may itself combine with the fragments of fuel and oxygen molecules, so that the fuel cannot be oxidized. Although the process is not completely understood, chain breaking is the most effective extinguishing action of dry chemical.

Uses of Dry Chemical

Monoammonium phosphate (ABC, multipurpose) dry chemical may, as its name implies, be used on class A, B and C fires and combinations of these. However, as noted above, ABC dry chemical may only control, but not extinguish, some deep-seated class A fires. Then an auxiliary extinguishment method, such as a water hoseline, is required.

All dry chemical agents may be used to extinguish fires involving

1. Flammable oils and greases
2. Electrical equipment
3. Hoods, ducts and cooking ranges in galleys and diet kitchens
4. The surfaces of baled textiles
5. Certain combustible solids such as pitch, naphthalene and plastics (except those that contain their own oxygen)
6. Machinery spaces, engine rooms and paint and tool lockers.

Limitations on the Use of Dry Chemical

There are limitations on the use of dry chemical.

1. The discharge of large amounts of dry chemical could affect people in the vicinity. The opaque cloud that is produced can reduce visibility and, depending on its density, cause breathing difficulty.
2. Like the other extinguishing agents that contain no water, dry chemical is not effective on materials that contain their own oxygen.
3. Dry chemical may deposit an insulating coating on electronic or telephonic equipment, affecting the operation of the equipment.
4. Dry chemical is not effective on combustible metals such as magnesium, potassium, sodium and their alloys, and in some cases may cause a violent reaction.
5. Where moisture is present, dry chemical may corrode or stain surfaces on which it settles.

Compatibility with Other Extinguishing Agents

Any dry chemical may be applied to a fire with any other dry chemical. However, different types of dry chemical should not be mixed in containers. Some have an acid base, and others an alkali base. Mixing could cause excess pressure in the container or cause the chemicals to lump.

Many foam extinguishing agents break down when attacked by dry chemical. AFFF may be used with dry chemical, since it was developed for use with potassium bicarbonate in a twinned system. In that system, hoses from an AFFF tank and a dry chemical tank led to twin nozzles. AFFF could be directed at the fire from its nozzle, and dry chemical from its nozzle, either individually or simultaneously.

Today, large combined agent systems are used to protect petroleum refineries and oil storage tank farms. On vessels with foam systems, only foam-compatible dry chemicals may be used. If a dry chemical is not listed in the Coast Guard *Equipment Lists* (CG190), the Coast Guard commandant should be consulted before it is stowed aboard ship.

Safety

Dry chemical extinguishing agents are considered nontoxic, but they may have irritating effects when breathed. For this reason a warning signal, similar to the one used in carbon dioxide systems, should be installed in any space that might be totally flooded with dry chemical. In addition, breathing apparatus and lifelines must be available in case crewmen must enter the space before it is entirely ventilated.

Dry chemical extinguishing agents are very effective on gas fires. However, as has been noted several times in this book, *gas flames should not be extinguished until the supply of fuel has been shut down upstream of the fire.*

DRY POWDERS

Dry powders were developed to control and extinguish fires in combustible metals, i.e., class D fires. As mentioned earlier, dry chemical and dry powders are not the same. Only dry powders are intended for combustible-metal fires, i.e., those involving magnesium, potassium, sodium and their alloys, titanium, zirconium, powdered or fine aluminum, and some lesser known metals.

Dry powders are the only extinguishing agents that can control and extinguish metal fires without causing violent reactions. Other extinguishing agents may accelerate or spread the fire, injure personnel, cause explosions or create conditions more hazardous than the original fire. Dry powders act mainly by smothering, although some agents also provide cooling.

Water and water-based agents such as foam should not be used on combustible-metal fires. The water may cause an explosive chemical reaction. Even when there is no chemical reaction, water droplets that move below the surface of the molten metal will expand with explosive violence and scatter molten material. However, water has been prudently used in some cases; for example, on large pieces of burning magnesium, water was applied to a portion not actually burning, to cool this part sufficiently so that the fire did not extend. In general, water should not be applied to molten metals themselves, but it can be used to cool down threatened areas.

Types of Dry Powders

Two commercially available dry powders are composed mostly of graphite. The graphite cools the fire and creates a very heavy smoke that helps smother the fire. These agents are effective on all metals listed above. They are applied with a scoop or shovel.

Dry powder with a sodium chloride (salt) base is propelled from portable extinguishers by carbon dioxide, and from large containers or fixed systems by nitrogen. The powder is directed over the burning metal; when it drops, it forms a crust on the metal and smothers the fire. Like the graphite types, it is effective on the combustible metals mentioned above.

Dry powder with a sodium carbonate base is intended for sodium fires. The powder may be scooped from buckets or propelled from a pressurized extinguisher. It forms a crust on the surface of the burning sodium to smother the fire.

There are a number of other extinguishing agents for combustible-metal fires. Most are specialized, intended for one or perhaps two kinds of metal. The National Safety Council, in Chicago, and the Manufacturing Chemists' Association, in Washington, D.C., issue data sheets concerning specific combustible metals. The data sheets include extinguishment methods and agents. It would be wise for owners (and masters) who might expect their ships to carry combustible metals to secure data sheets for these metals. The Coast Guard regulations require that appropriate extinguishing appliances be provided whenever a merchant vessel carries hazardous material whose extinguishment is beyond the firefighting capability of the ship's normal outfitting.

HALOGENATED EXTINGUISHING AGENTS (Halon)

Halogenated extinguishing agents are made up of carbon and one or more of the halogen elements: fluorine, chlorine, bromine and iodine.

Two halogen extinguishing agents are recognized for use in the United States, *bromotrifluoromethane* (more familiarly known as Halon 1301) and *bromochlorodifluoromethane* (Halon 1211). The NFPA has set up standards (No. 12A for 1301, and No. 12B for 1211) for systems using these agents. The Coast Guard *Equipment Lists* (CG190) include equipment for Halon 1301 systems and extinguishers, but not for Halon 1211. Thus, the permission of the Coast Guard commandant must be obtained before a Halon 1211 system or extinguisher is installed aboard a vessel.

Both Halon 1301 and Halon 1211 enter the fire area as a gas. Most authorities agree that the Halons act as chain breakers. However, it is not known whether they slow the chain reaction, break it up or cause some other reaction.

Halon 1301 is stored and shipped as a liquid under pressure. When released in the protected area, it vaporizes to an odorless, colorless gas and is propelled to the fire by its storage pressure. Halon 1301 does not conduct electricity.

Halon 1211 is also colorless, but it has a faint sweet smell. Halon 1211 is stored and shipped as a liquid and pressurized by nitrogen gas. Pressurization is necessary since the vapor pressure of Halon 1211 is too low to convey it properly to the fire area. It does not conduct electricity.

Uses of the Halons

The extinguishing properties of Halon 1211 and Halon 1301 allow their use on a number of different types of fire. These include

1. Fires in electrical equipment
2. Fires in engine rooms, machinery spaces and other spaces involving flammable oils and greases
3. Class A fires in ordinary combustibles. However, if the fire is deep seated, a longer soaking time may be needed, or a standby hoseline may be used to complete the extinguishment.
4. Fires in areas where articles of high value may be stored and thus damaged by the residue of other agents
5. Halon 1301 is recommended for fires involving electronic computers and control rooms. Halon 1211 carries no such recommendation.

There are few limitations on the use of Halons. However, they are *not* suited for fighting fires in *1)* materials containing their own oxygen and *2)* combustible metals and hydrides.

Safety

When inhaled, both Halon 1301 and Halon 1211 may cause dizziness and impaired coordination. These gases may reduce visibility in the area in which they are discharged. At a temperature slightly below 500°C or about 900°F the gases of both Halons will decompose. The normal vapors below that temperature are not considered very toxic; however, the decomposed gases may be very hazardous, depending on *a)* the concentration, *b)* the temperature and *c)* the amount that has been inhaled.

Halon 1211 is not recommended for the total flooding of confined spaces. If Halon 1301 is to be used for the total flooding of normally occupied spaces, an evacuation alarm must be provided. Personnel should leave the area promptly on hearing the alarm. Similarly, when a Halon 1301 extinguisher is used, those not directly involved in the operation should leave the area immediately. The extinguisher operator should step away as soon as the appliance is discharged. The area should be vented with fresh air before it is reentered. If it is necessary to remain in or enter an area where Halon 1301 has been discharged, breathing apparatus and lifelines should be used. The only valid reason for such entry would be to save life or to maintain control of the ship.

SAND

Sand has been used on fires since time immemorial. However, it is not very efficient when compared to modern extinguishing agents.

The function of sand is to smother an oil fire by covering its surface. But if the oil is more than an inch or so in depth, the sand will sink below the oil surface. Then, unless a sufficient amount of sand is available to cover the oil, it will be ineffective in extinguishing the fire. However, when properly applied, sand can be used to dam or cover an oil spill.

Sand must be applied to a fire with a scoop or shovel. Its minimal effectiveness may be further reduced by an unskilled user. After the fire, there is a cleanup problem. In addition to these deficiencies, sand is abrasive and has an ingenious way of getting into machinery and other equipment.

Title 46 CFR, Parts 34 and 95, lists requirements for sand as an extinguishing substance in the amount of 0.28 m^3 (10 ft^3) for spaces containing oil-fired boilers. However, an additional class B extinguisher may be substituted for the sand. The class B extinguisher is a good alternative to sand.

1. The extinguisher is more effective, pound for pound and cubic foot for cubic foot.
2. The extinguisher is easier to use.
3. The extinguisher has greater range.
4. Use of the extinguisher requires little or no cleanup.
5. The extinguisher occupies less space: 5.7×10^{-2} m^3 (2 ft^3) at most, as compared to 0.28 m^3 (10 ft^3) for sand.
6. The extinguisher is lighter in weight: 22.7 kg (50 lb) or less, as compared to 0.45 tonne (½ ton) for sand.

Suitable substitutes for the required sand are a 9.5-liter (2½-gal) foam extinguisher, a 6.8-kg (15-lb) carbon dioxide extinguisher, and a 4.5-kg (10-lb) dry chemical extinguisher.

It is difficult to smother combustible-metal fires with sand because the extremely hot temperature of the fire extracts oxygen from the sand. Any water in the sand will increase the intensity of the fire or cause such reactions as steam explosions; it would be very unusual to find completely dry sand aboard ship. Sand may be used to dam up running molten metal, but an approved dry powder should be used to extinguish the fire.

SAWDUST

Sawdust impregnated with soda is sometimes used to smother small oil fires. Like sand, it must be scooped up and placed at close range. The deficiencies of sawdust as an extinguishing agent are similar to those of sand. The alternative, a class B extinguisher, is more effective than sawdust for the reasons given in the discussion of sand. Although sawdust is considerably lighter than sand, the amount required—0.28 m^3 (10 ft^3)—weighs more than an extinguisher.

STEAM

Steam was one of the earliest extinguishing agents used aboard vessels. It was readily available for firefighting once the ship's boilers were lighted. Steam extinguishes fire by smothering, e.g., by forcing air away from the fire and by diluting the air around the fire. As long as the steam blanket is maintained, it will prevent reignition. However, there are several disadvantages in using steam, especially in comparison with other extinguishing agents.

Obviously steam is applied to the fire in the vapor state. Thus, most of its heat-absorbing ability is lost before it is applied, and it does little cooling. (Water fog, on the other hand, cools as it turns to steam.) Additionally, steam condenses when the supply is shut off. Its volume decreases substantially, and combustible vapors and air rush in to displace it. There is, then, a very good chance that the fire will reflash if it has not been completely extinguished and cooled. The temperature of the steam itself is high enough to ignite many liquid fuels. Finally, steam is hazardous to personnel; the heat it carries can inflict severe burns.

If a ship is equipped with a steam smothering system, the crew must, of course, use that system in case of fire. Some older ships may have fixed steam smothering systems for the protection of cargo; however, since January 1, 1962, such installations have not been allowed on new ships.

The use of the steam soot blowers on the boiler to extinguish uptake-type fires is extremely hazardous and should not be attempted. The high velocity and high temperature of the steam reacts with the powdered soot (mostly carbon) to form an explosive mixture. Several explosions have occurred as a result of this practice.

SHIPBOARD USE OF EXTINGUISHING AGENTS

Some extinguishing agents such as carbon dioxide and foam are required in ships. Some, like dry chemical and the halogenated agents, are approved for shipboard use in the Coast Guard *Equipment Lists* (CG190). Any extinguishing agent that is neither required nor specifically listed may be installed or carried if it is approved by the Coast Guard commandant. Sprinkler systems of any type must be approved by the Coast Guard commandant before they are installed. In short, the Coast Guard has the final say on fire extinguishing systems and appliances.

A fire extinguishing *system* consists of a supply of the extinguishing agent, an actuation device (manual or automatic), and the piping, valves and nozzles necessary to apply the agent. A fire *extinguisher* is a self-contained unit, portable or semiportable, consisting of a supply of the extinguishing agent, an expellant gas (if the apparatus is not pressurized) and a hose with a nozzle.

Officers and crewmen should familiarize themselves with the extinguishing agents, systems and appliances carried aboard their ships. They should be aware of the relative advantages of the various agents and the limitations on their use. For instance, they should remember that when a space is totally flooded with carbon dioxide, patience is not only a virtue but a necessity. They should also be aware of the toxic or suffocating properties of some agents used in total flooding systems, and the need for proper and sufficient ventilation before anyone enters a space that has been totally flooded with an extinguishing agent. A test of the atmosphere in such a space should be made with an oxygen indicator (*see* Chapter 16) to determine if the space is safe to enter. Officers and crewmen should realize that, although a space may look and smell clean or clear, it may contain sufficient carbon monoxide to render them helpless or insufficient oxygen to support life. Breathing apparatus and lifelines should be used when entering a compartment or tank whose contents are unknown. (*See* Chapters 8 and 9 for a discussion of the appliances, systems and equipment that use extinguishing agents.)

BIBLIOGRAPHY

National Fire Codes. NFPA. Boston, Mass, 1977

Fire Service Training. Ohio Trade & Industrial Ed. Services. Columbus, Ohio, 1977

Basic Fireman's Training Course. Md. Fire & Rescue Inst., Univ. of Md. College Park, Md, 1969

Fire Fighting—Principles & Practices. William Clark, New York, NY, 1974

Fire Control. California State Dept. of Ed. Sacramento, Ca, 1964

Fire Protection Guide on Hazardous Materials. NFPA. Boston, Mass, 1973

Fire Chiefs Handbook, 4th Ed. Dunn-Donnelly Pub. Company. New York, NY, 1977

Portable & Semiportable Fire Extinguishers

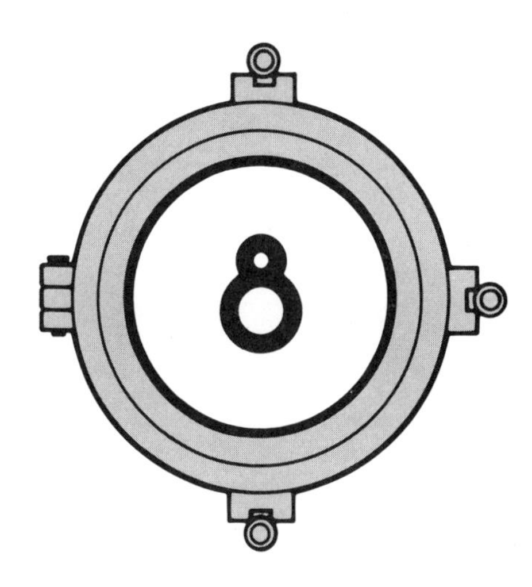

Since some fires start small, a fire discovered early and attacked quickly, usually can be extinguished easily. Portable fire extinguishers are used for a fast attack that will knock down the flames; semiportable extinguishing systems bring larger amounts of extinguishing agent to the fire. Both can be effective when used properly.

PORTABLE FIRE EXTINGUISHERS

Portable extinguishers can be carried to the fire area for a fast attack. However, they contain a limited supply of extinguishing agent. The agent is quickly expelled from the extinguisher; in most cases, continuous application can be sustained for only a minute or less. For this reason, it is extremely important to back up the extinguisher with a hoseline. Then, if the extinguisher does not have the capacity to put the fire out completely, the hoseline can be used to finish the job. However, a crewman who is using an extinguisher cannot advance a hoseline at the same time. Thus, the alarm must be sounded as soon as fire is discovered, to alert the ship's personnel to the situation.

There is a right way to use a portable fire extinguisher, and there are many wrong ways. Crewmen who have had little training with these appliances waste extinguishing agent through improper application. At the same time, untrained personnel tend to overestimate their extinguishing ability. Periodic training sessions, including practice with the types of extinguishers carried on board, are the best insurance against inefficient use of this equipment. Extinguishers that are due to be discharged and inspected may be used in these training sessions.

Classes of Fire Extinguishers

Every portable extinguisher is classified in two ways, with one or more letters and with a numeral. The letter or letters indicate the classes of fires on which the extinguisher may be used. These letters correspond exactly to the four classes of fires (*see* Chapter 5). Thus, for example, class A extinguishers may be used only on class A fires—those involving common combustible materials. Class AB extinguishers may be used on fires involving wood or diesel oil or both.

The numeral indicates either the relative efficiency of the extinguisher or its size. This does not mean the size of fire on which to use the extinguisher; rather, the numeral indicates how well the extinguisher will fight a fire of its class.

The National Fire Protection Association (NFPA) rates extinguisher efficiency with Arabic numerals. The Underwriters Laboratories (UL) tests extinguishers on controlled fires to determine their NFPA ratings. A rating such as 2A or 4A on an extinguisher would be an NFPA rating. (A 4A rating will extinguish twice as much class A fire as a 2A rating; a 20B rating will extinguish four times as much class B fire as a 5B rating.)

The Coast Guard uses Roman numerals to indicate the sizes of portable extinguishers. The numeral I indicates the smallest size, and V the largest. Thus, a BIII Coast Guard rating indicates a medium-sized extinguisher suitable for fires involving flammable liquids and gases. The Coast Guard ratings of the different types of extinguishers are given in Table 8.1.

Test and Inspection

The Coast Guard requires owners, masters or persons in charge to have portable and semiportable fire extinguishers and fixed fire-extinguishing systems tested and inspected "at least once in every twelve months." More detailed maintenance

Table 8.1. United States Coast Guard Extinguisher Classification.

		Water	*Foam*	*Dioxide*	*Chemical*
Type	*Size*	*Gallons*	*Gallons*	*Pounds*	*Pounds*
A	II	2½	2½	—	—
B	I	—	1¼	4	2
B	II	—	2½	15	10
B	III	—	12	35	20
B	IV	—	20	50	30
B	V	—	40	100	50
C	I	—	—	4	2
C	II	—	—	15	10

instructions are included with some of the discussions that follow.

Upon the completion of required tests, a tag should be placed on each extinguisher, showing the date and the person who completed the tests. Many ship owners have contracts with commercial fire protection companies to have their fire equipment tested and maintained. This does not relieve the master or officer-in-charge of fire protection from the responsibility of carrying out Coast Guard regulations regarding the maintenance of firefighting equipment.

General Safety Rules for Portable Extinguishers

1. When you discover a fire, call out your discovery, sound the fire alarm and summon help.
2. Never pass the fire to get to an extinguisher. A dead-end passageway could trap you.
3. If you must enter a room or compartment to combat the fire, keep an escape path open. Never let the fire get between you and the door.
4. If you enter a room or compartment and your attack with a portable extinguisher fails, get out immediately. Close the door to confine the fire and prepare to fight the fire while waiting for previously summoned help. Your knowledge of the situation will aid those responding.

WATER EXTINGUISHERS

Extinguishers that use water or a water solution as the extinguishing agent are suitable only for class A fires. There are five types of water extinguishers, but only two are currently produced. In 1969, the manufacture of the inverting types of extinguishers (the soda-acid, foam and cartridge-operated) was discontinued. However, since a large number of inverting extinguishers are still in use, they will be discussed along with the two currently produced types: the stored-pressure and pump-tank water extinguishers.

Soda-Acid Extinguisher

The soda-acid extinguisher (Fig. 8.1) comes only in a 9.5 liter (2½-gal) size that carries an NFPA rating of 2A. It weighs about 13.6 kg (30 lb) when charged, has a reach of from 10.7 m to 12.2 m (30–40 ft) and expends itself in about 55 seconds. The shell of the extinguisher is filled with a solution of 0.7 kg (1½ lb) of sodium bicarbonate in 9.5 liters (2½ gal) of water. The screw-on cap contains a cage that holds a 0.23-kg (8-oz) bottle, half filled with sulfuric acid, in an upright position. A loose stopper in the top of the acid bottle prevents acid from splashing out before the extinguisher is to be used.

Operation. The extinguisher is carried to the fire by means of the top handle. At the fire, the extinguisher is inverted, the acid mixes with the solium bicarbonate solution forming carbon dioxide gas and the pressure of the CO_2 propels the water out through the nozzle. The stream must be directed at the seat of the fire and moved back and forth to hit as much of the fire as possible. The nozzle should be aimed at the fire until the entire content of the extinguisher is discharged (Fig. 8.2). Remember, water is available for less than a minute!

The extinguishing agent, sodium bicarbonate solution mixed with acid, is more corrosive than plain water. The operator should avoid getting the agent on his skin or in his eyes, as the acid could cause burning. Moreover, soda-acid extinguishers must be carefully maintained. When the extinguisher is inverted, a pressure of 896 kilopascals (130 psi) or more is generated. If the container is corroded or otherwise damaged, this pressure could be sufficient to burst the container.

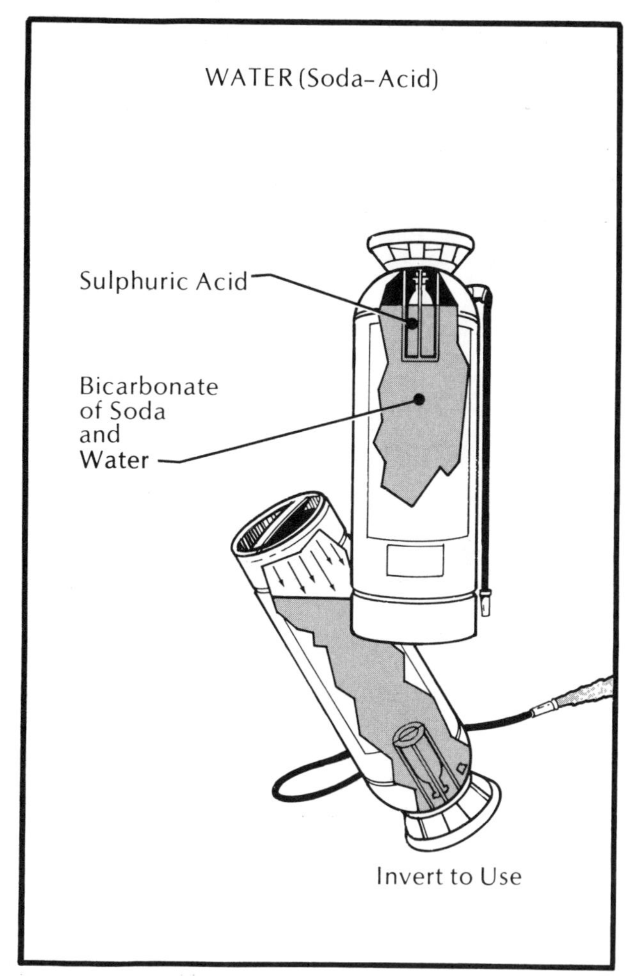

Figure 8.1. Soda-acid fire extinguisher used for class A fires only.

Figure 8.2. The soda-acid extinguisher is inverted, and the nozzle is swept back and forth across the base of the fire.

Maintenance. Soda-acid extinguishers should be stowed at temperatures above 0°C (32°F) to keep the water from freezing. They should be recharged annually and immediately after each use. During the annual recharging, all parts must be carefully inspected and washed in fresh water. The hose and nozzle should be checked for deterioration and clogging. The proper chemicals must be used for recharging. The sodium bicarbonate solution should be prepared outside the extinguisher, preferably with lukewarm fresh water. The recharging date and the signature of the person who supervised the recharging must be placed on a tag attached to the extinguisher.

Several times a year, each extinguisher should be inspected for damage and to ensure that the extinguisher is full and the nozzle is not clogged.

Cartridge-Operated Water Extinguisher

The cartridge-operated water extinguisher (Fig. 8.3) is similar in size and operation to the soda-acid extinguisher. The most common size is 9.5 liters (2½ gal), with an NFPA rating of 2A. It has a range of from 10.7-12.2 m (30–40 ft). The container is filled with water or an antifreeze

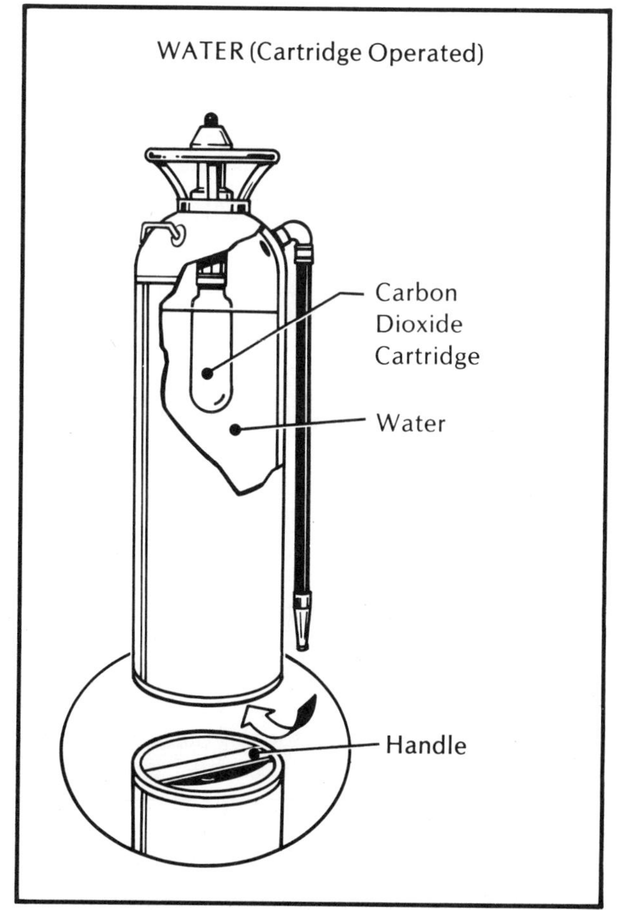

Figure 8.3. Cartridge-operated water extinguisher used for class A fires only.

solution. The screw-on cap contains a small cylinder of CO_2; when the cylinder is punctured, the gas provides the pressure to propel the extinguishing agent.

Operation. The extinguisher is carried to the fire, then inverted and bumped against the deck. This ruptures the CO_2 cylinder and expels the water. The stream should be directed at the seat of the fire. The nozzle should be moved back and forth, to quench as much of the burning material as possible in the short time available (Fig. 8.4). The discharge time is less than one minute. The entire content of the extinguisher must be discharged, since the flow cannot be shut off.

As with the soda-acid extinguisher, the container is not subjected to pressure until it is put to use. Thus, any weakness in the container may not become apparent until the container fails.

Maintenance. The pressure cartridge should be inspected and weighed annually. It should be replaced if it is punctured or if its weight is 14 gm (¼ oz) less than the indicated weight. The hose and nozzle should be inspected to ensure that they are clear. The container should be inspected for damage. Water should be added, if necessary, to bring the contents up to the fill mark.

Pin-Type Cartridge-Operated Extinguisher. A newer version of the cartridge-operated water extinguisher need not be inverted for use. Instead, a pin is pulled out of the cartridge, with the extinguisher upright. A lever is squeezed to discharge the extinguishing agent (water or antifreeze solution).

The cartridge is fitted with a pressure gauge. The gauge should be checked periodically to ensure that the cartridge pressure is within its operating range. Otherwise, maintenance is similar to that for the inverting-type cartridge extinguisher.

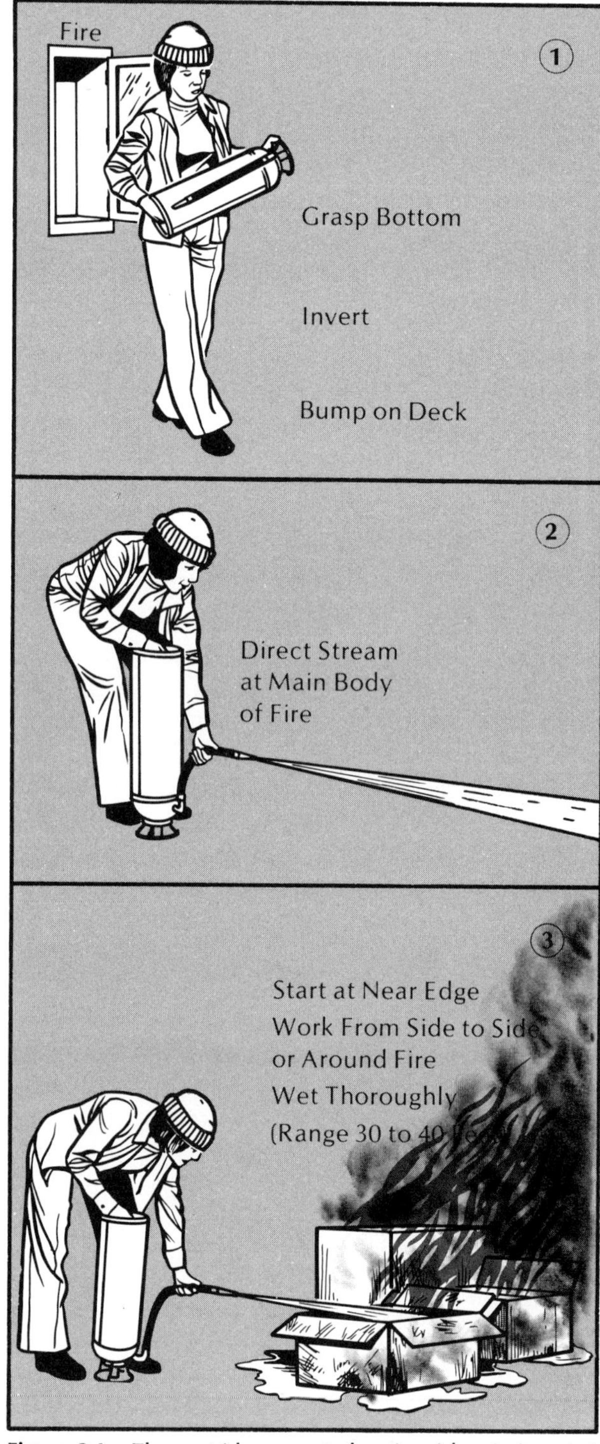

Figure 8.4. The cartridge-operated extinguisher is inverted and bumped on the deck. The stream is moved across the base of the fire.

Stored-Pressure Water Extinguisher

The stored-pressure water extinguisher (Fig. 8.5) is the most commonly used portable firefighting appliance. The 9.5-liter (2½-gal) size has an NFPA rating of 2A. It weighs about 13.6 kg (30 lb) and has a horizontal range of 10.7–12.2 m (35–40 ft). In continuous operation, it will expend its water in about 55 seconds. However, it may be used intermittently, to extend its operational time.

The container is filled with water or an antifreeze solution, to within about 15 cm (6 in.) of the top. (Most extinguishers have a fill mark stamped on the container.) The screw-on cap holds a lever-operated discharge valve, a pressure gauge and an automobile tire-type valve. The extinguisher is pressurized through the air valve, with either air or an inert gas such as nitrogen. The normal charging pressure is about 690 kilopascals (100 psi). The gauge (Fig. 8.5) allows the pressure within the extinguisher to be checked at any time. Most gauges are color coded to indicate normal and abnormal pressures.

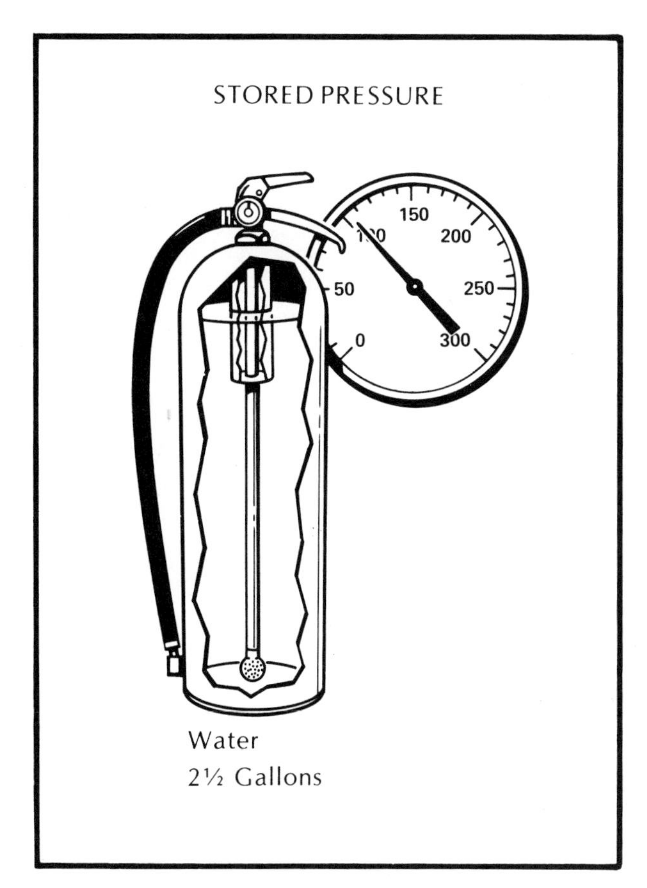

Figure 8.5. Stored-pressure water extinguisher used for class A fires only.

Operation. The extinguisher is carried to the fire, and the ring pin or other safety device is removed. The operator aims the nozzle with one hand and squeezes the discharge lever with the other hand. The stream should be directed at the seat of the fire. It should be moved back and forth to ensure complete coverage of the burning material. Short bursts can be used to conserve the limited supply of water.

As the flames are knocked down, the operator may move closer to the fire. Then, by placing the tip of one finger over the nozzle, the operator can obtain a spray pattern that will cover a wider area.

Maintenance. Inspect gauge for loss of pressure, check for leaks and check condition of hose and overall condition of the tank.

Pump-Tank Extinguisher

Pump tanks are the simplest type of water extinguishers. They come in sizes from 9.5–19 liters (2½ to 5 gal), with NFPA ratings of 2A to 4A. Ships do not carry pump-tank extinguishers and are not required to do so. However, in port, shoreside personnel often bring them aboard for fire protection during burning and welding operations.

The tank is filled with water or an antifreeze solution. A hand-operated piston pump is built into the extinguisher and is used to discharge water onto the fire. The pump is usually double acting, which means it discharges on both the up and down strokes. The range of the stream depends on the strength and ability of the operator, but is usually 9.2–12.2 m (30–40 ft). The 9.5 liter (2½-gal) size holds enough water for about 55 seconds of continuous operation.

Operation. The tank is carried to the fire and placed on the deck. It is steadied by placing one foot on the extension bracket. The operator uses one hand to operate the pump, and the other to direct the stream at the seat of the fire. If the operator must change position, he has to stop pumping and carry the pump tank to the new location.

If more than one person is at the fire scene, and there are no other extinguishers available, a joint operation is more effective. One person should direct the stream, and one should do the pumping. Other available personnel should bring additional water to keep the tank full. They should also relieve the pumpman periodically.

Maintenance. The pump-tank hose should be inspected periodically to ensure that it is clear. The efficiency of the pump should be checked by throwing a stream. The tank should be checked for corrosion and refilled to the fill mark.

Foam Extinguisher

Foam extinguishers (Fig. 8.6) are similar in appearance to those discussed previously, but they have a greater extinguishing capability. The most common size is 9.5 liters (2½ gal), with an NFPA rating of 2A:4B. This indicates that the extinguisher may be used on both class A and class B fires. It has a range of about 9.2–12.2 m (30–40 ft) and a discharge duration of slightly less than a minute.

The extinguisher is charged by filling it with two solutions that are kept separated (in the extinguisher) until it is to be used. These solutions are commonly called the A and B solutions; their designations have nothing to do with fire classifications.

Operation. The foam extinguisher is carried to the fire right side up and then inverted. This mixes the two solutions, producing a liquid foam and CO_2 gas. The CO_2 acts as the propellant and fills the foam bubbles. The liquid foam expands to about 8 times its original volume; this means the 9.5 liter (2½-gal) extinguisher will produce 68–76 liters (18–20 gal) foam.

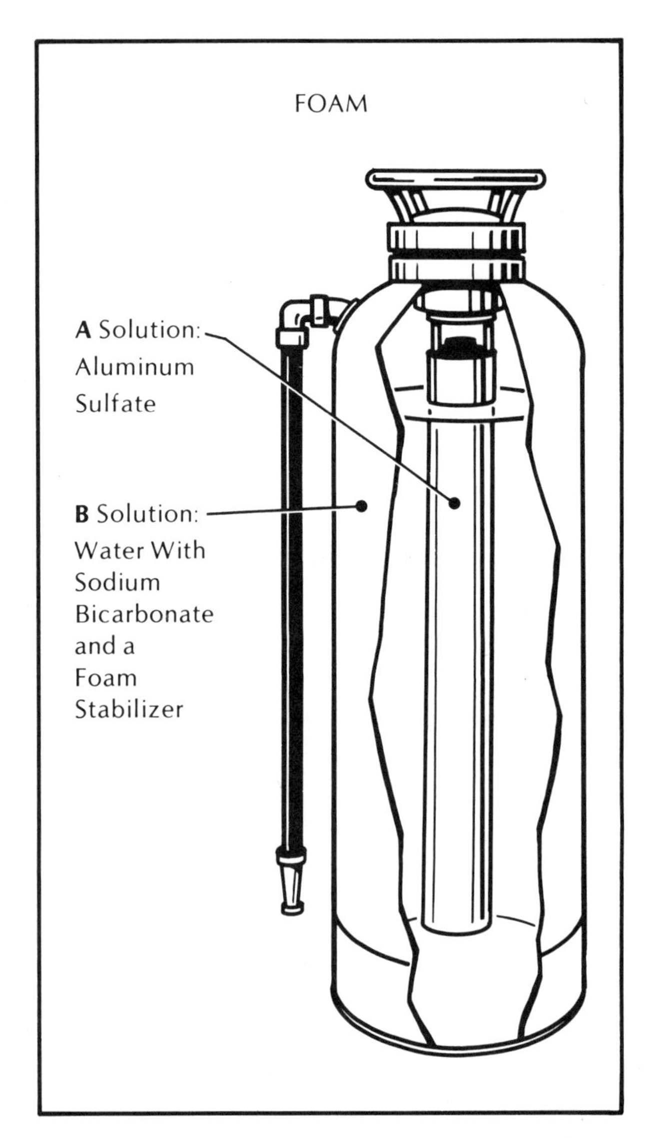

Figure 8.6. Cutaway of foam extinguisher used for class A and class B fires showing

The foam should be applied gently on burning liquids (Fig. 8.7). This can be done by directing the stream in front of the fire, to bounce the foam onto the fire. The stream also may be directed against the back wall of a tank or a structural member to allow the foam to run down and flow over the fire. Chemical foam is stiff and flows slowly. For this reason, the stream must be directed to the fire from several angles, for complete coverage of the burning materials.

For fires involving ordinary combustible materials, the foam may be applied in the same way, as a blanket. Or, the force of the stream may be used to get the foam into the seat of the fire.

Foam extinguishers are subject to freezing and cannot be stowed in low temperatures below 4.4°C (40°F). Once activated, these extinguishers will expel their entire foam content; it should all be directed onto the fire. As with other pressurized extinguishers, the containers are subject to rupture when their contents are mixed, and are a possible cause of injury to the operator. Maintenance consists mainly of annual discharging, inspection, cleaning and recharging.

CARBON DIOXIDE (CO_2) EXTINGUISHER

Carbon dioxide extinguishers are used primarily on class B and class C fires. The most common sizes of portable extinguishers contain from 2.3–9.1 kg (5–20 lb) of CO_2 not including the weight of the relatively heavy shell. The CO_2 is mostly in the liquid state, at a pressure of 5.86×10^6 pascals (850 psi) at 21°C (70°F). The 2.3 kg (5-lb) size is rated 5B:C, and the 6.8 kg (15-lb) size has a rating of 10B:C. The range varies between 1.8–2.4 m (3–8 ft), and the duration between 8–30 seconds depending on the size.

Operation

The extinguisher is carried to the fire in an upright position. (The short range of the CO_2 extinguisher means the operator must get fairly close to the fire.) The extinguisher is placed on the deck, and the locking pin is removed. The discharge is controlled either by opening a valve or by squeezing two handles together. Figure 8.8 shows the two-handle type.

The operator must grasp the hose handle, and not the discharge horn. The CO_2 expands and cools very quickly as it leaves the extinguisher. The horn gets cold enough to frost over and cause severe frostbite. When a CO_2 extinguisher is used in a confined space, the operator should guard against suffocation by wearing breathing apparatus.

Class B Fires. The horn should be aimed first at the base of the fire nearest the operator. The discharge should be moved slowly back and forth across the fire. At the same time, the operator should move forward slowly. The result should be a "sweeping" of the flames off the burning surface, with some carbon dioxide "snow" left on the surface.

Whenever possible, a fire on a weather deck should be attacked from the windward side. This will allow the wind to blow the heat away from the operator and to carry the CO_2 to the fire. Generally, CO_2 extinguishers do not perform well in a wind. The blanket of CO_2 gas does not remain on the fire long enough to permit the fuel to cool down.

Class C Fires. The discharge should be aimed at the source of a fire that involves electrical equipment. The equipment should be de-ener-

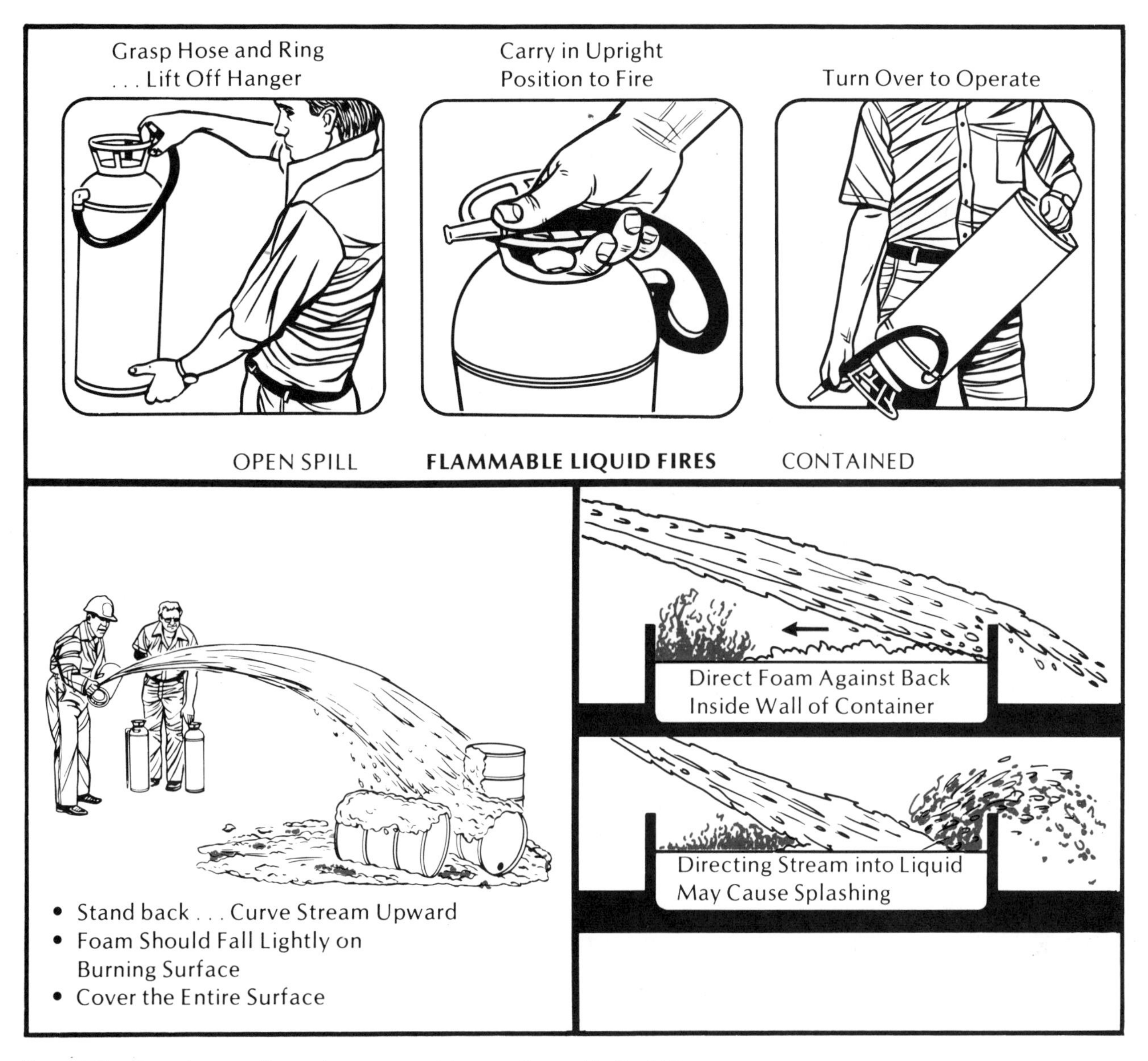

Figure 8.7. Steps in operating a foam extinguisher on flammable liquid fires.

gized as soon as possible to eliminate the chance of shock and the source of ignition.

Maintenance

CO_2 extinguishers need not be protected against freezing. However, they should be stowed at temperatures below 54°C (130°F) to keep their internal pressure at a safe level. (At about 57°C (135°F), the safety valves built into CO_2 extinguishers are activated at approximately 18.62×10^6 pascals (2700 psi), to release excess pressure.)

Several times each year, CO_2 extinguishers should be examined for damage and to ensure that they are not empty. At annual inspection, these extinguishers should be weighed. Any extinguisher that has lost more than 10% of its CO_2 weight should be recharged, *by the manufacturer*. A CO_2 extinguisher should also be recharged after each use, even if it was only partly discharged.

DRY CHEMICAL EXTINGUISHER

Dry chemical extinguishers are available in several sizes, with any of five different extinguishing agents. All have at least a BC rating; the monoammonium phosphate extinguisher carries an ABC rating. The different dry chemical agents have different extinguishing capabilities. If sodium bicarbonate is arbitrarily given an extinguishing capability of 1, then the relative capabilities of the other dry chemical agents are as follows.

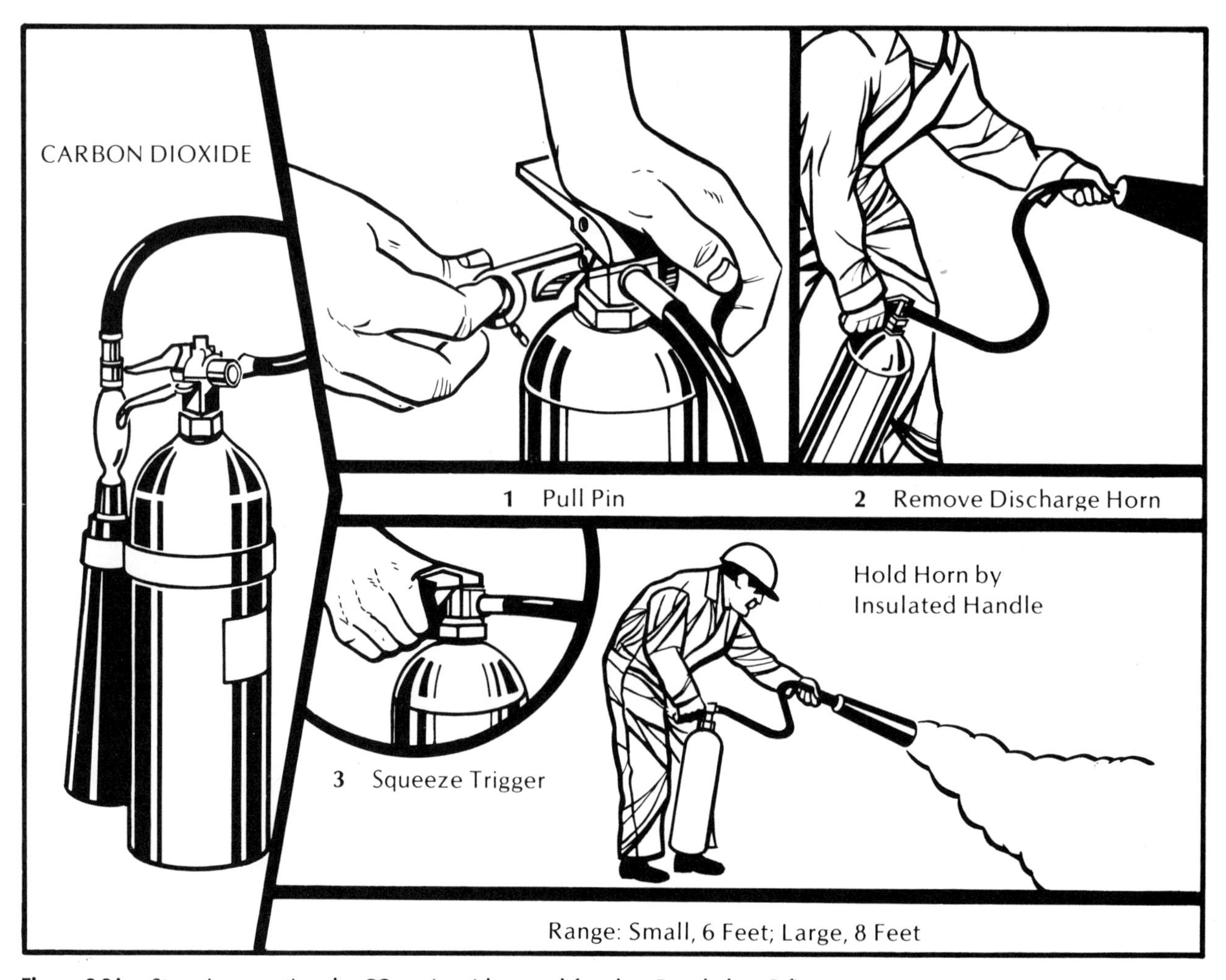

Figure 8.8A. Steps in operating the CO_2 extinguisher used for class B and class C fires.

Figure 8.8B. Extinguishing galley range fire with CO_2 portable extinguisher.

Table 8.2. Relative Extinguishing Capabilities of Dry Chemical Agents.*

Monoammonium phosphate (ABC)	1.5
Potassium chloride (BC)	1.8
Potassium bicarbonate (BC)	2.0
Urea potassium bicarbonate (BC)	2.5

*When sodium bicarbonate is classified as 1.

Thus, for example, potassium bicarbonate is twice as effective as sodium bicarbonate.

Cartridge-Operated Dry Chemical Extinguisher

Portable cartridge-operated, dry chemical extinguishers range in size from 0.91–13.6 kg (2–30 lb); semiportable models contain up to 22.7 kg (50 lb) of agent. An extinguisher may be filled with any of the five agents, and its rating will be based on the particular agent used. A small cylinder of inert gas is used as the propellant (Fig. 8.9). Cartridge-operated, dry chemical extinguishers have a range of from 3–9.1 m (10–30 ft). Units under 4.5 kg (10 lb) have a discharge duration of 8–10 seconds, while the larger extinguishers provide up to 30 seconds of discharge time.

Operation. The extinguisher is carried and used upright. The ring pin is removed, and the puncturing lever is depressed. This releases the propellant gas, which forces the extinguishing agent up to the nozzle. The flow of dry chemical is controlled with the squeeze-grip On–Off nozzle at the end of the hose. The discharge is directed at the seat of the fire, starting at the near edge. The stream should be moved from side to side with

Figure 8.9. Operating the cartridge-operated dry-chemical extinguisher.

rapid motions, to sweep the fire off the fuel. On a weather deck, the fire should be approached from the windward side if possible.

The initial discharge should not be directed onto the burning material from close range (0.91–2.4 m (3–8 ft)). The velocity of the stream may scatter the burning material.

If the propellant gas cylinder is punctured but the extinguisher is not put into use or is only partially discharged, the remaining gas may leak away in a few hours. Thus, the extinguisher must be recharged after each use or activation. However, the agent may be applied in short bursts by opening and closing the nozzle with the squeeze grips.

Dry chemical extinguishers extinguish class B fires by chain breaking, with little or no cooling. Thus, a reflash is possible if the surrounding surfaces are hot. Additional dry chemical or another appropriate extinguishing agent must be available as backup, until all sources of ignition are eliminated.

Dry chemical extinguishing agents may be used along with water. Some dry chemical extinguishers are filled with an extinguishing agent that is compatible with foam.

Stored-Pressure Dry Chemical Extinguishers

Stored-pressure dry chemical extinguishers are available in the same sizes as cartridge-operated types. They have the same ranges and durations of discharge and are used in the same way. The only difference is that the propellant gas is mixed in with the dry chemical in the stored-pressure type. And the extinguisher is controlled with a squeeze-grip trigger on the top of the container (Fig. 8.10). A pressure gauge indicates the condition of the charge.

Class A Extinguishment Using ABC Dry Chemical

Only one dry chemical extinguishing agent, monoammonium phosphate (ABC, multipurpose) is

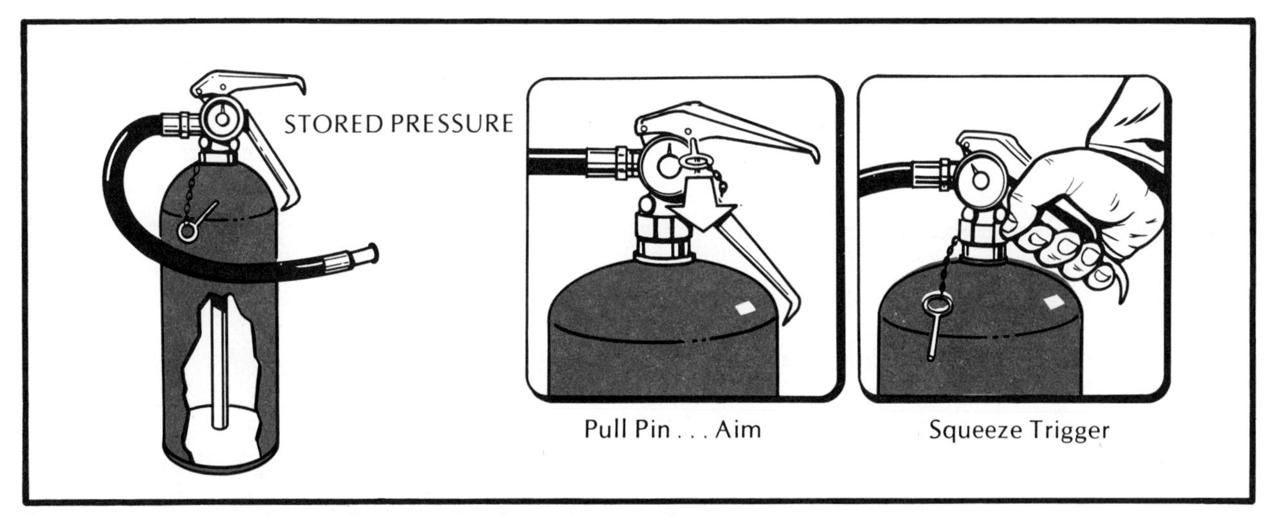

Figure 8.10. Operating the stored-pressure dry-chemical extinguisher.

approved for use on class A fires. This agent extinguishes fire by chain breaking, as do the other dry chemical agents. In addition, it softens and clings to the surfaces of burning materials to form a coating that deprives the fuel of air. As with the other agents, this dry chemical should be directed at the seat of the fire and swept from side to side to knock down the flames. However, once the fire has been knocked down, the operator should move close to the burning debris. Then all fuel surfaces should be thoroughly coated with the chemical agent. For this, the operator should use short, intermittent bursts.

Class B Extinguishment Using BC or ABC Dry Chemical

A flammable-liquid fire should be attacked as noted above. The agent should first be directed at the edge nearest the operator. The nozzle should be moved from side to side, with a wrist action, to cover the width of the fire. The operator should maintain the maximum continuous discharge rate, remembering that the extinguisher has a range of from 1.6–13.1 m (10–30 ft). The operator must be very cautious, moving in toward the fire very slowly. A liquid fire can flank an operator who moves in too rapidly, or reflash around an operator who is too close.

When all the flames are out, the operator should back away from the fire very slowly, being alert for possible reignition. Many types of flammable liquids will reflash under normal atmospheric conditions. A hot spot that the operator has missed could cause reignition, resulting in a duplicate of the original fire. For this reason it is always a good idea to have reserve units or additional extinguishers ready to move in to assist in the extinguishment of the fire.

In using dry chemical to approach a pressure gas fire to close off the fuel flow, the heat shield afforded by the dry chemical should be maintained constantly in front of the operator's face. When extinguishment is desired, the dry chemical stream must be directed into the gas stream nearly parallel to the gas flow, with approximately 10 degrees to the right or left side entry. If dry chemical is directed into the stream at too great an angle, the dry chemical will not penetrate the full stream and will be unsuccessful. Conversely, if the chemical stream does not have a slight right or left angle, the dry chemical will be deflected by the gas pipe.

Once the gas is shut off or extinguished, the operator should slowly back away. Remember, never extinguish a pressure gas fire unless by so doing the fuel flow can be controlled.

Class C Extinguishment Using BC or ABC Dry Chemical

When electrical equipment is involved in a fire, the stream of dry chemical should be aimed at the source of the flames. In small spaces, the smoke and the cloud produced by the dry chemical will limit visibility (and may cause choking). The chance of electrical shock is also increased. For this reason, electrical equipment that may be involved in a fire should be deenergized at its source, if at all possible, before any attempt is made to extinguish the fire.

Dry chemical extinguishing agents leave a coating on materials involved in the fire. This coating must be cleaned off electrical equipment before it can be used. Monoammonium phosphate (ABC) dry chemical leaves a sticky coating that is very difficult to remove. This coating also penetrates and sticks to circuit breakers and switching components, making them virtually useless. For that reason, ABC dry chemical is not recommended for use on electrical fires.

Dry chemical agents that contain sodium can contaminate or corrode brass and copper electrical fittings. Electric fires are best extinguished with carbon dioxide or Halon, which are "clean" extinguishing agents.

Maintenance of Dry Chemical Extinguishers

Dry chemicals and their propellants are unaffected by temperature extremes and may be stowed anywhere aboard ship. They do not deteriorate or evaporate, so periodic recharging is not necessary. However, the cartridges in cartridge-operated extinguishers should be inspected and weighed every 6 months. Cartridges that are punctured or weigh 14.2 gm (¼ oz) less than the indicated weight should be replaced. At the same time, the hose and nozzle should be checked to ensure that they are not clogged.

Stored-pressure extinguishers manufactured after June 1, 1965, have pressure gauges that indicate whether the internal pressure is within the operating range. These should be checked visually at intervals. (The gauge is located on the bottom of some extinguishers.)

DRY POWDER EXTINGUISHER

Dry powder (*not* dry chemical) is the only extinguishing agent that may be used on combustible-metal (class D) fires. The one available class D extinguisher is a 13.6 kg (30-lb) cartridge-operated model that looks very much like the cartridge-operated dry chemical extinguisher (Fig.

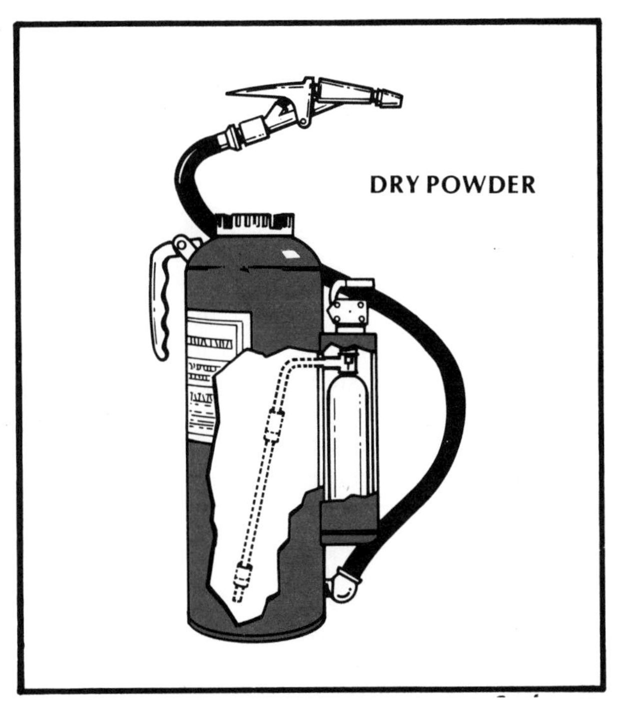

Figure 8.11. Dry-powder extinguisher used for class D fires only.

Figure 8.12. The dry-powder extinguisher is operated in an upright position. The agent must be applied gently, to maintain a crust on the burning metal.

8.11). One difference is that the class D extinguisher has a range of only 1.8–2.4 m (6–8 ft). The extinguishing agent is sodium chloride, which forms a crust on the burning metal.

Operation

The nozzle is removed from its retainer, and the puncture lever is pressed. This allows the propellant gas (CO_2 or nitrogen) to activate the extinguisher. The operator then aims the nozzle and squeezes the grips to apply the powder to the surface of the burning metal.

The operator should begin the application of dry powder from the maximum range 1.8–2.4 m (6–8 ft). The squeeze grips may be adjusted for the desired rate of flow, to build a thick layer of powder over the entire involved area. The operator must be careful not to break the crust that forms when the powder hits the fire (Fig. 8.12).

A large amount of dry powder is sometimes needed to extinguish a very small amount of burning metal. A brown discoloration indicates a hot spot, where the layer of dry powder is too thin. Additional agent should be applied to the discolored areas. When the fire involves small metal chips, the agent should be applied as gently as possible, so the force of the discharge does not scatter burning chips.

Class D dry powder also comes in a container, for application with a scoop or shovel (Fig. 8.13). Here, too, the agent should be applied very gently. A thick layer of powder should be built up, and the operator should be careful not to break the crust that forms.

Figure 8.13. Application of dry powder with a shovel or scoop.

HALON EXTINGUISHERS

Halon 1211 extinguishers are available in several sizes; Halon 1301 in only one size. All the Halon extinguishers look alike (Fig. 8.14) and are used in the same way.

Bromochlorodifluoromethane (Halon 1211) extinguishers contain from 0.91–5.44 kg (2–12 lb) of extinguishing agent and carry NFPA ratings of 5B:C to 10B:C. Their horizontal range is from 2.7–4.6 m (9–15 ft), and they discharge their contents in 9–15 seconds. Halon 1211 is more effective than CO_2, leaves no residue and is virtually noncorrosive. However, it can be toxic, and its vapors should not be inhaled.

Bromotrifluoromethane (Halon 1301) is available only in a 1.1-kg (2½-lb) portable extinguisher, with an NFPA rating of 5B:C. Its horizontal range is from 1.2–1.8 m (4–6 ft), and its discharge time is 8–10 seconds.

Both extinguishing agents are pressurized in a light weight steel or aluminum alloy shell. The cap contains the discharge control valve and discharge nozzle.

Operation

The extinguisher is carried to the fire, and the locking pin is removed. The discharge is controlled by squeezing the control valve–carrying handle. The Halon should be directed at the seat of a class B fire, and applied with a slow, side-to-side sweeping motion. It should be directed at the source of an electrical fire (Fig. 8.15).

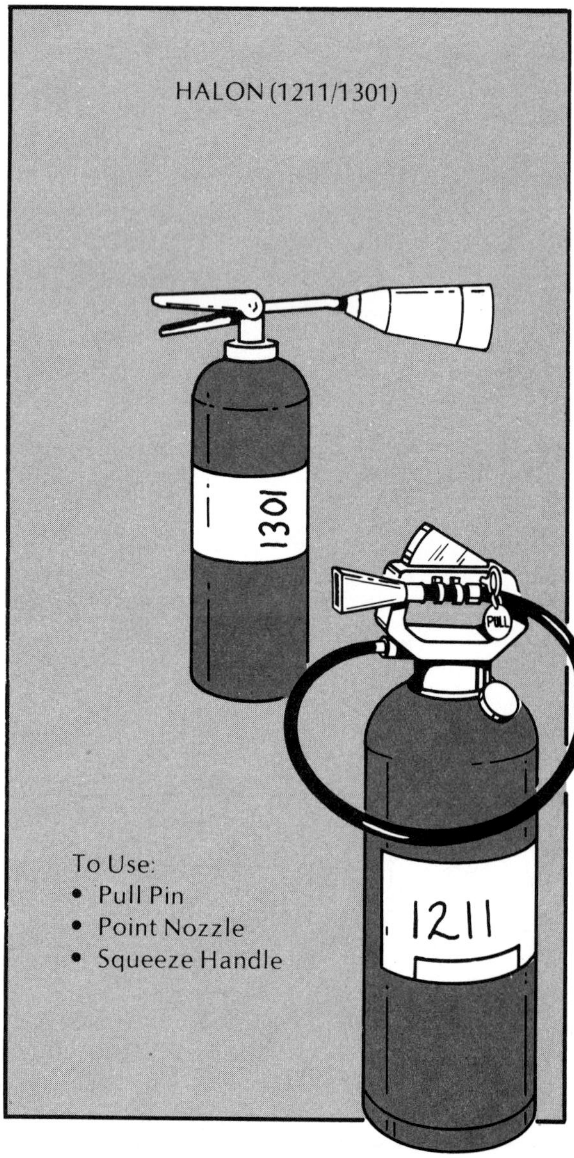

Figure 8.14. Halon extinguisher for use on class B and class C fires.

Figure 8.15. Operation of Halon extinguishers.

SEMIPORTABLE FIRE EXTINGUISHERS

A semiportable fire extinguisher (or extinguishing system) is one from which a hose can be run out to the fire. The other components of the system are fixed in place, usually because they are too heavy to move easily.

Semiportable systems provide a way of getting a sizable amount of extinguishing agent to a fire rapidly. This allows the operator to make a sustained attack. However, a semiportable system is also a semifixed system. One disadvantage is that the protected area is limited by the length of hose connected into the system. Extinguishing agents are applied to the fire in the manner described for portable extinguishers. The main differences are a slight increase in the effective range (from nozzle to fire) and the increased amount of extinguishing agent available.

Semiportable systems are usually set up to protect the same areas as fixed systems. Where possible, a fire is first attacked with the semiportable system. If this attack controls or extinguishes the fire, then the large fixed system need not be activated. Semiportable systems may also be used as primary extinguishing systems. Since they are initial attack systems, it is essential that they be backed up with additional firefighting equipment.

CARBON DIOXIDE HOSE-REEL SYSTEM

The carbon dioxide hose-reel system is employed in engine rooms and in spaces containing electrical equipment. The system consists of one or two CO_2 cylinders, a 1.27-cm (½-in.) diameter hose that is 15.2–22.9 m (50–75 ft.) in length, a reel for the hose and a CO_2 discharge horn with an On–Off control valve.

Operation

The system is activated manually, by use of a control lever mounted on top of the CO_2 cylinder. If the system uses two cylinders, only one lever need be operated; pressure from the first cylinder opens the valve of the second, so both will be used.

Here is the general procedure to be followed:

1. Activate the cylinders by removing the locking pin and operating the lever of the control cylinder (Fig. 8.16).
2. Run out the CO_2 hoseline to the fire area.
3. Open the horn valve by pushing the handle forward.
4. Direct the CO_2 at the near edge of the fire (Fig. 8.17). For a bulkhead fire, direct the CO_2 at the bottom and work up. As the flames recede, follow them slowly with CO_2.

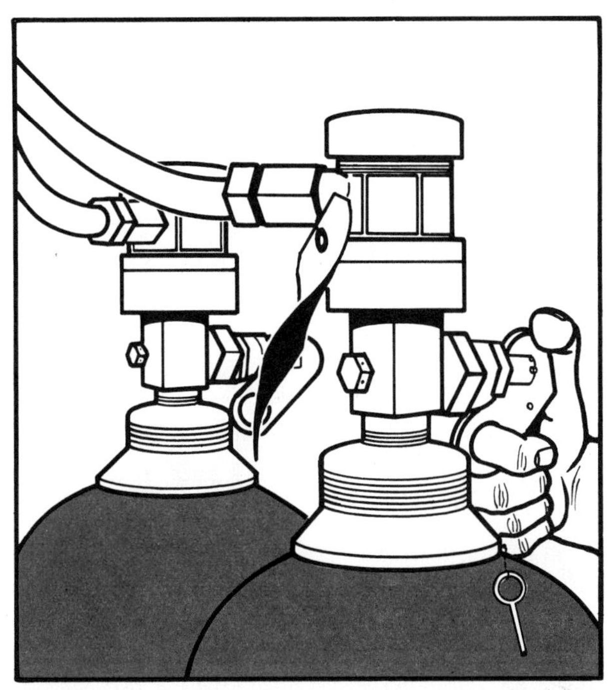

Figure 8.16. The activating lever on one CO_2 cylinder is operated after the locking pin is removed.

5. Continue the discharge until any smoldering materials are covered with snow.
6. To temporarily stop the flow of CO_2, close the horn valve by pulling the handle back.

To attack a bilge fire, it may be necessary to remove some floor plates to gain access to the fire. As few plates as possible should be removed. If it is necessary to drop the horn to attack an inaccessible fire, the horn valve may be locked in the open position. This is done by pushing the lock against the notch in the handle, with the handle forward (Fig. 8.18).

In an attack on an electrical fire, the gas should be directed into all openings in the involved equipment. After the fire is extinguished, the CO_2 discharge should be continued until the burned surfaces are covered with "snow." Although carbon dioxide is a poorer conductor than air, the equipment should be deenergized as soon as possible to prevent the fire from spreading.

DRY CHEMICAL HOSE SYSTEM

The dry chemical semiportable system consists of a storage tank containing the agent, pressurized cylinders containing nitrogen gas, a rubber hose and a nozzle with a control valve. The nitrogen is used as the propellant for the dry chemical. Systems employing sodium bicarbonate, potassium bicarbonate or potassium chloride can be located where class B and class C fires may be

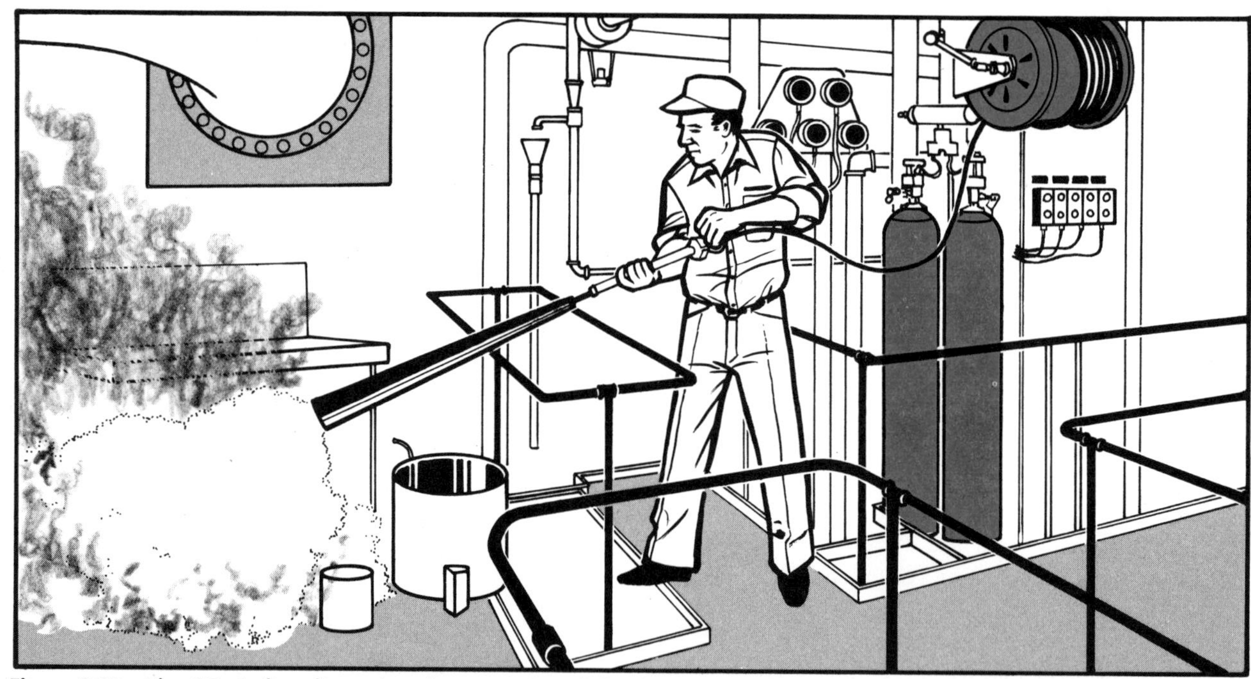

Figure 8.17. The CO_2 is first directed at the near edge of the fire. It is then directed at the receding flames until the fire is knocked down.

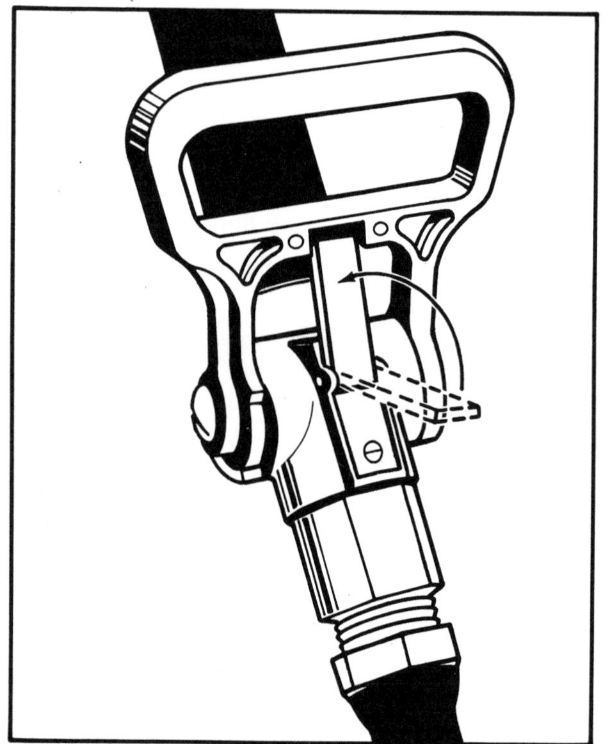

Figure 8.18. The CO_2 horn valve may be locked in the open position as shown.

expected. Systems employing monoammonium phosphate may be approved for any location on the ship. However, they should not be used to protect electrical gear, because of the sticky residue this dry chemical leaves.

Operation

The system is activated by pulling the release mechanism in the head of the nitrogen cylinder; some systems may also be activated by a remote-cable device (Fig. 8.19). When the system is activated, the nitrogen flows into the dry chemical tank. It fluidizes the chemical and propels it into the hoseline, up to the nozzle. The hose is run out to the fire attack position, and the nozzle is opened to commence the attack. The full length of the hose should be run out to ensure an even, continuous flow of extinguishing agent.

HALON HOSE-REEL SYSTEM

The semiportable Halon system is very similar to the carbon dioxide system and is employed to combat class B and class C fires. Most semiportable systems use Halon 1301. The system consists of one or two pressurized cylinders containing the extinguishing agent, a hoseline and a nozzle with an On–Off control valve.

Operation

The system is activated by operating a release mechanism at the top of the cylinder, similar to the CO_2 release device. If two cylinders are used, they are both opened when the pilot cylinder is activated. When the agent is released, it travels through the hose up to the nozzle. The hose is then run out to the fire, and the agent is applied as required.

PORTABLE FOAM SYSTEMS

A foam system using an in-line proportioner or a mechanical foam nozzle with pickup tube can be

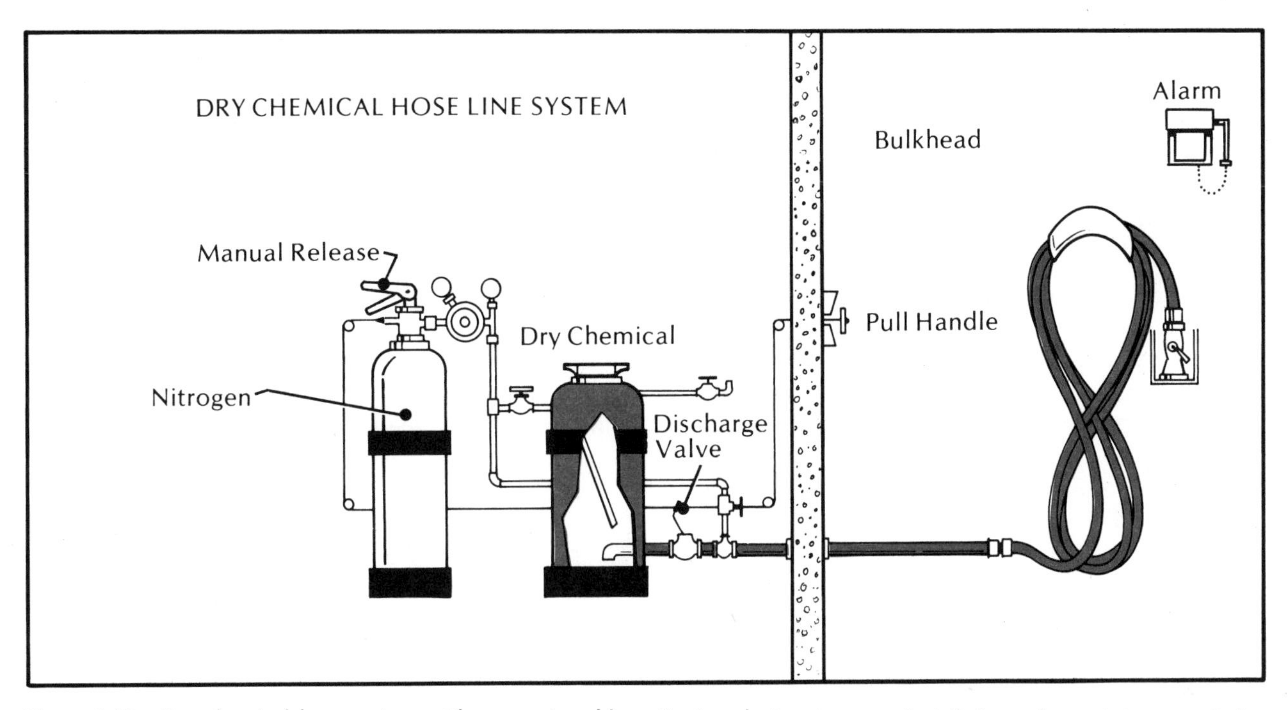

Figure 8.19. Dry chemical hose system with a remote-cable activating device. In some installations, the entire system is located in one space.

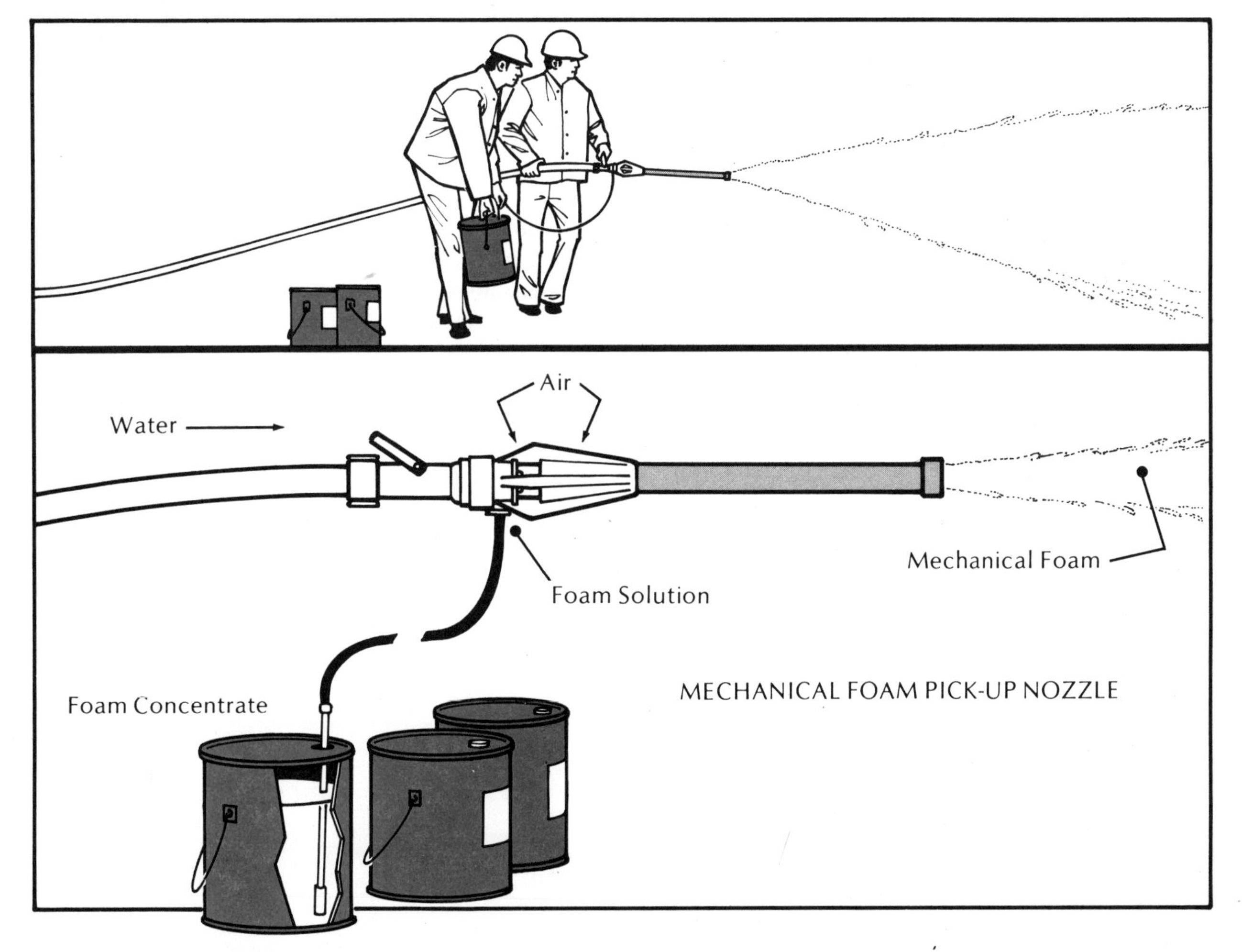

Figure 8.20. How foam is produced by the mechanical foam nozzle with pickup. The nozzle itself is a very efficient foam producer. However, the nozzleman's movements are restricted by the need to keep the pickup tube in the foam-concentrate container.

carried to various parts of the ship. The foam system is used with the ship's firemain system. It is an efficient method for producing foam, but it requires more manpower than semiportable systems employing other extinguishing agents.

Mechanical Foam Nozzle with Pickup Tube

In use, the mechanical foam nozzle with pickup tube is attached to a standard hoseline from the firemain system. It draws air in through an aspirating cage in its hoseline end. At the same time it introduces mechanical foam concentrate into the water stream through a pickup tube (Fig. 8.20). When the air and foam solution mix, foam is discharged from the nozzle.

One type of nozzle consists of a 533-mm (21-in.) length of flexible-metal or asbestos-composition hose, 51 mm (2 in.) in diameter, with a solid metal outlet. A suction chamber and an air port in the hoseline end form the aspirating cage. The pickup tube is a short piece of 16-mm (5/8-in.) metal pipe with a short piece of rubber hose on one end. It is used to draw up the contents of a 19-liter (5-gal) container of foam concentrate. The pickup tube operates on suction created in the suction chamber of the nozzle.

Operation. The mechanical foam nozzle is screwed onto the fire hose, and the pickup tube is screwed into the side port in the base of the nozzle. The metal pipe at the end of the pickup tube is inserted into the foam-concentrate container. When water pressure is applied to the hose, foam concentrate is drawn up to the nozzle, where it mixes with the air and water. The resulting foam is applied in the usual manner. As Figure 8.20 shows, the mobility of the foam nozzle is improved if one firefighter operates the nozzle while another follows with the concentrate container.

Portable In-Line Proportioner

The portable in-line foam proportioner, or *eductor,* allows the nozzlemen more freedom of movement than the nozzle with pickup tube. The proportioner may be installed anywhere in the hoseline, between the firemain and the foam nozzle. It, too, feeds mechanical foam to the nozzle, but it may be placed at a convenient distance from the heat of the fire (Fig. 8.21).

The in-line proportioner is a light weight venturi device. It uses the water-stream pressure to draw foam concentrate from a 19-liter (5-gal) container, through a pickup tube, and into the water stream, in the proper proportion.

Operation. The male end of the hoseline feeding water to the proportioner is screwed into the female (gauge) end of the proportioner. The pickup tube is screwed into the top center of the proportioner. The female end of the firefighting hose is screwed into the male end of the proportioner. The male end of the firefighting hose is advanced to the fire, and the mechanical foam nozzle is screwed on. The firefighting hose should not be longer than 45.7 m (150 ft) from proportioner to nozzle.

The hoselines are now charged from the firemain. When the water pressure on the inlet side of the proportioner reaches 448 kilopascals (65 psi) as shown on the gauge, the suction end of

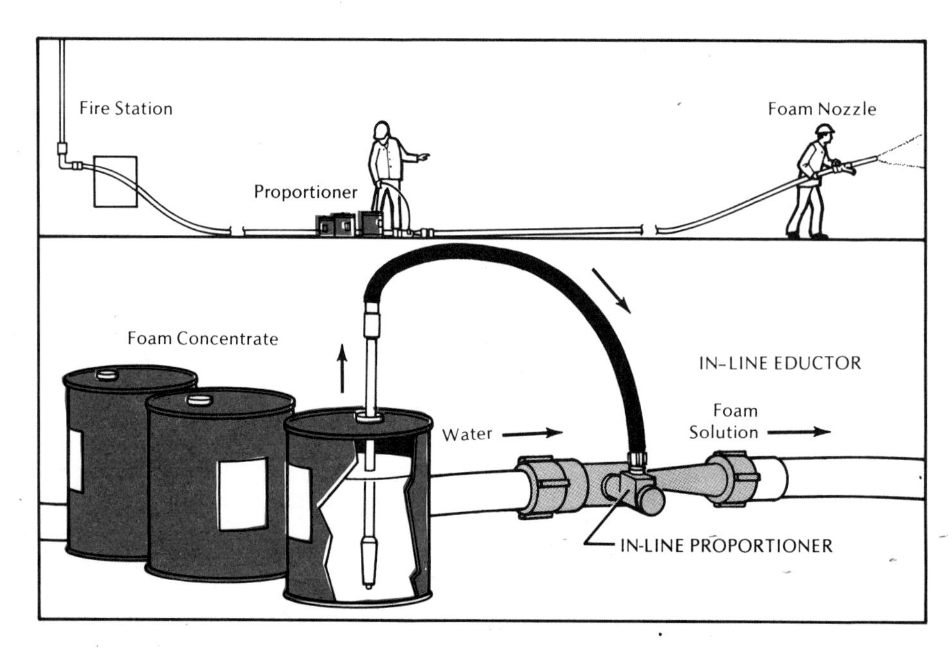

Figure 8.21. Production of mechanical foam by an in-line proportioner, or eductor. The proportioner can be placed in the hoseline away from the fire, so the nozzleman has more mobility.

the pickup tube is inserted into the foam-concentrate container. Mechanical foam is discharged from the nozzle and directed onto the fire.

Foam Supply

Whether a mechanical nozzle with pickup tube or a proportioner is being used, extra containers of foam concentrate should be opened and kept on hand. This will allow the pickup tube to be quickly transferred from an empty container to a full one, so there is no break in the foam discharge. The 19-liter (5-gal) containers of foam concentrate are used up quickly. At 345 kilopascals (50 psi) water pressure, one 19-liter (5-gal) container lasts approximately 2½ minutes; at 689 kilopascals (100 psi) water pressure, a container lasts about 1½ minutes.

BIBLIOGRAPHY

Instructor's Guide, Fire Service Extinguishers. G. Post, J. Smith, R. J. Brady visual teaching aid program, unit 3, 1978, Bowie, Md.

IFSTA Fire Extinguisher, Oklahoma State University, Stillwater, Okla. NFPA 14th Ed.

Fixed Fire-Extinguishing Systems

The primary objective of firefighting is quick control and extinguishment. This objective can be achieved only if the extinguishing agent is brought to the fire rapidly and in sufficient quantity. Fixed fire-extinguishing systems can do exactly that. Additionally, some of these systems are also capable of applying the agent directly to the fire—without the assistance of crew members.

The Coast Guard regulates the installation of fixed firefighting systems aboard U.S. vessels. In its Navigation and Vessel Inspection Circular 6-72, *Guide to Fixed Fire-fighting Equipment aboard Merchant Vessels,* the Coast Guard states:

> Fire extinguishing systems should be reliable and capable of being placed into service in simple, logical steps. The more sophisticated the system is, the more essential that the equipment be properly designed and installed. It is not possible to anticipate all demands which might be placed upon fire extinguishing systems in event of emergency. However, potential casualties and uses should be considered, especially as related to the isolation of equipment, controls, and required power from possible disruption by a casualty. Fire protection systems should, in most cases, serve no function other than fire fighting. Improper design or installation can lead to a false sense of security, and can be as dangerous as no installation.
>
> Fixed extinguishing equipment is not a substitute for required structural fire protection. These two aspects have distinct primary functions in U.S. practice. Structural fire protection protects passengers, crew, and essential equipment from the effects of fire long enough to permit escape to a safe location. Firefighting equipment, on the other hand, is for protection of the vessel. Requirements for structural fire protection vary with the class of vessel and are the most detailed for passenger vessels. However, approved fixed extinguishing systems are generally independent of the vessel's class.

Coast Guard regulations ensure that shipboard firefighting systems are properly designed and installed to provide reliable protection for the ship and its crew. The material presented in this chapter reflects Coast Guard thinking on the subject. However, the chapter is not intended to be a digest of their regulations, but rather a discussion of fire protection systems aboard U.S. flag vessels. Title 46 CFR contains specific fire protection requirements for ships, based on age, tonnage, service and other factors.

DESIGN AND INSTALLATION OF FIXED SYSTEMS

Fire extinguishing systems are designed and installed in a ship as a part of its original construction. The ship's master, officers and crew members rarely have any influence on the type of firefighting systems employed. Marine and fire protection engineers generally make these decisions to conform with Coast Guard Regulations. The crew's duties require them to learn how the systems operate, perform proper maintenance and conduct required tests and inspections.

Many factors must be analyzed when a fixed extinguishing system (or combination of systems) is installed on a ship. A study is made of the overall ship design and the potential fire hazards. Among the things considered are

- Fire classes (A, B, C and D) of potential hazards
- Extinguishing agent to be employed
- Locations of specific hazards
- Explosion potential
- Exposures
- Effects on the ship's stability
- Methods of fire detection
- Protection of the crew.

Generally the fire class of the hazard determines the type of system to be installed. Aboard ship, there are exceptions to this rule. For example, ship's spaces are normally protected against class A fires by systems using water as the extinguishing agent. The cooling effect of water makes it the logical extinguishing agent for fires involving ordinary combustible materials. Yet water is not used to protect cargo holds, even though they contain class A materials. Water would not be effective in a hold because the closely packed cargo would probably prevent the water from reaching the seat of the fire, and the use of excessive amounts of water could cause the ship to lose stability and develop a list. The system used for cargo hold protection is the carbon dioxide flooding system, which controls and extinguishes fire by smothering.

Shipboard extinguishing systems are thus designed to be consistent with both the potential fire hazards and the uses of the protected space. Generally,

- Water is used in fixed systems protecting areas containing ordinary combustibles, such as public spaces and passageways.
- Foam or dry chemical is used in fixed systems protecting spaces subject to class B (flammable liquid) fires. Flammable gas fires are not extinguished by fixed systems; controlled burning is recommended until the fuel source can be shut off.
- Carbon dioxide, Halon or a suitable dry chemical is used in fixed systems that protect against class C (electrical) fires.
- No fixed extinguishing system is approved for use against class D fires involving combustible metals.

The design of shipboard extinguishing systems is also consistent with the ship's purpose: a cargo vessel, tanker, grain ship, LNG carrier or passenger vessel. Each system is tailored to the configuration of the ship and the spaces to be protected. Because of the many variables that must be considered, the selection, design and installation of an automatic fire extinguishing system is a highly complex process. It requires expertise in a variety of technical disciplines. The unauthorized alteration or jury-rigging of a fire extinguishing system could render it incapable of controlling a fire.

United States ships use seven major types of fixed fire-extinguishing systems:

1. Fire-main systems
2. Automatic and manual sprinkler systems
3. Spray systems
4. Foam systems
5. Carbon dioxide systems
6. Halon 1301
7. Dry chemical systems

The first four systems use liquid extinguishing agents; the next two use gaseous agents; the last uses solid agents. Each of these systems is discussed in the sections that follow.

FIRE-MAIN SYSTEMS

The fire-main system is the ship's first line of defense against fire. It is required no matter what other fire extinguishing systems are installed. Every crew member can expect to be assigned to a station requiring knowledge of the use and operation of the ship's fire main.

The fire-main system supplies water to all areas of the vessel. Fortunately, the supply of water at sea is limitless. The movement of water to the fire location is restricted only by the system itself, the effect of the water on the stability of the ship and the capacity of the supply pumps.

The fire-main system is composed of the fire pumps, piping (main and branch lines), control valves, hose and nozzles. The fire pumps provide the power to move water through the piping to fire stations located throughout the vessel. The valves, hose and nozzles are used to control the firefighting water and direct it onto the fire.

Hydrants and Piping

The piping directs firefighting water from the pumps to hydrants at the fire stations. The piping must be large enough in diameter to distribute the maximum required discharge from two fire pumps operating simultaneously. The water pressure in the system must be approximately 345 kilopascals (50 psi) at the two hydrants that are highest or furthest (whichever results in the greatest pressure drop) for cargo and miscellaneous vessels, and 517 kilopascals (75 psi) for tank vessels. This requirement ensures that the piping is large enough in diameter so that the pressure produced at the pump is not lost through friction in the piping.

The piping system consists of a large main pipe and smaller branch lines leading off to the hydrants. The main pipe is usually 102–152 mm (4–6 in.) in diameter. The branch lines are generally 37–64 mm (1½–2½ in.) in diameter. Although the smaller branch lines reduce the flow of water, they make it easier to maintain the required pressure at the fire stations. Branch lines may not be connected into the fire-main system for any purpose other than firefighting and deck washing.

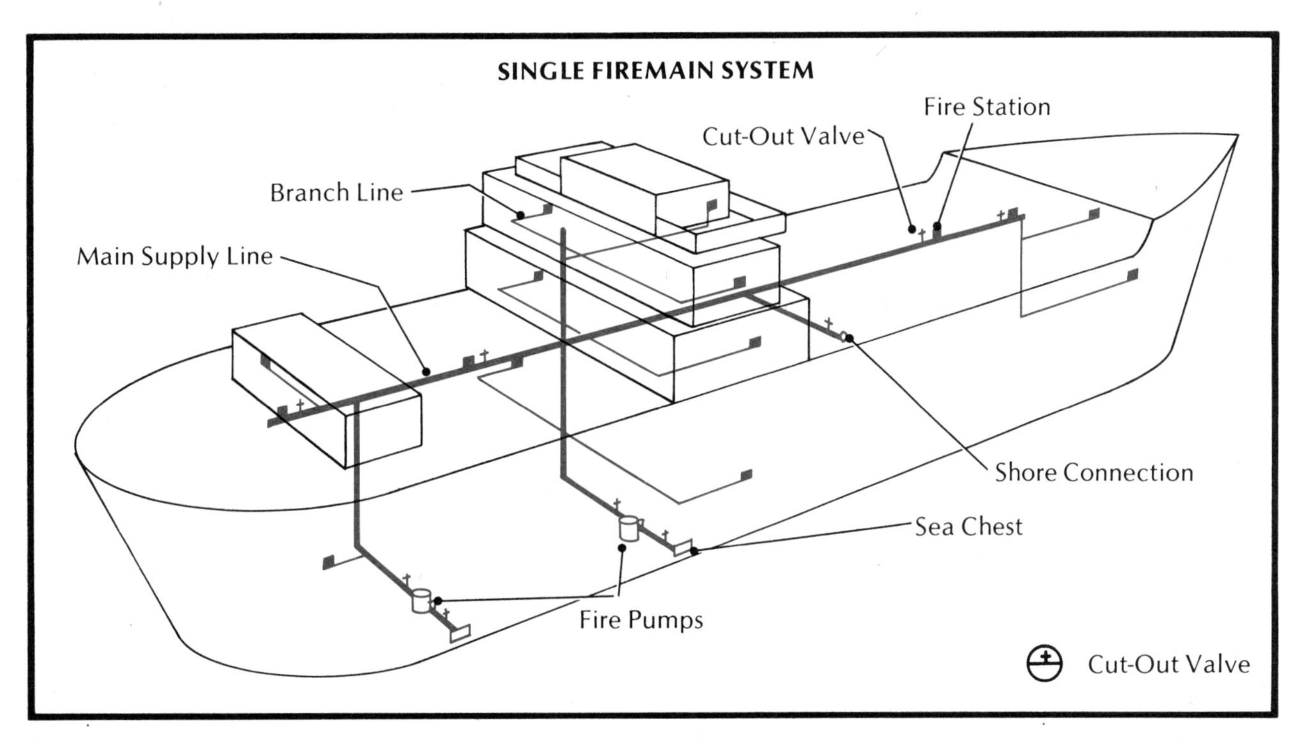

Figure 9.1. Typical single main system.

All sections of the fire-main system on weather decks must be protected against freezing. For this purpose, they may be fitted with isolation and drain valves, so that water in the piping may be drained in cold weather.

There are two basic main-pipe layouts, the single main and the horizontal loop.

Single Main System. Single main systems make use of one main pipe running fore and aft, usually at the main deck level. Vertical and horizontal branch lines extend the piping system through the ship (Fig. 9.1). On tankers, the main pipe usually runs the length of the vessel, down its centerline. On grain vessels of the Great Lakes

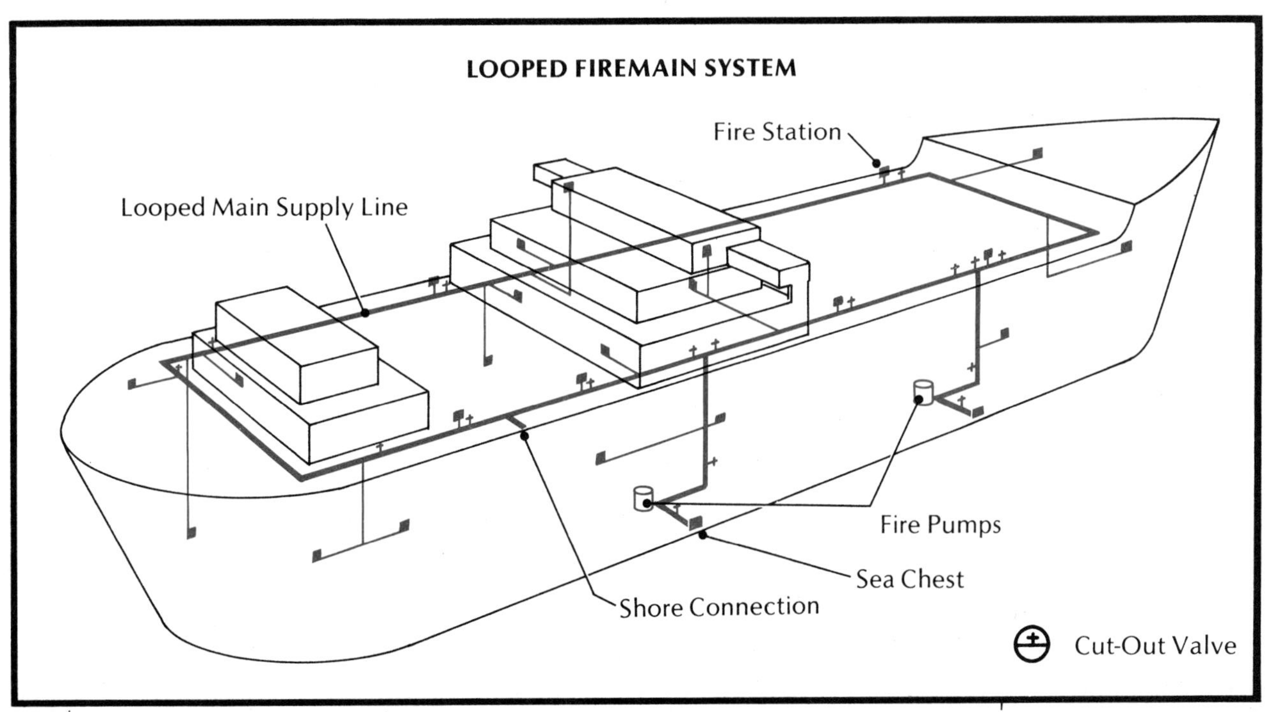

Figure 9.2. Typical horizontal loop fire-main system.

or similar configuration, the main pipe is located along the port or starboard edge of the vessel's main deck. A disadvantage of the single main system is its inability to provide water beyond a point where a serious break has occurred.

Horizontal Loop System. The horizontal loop system consists of two parallel main pipes, connected together at their furthest points fore and aft to form a complete loop (Fig. 9.2). Branch lines extend the system to the fire stations. In the horizontal loop system, a ruptured section of the main pipe may be isolated. The system can then be used to deliver water to all other parts of the system. Isolation valves are sometimes located on the main pipeline, forward of each hydrant location; they are used to control the water flow when a break occurs in the system. Some single loop systems have isolation valves for the fore and aft decks only.

Shore Connections. At least one shore connection to the fire-main system is required on each side of the vessel. Each shore connection must be in an accessible location and must be fitted with cutoff and check valves.

A vessel on an international voyage must have at least one portable international shore connection (Fig. 9.3) available to either side of the vessel. International shore connections may be connected to matching fittings that are available at most ports and terminals throughout the world. They enable the crew to take advantage of the pumping capability of the shore installation or fire department at any port. The required international shore connections are permanently mounted on some vessels.

Figure 9.3. Portable international shore connection for ship's fire-main system.

Fire Pumps

Fire pumps are the only means for moving water through the fire-main system when the vessel is at sea. The number of pumps required and their capacity, location and power sources are governed by Coast Guard regulations. In brief, the minimum requirements are as follows.

Number and Location. Two independently powered fire pumps are required on a tank ship 76 m (250 ft) or more in overall length or 1016 metric tons (1000 gross tons) and over on an international voyage. A cargo or miscellaneous vessel of 1016 metric tons (1000 gross tons) and over also requires at least two independently powered fire pumps, regardless of its length. All passenger vessels up to 4064 metric tons (4000 gross tons) on international voyages must have at least two fire pumps, and those over 4064 metric tons (4000 gross tons) must have three pumps, regardless of their lengths.

On vessels that are required to carry two fire pumps, the pumps must be located in separate spaces. The fire pumps, sea suction and power supply must be arranged so that fire in one space will not remove all the pumps from operation and leave the vessel unprotected. Any alternative to the two separate pump locations requires the approval of the Commandant, U.S. Coast Guard, and the installation of a CO_2 flooding system to protect at least one fire pump and its power source. This arrangement is permitted only in the most unusual circumstances. Generally, it is used only on special ships, where safety would not be improved by separating the pumps.

The crew is not usually responsible for ensuring that their ship carries the correct number of pumps, located and powered as required. Ships are designed, built and, when necessary, refitted to comply with Coast Guard regulations. However, the crew is directly responsible for keeping the pumps in good condition. In particular, engineering personnel are usually charged with the responsibility for maintaining and testing the ship's fire pumps, to ensure their reliability during an emergency.

Water Flow. Each fire pump must be capable of delivering at least two powerful streams of water from the hydrant outlets having the greatest pressure drop, at a pitot-tube pressure of 517 kilopascals (75 psi) for tanker vessels and 345 kilopascals (50 psi) for passenger and cargo vessels. These requirements match those for the fire-main piping. They must be met when the system is tested.

Safety. Every fire pump must be equipped with a relief valve on its discharge side. The relief valve should be set at 862 kilopascals (125 psi), or at 172 kilopascals (25 psi) above the pressure necessary to provide the required fire streams, whichever is greater. A pressure gauge must also be located on the discharge side of the pump.

Other firefighting systems (e.g., a sprinkler system) may be connected to the fire-main pumps. However, the capacity of the fire pumps must then be increased sufficiently so they can supply both the fire-main system and the other system with the proper water pressure *at the same time.*

A pump that is connected to an oil line should not be used as a fire pump. The pump could possibly pump a flammable liquid, rather than water, through the fire main. In addition, the pump could contaminate the system with oil, which would clog applicator and nozzle openings. Oil in the water would also rot the linings in hoses.

Use of Fire Pumps for Other Purposes. Fire pumps may be used for purposes other than supplying water to the fire main. However, one of the required pumps must be kept available for use on the fire main at all times. This does not mean that one pump must be reserved exclusively for the fire main. The reliability of fire pumps is probably improved if they are used occasionally for other services and are then properly maintained. When control valves for other services are located at a manifold adjacent to the pump, any other service may be readily secured if the valve to the fire main must be opened.

The fire-main piping is a tempting source of "free" water, already installed in most spaces. However, improper or careless use quickly reduces the reliability of the system. If the fire-main pumps are used for purposes other than firefighting, deck washing and tank cleaning (on tankers), connections should be made only to a discharge manifold near the pump. The fire-main piping may be used only when specific exceptions are granted. The fire main may be used for deck washing and tank cleaning simply because, in those cases, someone knows that the system is in use, and crewmen are usually in attendance.

Connections to the fire main for low-water-demand services in the forward portion of the vessel (such as anchor washing, forepeak eductor or chain-locker eductor) have frequently been allowed. In such cases, each fire pump must be capable of meeting its water flow requirements with the other service connection open. This ensures that the effectiveness of the fire-main system is maintained if the other service connection is accidentally opened.

Fire Stations

The purpose of the fire-main system is to deliver water to the fire stations that are located throughout the ship. A fire station consists basically of a fire hydrant (water outlet) with valve and associated hose and nozzles. It is important that all required firefighting equipment be kept in its proper place.

Fire stations and hoses must be highly visible and easily put into service. However, this visibility makes them vulnerable to misuse and damage. One type of misuse is washing down decks and bulkheads. The valve or piping can be damaged if it is used as a cleat for typing a line. Hydrant valve stems can also be damaged during the handling of cargo or the moving of heavy materials through passageways. Hydrants located on weather decks may become corroded or encrusted with salt, causing their valves to freeze in position and become inoperable. When a section of hose or a nozzle is borrowed for use at another station, at least one fire station is made useless as a firefighting unit. Couplings and hoses that are abused (as by being dropped or dragged on the deck) may fail in use or, at least, become difficult to connect.

Crew members should make every effort to protect all parts of the fire-main system and avoid unauthorized use of the system. Weekly visual inspection of fire stations should be a standard procedure to ensure that all required equipment is in its proper place.

Different hydrants should be opened during succeeding weekly fire drills to ensure that water is allowed to flow from each hydrant at least once every 2 months. This will reduce crusting and rust. Whenever the opportunity arises, the fire-main system should be flushed out with fresh water to destroy any marine growth in the lines.

Fire Station Locations. Fire stations are located to ensure that the water streams from at least two hydrants will overlap. U.S. Coast Guard regulations specify hydrant locations as follows:

- Fire hydrants shall be sufficient in number and so located that any part of the vessel, other than main machinery spaces, is accessible to persons on board while the vessel is being navigated, and all cargo holds may be reached with at least two streams of water from separate outlets. At least one of these streams shall be from a single length of hose.
- In main machinery spaces, all portions of such spaces shall be capable of being reached by at least two streams of water, each of

which shall be from a single length of hose and from separate outlets.

- Fire stations should be numbered sequentially as required by regulations on all vessels to be certified by the Coast Guard.

If deck cargo is carried, it must be stowed so that it does not block access to the fire station hydrant.

Hydrants. The fire station hydrant (Fig. 9.4) has three major components: *1)* a control valve; *2)* the hose connection, either 38.1 or 63.5 mm (1½ or 2½ in.) with appropriate threads; and *3)* a hose rack.

Regulations require that:

- Each fire hydrant outlet must have a valve that allows the hose to be removed while there is pressure in the fire-main system.
- The fire hydrant outlet may be in any position, from horizontal to pointing vertically downward. It should be positioned to minimize the kinking of the fire hose.
- The threads on the fire hydrant outlet must be National Standard fire-hose coupling threads. These standard threads allow all approved hose to be attached to the hydrant.
- On interior hydrants in certain passenger vessels, a 63.5-mm (2½-in.) outlet may be wyed for two 38.1-mm (1½-in.) hoses with a wye gate connection.
- A rack must be provided for the proper stowage of the fire hose. The hose must be stowed in the open or where it is readily visible.

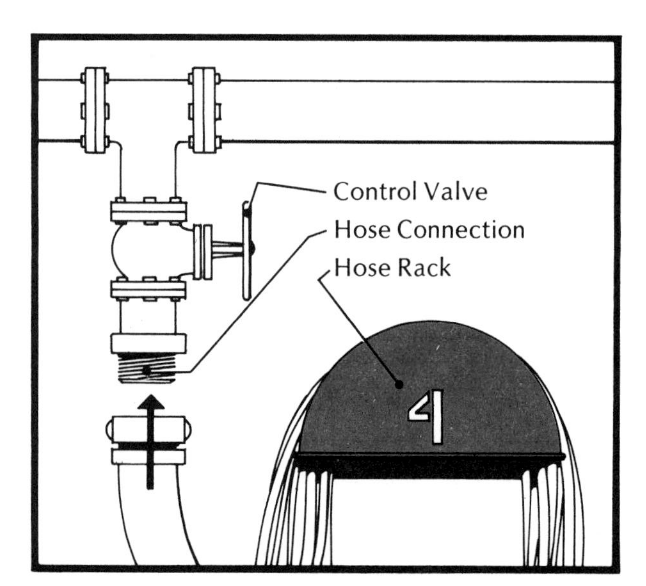

Figure 9.4. The three required components of a fire station hydrant.

All water enters the fire-main system through the sea chest, which is frequently covered with marine growth. It would thus be a good practice to fit all hydrant outlets with self-cleaning strainers. These strainers remove matter that might clog the nozzle, particularly the fine holes in combination nozzles and low-velocity applicators (*see* Fig. 9.5). Combination nozzles installed since 1962 must allow the free flow of foreign matter through nozzle orifices up to 9.53 mm (⅜ in.) in size. On vessels that are not required to carry such combination nozzles, self-cleaning strainers should be installed on the hydrant, or combination nozzles with internal strainers must be used.

Fire Hose, Nozzles and Appliances. The efficiency of a fire station depends largely on the equipment stowed at the station and its condition. A single station should have the following equipment.

Hoses. A single length of hose of the required size, type and length: 63.5-mm (2½-in.) diameter hose is used at weather-deck locations; 38.1-mm (1½-in.) diameter hose is used in enclosed areas. The hose must bear the Underwriters Laboratory (UL) label or comply with federal specification JJ–11–571 or ZZ–11–451a. Unlined hose may not be used in machinery spaces. The hose couplings must be of brass, bronze or a similar metal and be threaded with National Standard fire-hose coupling threads.

The hose must be 15 m (50 ft) in length, except on the weather decks of tankers. There, the hose must be long enough to permit a single length to be goosenecked over the side of the tank ship. Goosenecking is directing a stream of water over the vessel's side, perpendicular to the water surface.

The fire hose must be connected to the hydrant at all times, with the appropriate nozzle attached. However, when a hose is exposed to heavy weather on an open deck, it may be temporarily removed from the hydrant and stowed in a nearby accessible location. Fire hose may also be temporarily moved when it might be damaged by the handling of cargo. (When fire hose is removed, the exposed threads of the hydrant should be covered with a thin coating of grease and a protective screwcap. If a screwcap is not available, a heavy canvas, lashed over the threads, gives some protection.)

Fire hose may not be used for any purpose other than firefighting, testing and fire drills.

Nozzles. A nozzle, preferably of the combination type, so that water flow may be controlled

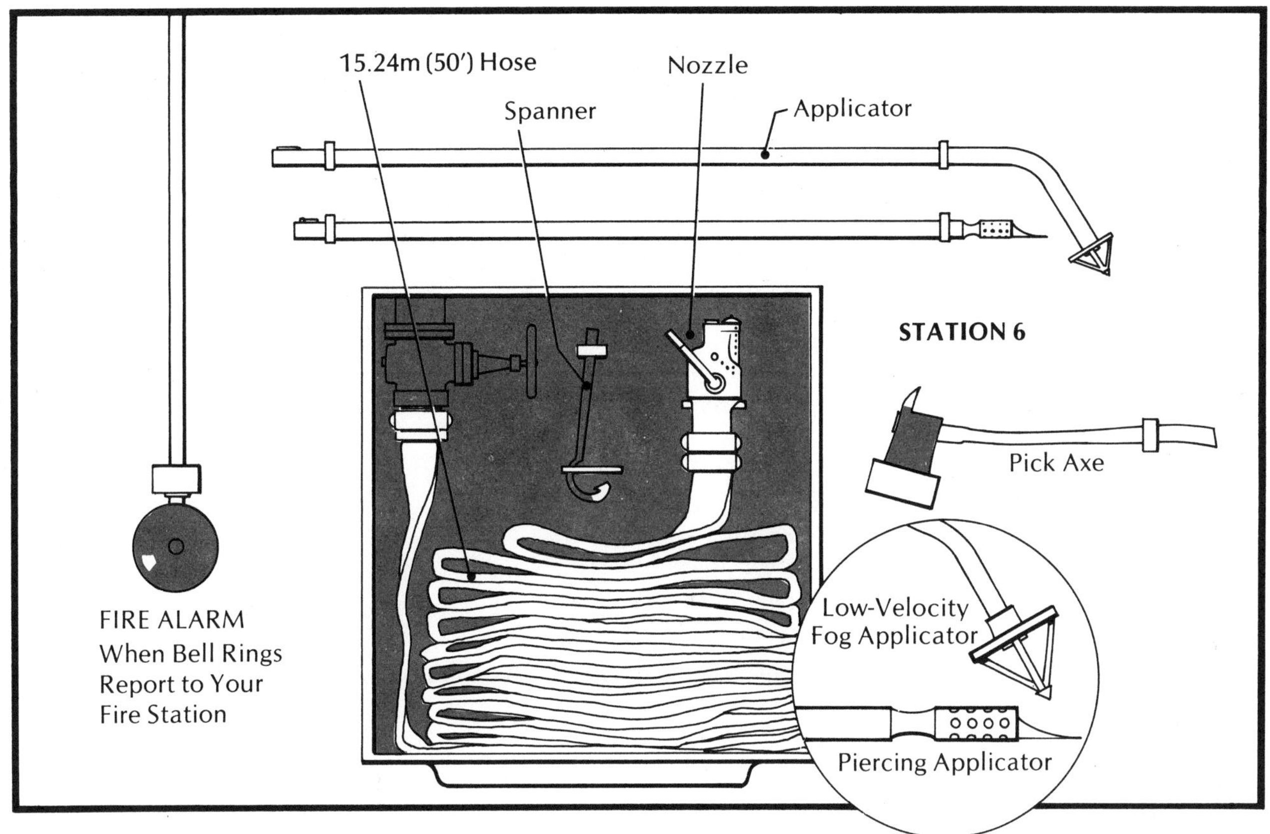

Figure 9.5. Shipboard fire station equipment.

must be connected to the hose at all times. The following regulations apply to vessels contracted or built after May 26, 1965: Tank vessels must be equipped with combination nozzles throughout. Cargo and miscellaneous vessels must be equipped with combination nozzles in machinery spaces and may use smooth-bore solid-stream nozzles in other spaces.

The combination nozzle must be fitted with a control that permits the stream to be shut off and to be adjusted for solid stream or high-velocity fog. On a 63.5-mm (2½-in.) combination nozzle, the solid-stream orifice must be at least 22.2 mm (⅞ in.) in diameter; on a 38.1-mm (1½-in.) nozzle, the opening must be at least 15.8 mm (⅝ in.) in diameter.

At this writing, the Coast Guard is considering a regulation that would eliminate the use of smooth-bore solid-stream nozzles on U.S. flag vessels. If the regulation is put into effect, the Coast Guard may allow the continued use of smooth-bore nozzles on ships that currently carry them; however, approved nozzles would have to be substituted when the smooth-bore nozzles are replaced in normal service.

Fog applicator. A low-velocity fog applicator for use with the required combination nozzle must be provided at each station. On exterior decks, applicators should be 3.0–3.6 m (10–12 ft) in length. In machinery spaces, applicators are limited to 1.8 m (6 ft) in length. Where combination nozzles are not required but are installed, the low-velocity applicator need not be furnished.

On container ships, a bayonet-type applicator should be provided. This applicator is similar to the fog applicator, but it has a sharp tip that can cut and penetrate the metal skin of a container.

Other Useful Tools. A *spanner wrench* whose size matches the hose coupling, or an adjustable spanner wrench. Depending on the location of the fire station, a *pickhead axe* may also be required. A fully equipped fire station is shown in Figure 9.5.

Fire Hose

A fire hose is a flexible tube that is used to transport water from the hydrant to the fire. Most of the hose in use is lined to stand up under high water pressure and minimize frictional loss. The lining is usually constructed of a rubber or synthetic material. Its inner surface is very smooth, so water will flow through it with a minimum of friction. The outer covering of the hose is a jacket of heavy cloth or synthetic material. The hose has a male coupling at one end and a female coupling at the other; these couplings are some-

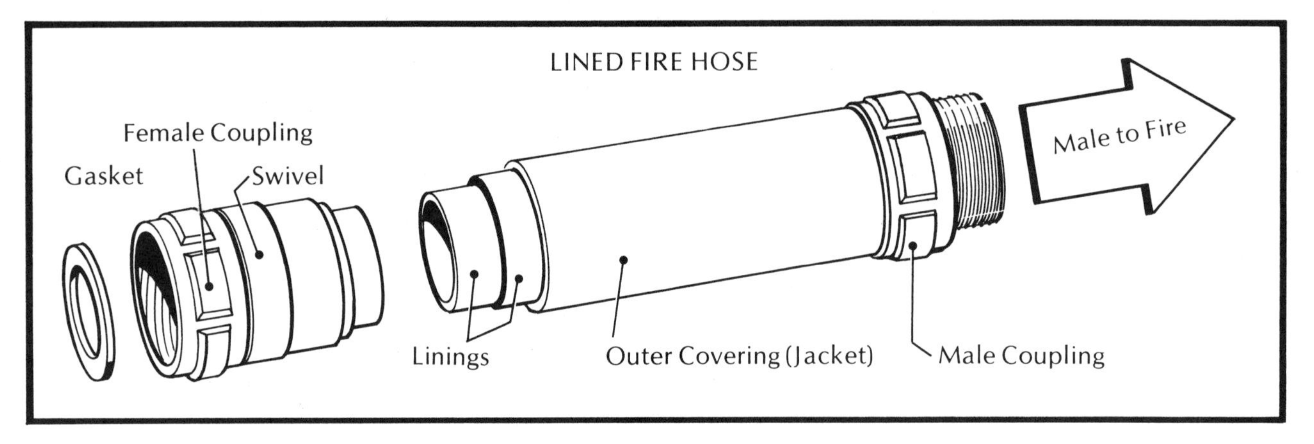

Figure 9.6. Lined fire hose and hose couplings.

times called *butts*. The female coupling is attached to the hydrant, and the male coupling to the nozzle. (Fig. 9.6).

The fire hose is the most vulnerable part of the fire-main system. It is easily damaged through misuse. Failure to remove dirt, grease, abrasives and other foreign substances from the outer surface of a hose can cause it to fail under pressure. Fire hose may be cleaned by washing with fresh water and mild detergent, using a soft brush. Abrasive cleaners should not be used as they cut into the outer covering of the hose and weaken it.

Hose that is dragged across metal decks can be permanently damaged. The jacket may be cut, or the coupling threads may be bent or broken. Failure to drain hose thoroughly, prior to racking, allows trapped moisture to cause mildew and rot; possibly resulting in failure under pressure. In addition, cold, heat and seawater tend to weaken the hose.

Fire hose should be inspected visually each week. Every hose on board should be tested monthly, through actual use under the pressure required to produce a substantial water stream. This can de done by alternating weekly fire drills from station to station, or through a rotating testing schedule.

Fire hose should be taken from the rack periodically and visually inspected for dry rot and other damage. If the hose is sound, it should be replaced on the rack with the bight folds at different locations. This prevents cracking of the hose liner and the friction loss that is caused by deep bends.

Racking and Stowage Procedures. Most shipboard racks for the stowage of hose at fire stations require that the hose be faked. The procedure should include the following steps:

1. Check the hose to make sure it is completely drained. Wet hose should not be racked.
2. Check the female coupling for its gasket.
3. Hook the female coupling to the male outlet of the hydrant. (The hose should always be connected to the hydrant.)
4. Fake the hose so that the nozzle end can be run out to the fire (*see* Fig. 9.5).
5. Attach the nozzle to the male end of the hose, making sure a gasket is in place.
6. Place the nozzle in its holder or lay it on the hose, so that it will not come adrift.

There are several different types of hose racks. One type consists of a half round plate, over which the hose is faked. A horizontal bar swings into position, holding the hose snug. Reels are used in engine rooms. They are also used for rubber hose, such as that found on a semiportable CO_2 extinguisher.

Rolling Hose. After spare hose is used, it should be rolled and replaced in stowage. The hose must first be drained and dried. Then it should be placed flat on the deck with the female coupling against the deck. The hose is next folded back on itself, so the male coupling is brought up to about 1.2 m (4 ft) from the female couplings. The exposed thread of the male coupling should be layered between the hose when the roll is completed. The roll should be tied with small stuff to keep it from losing its shape.

Nozzles and Applicators

Two types of nozzles are used on merchant marine vessels, combination nozzles and smooth-bore nozzles. Both have been mentioned earlier in this chapter. (*See* Chapter 10 for a description of their operation.)

Nozzles are quite rugged but are still subject to damage. For example, the control handle can become stuck in the closed position, owing to the corrosive action of seawater. Combination nozzles and applicators are often clogged by minute pieces of dirt that enter and collect around openings. Periodic testing and maintenance will help detect and correct deficiencies.

The combination nozzle has a spring latch that allows the high-velocity tip to be released (Fig. 9.7). The latch often freezes into position from misuse. During inspections and drills, the tip should be released and the applicator inserted into position for proper operation. The high-velocity tip should be attached to the nozzle by a substantial chain, so that it cannot be completely separated from the nozzle.

Applicators (*see* Fig. 9.5) are strong, but not strong enough to be used as crowbars, levers or supports for lashing. If misused, the applicator can be crimped or bent along its length. The bayonet end can be damaged so that it cannot fit in the nozzle receptacle. Applicators should be stowed in the proper clips at the fire station, and used for firefighting and training only. When stowed, applicator heads should be enclosed in sock-type covers to keep foreign matter out. (*See* Chapter 10 for a description of the proper use of applicators.)

Spanner Wrenches

A spanner wrench is a special tool designed specifically for tightening or breaking apart fire-hose connections. The spanner should match the hose size and butt configuration. Hose-butt lug designs change over the years, making some spanner wrenches obsolete. When new hose is ordered, the available spanner wrenches must be compatible with the new hose couplings, or new spanner wrenches must also be ordered.

Most hose connections can be made handtight and do not require excessive force.

Wye Gates and Tri-Gates

It is sometimes advantageous to have two smaller 38.1-mm (1½-in.) hoselines available, rather than one large 63.5-mm (2½-in.) line. Devices called wye gates and tri-gates are used to reduce the hoseline size and separate the lines.

A wye gate is a connector in the shape of a "Y" (Fig. 9.8). It has one female 63.5-mm (2½-in.) inlet butt and two 38.1-mm (1½-in.) male outlet butts. The large inlet butt is attached to a 63.5-mm (2½-in.) fire hydrant outlet so that water flows out through the two smaller outlets. A hose may be attached to each outlet.

The device is gated, which means it has valves that can be used to shut off the flow of water. The two valves, or gates, are independent of each other, so that one can be closed while the other

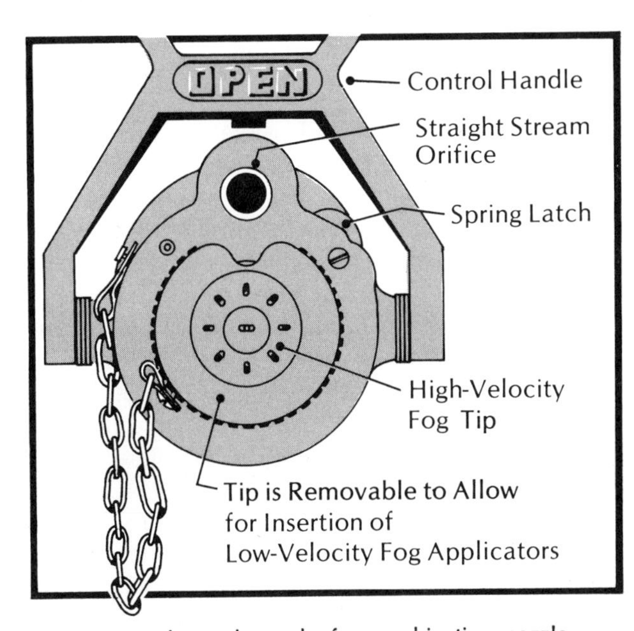

Figure 9.7. The outlet end of a combination nozzle.

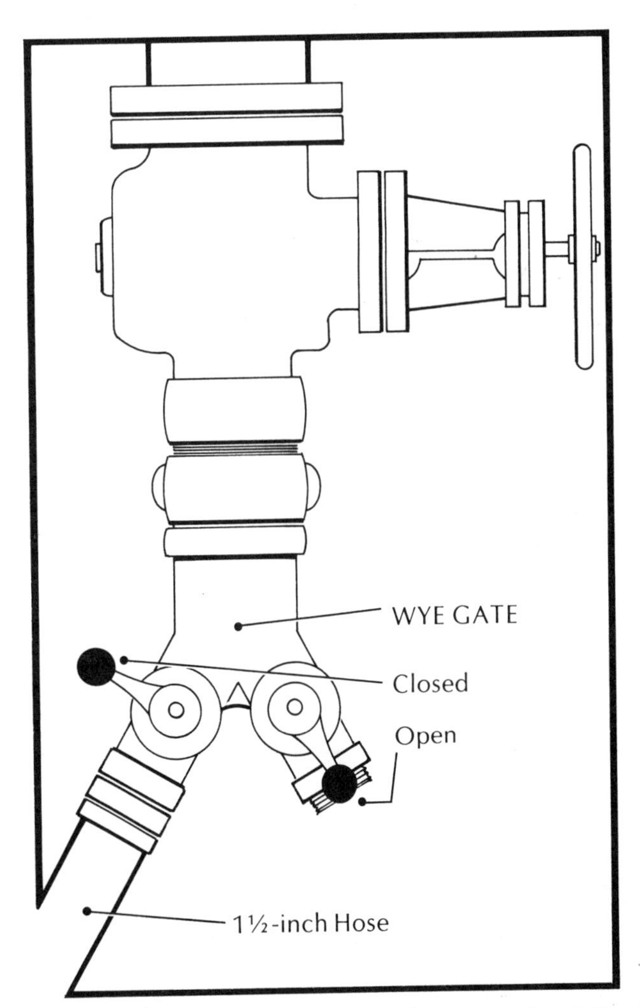

Figure 9.8. Wye gate attached to a hydrant outlet. *Left,* closed valve; *right,* open valve.

is open. The gates are opened or closed with a quarter turn. When a gate handle is parallel with the waterway, it is in the open position (Fig. 9.8).

If two hoses are connected to a wye gate, both gates should be closed when the hoses are not in use. If one hose is connected, its gate should be closed. The other gate should be open, allowing leakage from the hydrant to drip out the opening.

The U.S. Coast Guard permits the use of wye gates at fire hydrants under certain conditions. The tri-gate is similar to the wye gate, but it provides three 38.1-mm (1½-in.) outlets. While these devices may allow additional lines to be directed onto a fire, they result in a large pressure drop at the nozzle. Even the reduction of a 63.5-mm (2½-in.) line to two smaller lines could drop the water pressure to the point where both streams are ineffective for firefighting. It is better to have one good firefighting stream that can penetrate into the fire than two poor streams.

WATER SPRINKLER SYSTEMS

United States ships are constructed in accordance with Method I of the Safety of Life at Sea (SOLAS) convention. Method I calls for fire protection through the use of noncombustible construction materials, rather than reliance on automatic sprinkler systems. For this reason, sprinkler systems are not widely used on U.S. merchant vessels. They are generally used only to protect living quarters, adjacent passageways, public spaces, and vehicular decks on roll-on/roll-off (ro-ro) vessels and ferryboats.

Sprinkler systems may extinguish fire in these spaces. However, their primary function is to protect the vessel's structure, limit the spread of fire and control the amount of heat produced. They also protect people in these areas and maintain escape routes.

Components of Sprinkler Systems

All sprinkler systems consist of piping, valves, sprinkler heads, a pump and a water supply.

Piping. The piping must comply with standards developed for such systems. The piping size and layout are chosen to deliver the proper amount of water to the sprinkler heads. The main supply line from the pump carries the water to branch lines. The diameters of the branch lines decrease as they extend further from the source of the water. The branch lines supply the water to the sprinkler heads.

Valves. Valves are located at the pump manifold and outside the protected spaces. They should be readily accessible in case of fire. Control valves should be clearly marked as to their function, e.g., "Control Valve for Automatic Sprinkler System." They should also be marked as to their normal position: "Keep Open at All Times" or "Close Only to Reset the System." If the sprinkler system is divided into separate zones, the control valves should be clearly identified with their zone numbers.

Sprinkler Heads. The heads are actually valves of special design. They release water from the system and form the water into a cone-shaped spray. A sprinkler head is made up of a threaded frame (for installation in a branch pipe), a waterway and a deflector for forming the water spray pattern. Sprinkler heads for automatic systems may be equipped with a fusible link. The link keeps the head closed normally. Heads for manual systems are open normally; they do not include a fusible link.

Coast Guard regulations require that each protected space have sufficient heads located so that no part of an overhead or vertical projection of a deck is more than 2.1 m (7 ft) from a sprinkler head.

Fusible Links. A fusible link is a pair of levers, held within the sprinkler head frame by two links. The links are connected by eutectic alloy or a similar low melting-point metal (Fig. 9.9). The levers hold a valve cap in place over the sprinkler head outlet, preventing the flow of water. Since the sprinkler head is closed, the piping may be charged with water up to the head. (*See* Chapter 6 for a discussion of other types of heat detectors.)

When heat from a fire increases the temperature of the eutectic alloy enough to melt it, the links come apart. This releases the levers, opening the sprinkler head waterway (Fig. 9.9).

Temperature Ratings of Fusible Links. Sprinkler heads on some ships may be color coded to indicate the temperature at which the fusible metal (solder) will melt and activate the head. Table 9.1 gives the standard operating temperatures of sprinkler heads and the corresponding color codes. The color is painted on the frame arms of the sprinkler head. No other part of a sprinkler head should be painted—especially not the fusible element. The paint would insulate the fusible metal from the heat of the fire and keep it from melting at its operating temperature.

The sprinkler heads normally used on ships are ~~these~~ that operate 57.2°C–73.8°C/or 100°C (135°F–165°F/or 212°F) (uncolored or white). Heads with lower operating temperatures are

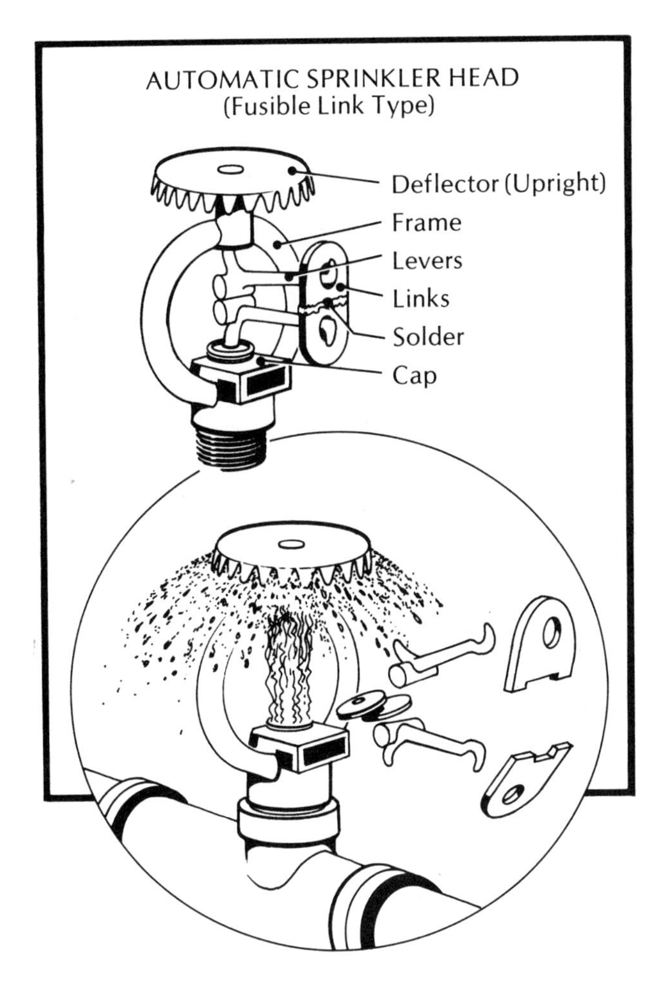

Figure 9.9. **A.** Heat from fire melts solder, allowing links to separate. **B.** The levers come apart and **C.** water pressure pushes the valve cap off the sprinkler outlet. **D.** Water flows up against the deflector, forming a spray that falls onto the fire.

Table 9.1. Operating Temperatures and Color Coding of Fusible Metal links for Sprinkler Heads.*

Operating temperature °C (°F)	*Color code*
57.2 (135); 65.5 (150); 71.1 (160); 73.8 (165)	Uncolored
79.4 (175); 100 (212)	White
121 (250); 138 (280); 141 (286)	Blue
163 (325); 171 (340); 177 (350); 182 (360)	Green
232 (450); 260 (500)	Orange

*Color coding may be found on some vessels.

used in spaces where normal temperatures can be expected, such as living spaces. Higher temperature heads are used where temperatures above normal are expected, such as galley areas. A sprinkler head must always be replaced with a head that has the same temperature rating. A higher temperature head will not protect the space properly; a lower temperature head could be operated by a heat source other than a hostile fire.

Spray Patterns. The spray deflector on a sprinkler head is designed to direct the water in a specific direction. The *upright* deflector in Figure 9.9 directs water down toward the deck. The *pendant* deflector also directs water downward, but it is placed differently in the piping. The *sidewall* deflector directs water away from bulkheads, toward the center of the protected space. The position in which a sprinkler head should be installed is stamped on the frame or deflector. Pendant heads should not be installed as replacements for upright heads, and vice versa. Improper installation can destroy the firefighting capability of a sprinkler head.

Automatic Sprinkler Systems

Automatic sprinkler systems are not used extensively on U.S. merchant ships. The automatic sprinkler makes use of closed sprinkler heads (Fig. 9.9), so the piping can be charged with water. The fusible links serve as the fire detectors and the activating devices. A pressure tank serves as the initial water source. The pressure tank is partially filled with fresh water (usually to two thirds of its capacity). The remainder of the tank is filled with air under pressure. The air pressure propels the water to and through the sprinkler heads when they open. The pressure tank must hold enough water to fill the piping of the largest zone, and in addition, force out at least 757 liters (200 gal) at the least effective head in the zone at a pitot tube pressure of at least 103.42×10^3 pascals (15 psi). Fresh water is used in the system to avoid the breakdown of metal by electrolysis.

How the System Works. Heat from the fire melts the fusible links of one or more sprinkler heads. The heads open, allowing water to flow. The initial supply of water comes from the piping, and then from the pressure tank. As water flows out of the tank, its pressure is reduced. This pressure drop causes a pressure-sensitive switch to electrically activate the sprinkler water pump and the alarm bells. The sprinkler pump takes over as the water source, supplying water from a fresh water holding tank (Fig. 9.10). Check valves in the piping ensure that the water flows from the pump to the sprinkler heads, rather than into the pressure tank. When the holding tank water supply is depleted, the pump suction must be manually shifted to seawater.

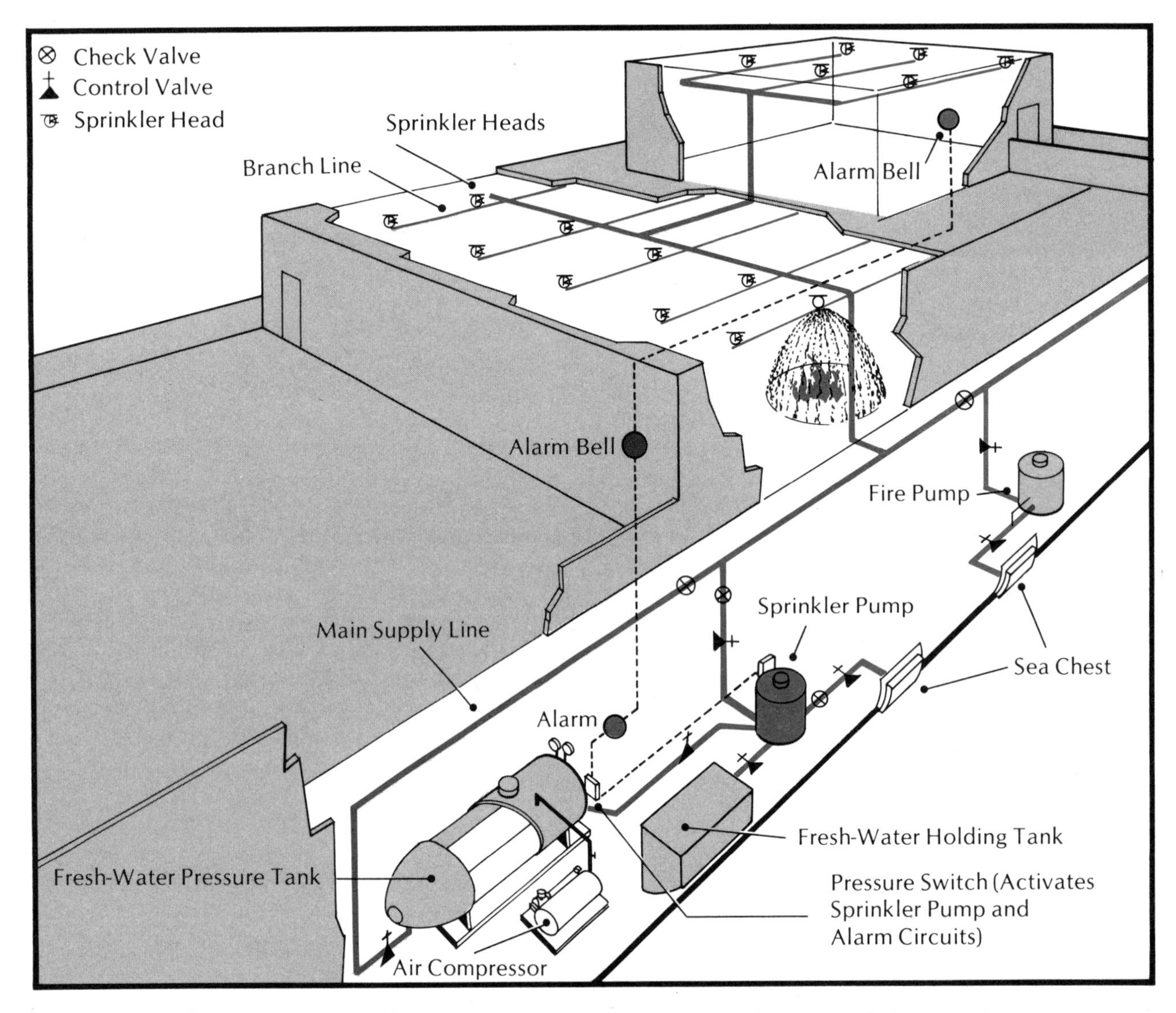

Figure 9.10. Shipboard automatic sprinkler system. The sprinkler pump is started automatically by a switch in the pressure tank.

Crewmen should not depend on an automatic sprinkler system as the sole method of extinguishment. As in all fire attack operations, the initial attack (by the sprinkler system) should be backed up with charged hoselines.

An activated sprinkler system should not be shut down until the fire is at least knocked down and hoselines are in position to extinguish any remaining fire. It is important to prevent unnecessary water accumulation, but the primary objective is to get the fire out. If an automatic sprinkler system is shut off too soon, heat from the continuing fire can cause many more sprinkler heads to open. The additional open heads can put an excessive load on the system, beyond the capability of the sprinkler pump. The result would be reduced pressure in the system and insufficient water flow from the sprinkler heads. The heads then would not be able to form the spray pattern necessary to achieve extinguishment.

After the fire is extinguished, the sprinkler system should be restored to service. The sprinkler heads that were opened should be replaced with heads of the same temperature rating and deflector type. A supply of heads of the proper types should be kept on board for this purpose. The pressure tank should be refilled and pressurized, and the valves reset.

Manual Sprinkler Systems

The manual sprinkler system differs from automatic systems in two respects: *1)* the sprinkler heads are normally open and *2)* the piping does not normally contain water. Water is supplied to the manual system by the ship's fire pumps; no pressure tank is required.

The system is composed of the piping, open sprinkler heads, control valves, fire pumps and water supply. It may be used along with a fire detection system. However, the fire detectors do not activate the system automatically; they sound alarms so that the system can be put into action manually.

How the System Works. When fire is discovered or the alarm is sounded, the fire pumps are started. A control valve is manually opened to allow water to flow into the system. The control valve is located either at the fire-pump manifold or near (but not in) the protected area. Water is discharged out of all the sprinkler heads, so the entire area is covered with water spray. The area is thus saturated with a large volume of water, capable of knocking down a sizable fire.

With manual systems, there is a delay in getting water onto the fire, and then an excessive amount of water is applied, well beyond that needed for extinguishment. Manual systems are, however, effective protection for vehicular decks on ro-ro vessels and ferryboats. The large amount of water is effective in knocking down the fire and protecting the vessel and exposed vehicles. It will also dilute and carry off flammable liquids, if they are involved. Manual sprinkler systems are also used in cargo spaces that are accessible to the crew when the vessel is under way.

Free surface water is a constant threat to a ship's stability. A sprinkler system with open heads can easily flow 1.89 m^3/min (500 gal/min). Where large systems are used, as on ro-ro vessels, provisions are made to drain the water off through scuppers or internal drains. Scuppers drain the water overboard; internal drains direct the water to bilges in the ship's bottom. However, bilges can overflow during firefighting operations, spilling the water into the cargo holds. Therefore, the vessel's bilges should be pumped out while water is draining into them; otherwise, the free surface water would simply be moved from the fire area to the bilge, possibly resulting in a serious list.

Reliability of Sprinkler Systems

Land-based sprinkler systems are very reliable. In most instances, fires are controlled or extinguished as soon as one or two heads are opened. The water supply to these systems is clean and free of debris. However, shipboard systems are not as reliable, because they are supplied with water through sea chests. In most instances, this water contains solid matter of sufficient size to clog the system, especially at the sprinkler head openings.

To help ensure some measure of reliability, sprinkler systems must be tested periodically. The testing procedure must conform with Coast Guard regulations.

Zoning of Sprinkler Systems

When a large portion of a passenger ship is to be protected by sprinklers, several small subsystems, rather than one large system, are used. The subsystems are placed within spaces separated by fire-retarding bulkheads; these spaces between bulkheads are called *fire zones*. Fire zones extend across the beam of the vessel and are confined between main vertical zones (class A bulkheads). Zones may not exceed 40 m (131 ft) in length. The bulkheads are fitted with doors of the same fire-retarding capability as the bulkheads. (Class A bulkheads must be capable of retarding the passage of smoke and flame for 1 hour.)

This arrangement has two advantages: *1)* No subsystem is very large, so it will not overtax the water supply and *2)* when the bulkhead doors are closed, the fire is confined in an area that is designed to keep it from spreading.

Two separate subsystems may be installed within a single fire zone, e.g., one to protect the port size, and the other the starboard side. However, the coverage of the two subsystems must overlap, to ensure full protection of the fire zone. The sprinkler systems on ro-ro vessels are zoned but do not require bulkheaded divisions.

Sprinkler Zone Chart. A chart must be posted (in the wheelhouse or control station, adjacent to the detecting cabinet) showing the arrangement of the fire zones, their identification numbers and the sprinkler system layout within the zones. The chart must also show the piping system, including the locations of control valves and water supply pumps (Fig. 9.11).

WATER SPRAY SYSTEMS

Water spray systems are similar to sprinkler systems but make use of a different type of head and a different piping arrangement.

Spray Heads

Spray heads are open heads (Fig. 9.12) that shape the discharged water into a spray pattern. However, unlike some sprinkler heads, which discharge hollow spray patterns, spray heads discharge a solid cone of water, giving them superior cooling capabilities. In addition, a spray head can be aimed to hit a specific target area.

Figure 9.11. Zone valves and zone chart for a shipboard sprinkler system.

Water Supply

Water can be supplied to the spray system by a separate pump or by one of the ship's fire pumps. A fire pump may be used if it can adequately supply both the fire main and the spray system when both are in operation at the same time. An extensive spray system needs a substantial water supply and would most likely require a pump other than a fire pump.

The spray system piping is normally empty, because the spray heads are open. When fire is discovered, the system is activated manually by opening the proper valves and starting the water pump. The spray heads provide a very finely divided water spray that blankets the protected area.

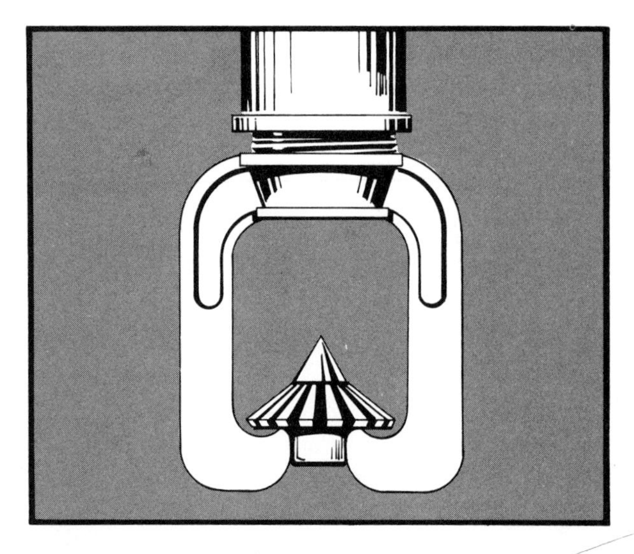

Figure 9.12. Typical spray head. There is no fusible link; the head is open at all times.

Applications

Water spray systems are used to protect the piping and exposed sections of storage tanks on vessels transporting cryogenic gases such as LNG. They are also used to protect loading stations and manifolds. In the event of a gas leak with fire, their primary functions are to cool exposed tanks and piping and to confine the fire until the leak can be stopped. If a leak occurs without fire, the water spray could be effective in diluting the leaking vapors. The water spray also helps protect metal surfaces directly exposed to the leak from fracture, and it can be used to dissipate the vapors under the right conditions.

In addition, water spray can be used to protect the superstructure of the vessel from radiant heat in the event of a massive fire. When used for this purpose, the spray is cascaded directly onto the surfaces of bulkheads and decks (Fig. 9.13), taking the most advantage of the cooling capabilities of the water.

The water spray is not normally used to extinguish the fire, dry chemical extinguishing units are employed on LNG and LPG vessels for this purpose. Crewmen should, however, realize that extinguishing such a fire may create a greater hazard—a flammable vapor cloud. It is usually best to allow the LNG to burn under controlled conditions until the fuel is exhausted.

FOAM SYSTEMS

Foam is used mainly in fighting class B fires, although low-expansion foam (with a high water content) can be used to extinguish class A fires. Foam extinguishes mainly by smothering, with some cooling action. (*See* Chapter 7 for a discussion of foam as an extinguishing agent.)

Foam may be generated chemically or mechanically. *Chemical* foam is produced by chemical reactions taking place in water. The foam bubbles are filled with CO_2. *Mechanical* foam is produced by first mixing foam concentrate with water to produce a foam solution, then mixing air with the foam solution. The bubbles are thus filled with air.

Foam systems are acceptable as fire protection for boiler rooms, machinery spaces and pump rooms on all vessels. Mechanical foam systems may be installed in these spaces instead of other approved systems such as CO_2. Deck foam systems must be installed on tankers constructed after January 1, 1970, as fire protection for flammable-liquid cargo. Some older vessels may have foam systems protecting flammable-liquid cargo holds; foam systems are no longer employed for this purpose.

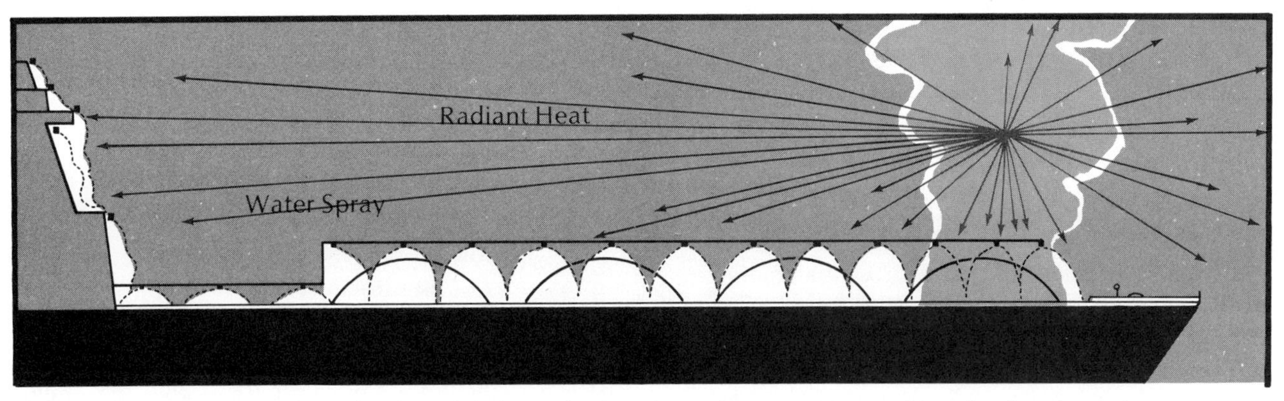

Figure 9.13. LNG and LPG burn cleanly with little smoke, but do produce massive quantities of radiant heat. The spray system can provide protection by cooling exposed decks, tanks, pipelines and superstructure through a continuous cascade of water.

Foam systems must meet Coast Guard requirements. Other guidance regarding fixed systems can be found in the recommendations of the Intergovernmental Maritime Consultative Organization (IMCO). The Coast Guard regulations are usually consistent with, but more stringent than, IMCO recommendations.

Chemical Foam Systems

Chemical foam is produced by the reaction of bicarbonate of soda with aluminum sulfate (or ferric sulfate). A *foam stabilizer* is added to improve its extinguishing properties. Chemical foam has more body than mechanical foam and will build a stouter blanket.

Continuous-Type Generator. A continuous-type chemical foam generator is shown in Figure 9.14. The generator may be fixed or portable. It consists of a hopper with a foam ejector at the bottom; its function is to dissolve the dry foam chemicals in a stream of water. The generator inlet is connected to a hoseline or piping to the fire main; the outlet is connected to a 63.5-mm (2½-in.) hoseline. After water at 517–689 kilopascals (75–100 psi) has been started through the generator, the mixture of dry foam chemicals is poured into the hopper. The chemical reaction takes place downstream of the ejector.

The temperature of the water governs the speed of foam production (it is slower at lower temperatures), and the length of the outlet hose should be varied accordingly. At temperatures above 32.2°C (90°F), 15.24 m (50 ft) of hose is adequate; from 10°C to 32.2°C (50°F to 90°F), 30.48 m (100 ft) of hose should be used; at temperatures below 10°C (50°F), 45.72 m (150 ft) is required. The hose should have a 38.1-mm (1½-in.) diameter nozzle for the most effective foam discharge.

The continuous-type generator uses foam chemical at a rate of about 45.4 kg/min (100 lb/min) with either fresh or salt water at 21.1°C (70°F). Since 0.45 kg (1 lb) of foam powder produces about 30 liters (8 gal) of foam, the unit produces about 3000 liters/min (800 gal/min) of foam. In one minute, this quantity of foam can cover an area of 37 m^2 (400 ft^2) to a thickness of 76.2 mm (3 in.). This area is equivalent to a square, 6.1 m (20 ft) on each side.

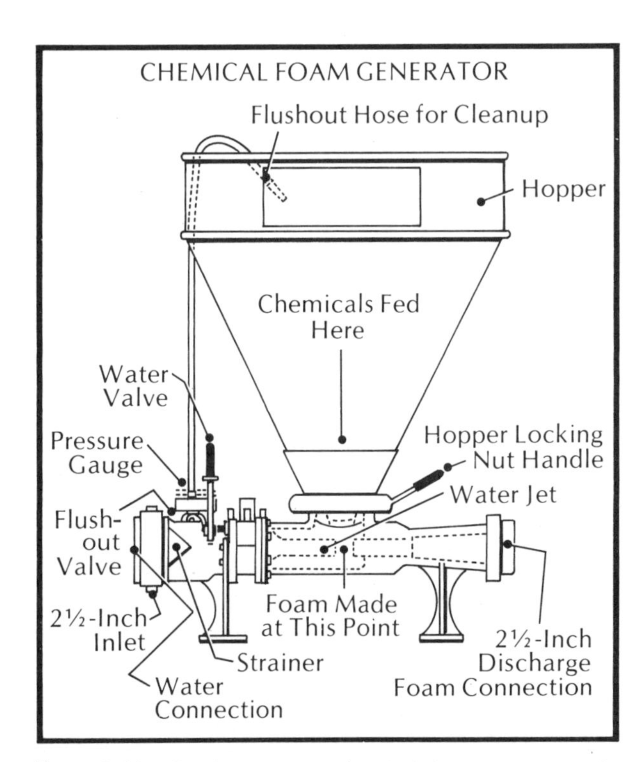

Figure 9.14. Continuous-type chemical foam generator. **A.** The generator is connected to a hoseline or piping to the fire main. **B.** Foam chemicals are discharged into water flowing through the bottom of the generator. **C.** Chemical foam is produced downstream of the generator.

The continuous-type foam generator is also available with two separate hoppers. The dry foam chemicals are dissolved separately in two streams of water. The two solutions are brought together at the discharge outlet to produce foam.

Hopper-Type Generator. In the large hopper-type generator, the chemicals are stored separately, in powder form, in twin compartments in the hopper. The chemicals are thus always ready for use. This type of generator is usually equipped with a mechanical turning device that should be turned occasionally to keep the chemicals from settling and packing. When the generator is to be used, control levers are operated to release the chemicals into the water stream.

Extra chemicals are usually stored in containers holding 22.7 kg (50 lb) each. These containers must be kept sealed and must be stored in a cool dry place until used.

Twin-Solution Generator. In the twin-solution foam generator, a solution of bicarbonate of soda and foam stabilizer is contained in one cylinder, and an aluminum sulfate solution in the other. The contents of the two cylinders are pumped separately to discharge outlets. At the outlets, the solutions mix to produce and discharge foam on the protected area.

Mechanical versus Chemical Foam

Chemical foam generators have been replaced by mechanical foam generators on almost all vessels. Chemical foam systems are no longer approved for installation by the U.S. Coast Guard. However, older ships may still use chemical foam systems provided they are in good condition and operate properly. Whichever system is installed aboard a vessel, crewmen should be well trained in its use. Periodic drills build efficiency and help to avoid mistakes when a fire occurs.

Mechanical Foam Systems

Mechanical foam concentrate is available in 3% and 6% concentrations (*see* Chapter 7). It may be mixed with either fresh or salt water to produce foam solution:

- 12 liters (3 gal) 3% concentrate, mixed with 367 liters (97 gal) of water produces 379 liters (100 gal) foam solution.
- 23 liters (6 gal) 6% concentrate, mixed with 356 liters (94 gal) water, produces 379 liters (100 gal) foam solution.

When the foam solution is mixed with air, it expands. The expansion ratio of the foam indicates the proportions of air and water it contains. Thus, for example, a 4:1 foam expansion ratio is defined as the quantity of moisture contained in a given quantity of foam. In 1000: 1 high expansion foam there is one gallon of moisture in 1000 gallons of the high expansion foam. A 100:1 expansion ratio means the foam contains 99 volumes of air for each volume of water. The air is introduced into the foam solution at a foam spray nozzle, monitor or turret nozzle.

In fixed foam extinguishing systems the air-to-water ratio is set to obtain the desired foam properties. In general, the lower the expansion ratio,

- The wetter the foam
- The more fluid the foam
- The heavier the foam
- The more heat resistant the foam (for a given type of concentrate)
- The less the foam adheres to vertical surfaces
- The more electrically conductive the foam
- The less the foam is subject to movement by wind.

The foams used in engine rooms should be soupy, mixed at about a 4:1 expansion ratio. A mixture of this consistency is capable of flowing rapidly around obstructions. It has good cooling qualities (for foam) and resists heat (holds up longer). However, the actual ratio may vary. Because a 4:1 ratio foam is loaded with water, there is a fairly rapid water runoff. It is difficult to build up a deep layer of 4:1 foam, unless it is confined in a limited area.

High-expansion foam, with a ratio of 100:1 or more, is stiff, does not flow rapidly and is easily pushed around by the wind in open areas. It builds up a deep layer of foam rapidly—up to 0.9 m (3 ft) in open, unconfined areas, and up to 5.9 m (20 ft) in a confined space. High-expansion foam is a poor conductor of electricity since it contains little water.

Low-Expansion Mechanical Foam System

One low-expansion foam system used on ships is the *balanced-pressure proportioning system.* The system gets its name from the action of the proportioning device: The water and the foam concentrate are pumped into the proportioner separately, under pressure. Monitoring devices in the proportioner regulate (balance) these two flows to produce the desired foam solution.

A typical system is shown in schematic form in Figure 9.15. The major components are

- A water supply
- The fire pump
- The foam-concentrate pump

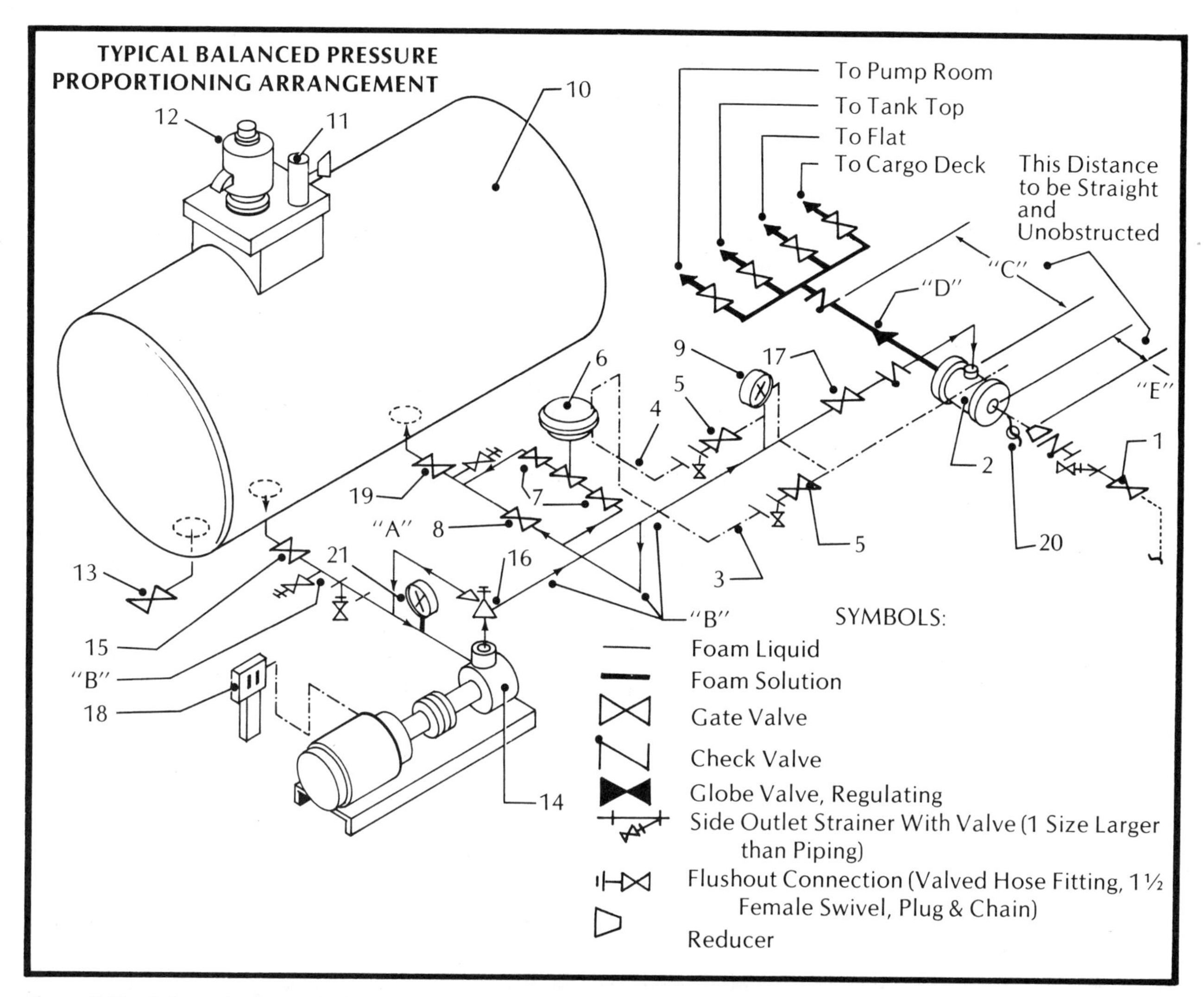

Figure 9.15. Schematic diagram of a typical balanced-pressure proportioning system. **A.** Water supply valve (normally closed). **B.** Ratio-flow proportioner. **C.** Water balance line. **D.** Foam concentrate balance line. **E.** Balance line valves (normally open). **F.** Diaphragm control valve (automatic bypass). **G.** Block valves (normally open). **H.** Regulating globe valve (manual bypass; normally closed). **I.** Water and foam concentrate pressure gauge. **J.** Foam-concentrate storage tank. (Courtesy National Foam System, Inc.)

- A holding tank for the foam concentrate
- The proportioning device
- The discharge foam spray nozzles or monitors
- The piping, control and check valves.

How the System Works. The system must be activated manually when fire is discovered in the protected area. First the water and foam-concentrate supply pumps are started. Then the proper control valves are opened to allow the water and foam concentrate to flow to the proportioner. If central foam-producing equipment furnishes foam solution to more than one piping system, then the control valve to the proper system must be opened. It is important that the crew member reporting the fire give the fire's exact location, so the proper valves may be opened without delay.

When the necessary valves are open and the pumps are activated, foam concentrate and water are pumped into the proportioner and mixed in preset proportions. The foam solution then flows through piping to the desired location for discharge. In a fixed system, the foam is discharged through nozzles located in the protected area. As the solution flows into each nozzle, it passes through an aspirator and is mixed with air to form the foam bubbles. In most fixed systems the nozzles are aimed at a bulkhead or metal deflector, so the foam flows gently onto the surface of the burning liquid. All the spray nozzles in the system discharge foam at the same time, to cover the area rapidly with a blanket of foam.

The system will continue to operate and produce foam until the foam-concentrate supply in the storage tank is depleted. When this occurs, water will continue to flow through the system and out the foam nozzles. If the water is allowed to flow more than 2 or 3 minutes, it will start to dilute the foam blanket and break it down. Therefore, it is important to shut off the system when it is no longer producing foam.

Foam-Concentrate Supplies. The rate of application affects the ability of foam to control flammable-liquid fires; it is essential that the required amount of foam be discharged in 3 minutes. The pumps, piping and nozzles are designed to do this. However, sufficient foam concentrate must be carried to produce the required amount of foam solution, which is 6.5 liters/min per m^2 (1.6 gal/min/per 10 ft^2) of protected area. The total available supply of foam concentrate must at least be sufficient for the space requiring the greatest amount. In conjunction with the deck foam system which requires a 20-minute supply, this will automatically cover the requirements for other spaces.

Most foam concentrates have a storage life of 5–20 years, depending on the manufacturer. However, the concentrate must be stored properly, on sturdy racks, where the containers will not become damaged. The storage space should be ventilated and fairly dry, with an ambient temperature not exceeding 38°C (100°F). The foam should be kept away from steam pipes and hot bulkheads. Excessive temperatures deteriorate the liquid concentrate and reduce its foam-making capability.

Foam-Concentrate Tanks. In the diaphragm-tank foam system shown in Figure 9.16, the foam-concentrate tank is fitted internally with a flexible rubber diaphragm. The diaphragm is one half the size of the tank and is fastened to a metal lip around the tank midsection. As the tank is filled with foam concentrate, the rubber diaphragm is pressed against the walls of the tank. When the foam system is activated, the foam system pump supplies water to both the proportioner and the tank at a predetermined pressure. The water enters the tank on the diaphragm side. It pushes against the diaphragm with enough pressure to force foam concentrate out of the tank and into the foam proportioner. An adjustable metering valve provides a measured flow of foam concentrate to the proportioner; this assures the proper proportions of water and concentrate to produce the foam solution. After operation, the water is drained from the tank. The tank is then refilled with the appropriate type and amount of foam concentrate. The system is very reliable and does not require the separate foam-concentrate pump.

Another type of foam-concentrate tank is shown in Figure 9.17. Here, water moving

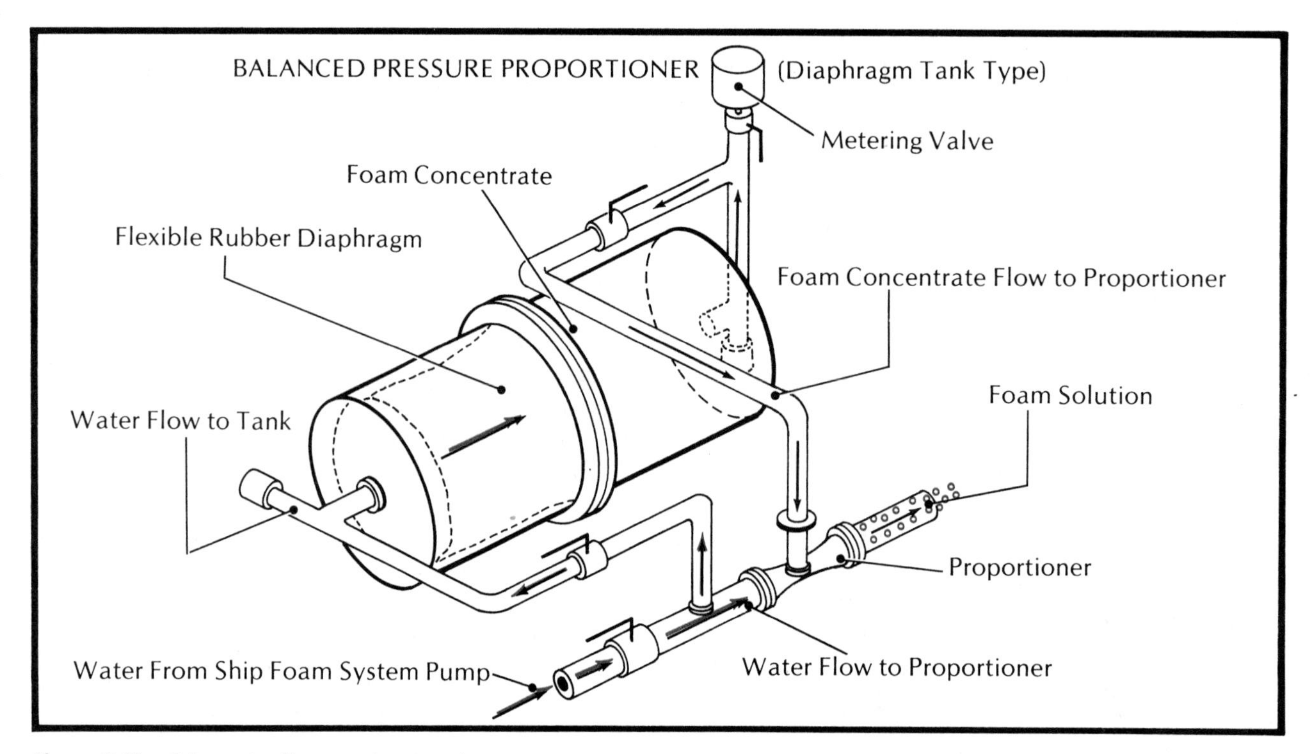

Figure 9.16. Schematic diagram showing the operation of a diaphragm-type foam concentrate tank. (Courtesy Rockwood System Corporation)

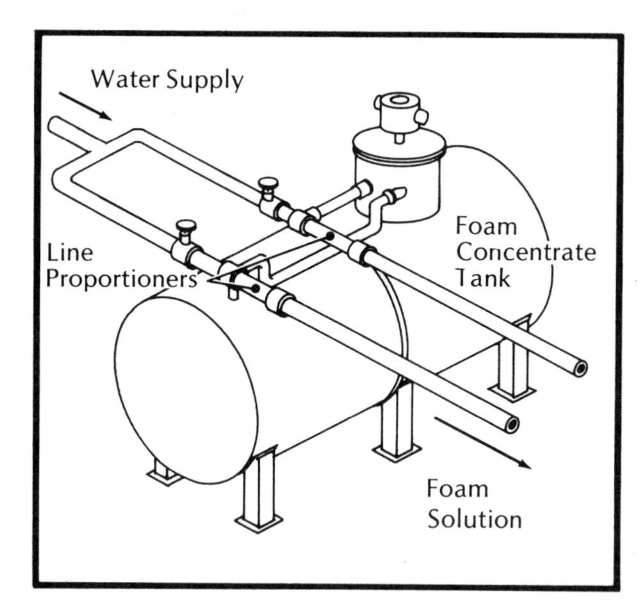

Figure 9.17. Venturi-affect foam tank with dual proportioners. This type of tank does not require a separate foam concentrate pump. (Courtesy National Foam System, Inc.)

through the proportioner creates a slight vacuum. The vacuum draws a metered amount of foam concentrate up from the tank and into the water stream. Note the two separate water lines, with proportioners in Figure 9.17. Each line has its own water control valve and can serve a different foam piping system.

Foam-concentrate storage tanks must be kept filled, with liquid halfway into the expansion dome to ensure prolonged storage life. The tank should be kept closed to the atmosphere, except for the pressure vacuum vent. When a tank is partially empty, there is a larger liquid surface area to interact with air. This allows excessive evaporation and condensation, which degrade the foam concentrate and permit corrosion of the tank shell.

Nozzle Placement. Once the foam solution is produced, it can be piped to supply a deck-system turret, handline nozzles or fixed marine floor or overhead spray deflectors (Fig. 9.18). The supply can also service high-expansion foam devices if proper concentrate is used.

Nozzles are located so that no point in the protected area is more than 9 m (30 ft) from a nozzle. If there is an obstruction to the flow of foam, additional nozzles must be employed.

The nozzles protecting boiler flats are positioned to spread the foam under the floor plates on the lower boiler flat. The foam will then follow the path of a fuel-oil spill, providing it does not have to travel very far from the nozzle. In machinery spaces the nozzles are placed to protect the bilge. U.S. Coast Guard regulations require that the distance from the bilge nozzles to the bilge be no less than 152 mm (6 in.). When the foam system is used to protect an oil-fired boiler installation on a boiler flat that can drain to the lower engine room, both spaces should be protected simultaneously. Nozzles should be located near the boiler flat and near the floor plates.

Hydrant Requirement—Additional Protection. According to Coast Guard regulations, "two additional fire hydrants are required outside of machinery spaces to extinguish residual fires above the floor plates." These two hydrants are additional to those required for the fire-main system.

If the foam system is blanketing a fire below the floor plates, fire above the floor plates should be attacked with the low-velocity applicator. Water usage should be kept to a minimum to prevent excessive drainage into the bilge and dilution of the foam. A straight stream should definitely not be used. It could "dig into" the foam blanket, break it up and allow the fire to reflash.

Valves and Piping. A diagram of the piping system and control valves should be posted in the foam supply room. It should show which valves are to be opened in the event the system must be activated. The diagram should explain thoroughly and clearly all the steps necessary to put the system into operation. Color coding the valves aids in identification, e.g., all valves that are to be opened when a fire alarm is received might be painted some distinctive color. Each valve could also be labeled as to its function; this would be of help in operating, restoring and maintaining the system.

Deck Foam Systems (Tankers)

Deck foam systems are required on all tank vessels by the 1970 Tank Vessel Regulations. The foam system replaces the fixed-pipe, inert-gas smothering system, for improved fire protection. With a fixed-pipe, inert-gas system, the rupture of a key inert-gas line would make it impossible to get inert gas to the fire. The rupture of a tank would make it impossible to maintain an inert-gas concentration.

The deck foam system is intended to protect any deck area with foam applied from stations (monitors or hose stations) located aft of the area. At least 50% of the required rate of application must come from mounted devices (deck foam monitors). Mounted appliances have greater capacity and range, require fewer personnel, and can be put into operation in a much shorter time than handheld devices. Title 46 CFR 34.20 re-

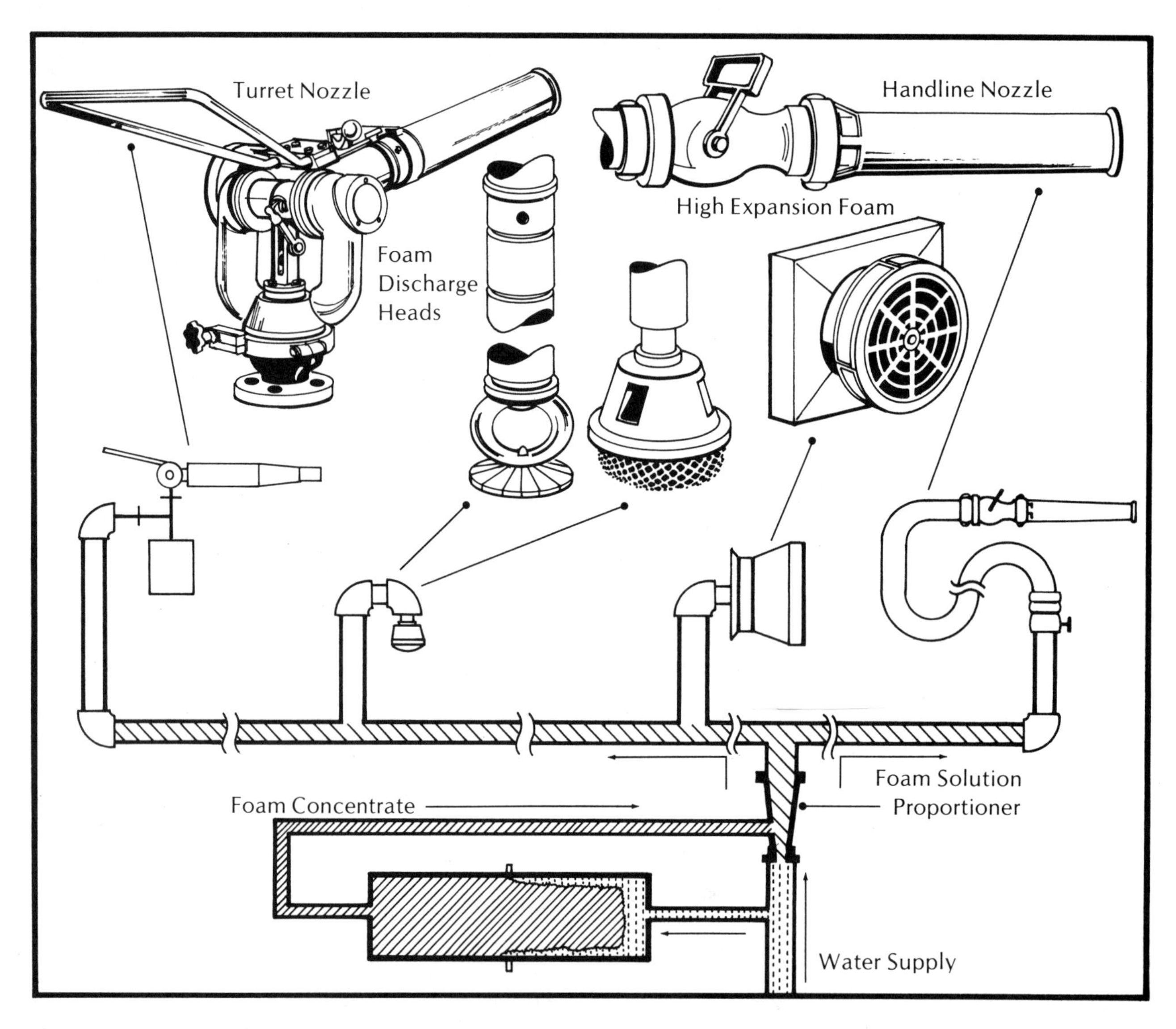

Figure 9.18. Fixed mechanical foam systems can supply turret and handline nozzles, foam discharge heads and high-expansion foam devices. (Courtesy Rockwood Systems, Inc.)

quires that each foam monitor have a capacity of at least 3 liters/min per m^2 covered (0.073 gal/min per ft^2 covered). At least one handheld device must also be provided at each foam station for flexibility during the final stages of extinguishment. The system piping and the foam stations must be arranged so that a ruptured section of piping may be isolated during a fire. With this arrangement, it is possible to fight a fire effectively by working forward from the after house (assuming the machinery and foam generating equipment is aft).

The system is supplied from a central station that houses the foam-concentrate tank, proportioning unit, foam pump and control valves. Piping carries the foam solution from the central station to foam stations located on deck, above the cargo tanks. Each foam station is equipped with a foam turret nozzle and may have one or two foam-dispensing handlines. The stations are generally located so that the foam pattern from each station overlaps the foam patterns from adjacent stations.

How the System Works. The foam system and each foam station are activated manually. The first step is to activate the foam pumps and open the proper valves in the foam supply room. This starts the flow of foam solution to the fire station through the main piping. The turret nozzle is put into operation by opening a valve that is usually located in the supply pipe at the base of the turret. The handlines at the foam station also must be put into operation manually. When the foam solution passes into the foam turret or hand nozzle, air is drawn in; it mixes with the foam solution to produce low-expansion mechanical foam.

Rate of Foam Flow. The required foam solution rate is 0.65 liter/min per m^2 (0.016 gal/min per ft^2) of the entire tank surface, for 15 minutes. The *entire tank surface* is defined as the maximum beam of the vessel times the longitudinal extent of the tank spaces. The required rate is based on the typical T-2 tanker configuration shown in Figure 9.19. *Note:* The term "water rate" is used by USCG; its meaning is synonymous with that of "foam solution rate" for 3% and 6% concentrate systems.

For the usual petroleum products, the foam solution rate must be at least 0.65 liter/min per m^2 (0.016 gal/min per ft^2) of cargo area or 9.8 liters/min per m^2 (0.24 gal/min per ft^2) of the horizontal sectional area of the single tank having the largest area, whichever is greater. The quantity of foam available must be sufficient for 15 minutes of operation, or 20 minutes of operation without recharging on installations after January 1, 1975. The *cargo area* is defined as the maximum beam of the vessel times the longitudinal extent of the tank spaces.

The foam solution rate was determined by assuming fire in hold no. 3C, and probable and possible fire areas as shown in Figure 9.19. The total possible fire area is approximately one third of the total tank area of the vessel. The time of application of the required rate (15 minutes) was based upon two considerations:

1. If the fire were to burn longer than 15 minutes, it is improbable that it could be contained and extinguished by the vessel's crew.
2. The previous requirement for fixed systems protecting cargo tanks was to apply foam for 5 minutes per tank. Since the possible fire area covers three tanks, the required application time is $3 \times 5 = 15$ minutes.

Fire extinguishment does not always depend on the thickness of the foam blanket. What is important is to maintain an effective vaportight cover on the fire. Some foams require thick blankets to accomplish this. Other foams can do an equally effective job with thinner blankets. Thus, the rate of application, as related to some standard rate, is the important factor.

Pumps. The use of the deck foam main must not interfere with simultaneous use of the fire main. The ship's fire pumps may be used to provide water for foam generation if, in a fixed system, the pumps are located outside the protected space. If the foam system water supply is taken directly from the fire main, a single fire pump must be capable of meeting the fire-main and foam system requirements simultaneously.

The foam system piping may not be used for any other purpose. If it were, complex operating instructions would be required. This would make the foam system something other than a versatile fire protection system that can be put into immediate use. In addition, there would be a possibility of pumping the foam out through the ballast connection rather than the monitors and handline nozzles.

CARBON DIOXIDE SYSTEMS

Carbon dioxide (CO_2) systems are used to protect cargo spaces, pump rooms, generator rooms, storage spaces such as paint and lamp lockers, galley ranges and duct systems. They are also used in engine rooms and to protect individual generators.

As an extinguishing agent, CO_2 is especially adaptable to shipboard use: It will not damage expensive cargo or machinery. It leaves no undesirable residue to be cleaned off equipment and

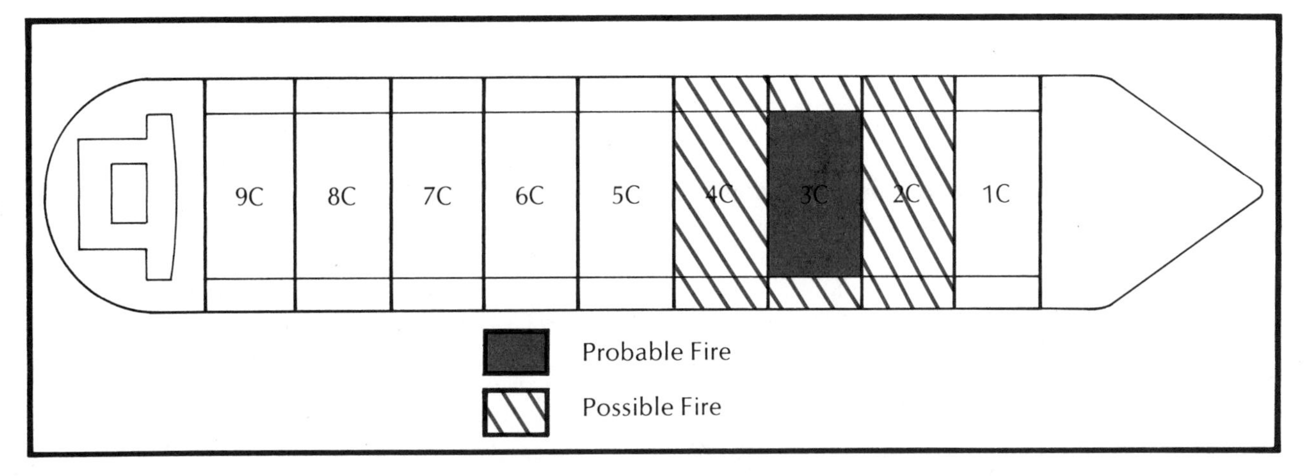

Figure 9.19. Tanker configuration used to determine the required rate of foam flow.

decks. It does not conduct electricity, and so can be used on live electrical equipment. It is released as a liquid under pressure and expands to a dense gas at atmospheric pressure. It will remain at the lower levels of a space until it diffuses with time and a temperature rise.

There are some disadvantages to CO_2. The amount that can be carried on a ship is limited, because it must be stored in cylinders under pressure. CO_2 has little cooling effect on materials that have been heated by the fire. Instead, CO_2 extinguishes fire by smothering, i.e., by displacing the oxygen content in the surrounding air to 15% or lower. Thus, materials that generate their own oxygen as they burn cannot be extinguished by CO_2.

CO_2 is hazardous to humans. The minimum concentration sufficient to extinguish fire does not reduce the oxygen content of the air to a hazardous level. However, when inhaled, the CO_2 raises the acidic level of the blood. This prevents the hemoglobin from absorbing oxygen in the lungs, which can lead to a respiratory arrest. Thus, it is extremely dangerous to enter any compartment in which CO_2 has been discharged, without proper breathing apparatus. This applies even to supposedly short periods of time, e.g., a crewman might be tempted to hold his breath while darting into a compartment to rescue a person lying unconscious on the floor.

CO_2 is especially effective against fires involving flammable liquids. It will also control fires involving class A combustibles in confined spaces.

Types of Marine Systems

Two fixed CO_2 systems are used for the vessel's protection: The total-flooding system for machinery space and the cargo system. A total-flooding system for machinery space is activated only as a last resort, after all other extinguishing methods have been tried and have failed to control the fire. This system for machinery spaces expels 85% of its total CO_2 capacity within 2 minutes to achieve rapid saturation of the air with CO_2 and quick extinguishment. This rapid release of the CO_2 is necessary in spaces such as engine rooms, where fast-burning flammable liquids must be extinguished quickly. Smaller versions of the total-flooding system are used in generator rooms, pump rooms and paint lockers. The systems designed for these spaces may be supplied by the main system, or they may be complete, independent systems.

The cargo system is not activated immediately upon discovery of the fire. The involved space (usually a cargo hold) is first sealed. Then the agent is introduced into the space at a preset rate, to reduce and maintain the oxygen content at a level that will not support combustion. Cargo systems are used in a break-bulk, ro-ro and stacked-container cargo holds. The cargo tanks aboard cargo and passenger vessels may be protected by a type of CO_2 cargo system. Tank vessels contracted prior to January 1, 1962, may have CO_2 systems in their cargo tanks. Tank vessels contracted on or after January 1, 1970, must be equipped with a deck foam system and may have an approved inert-gas or water spray system for cargo tank protection.

All CO_2 systems consist basically of piping, discharge nozzles of a special configuration, valves and CO_2 cylinders. The cylinders are arranged to discharge their contents into the system through a manifold. The CO_2 is also used to activate alarm devices and pressure switches that shut down ventilation systems. Total-flooding systems and cargo systems are activated manually. Smaller systems (those using less than 136 kg (300 lb) of CO_2) for paint lockers and other small spaces may be automatically activated by heat sensitive devices or may be operated manually.

Actuating a Typical Total-Flooding System

The total-flooding system is actuated manually by pulling two cables. The cable pulls are housed in pull boxes. They are connected through corner pulleys to controls in the CO_2 room. Coast Guard regulations require that the pull boxes be located outside the area being protected, for example, outside an engine room doorway that would be a normal route of escape. The cable pulls are protected by glass to prevent tampering. To operate the glass must be broken with the hammer that is provided. Then each cable must be pulled straight out. *The cables must be pulled in the required sequence.* Instructions explaining how to actuate the system should be posted over the pull boxes (Fig. 9.20).

One cable is connected to the control heads on the pilot cylinders; the other is connected to a control head mounted on the pilot port valve. When both cables are pulled, CO_2 discharges from the two pilot cylinders and opens the pilot port valve. The CO_2 is delayed from discharging into the fire area by a stop valve. During the delay, the CO_2 is routed through the pilot port valve to the discharge delay device, into piping where it actuates pressure switches. About 20 seconds is required for the CO_2 to pass through the discharge delay device. During this time, ventilation systems are shut down and alarm de-

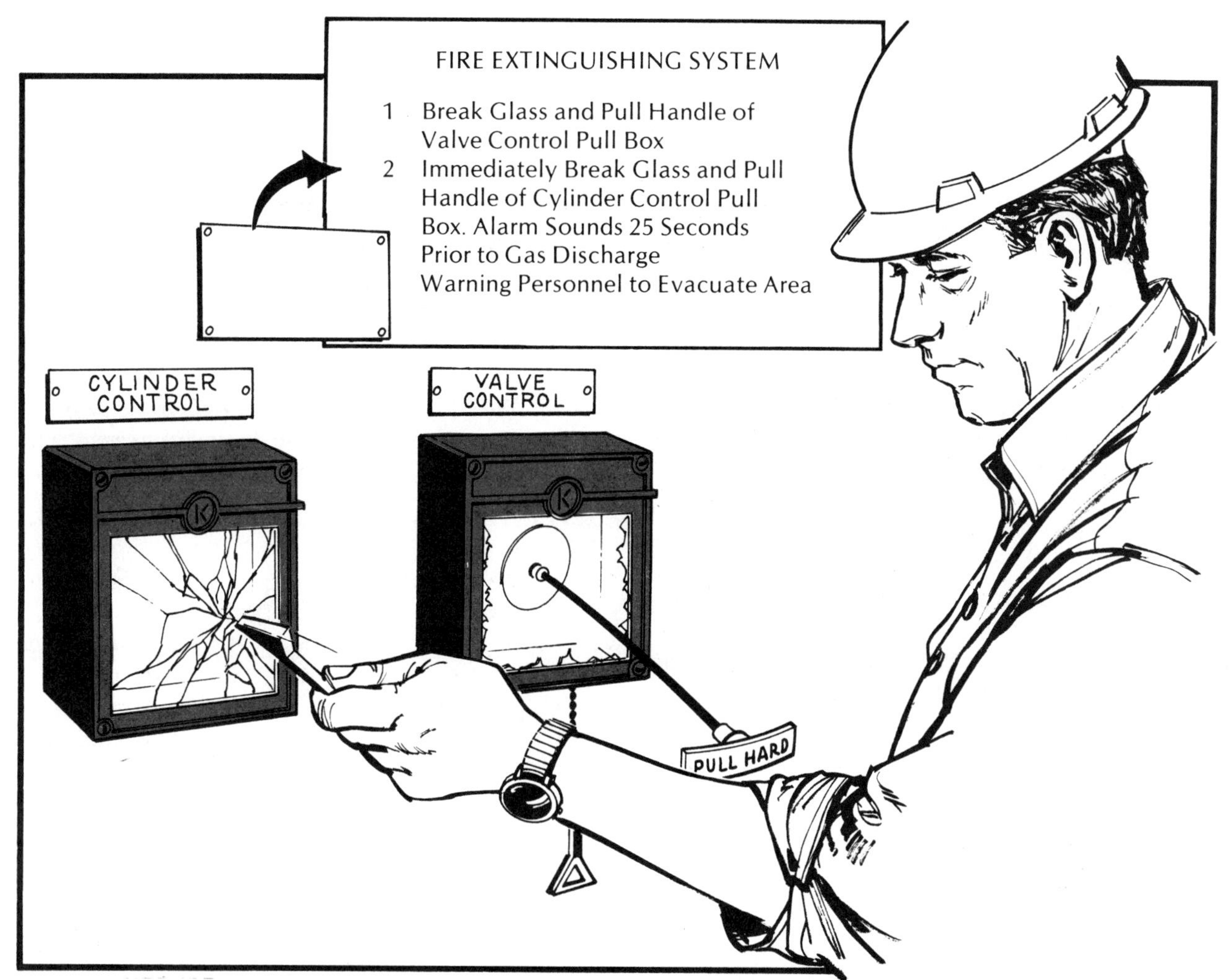

Figure 9.20. The pull cables used to activate the total-flooding CO_2 system. The cables must be pulled in the proper order (valve control first) as noted in the posted instructions.

vices are actuated. After the 20-second delay, the CO_2 acts on a pressure control head mounted on the stop valve. The valve opens, permitting the CO_2 to discharge into the protected space.

Carbon Dioxide Warning Alarm. An approved audible alarm must be installed in every space protected by a CO_2 extinguishing system (other than paint and lamp lockers and similar small spaces) and normally accessible to persons on board while the vessel is being navigated. The alarm must be arranged to sound automatically for at least 20 seconds prior to the discharge of CO_2 into the space. It must not depend on any source of power other than the CO_2. The alarm must be conspicuously and centrally located and marked "WHEN ALARM SOUNDS VACATE AT ONCE. CARBON DIOXIDE IS BEING RELEASED" (Fig. 9.21).

Actuation Procedure. In case of fire in the engine room, once the decision to release the fixed CO_2 system is made, the following procedure should be followed:

The alarm is a warning that the carbon dioxide system has been activated. Once it sounds, you have about 20 seconds to get out of the space. *Do not delay—leave immediately.* If you delay, the CO_2 will flood the space and reduce the oxygen content below the level required to sustain life. *Failure to evacuate immediately could result in loss of life.*

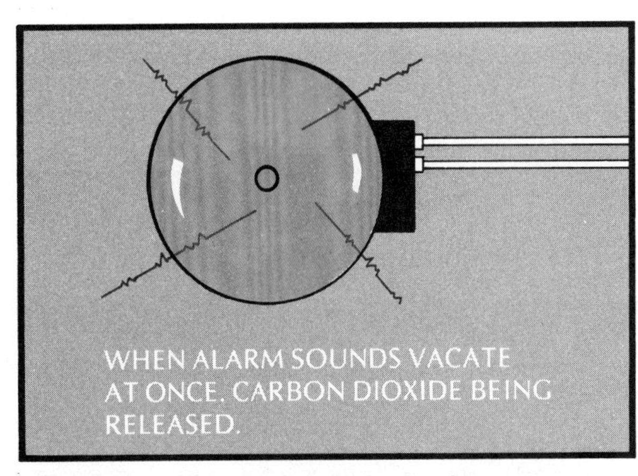

Figure 9.21. Carbon dioxide alarm and posted warning.

1. Warn personnel to evacuate the space.
2. Close all doors, hatches and other openings.
3. Secure main and auxiliary machinery.
4. Go to the pull boxes at the engine room exit.
5. Break the glass and pull the handle of the pull box marked "valve control" (Fig. 9.20).
6. Immediately break the glass and pull the handle of the pull box marked "cylinder control" (Fig. 9.20).

The system may also be activated from within the CO_2 room. Here the procedure is to remove the locking pins and operate the levers of the control heads mounted on the two pilot cylinders and the pilot port valve. The discharge delay may be bypassed by removing the locking pin and operating the lever of the pressure control head mounted on the stop valve. (*See* Chapter 10 for a discussion on using CO_2 total-flooding system to combat an engine room fire.)

Combination Smoke Detection—Carbon Dioxide System

Smoke detection systems are often installed in cargo spaces, along with carbon dioxide extinguishing systems. When the presence of fire is detected by the smoke detection system, it sounds alarms in the CO_2 room, bridge and engine room. The line number indicator locks on the monitoring line from which smoke is detected. The CO_2 extinguishing system must then be activated manually. (*See* Chapter 6 for a discussion of smoke detection systems.)

Recheck and Initial Discharge. The following procedure should be used when the smoke detection system signals a cargo space fire:

1. Confirm the presence of smoke. To do so, depress the recheck (or reset) button on the main detecting cabinet, and note whether the indicator locks on the same line again. Then visually inspect all monitoring lines for smoke.
2. Check the number of the line showing smoke against the line index chart, to determine which space is involved.
3. Make certain no one is in the space.
4. Shut off all mechanical ventilation; seal all ventilators, ports, sounding pipes and hatches leading to the space.
5. Refer to the line index chart or profile chart for the number of cylinders to be discharged into the involved space.
6. Open the three-way valve whose number (on the handle) is indicated on the line index chart. To do so, pull the handle down so the word "extinguishing" is visible. When more than one line is installed in a space, open all the three-way valves for that space (Fig. 9.22).
7. Discharge the required number of cylinders, in pairs. To do this, remove the locking pin and operate the lever of the control head mounted on one cylinder of each pair. *Caution: Do not operate the control heads mounted on the pilot cylinders until all other cylinders have been discharged.* Operating the pilot cylinders first will activate all the cylinders connected to the manifold (Fig. 9.23).

If fire occurs in two spaces simultaneously, only one three-way valve is opened; the required number of clinders is discharged into the space served by that valve. When this first space has been charged with CO_2, the first valve is closed. The second three-way valve is then opened, and the required number of cylinders is discharged into the second space. Carbon dioxide should be discharged into the *lowest* involved space first, then into the next higher space, and so on.

Delayed Discharge. To maintain the proper CO_2 level, additional cylinders should be discharged into the involved space at intervals ranging from 30 minutes to 6 hours. If the smoke increases in intensity or the surrounding plates and bulkheads get hotter, shorter intervals are indicated; if conditions are favorable, longer intervals are acceptable. The number of cylinders to be discharged and the intervals are shown on the line index chart and profile chart, as delayed discharges.

Because the supply of CO_2 is limited, it should be used carefully. The distance of the ship from port and the possibility of obtaining additional CO_2 at a port should be taken into account. Cargo hold fires usually are not extinguished quickly; they often require days to extinguish. (*See* Chapters 3 and 10 for detailed discussions of cargo hold fires.)

Lash and Seabee Barge-carrying Ships. Combination smoke detection–carbon dioxide systems are used in the cargo holds of Lash and Seabee barge carrying ships. When fire is detected in the hold of a Lash vessel carrying barges, the CO_2 is released into the entire hold—not into individual barges. On Seabee vessels, a smoke detector monitoring line is attached to each barge. When fire occurs, the smoke detector identifies

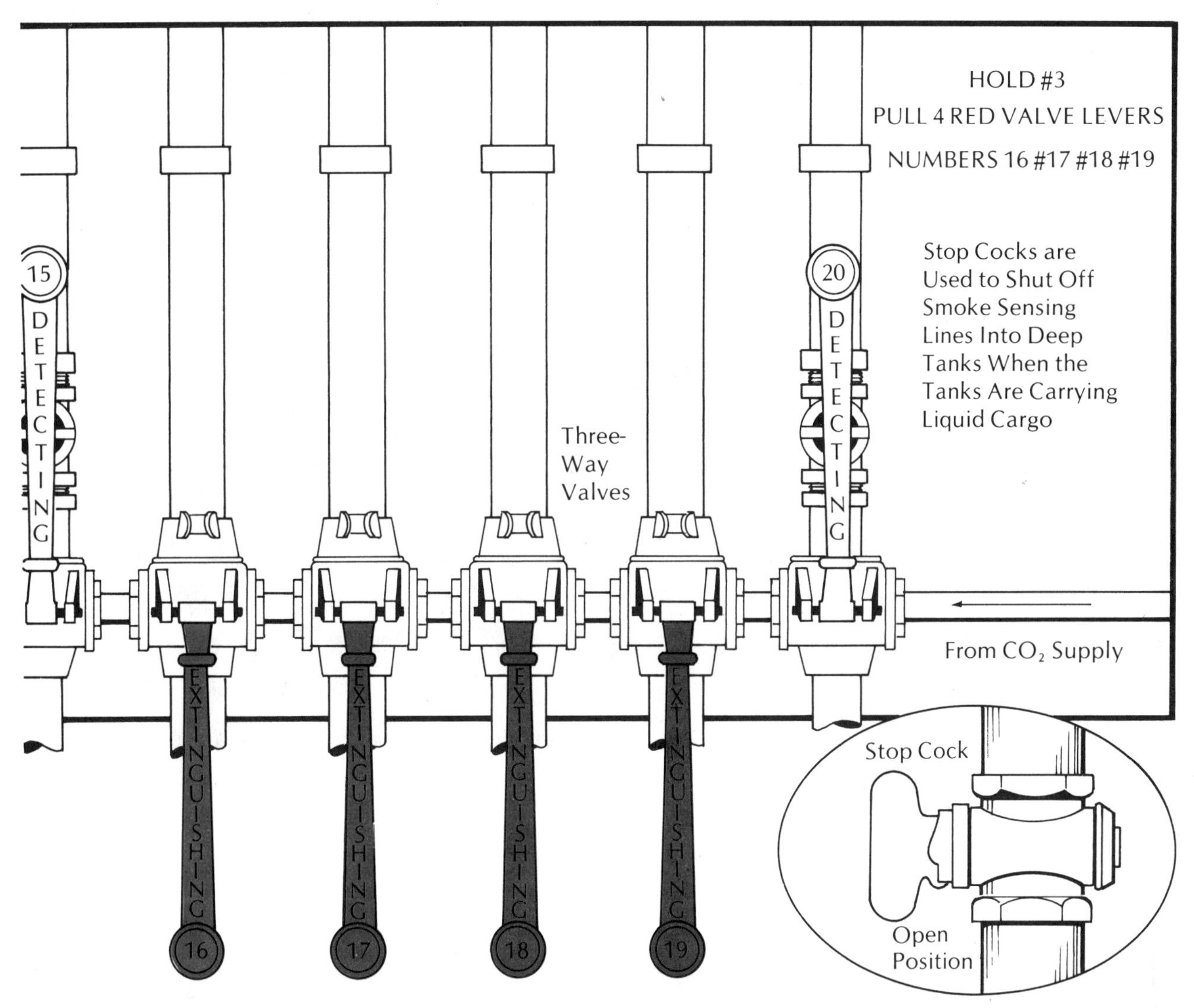

Figure 9.22. The proper three-way valve (or valves) must be pulled down to route CO_2 to the involved space. Here, the involved space is served by four lines.

the barge. CO_2 is manually released into the involved barge through the smoke detector monitoring line attached to that barge. The specific amount of CO_2 to be released is given in the CO_2 discharge instructions.

Independent Carbon Dioxide Systems

The paint locker, lamp locker, engineer's paint locker and generator rooms can be protected by independent fixed CO_2 systems, i.e., each system has its own CO_2 supply, independent of other CO_2 systems. If the space requires less than 136 kg (300 lbs) of CO_2, the cylinders may be installed inside each protected space provided they are capable of automatic operation. These systems are manually operated, in addition to being activated automatically by heat detectors.

Independent Automatic System. An independent automatic system is composed of heat detectors, one or more CO_2 cylinders, piping, valves and discharge nozzles. The heat detectors are usually of the pneumatic type. They are located on the overhead of the protected space and are connected by tubing to a pneumatic control head mounted on the pilot cylinder. Air contained in the heat detector expands as the temperature rises; the resulting pressure increase is transmitted to the control head through the tubing. The control head is vented so that normal increases in temperature will not activate the system. However, the sudden increase in pressure caused by a fire cannot be vented off fast enough. This increased pressure operates a diaphragm-lever arrangement in the control head and releases the contents of the CO_2 cylinder. The carbon dioxide then discharges into the space through approved nozzles. When the system goes into operation, it sounds an audible alarm to alert crew members to begin evacuation immediately.

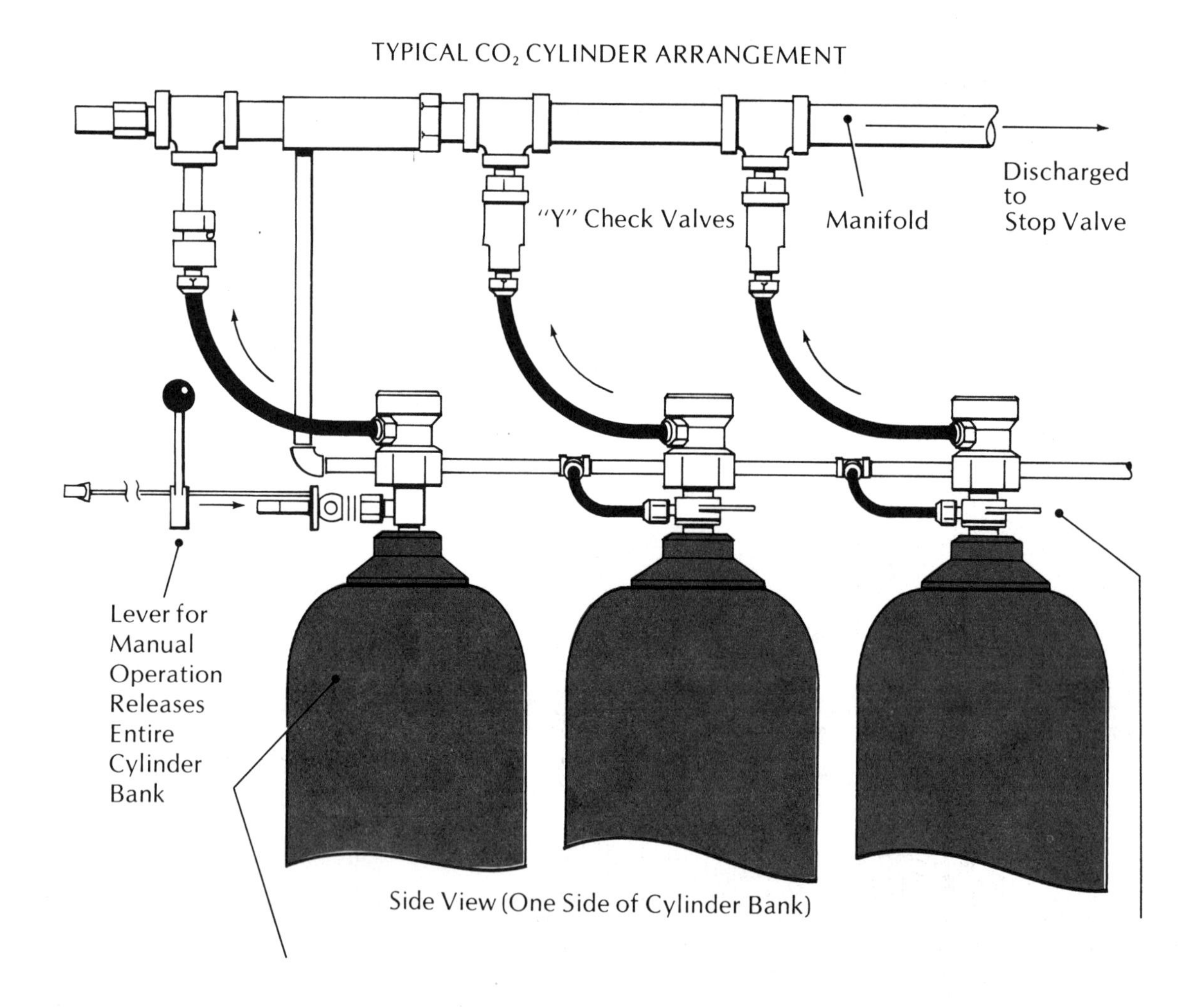

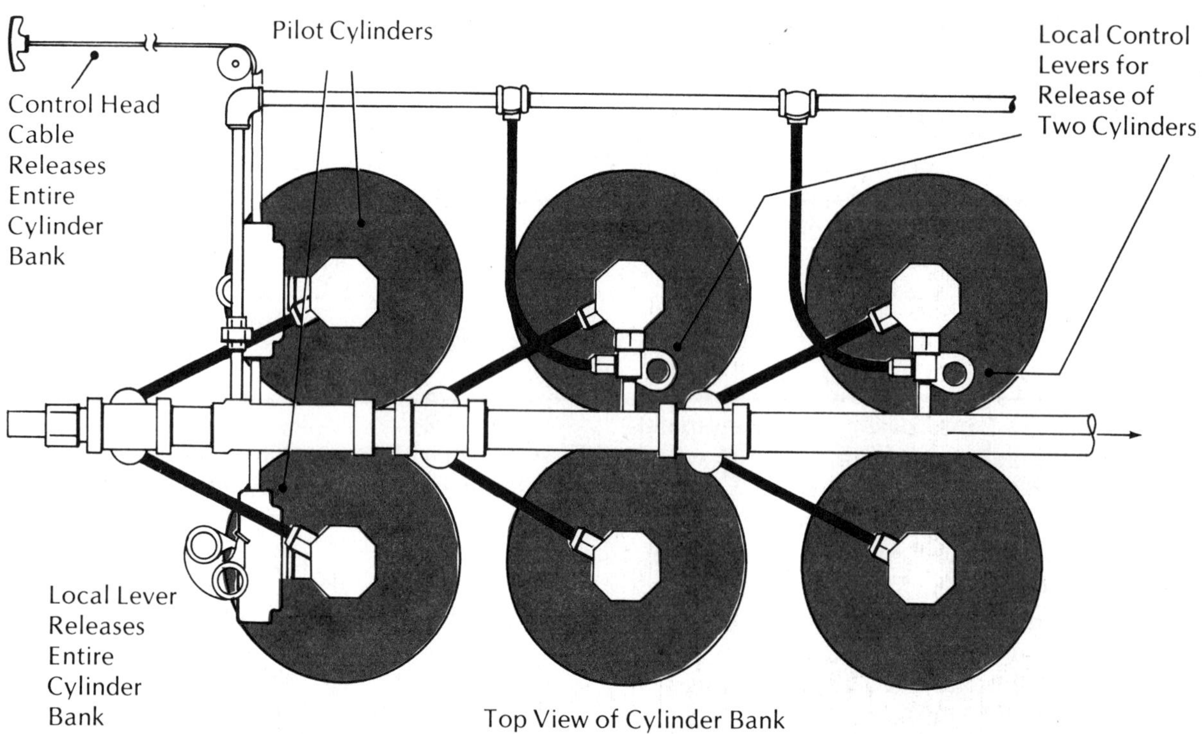

Figure 9.23. The pilot CO_2 cylinders must be activated last when CO_2 is routed to a cargo space fire. Otherwise, all the cylinders in the bank will be opened at once. (Courtesy Walter Kidde and Company)

Independent Manual System. To activate an independent manual system, one or two pull cables (depending on the manufacturer) must be operated. The cables are, as usual, located in pull boxes outside the protected space.

To activate a manual system

1. Make sure no one is in the protected space.
2. Close all doors, hatches and other openings to the space.
3. Operate the pull cable (or cables) according to the posted instructions.

The involved space should remain buttoned up for several hours, if possible. Then, if no heat buildup is evident on decks or bulkheads, a hoseline, charged with water and ready for use, should be positioned outside the door. The door should then be opened slightly. If fire is not evident, the door should be left open, but no one should enter the space. First, a breathable oxygen level must be allowed to build up. If the CO_2 has extinguished the fire, the atmosphere in the space may be tested for oxygen content after some time has passed. It is important not to rush in before making sure that the space will support life.

Carbon Dioxide Protection for Rotating Electrical Equipment. Carbon dioxide systems are used to protect generators with CO_2 discharge nozzles located inside the casing. The piping leads from the casing to a cylinder of CO_2. Most units of this type are activated manually. The need for evacuation of personnel should be evaluated when the system is designed, by determining the degree of oxygen depletion caused when CO_2 is discharged. Since the CO_2 does not flood the engine room, evacuation may not be required.

Inspection and Maintenance of Carbon Dioxide Systems

Carbon dioxide systems are reliable when they are maintained properly. Almost all malfunctions are due to neglect. When CO_2 systems have failed to control or extinguish fire, it was usually because they were used incorrectly, owing to a lack of knowledge. These fire extinguishing systems require only normal care to ensure proper operation when they are needed. However, they should be inspected on a regular basis, to combat the tendency to neglect emergency equipment of this kind.

Monthly Inspection. At least once a month, each fixed CO_2 system should be checked to ensure that nothing has been stowed so as to interfere with the operation of the equipment or with access to its controls. All nozzles and piping should be checked for obstruction by paint, oil or other substances. The semi-portable hose-reel horn valve should be operated several times. Any damaged equipment must be replaced immediately.

Annual Inspection. It is recommended that a qualified fire protection technician or engineer make the annual inspection. Each year all the cylinders should be weighed, and the weights recorded on the record sheet. If a weigh bar is not installed above any cylinder, the cylinder must be placed on a scale for weighing. The full and empty weights of each cylinder are stamped on the cylinder valve. A cyclinder is considered satisfactory if its weight is within 10% of the stamped full weight of the charge.

Removing Charged Cylinders. When charged cylinders are to be removed from service, the discharge must be disconnected first. This eliminates the possibility of accidentally discharging the cylinders. Here is the recommended procedure for a typical system:

1. Remove the discharge heads from all cylinder valves by loosening the mounting nuts, which have right-hand threads. On installations of more than one cylinder, allow the discharge heads to hang on the loops.
2. Remove all the control heads from the cylinder valves by loosening the right-hand-threaded mounting nuts.
3. Screw a large top protection cap onto the threads on top of the cylinder valve. Screw a side protection cap onto the cylinder-valve control-head outlet.
4. Remove the cylinder rack.
5. Remove the cylinder. It is recommended that the cylinder cap be screwed on to prevent damage to the cylinder valve during removal.

Installing Charged Cylinders. When charged cylinders are placed in service, the discharge heads are replaced last. Here is the recommended procedure for a typical system:

1. Place the fully charged cylinder in the cylinder rack before removing the cylinder cap.
2. Install the cylinder rack, and handtighten the bolts so that the cylinder may be rotated in place.
3. Remove the cylinder cap and the top and side protection caps.

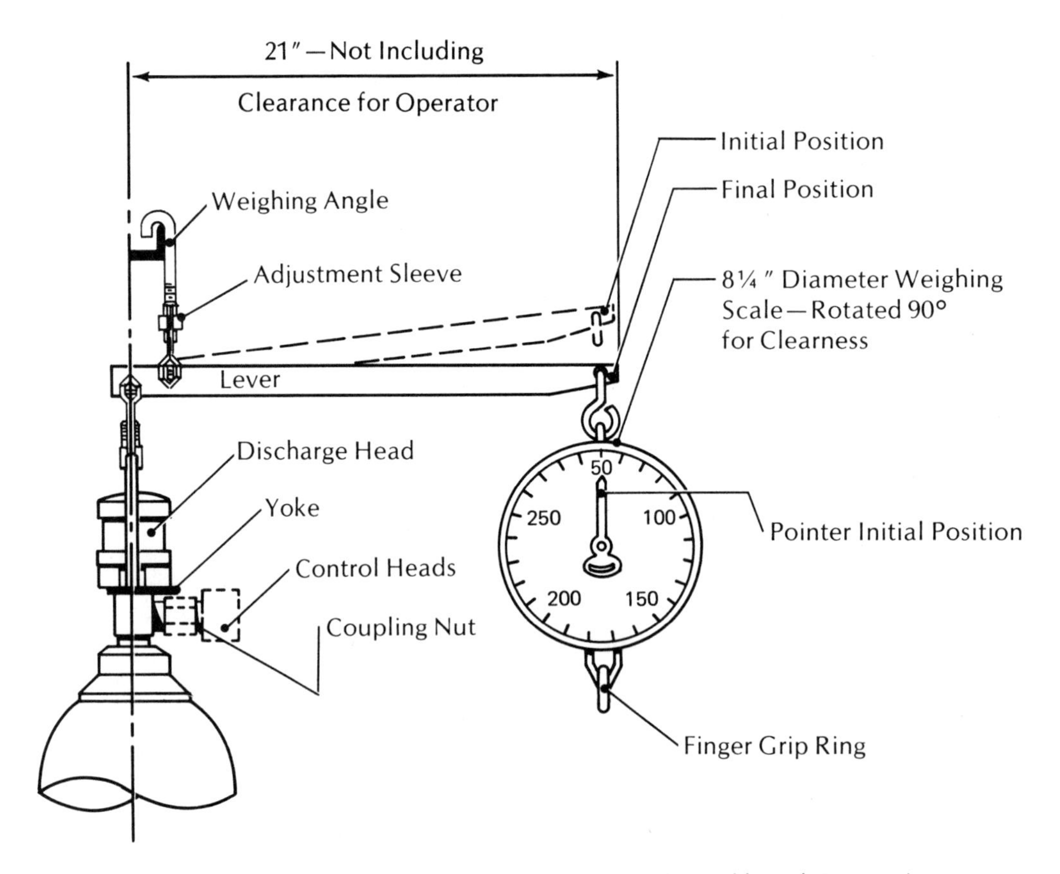

Figure 9.24. Carbon dioxide cylinder weigh bar and its use. (Courtesy Walter Kidde and Company)

4. Turn the cylinder so that the control-head outlet points in the proper direction. Tighten the cylinder rack bolts securely.
5. Make certain that all control heads have been reset, as follows:
   a. *Cable-operated control head:*
      (1) Remove the cover from the control head.
      (2) Make sure the plunger is retracted below the surface of the control-head body. Then engage a few threads of the mounting nut onto the cylinder valve.
      (3) Retract the actuating roller as far as possible from the direction of pull.
      (4) Replace the cover and locking pin, and install a new seal wire. (Note: When two control heads are connected in tandem, make certain both are completely reset before assembling them to the cylinder valves.)
   b. *Lever-operated control head:*
      (1) Return the lever to the set position, with the plunger fully retracted into the control-head body.
      (2) Replace the locking pin, and install a new seal wire.
   c. *Pneumatic control head:*
      (1) Insert a screwdriver into the reset stem. Turn it clockwise until the stem locks in position with the arrow on the reset stem lined up with the set arrow on the nameplate. The plunger should be fully retracted into the control head body.
      (2) Replace the locking pin, and install a new seal wire.
6. Reinstall the control head on the cylinder valve, tightening the mounting nut securely.
7. Connect the discharge head to the cylinder valve, tightening the mounting nut with a wrench that is at least 457 mm (18 in.) long.

Replacing Damaged Discharge Nozzles. If a discharge nozzle in a total-flooding system must be replaced, it should be replaced with a nozzle of the same size and discharge rate. Each discharge nozzle is installed to achieve a set discharge rate, and to ensure saturation of the space it protects in a certain length of time. The wrong nozzle can destroy the effectiveness of the system in the affected space.

MARINE HALON 1301 SYSTEM

A halogenated extinguishing agent, Halon 1301, has been accepted by the Coast Guard for limited use in fixed firefighting systems aboard U.S. ships. Halon 1301 is a very efficient extinguishing agent for fires involving flammable liquids and gases and live electrical equipment. It is a clean agent; its residue does not contaminate electrical contacts or circuits. It is a nonconductor of electricity.

Halon 1301 is a colorless, odorless gas. It may be toxic when exposed to flames. (This is taken into consideration in the engineering of Halon 1301 systems.) When the flames are extinguished quickly, a minimal amount of toxic material is produced. Slow extinguishment allows increased production of toxic materials at levels that could be dangerous to personnel.

The effectiveness of Halon 1301 as an extinguishing agent comes from its ability to chemically interrupt the combustion process. When applied in the proper concentration and at the proper delivery rate (usually in less than 10 seconds), it extinguishes flames very rapidly. (*See* Chapter 7 for a discussion of the other properties of Halon 1301.)

Halon 1301 System Requirements

To be acceptable for use on U.S. ships, a Halon 1301 system must be at least as reliable and effective as the system it replaces. Most of the Halon 1301 systems approved by the U.S. Coast Guard protect machinery spaces, turbine enclosures and pump rooms, where the usual petroleum products may be found. Halon 1301 systems are not yet approved for installation in the holds of ships carrying general cargo (usually class A materials.)

The spaces for which Halon 1301 systems have been approved are those normally protected by CO_2 systems. Thus, Halon 1301 systems must meet all the design requirements for CO_2 total-flooding systems. These include

1. Evacuation of all personnel from the protected space before the extinguishing agent is discharged. (A warning alarm, audible above operating machinery, is required. Personnel must be protected from both Halon 1301 and its toxic decomposition products.)
2. Stowage of the extinguishing agent outside the protected space except for space less than 169.9 m^3 (6000 ft^3) and modular systems.
3. Performance of two separate actions to activate the system. (Two pull boxes are used; one activates the pilot cylinders and one controls the stop valve, as in the CO_2 system.)
4. Manual activation of the system, except for spaces with a volume less than 169.9 m^3 (6000 ft^3). (Systems for these smaller spaces may be activated automatically, and the extinguishing agent may be stowed within the protected space. However, an automatic system must also be capable of manual operation.)
5. Posting of detailed instructions for activating alternate means of discharging the system at the remote release station.

Two types of Halon 1301 fire extinguishing systems have been approved by the Coast Guard. One, the *preengineered* type, includes a system approved for limited installation in unmanned spaces on hydrofoil craft (maximum volume of 63.7 m^3 (2250 ft^3) and uninspected pleasure craft. The other, the *engineered* type, includes systems installed on merchant vessels.

Engineered Halon 1301 Systems

The engineered system is a total-flooding system. The Halon 1301 extinguishing agent is contained in cylinders in liquid form. It is pressurized with dry nitrogen to a pressure of 2482 or 4137 kilopascals (360 or 600 psi) at 21°C (70°F). The cylinders are stowed outside the protected space, in an area whose temperature is maintained between −29°C and 54.4°C (−20°F–130°F). The bulkheads if contiguous separating the cylinders from the protected space must be A-60 bulkheads.

The cylinders are connected to a manifold that, in turn, is connected to piping leading to the protected space. All cylinders on a common manifold must be of the same size and must contain the same quantity of Halon 1301. This ensures equal flows from all cylinders. The cylinders should be adequately supported. Each cylinder must have a pressure relief device and a pressure gauge. Figure 9.25 shows one of the general arrangements of Halon 1301 cylinders approved by the U.S. Coast Guard.

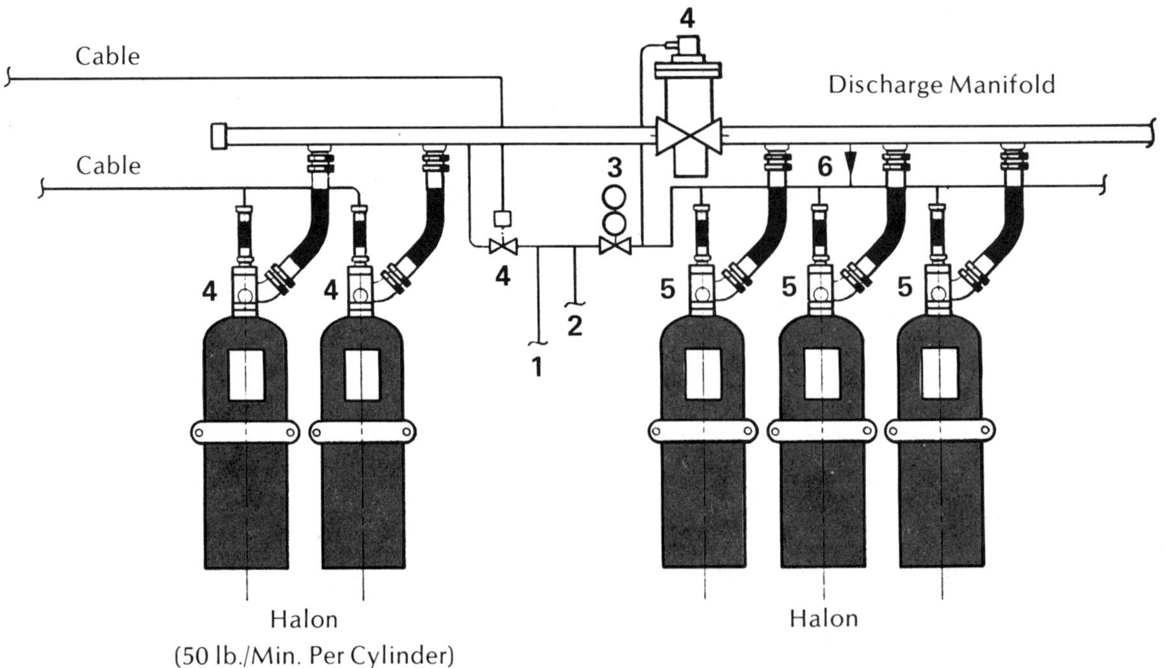

Figure 9.25. A U.S. Coast Guard approved arrangement for Halon 1301 cylinders in a manually operated engineered extinguishing system. *1)* Line to pressure-operated alarm. *2)* Line to pressure-operated switches. *3)* Time delay. *4)* Valve capable of manual operations. *5)* Pressure-operated valve (cannot be operated manually). *6)* Check valve.

Extinguishing-Agent Discharge Requirements. Enough Halon 1301 must be available to provide a minimum concentration of 6% of the gross volume of the protected space. A concentration of 7% may be required for the effective extinguishment of fires involving most marine fuels.

Both liquid and gaseous Halon 1301 flow through the piping to protected areas. Thus, the flow rates and piping sizes must be carefully computed when the system is designed. In addition, the system must be "balanced" so that all spaces that may require simultaneous discharge are adequately served.

Controls. A remote release (pull-box) station is required for each protected space. It should be located close to one exit from that space. Posted instructions at the release station should describe how to activate the system from the station. They should also describe an alternative means of activating the system in case the remote release fails. The instructions should be in large print and easily understood.

A warning device, actuated by pressure from the Halon 1301 system, must sound an alarm when the agent is about to be discharged into the protected space. The discharge must be delayed, to give personnel sufficient time to evacuate the space before the Halon 1301 is released. A sign should be posted at the warning device, explaining its purpose. In addition, a sign must be posted at each entrance to each protected space. The sign must warn crew members not to enter the space without breathing apparatus after the system has been activated.

A schematic diagram of the entire system should be posted in the Halon 1301 storage room. Each section of the system should be numbered, color coded or identified by name. The valves controlling these sections should be similarly identified. Instructions for activating the system should be posted in the storage room and at each pull box or stop valve. Again, the instructions should be in large print and easily understood. They should indicate which valves must be operated to activate each section of the system.

Ventilation. If a protected space is ventilated mechanically, the ventilation system must be automatically shut down by the release of the Halon 1301. Time must be allowed for fans and motors to stop rotating before the agent is released into the space. There must be some provision for sealing off points where the Halon 1301 could escape from the protected space. If this is not done, the concentration of the agent can be reduced below the effective level. Additional Halon must be provided, to make up for any leakage.

If a diesel or gasoline engine draws air from the protected space, the engine must be shut down before the extinguishing agent is released. Otherwise, the Halon 1301 would be decomposed by the high pressure and temperature within the engine. An automatic shutoff, activated by the extinguishing system, is required.

Inspection and Maintenance. If any system is to perform properly in an emergency, it must be inspected at intervals and maintained as necessary. Halon 1301 systems should be checked as follows

1. The cylinders should be weighed periodically. A weight loss of 5% or more indicates that the affected cylinder should be replaced or recharged.
2. The cylinder pressures should also be checked periodically. Table 9.2 gives normal pressures for a range of ambient temperatures. A pressure loss of 10% or more (for a given temperature) indicates that the affected cylinder must be recharged or replaced.
3. Remote release levers, cables and pulleys should be checked to ensure smooth operation.
4. Automatic switches and warning alarms should be checked to verify that they are operating properly.
5. Halon 1301 cylinders should be hydrostatically tested every 12 years.

DRY CHEMICAL DECK SYSTEMS

Ships carrying liquefied gases in bulk are now being fitted with a dry chemical fire extinguishing system to conform with IMCO and U.S. Coast Guard recommendations and regulations. The system is used to protect the cargo deck area and all loading-station manifolds on the ship (*see* Fig. 10.19). Each deck system is actually made up of several independent skid-mounted units (Fig. 9.26). The units are placed on deck so that they protect overlapping areas. The units are self-contained firefighting systems that use dry chemical.

Components of the Skid-Mounted Unit (A Typical System)

Each unit consists of a large capacity storage tank holding up to 1361 kg (3000 lb) of dry chemical 11.3 m^3 (400-ft^3) capacity nitrogen cylinders (6–8 per skid) and 30.5–45.7 m (100–150 ft) of lined, round rubber hose on reels. The unit can be fitted with a turret nozzle and several handlines. In some systems, handlines are used exclusively; in this case, up to six handline stations can be supplied, by each unit. Generally, the hose reels are mounted on the unit. However, in some installations, remote handlines are connected to the unit via piping. The hoselines are equipped with special On–Off controlled nozzles.

IMCO requires that monitor turrets have a discharge rate of not less than 10 kg/sec (22 lb/sec), and handline nozzles not less than 3.5 kg/sec (7.7 lb/sec). The maximum nozzle discharge rate is set by the requirement that one man be able to control the handline. The U.S. Coast Guard has adopted the recommendations of IMCO as at least its minimum standards.

The monitor-turret range required by IMCO is based on the discharge rate:

Monitor-turret maximum capacity kg/sec (lb/sec)	Maximum reach m (ft)
10 (22)	10 (33)
25 (55.4)	30 (99)
45 (99)	40 (132)

A handline is considered to have a range equal to its length. The actual coverage is affected when the target is above the nozzle. Wind conditions also affect coverage.

Table 9.2. Normal Halon 1301 Extinguishing System Cylinder Pressure as Related to Temperature.

Temperature °C (°F)	Cylinder pressure: 2482-kilopascals (360-psi) system	Cylinder pressure: 4137-kilopascals (600-psi) system
4.4 (40)	1896 (275)	3447 (500)
10 (50)	2068 (300)	3654 (530)
16 (60)	2275 (330)	3896 (565)
21 (70)	2482 (360)	4137 (600)
27 (80)	2723 (395)	4413 (640)
32 (90)	2965 (430)	4688 (680)
38 (100)	3241 (470)	5033 (730)

Figure 9.26. Typical skid-mounted dry chemical deck unit. (Courtesy Ansul Company)

A sufficient quantity of dry chemical should be stored on each unit to provide at least 45 seconds of continuous discharge through all its monitors and handline.

How the System Works

When fire is discovered, each skid-mounted unit is activated manually. The nitrogen cylinder valve is opened to release the nitrogen propellent. The nitrogen flows into the dry chemical storage tank through a perforated aerating tube. The holes in the tube are covered with rubber, so that nitrogen can flow into the tank but dry chemical cannot enter the tube. The action is similar to that of a check valve. The nitrogen cylinder valve is calibrated to release dry chemical to the nozzles at the proper rate. The hose should always be stowed with the nozzle in the closed position, since the dry chemical flows to the nozzle as soon as the nitrogen is released.

Activating and operating details vary with different units and manufacturers. In each case, the manufacturer's instructions should be followed carefully. Most important, and common to all systems, is the need to get the skid units into operation quickly when fire occurs. For this, standard activation procedures must be practiced. Emergency procedures, which are used in case the nitrogen valve does not operate properly, must also be practiced.

Firefighting Operations

If a handline is to be used, its entire length should be pulled off the reel. This ensures a smooth flow of dry chemical and provides enough hose for maneuvering at the fire. The handline-nozzle lever has two operating positions: When the lever is pushed forward, the nozzle is closed. When the lever is pulled back, the nozzle is open.

The flow of dry chemical into the turret nozzle is controlled by a turret control valve. The valve must be opened by the turret nozzleman when he is in position. (Some turret nozzles can be activated and controlled from a remote station. Remote operating procedures are given in the manufacturer's instruction manual.) Streams from both handlines and turrets are directed onto fires in the same way.

To fight a spill fire, the stream should be aimed at the base of the fire and moved back and forth in a sweeping action. When a turret and handlines are used together, the turret stream should be used to quickly knock down the bulk of the flames. The handlines should be directed at the flanks of the fire.

Hoselines from the fire-main system should be run out and charged with water. However, water streams should not be directed into the fire unless it is absolutely necessary. Initially, they should be used to protect personnel from the radiant heat, which can be very intense. If possible, the flames should be extinguished or confined to a small area by the use of dry chemical only. Then, water streams can be directed into the area to cool hot surfaces.

In combating a combustible liquid or gas pressure fire, the streams should be directed into the source of the escaping fuel. The velocity of the fuel will carry the extinguishing agent to the flames. It is important, however, to remember that gas leak fires are usually not extinguished, but only controlled until the leak can be stopped. The flames may have to be extinguished if they block the path to the shutoff valve or when lives are in jeopardy. If the flames are extinguished, the area should be kept saturated with water fog until the leak is stopped.

Blowdown and Recharging

After the fire is completely extinguished and the master has declared the area safe, the skid units should be restored to "standby" condition. Dry chemical should be blown out of all handlines and piping. Otherwise, it will cake up within the lines and restrict the flow of agent during the next use. The dry chemical tank should be refilled with the proper agent. The nitrogen cylinders should be replaced, and the remote and pneumatic actuators reset.

It is very important that this blowdown and recharging procedure be performed exactly as described by the manufacturer. Some parts of the procedure can be dangerous if the operating manual is not followed carefully.

Inspection and Maintenance

Every skid-mounted unit should be checked each week to ensure that it is in operation condition. The weekly inspection should include the following:

1. Check the dry chemical tank and all components subjected to the weather for mechanical damage and corrosion.
2. Check the readability of the plates that give operating instructions.
3. Ensure that the cylinder gauges register properly, according to the operating manual.
4. Check the dry chemical level to ensure that the tank is filled properly. The fill cap, if provided, should be hand-tightened only when it is replaced.
5. Check that all reels are in the unlocked position. Pull several feet of hose off each reel, to ensure that the reel moves freely.
6. Check the handline nozzles for obstructions, and operate their levers to check for free movement. In replacing a hose on the reel, make sure the nozzle is secured and the lever is in the closed position.
7. Make sure all tank valve handles have their ring pins and are sealed in the operating position.

Other inspection and maintenance steps may be detailed in the operating manual. Replacement parts should be installed in strict compliance with the manufacturer's instructions.

GALLEY PROTECTION

Three areas within the galley are especially subject to fire. These are

1. The cooking area, including the frying griddles, broilers, deep-fat fryers and ovens.
2. The area immediately behind the filter screens, called the "plenum"
3. The duct system that vents heated gases.

Fires in the cooking area can be serious. However, since they are out in the open, they usually can be extinguished completely. Fires in the plenum and the duct system are of most concern. Even after such fires are apparently extinguished, there may still be some fire hidden from view, or the fire may have extended out of the duct and into nearby compartments. For this reason, automatic fire extinguishing systems should be installed to protect all parts of the range, plenum and ducts. Three types of systems are discussed in this section.

Fire Prevention

Ventilation plays a critical part in protecting the range area against fire. The function of the exhaust blower is not simply to remove odors; its main function is to move sufficient air to keep the entire facility at a safe operating temperature. The blowers and the airflows (under the hood and in the ducts) are designed to keep the temperature from exceeding 93°C (200°F).

Maintenance and cleanliness are also very important. Electric circuits and gas lines must be kept in good condition. Filters should be cleaned often. The filters remove grease and oils from the smoke generated by the cooking process. When these substances remain on the filters, they react chemically. The reactions produce flammable substances, which are a fire hazard. In addition, clogged filters restrict the flow of ventilating air, so that the temperature rises above the safe level. The result is usually a fire.

Dry Chemical Galley Range System

The dry chemical galley system is composed of a pressurized dry chemical cylinder, piping, nozzles and detectors. There are two sets of piping, one in front of the filters over the cooking surface, and one behind the plenum and in the ducting (Fig. 9.27). The proper types and locations of piping and nozzles are determined by fire protection engineers. The nozzles are covered with blowout seals so that they cannot become clogged with grease.

Several types of fire detectors can be used. The system in Figure 9.27 has several fixed temperature fusible links connected by stainless steel cable. The links and cable are connected to a stretched spring. If any fusible link melts, the cable releases the spring, which closes an electric circuit. The resulting current opens the valve of the dry chemical cylinder, releasing its entire contents into the system.

Such galley range systems can usually be activated manually, as well as automatically. Manual controls are normally located near the range and at a remote location, preferably near an exit door. Either manual or automatic activation should also trigger alarms in the galley and in the engine room.

The exhaust blower should not be shut down when the system is activated. It helps to distribute the dry chemical through the ducting, thereby increasing the fire coverage. However, the burner gas or electrical system should be shut down when the extinguishing system is activated.

Carbon Dioxide Galley System

The carbon dioxide system may be used to protect galley ranges, deep-fat fryers and ducting. The system is composed of one of two CO_2 cylinders, detectors, piping and discharge heads. Most automatic CO_2 galley systems use fusible-link detectors. As with all shipboard systems, a manual activating device must be provided; it is usually placed at an exit from the galley.

Fires in the enclosed ducting or plenum are extinguished mainly by smothering. However, a range top or deep-fat fryer fire cannot be extinguished by smothering alone, because the fire is

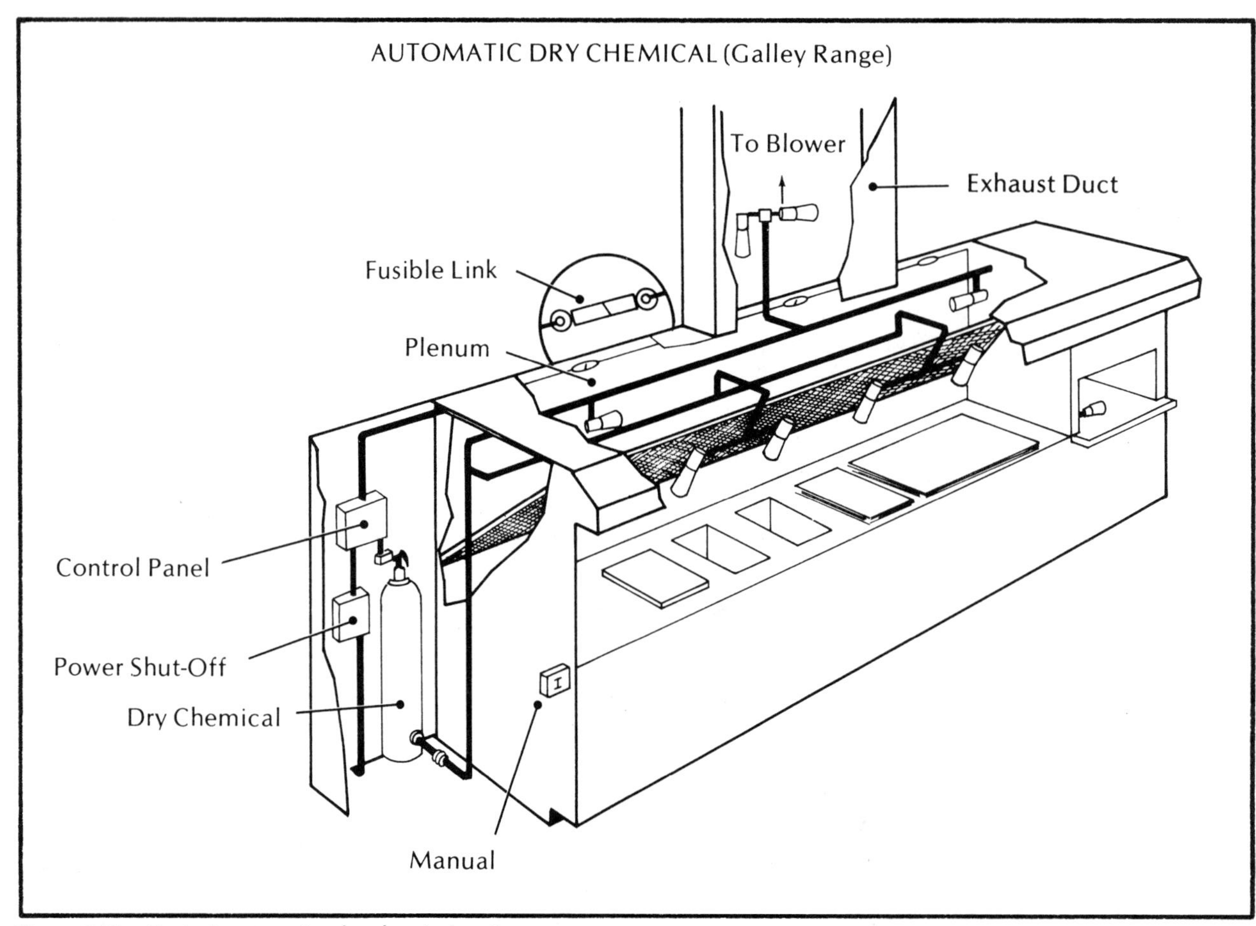

Figure 9.27. Typical automatic dry-chemical galley range system.

not confined. The required CO_2 concentration cannot be maintained in these cases. Rather than smothering alone, the CO_2 system extinguishes fire by a combination of oxygen dilution and high-velocity discharge of the gas. (The latter action may be compared to blowing out the flame of a match with your breath.)

The position and height of the nozzle over the stove, its angle, the velocity of the CO_2 discharge and the number of nozzles installed are quite critical to the effectiveness of the system. It is essential that the nozzles are not moved or tampered with in any way.

When activated, some CO_2 systems automatically shut off the supply of power or fuel to the galley range; others do not. This supply must be shut off in case of fire. If it is not done automatically, then galley personnel should do so manually, as part of their emergency procedures.

Galley Ventilator Washdown System

The main function of the ventilator washdown system is to prevent ducting fires by keeping the galley ductwork clean. In the event of a range fire, the system also protects the ductwork from fire spread. It is, thus, both a fire prevention and firefighting system that can help eliminate a troublesome fire hazard.

Ductwork Cleaning Action. When the range is being used for cooking, grease-laden air from the cooking area enters the ventilator duct. The air is forced to curve back and forth around several baffles at high speed (Fig. 9.28). This zigzag motion tends to throw the grease and any lint and dust out of the airstream and onto the ducting and baffles. From there, it flows to a grease-collecting gutter. When the day's cooking is done, the grease is automatically washed out of the collecting gutter. Nozzles in the ducting spray a solution of detergent in hot water onto the gutter. The scrubbing action of this spray washes the grease through a drain to a holding tank. The net result is ductwork that is free of fuel for a fire.

Ductwork Fire Protection. If fire occurs on the range, heat detectors actuate a damper and turn on the washdown spray system. These actions prevent the fire from entering the ductwork. Activation of a thermostat causes the following (Fig. 9.29):

1. The fire damper baffle is closed. This stops the natural draft through the ventilator duct and prevents flames and hot gases from entering the duct.
2. The exhaust blowers are automatically shut down.

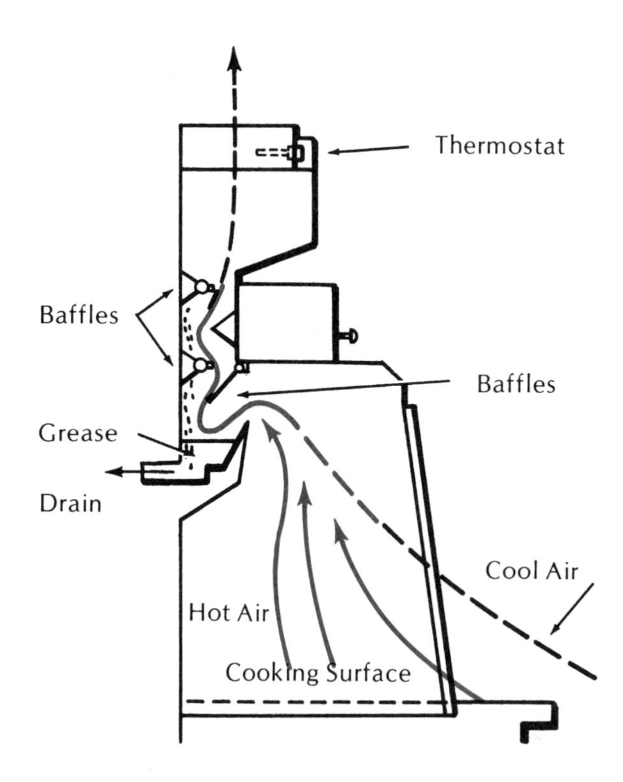

Figure 9.28. Galley ventilator washdown system. During normal operation, grease-laden air passes around a series of baffles, where the grease is removed.

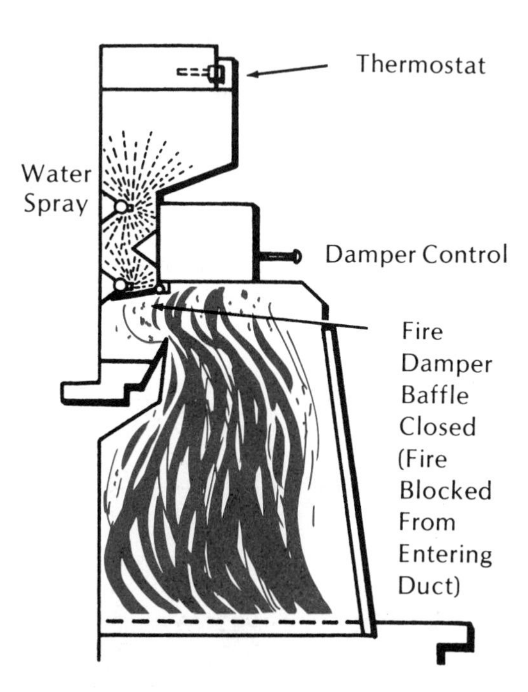

Figure 9.29. The fire damper system. **A.** Normal operation. **B.** During a fire. When the thermostat senses fire, the fire damper baffle is closed and the spray nozzles are activated. This keeps fire out of the ductwork. (Courtesy Gaylord Industries, Inc.)

3. Water is sprayed through the interior of the ventilator duct, to smother any fire that may have extended into the duct. The cleaning system spray nozzles are used for this purpose. The water spray continues as long as the temperature in the duct is over 121°C (250°F). When the duct cools below that temperature, the spray system goes through one normal wash cycle and then shuts off automatically.

After the fire is extinguished, the system is put back into operation by resetting the fire damper baffle and turning on the exhaust blowers.

INERT GAS SYSTEM FOR TANK VESSELS

Although the inert gas system is not a fire extinguishing system, it is designed to prevent fires and explosions. With few exceptions, every tank ship of 100,000 or more dead weight tonnage, and with a keel-laying date of January 1, 1975, or later, must have an inert gas system. The system must be capable of supplying to the cargo tanks a gas mixture with an oxygen content of 5% or less by volume. It must be operated as necessary to maintain an inert atmosphere in the cargo tanks except during gas freeing operations. The system must eliminate fresh air in the cargo tanks except when they are being freed of gas. It must be capable of maintaining an inert atmosphere in tanks that are being mechanically washed.

The inert gas system is composed of a gas generator, a scrubber, blowers, distribtuion lines, valves, instrumentation, alarms and controls.

Inert Gas Generator

The inert gas generator may be an automatic oil-fired auxiliary burner. It must be capable of supplying inert gas at 125% of the combined maximum rated capacities of all cargo pumps that can be operated simultaneously. It must also be able to maintain a gas pressure of 100 mm (4 in.) of water on filled cargo tanks during loading and unloading operations.

Gas Scrubber

If the inert gas produced by the generator is heated or contaminated, scrubbers are required. The scrubbers (or other similar devices) must be installed to cool the gas and reduce its content of solid and sulphurous combustion products. Water for the scrubbers must be supplied by at least two sources. The use of this water must not interfere with the simultaneous use of any shipboard firefighting system.

Blowers

The system must include at least two independent blowers. Together, the blowers must be capable of delivering inert gas at 125% of the combined maximum rated capacities of all cargo pumps that can be operated simultaneously. They must be designed so that they cannot exert more than the maximum design pressure on the cargo tanks.

Gas Distribution Line and Valves

The inert gas main must be fitted with two non-return devices, one of which must be a water seal. This water seal must be maintained at an adequate level at all times.

There must be an automatic shutoff valve fitted to the gas main where it leaves the production plant. Every shutoff valve must be designed to close automatically if the blowers fail.

Stop valves must be fitted in each branch pipe at each cargo tank. These valves must give a visual indication as to whether they are opened or closed.

Instrumentation

Sensors must be fitted downstream of the blowers and connected to the following instruments:

1. An oxygen-concentration indicator and permanent recorder
2. A pressure indicator and permanent recorder
3. A temperature indicator.

Each of these instruments must operate continuously while inert gas is being supplied to the tanks. Readouts of oxygen concentration, pressure and temperature must be provided at the cargo control station and at the location of the person in charge of the main propulsion machinery.

Each ship that has an inert gas system must carry portable instruments for measuring concentrations of oxygen and hydrocarbon vapor in an inert atmosphere.

Alarms and Controls

Every inert gas system must include the following alarms and automatic controls; the alarms must sound at the location of the controls for the main propulsion machinery:

1. An alarm that gives audible and visual warnings when the oxygen content of the inert gas exceeds 8% by volume.
2. An alarm that gives audible and visual warnings when the gas pressure in the inert gas main downstream of all nonreturn devices is less than 101.6 mm (4 in.) of water.

3. An alarm that gives audible and visual warnings (and a control that automatically shuts off the blowers) when the normal water supply at the water seal is lost.
4. An alarm that gives audible and visual warnings (and a control that shuts off the blowers) when the temperature of the inert gas being delivered to the cargo tanks is higher than 65.6°C (150°F).
5. An alarm that gives audible and visual warnings (and a control that automatically shuts off the blowers) when the normal cooling water supply to any scrubber is lost.

STEAM SMOTHERING SYSTEMS

Steam smothering systems for firefighting are not installed on U.S. ships contracted on or after January 1, 1962. Vessels equipped with these systems may continue to use them. The systems may be repaired or altered, provided that the original standards are maintained.

Steam for a smothering system may be generated by the main or auxiliary boilers. The steam pressure should be at least 690 kilopascals (100 psi). The boilers should be capable of supplying at least 1.3 kg of steam per hour per m^3 (1 lb of steam per hour per 12 ft^3) of the largest cargo compartment.

Piping

The steam supply line from the boiler to any manifold must be large enough to supply all the branch lines to the largest compartment and all adjacent compartments. The distribution piping from the manifold to the branch lines must have a cross-sectional area approximately equal to the combined cross-sectional areas of all the branch lines it serves. There must be provisions for draining the manifolds and distribution lines to prevent them from freezing.

The steam piping must not run into or through spaces that are accessible to passengers or crew members while the vessel is being navigated. The piping may, however, run through machinery spaces and corridors. Wherever possible, the piping for dry-cargo spaces, pump rooms, paint and lamp lockers and similar spaces must be independent of the piping for bulk cargo tanks.

Valves and Controls

The steam supply line to each manifold must be fitted with a master valve at the manifold. The branch line to each compartment must be fitted with a shutoff valve. The valve must be clearly marked to indicate the protected space.

On vessels the valves leading to cargo tanks must be open at all times. Thus, in case of fire, it is only necessary to open the master valve to ensure a flow of steam into each tank. The valves leading to tanks not involved in the fire may then be closed. On cargo vessels, the master valve is always open, and the valves leading to individual compartments are closed.

All controls and valves for operating the steam smothering system must be located outside the protected space. They may not be located in any space that might be cut off or made inaccessible by fire in the protected space. The control valves for the pump room extinguishing system must be located next to the pump room exit.

Steam Outlets

In the pump room, the steam outlets must be located just above the floor plates. In cargo holds, the outlets must be placed in the lower portion of each cargo hold or 'tween deck.

BIBLIOGRAPHY

Navigation and Vessel Circular No. 6-72 (Change 1). United States Coast Guard. February 28, 1977

Fire Fighting Manual For Tank Vessels. CG-329. United States Coast Guard. January 1, 1974

National Fire Protection Association Standard 12-A, Boston, Mass., 1977

National Fire Protection Association Handbook. 14th ed. Boston, NFPA, 1976

A Manual for the Safe Handling of Inflammable and Combustible Liquids and other Hazardous Productions. CG-174. United States Coast Guard. June 1, 1975

Bahme, CW: Fire Officer's Guide to Extinguishing Systems. Boston, NFPA, 1970

Bryan, JL: Fire Suppression and Detection Systems. Los Angeles, Glencoe Press, 1974

Haessler, WM: What You Should Know About Carbon Dioxide Fire Extinguishing Systems. Norris Industries Fire and Safety Equipment Division. Newark, N.J., 1969

Kidde Instruction Manuals for Smoke Detecting Systems and Carbon Dioxide Fire Extinguishing System. Walter Kidde and Co., Inc. Belleville, N.J.

Foam Protection for Tank Vessels (Section X), National Foam System, Inc. Lionville, Pa.

Design Manual for SK-3000 Dry Chemical Extinguishing System. The Ansul Company. Marinette, Wis., 1976

Navigation and Vessel Inspection Circular No. 6-72: Guide to Fixed Fire-Fighting Equipment Aboard Merchant Vessels. U.S. Coast Guard. Washington, D.C., 1972

Ansul Company Operating and Maintenance Manuals. Marinette, Wisc.

National Foam Systems, Inc. Technical Bulletins, Operating and Maintenance Manuals. Lionville, Pa.

Combating the Fire

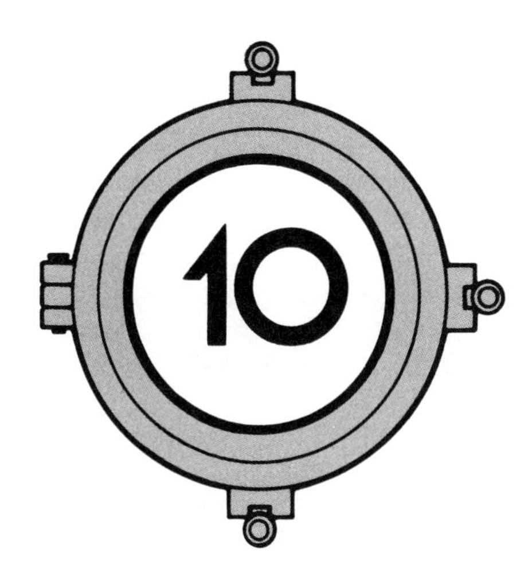

Ship fires are among the most difficult to control. The variety of fuels aboard ship and the ways in which their combustion products can hamper firefighting operations have already been discussed. In addition, the ship's configuration complicates extinguishment. If the fire is located in a below-deck compartment, it will be surrounded by steel decks and bulkheads; the space will be difficult, if not impossible, to ventilate. Materials burning in a lower cargo hold may be impossible to reach, since everything stowed above the fire would have to be removed. This is very impractical, especially if the ship is at sea. Fires located on weather decks may be easier to reach, but firefighting operations could be complicated by adverse wind conditions.

What is actually burning determines the appropriate type of extinguishing agent, but the location of the fire dictates the method of attack. In some instances the fire location determines both the extinguishing agent and the attack method. Cargo hold fires are an obvious example; they are fought with CO_2 rather than water, even when class A fuels are involved. The method of attack is an indirect one that is somewhat unique to cargo hold fires.

An important question is: How should the crew attack a certain type of fire in a certain location aboard ship? Part of the answer has been presented in the last two chapters; part is presented in this chapter. No one answer or set of answers will fit every ship exactly. Instead, the master must answer that question for his own ship; and, based on the ship's configuration, her crew size and the firefighting equipment she carries, a *prefire plan* must be developed for each space on the ship. A prefire plan is exactly what its name indicates—a plan for fighting fire that is worked out before a fire actually occurs. The concept of prefire planning is hard to disagree with, yet many ships do not have such a plan.

The firefighting procedures discussed in this chapter require teamwork, which can only be developed through constant drill. Prefire plans should be the basis for weekly fire drills and, thus, for developing coordination among personnel. This is especially important on ships with a large turnover of personnel.

INITIAL PROCEDURES

The initial procedures are those that must be performed before actual firefighting operations begin. The most important are, obviously, sounding the alarm and reporting the location of the fire.

Sounding the Alarm

The crew member who discovers the fire or the indication of fire must sound the alarm promptly. This point has been stressed in previous chapters, but it bears repeating. A delay in sounding the alarm usually allows a small fire to become a large fire. Once a fire gains intensity, it spreads swiftly. No crew member—no one on a ship—should ever attempt to fight a fire, however small it may seem, until the alarm has been turned in. Of course, if two or more people discover the fire, only one is required to sound the alarm. The others should stay and attempt to extinguish the fire with available equipment. A small fire in a metal wastebasket could be covered with a noncombustible lid, if readily available, before the crewman who discovered it leaves to report the fire.

All fires must be reported, even if self-extinguished (that is, the fire goes out by itself from lack of fuel or oxygen). The resulting investigation could uncover defects or conditions which, when corrected, would prevent future fires.

Reporting the Fire Location

The crewman who sounds the alarm must be sure to give the exact location of the fire, including compartment and deck level. This is important for several reasons. First, it confirms the location for the ship's fire party. Second, it gives them information regarding the type of fire to expect. Third, the exact location may indicate the need to shut down certain ventilation systems. Finally, it indicates what doors and hatches must be closed to isolate the fire.

Precautionary Measures. If flames can be seen, the location of the fire is obvious. However if only smoke is evident, the fire may be hidden behind a bulkhead or a compartment door. Then, certain precautions must be observed during attempts to find its exact location.

Before a compartment or bulkhead door is opened to check for fire, the door should be examined. Discolored or blistered paint indicates fire behind the door. Smoke puffing from cracks at door seals or where wiring passes through the bulkhead is also an indication of fire (Fig. 10.1). The bulkhead or door should be touched with a bare hand. If it is hotter than normal, it is probably hiding the fire.

Once a hidden fire has been located, the door to the area should *not* be opened until help and a charged hoseline are at hand. A fire burning in an enclosed space consumes the oxygen within that space. The fire seeks additional oxygen, and a newly opened door presents it with a generous supply. When the door is opened, air is pulled through the opening to feed the fire. As a result, the fire usually grows in size with explosive force. Flames and superheated gases are then forced violently out through the opening. Anyone in its path could be severely burned. If the fire is not attacked with a hoseline, it can travel through the area uncontrolled. The longer the fire has been burning, undetected, the more dangerous the situation will be. Therefore, cool the door with water before opening. Have everyone stand clear of the door to the side opposite the hinges. Always open the door from a position clear of the opening and opposite the hinges.

FIREFIGHTING PROCEDURES

Fire travels via the radiation, conduction and convection of heat (*see* Chapter 4). For the most part, these processes will extend the fire laterally and upward: laterally along passageways and ducting, and upward through hatches and stairways. In certain situations, fire will also travel downward, through ducting or through deck plates (by conduction). Burning embers, dropping from one deck to another, provide a more dangerous method of downward fire extension (Fig. 10.2).

Every fire will extend to new sources of fuel and oxygen if these sources are available. In this respect, all fires are similar. However, the path through which a particular fire extends will depend on the location of the fire and the construction features of surrounding spaces. These factors must, therefore, be taken into account when the

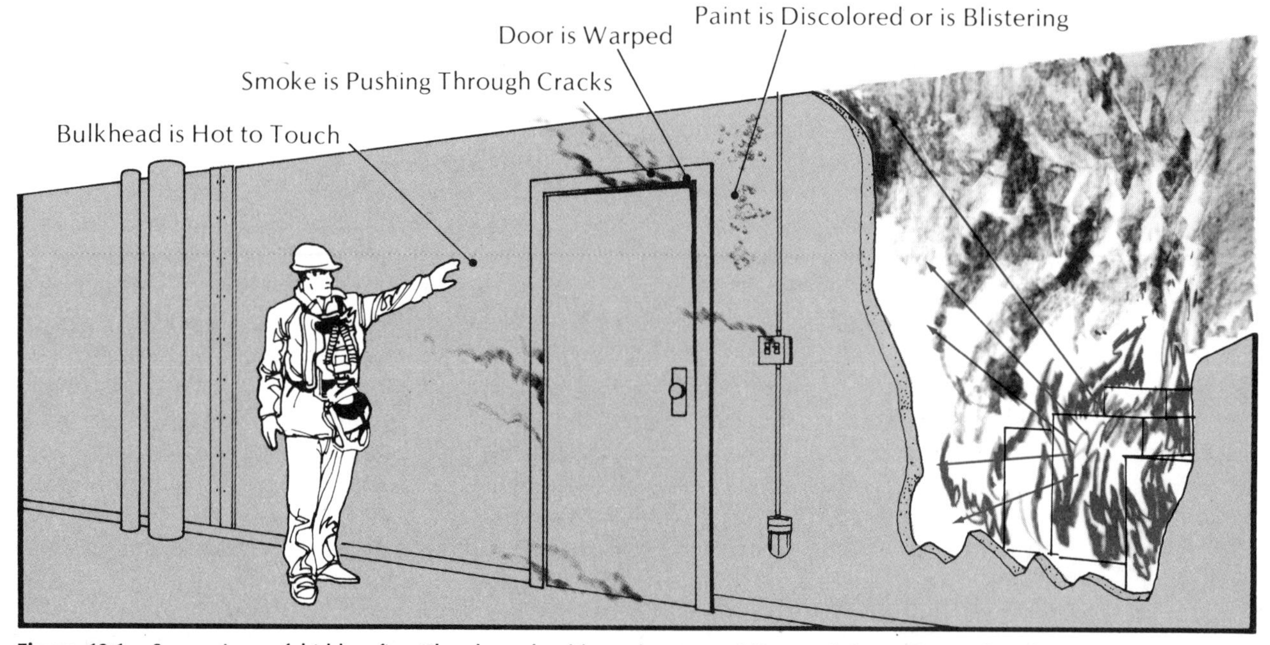

Figure 10.1. Some signs of hidden fire. The door should *not* be opened if any of these signs is found.

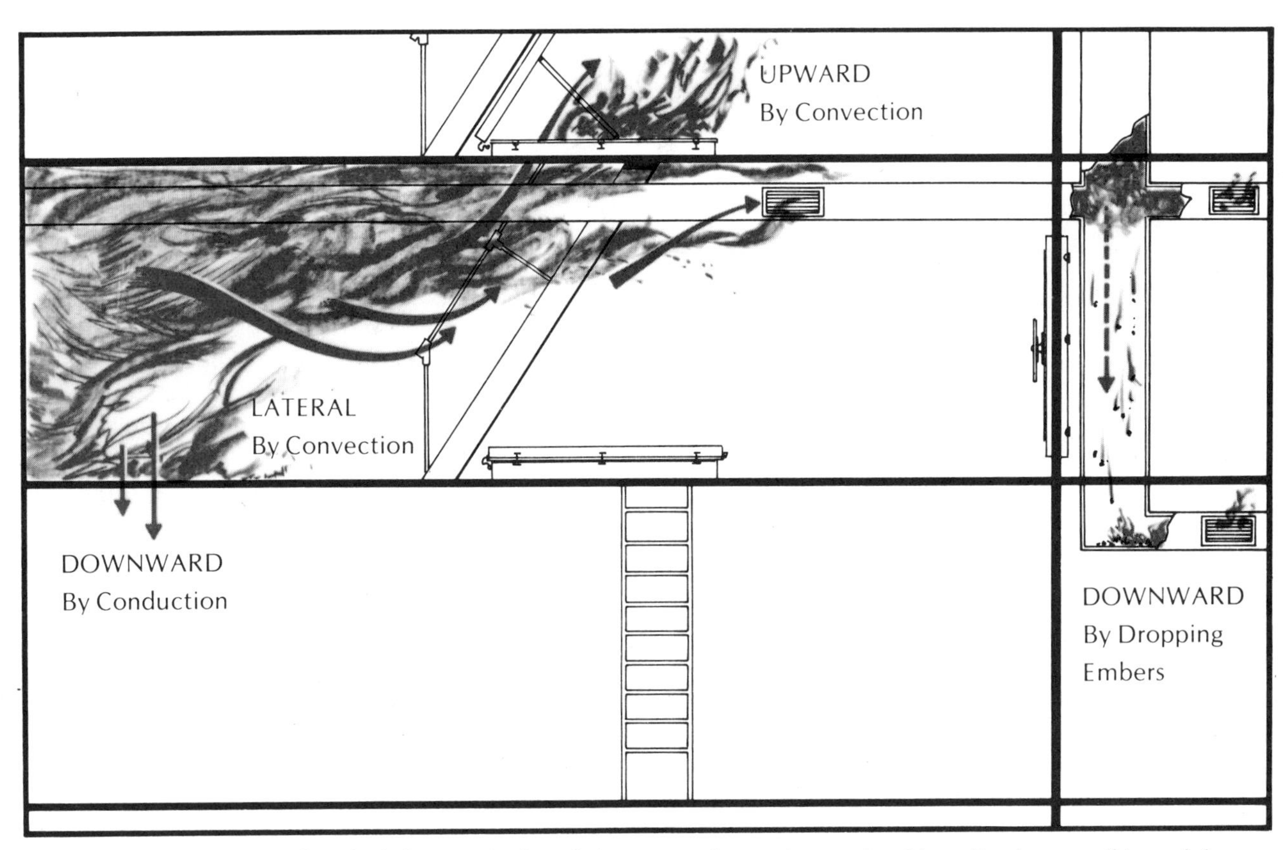

Figure 10.2. The processes by which fire travels aboard ship. Fire will spread upward and laterally where possible, and downward in some situations.

fire is attacked. In addition, the fuel and its combustion products will affect firefighting operations.

For these reasons, no fire can be fought routinely, although all fires must be fought systematically. The procedures described in the next several sections should be part of every firefighting operation. The particular fire situation will dictate the order in which they are to be performed, whether some must be performed simultaneously, and the amount of effort that should be devoted to each procedure.

Sizeup

Sizeup is the evaluation of the fire situation. The on-scene leader should determine, as quickly as possible,

1. The class of fire (what combustible materials are burning)
2. The appropriate extinguishing agent
3. The appropriate method of attack
4. How to prevent extension of the fire
5. The required manpower and firefighting assignments.

A small fire might be extinguished by the first few crewmen to arrive; they would probably perform a partial sizeup and begin the attack instinctively. Larger fires would require a coordinated attack, efficient use of manpower and equipment, thus, a more thorough assessment. During sizeup, or as soon thereafter as possible, communications and a staging area should be set up.

Communications. Communications with the master should be established by phone or by messenger. Communications with firefighting teams must also be established and maintained. Messengers would be best for this purpose, since telephone lines might be destroyed by the fire, and firefighters would be moving constantly. An internal two-way radio system, if available, could be used to coordinate firefighting efforts.

Staging Area. The staging area should be established in a smokefree area, as near as possible to the fire area. An open deck location, windward of the fire, would be ideal. However, if the fire is deep within the ship, the staging area should be located below deck. A location near a ship's telephone, if feasible, would be helpful in establishing communication links. However, the staging area should not be located where it might be endangered by the spread of fire.

All the supplies needed to support the firefighting effort should be brought to the staging area. These would include backup supplies of hose, nozzles and axes; spare cylinders for breathing apparatus; and portable lights. The staging area should also be used as the first aid station. The equipment required to render first aid to injured crewmen should be set up there.

Attack

The attack should be started as soon as possible, to gain immediate control and to prevent or minimize the extension of fire to exposures. (Exposures are the areas of the ship that are adjacent to the fire area on all four sides and above and below.) The attack will be either direct or indirect, depending on the fire situation. Direct and indirect attacks differ widely in how they achieve extinguishment; both are efficient when properly employed.

Direct Attack. In a direct attack, firefighters advance to the immediate fire area and apply the extinguishing agent directly into the seat of the fire. There may be no problem in getting to the immediate fire area if the fire is small and has not gained headway. However, as a fire increases in intensity, the heat, gases and smoke increase the difficulty of locating and reaching the seat of the fire. Once the fire has gained headway, a direct attack should be coupled with venting procedures (*see* section on Ventilation).

Indirect Attack. An indirect attack is employed when it is impossible for firefighters to reach the seat of the fire. Generally this is the case when the fire is in the lower portions of the vessel. The success of an indirect attack depends on complete containment of the fire. All possible avenues of fire travel must be cut off by closing doors and hatches and shutting down ventilation systems. The attack is then made from a remote location.

One technique involves making a small opening into the fire space, inserting a nozzle, and injecting a spray (fog) pattern into the space. Heat converts the fog to steam, which acts as a smothering agent. Two things are essential for a successful attack of this type. First, the fire must be completely enclosed, so that the steam will reduce the oxygen content of the air around the fire. Second, the fire must be hot enough to convert the water to steam.

Another indirect technique employs a smothering agent such as carbon dioxide. The use of this technique in fighting cargo hold fires has already been mentioned. It is discussed in detail later in this chapter.

Ventilation

Ventilation is the action taken to release combustion products trapped within the ship and vent them to the atmosphere outside the ship. Most fire fatalities do not result from burning, but rather from asphyxiation by combustion gases or lack of oxygen. Before smoke or heat becomes apparent, deadly carbon monoxide and other noxious gases seep into compartments. People who are asleep are easily overcome by these gases. However, if the fire is vented promptly and properly, the smoke, heat and gases can be diverted away from potential victims and from uninvolved combustibles.

Ventilation is used only when a direct attack is made on the fire. During an indirect attack the fire area must be made as airtight as possible, to keep oxygen out and the extinguishing agent in.

Vertical Ventilation. The smoke and hot gases generated by the fire should be vented to the outside air if possible. As a fire intensifies, the combustion gases become superheated; if they are ignited, they will spread the fire very quickly. In the ideal situation, the gases are released at a point directly above the fire, as the extinguishing agent is brought to bear on the fire (Fig. 10.3). This ideal *vertical ventilation* is just about impossible to achieve aboard ship, since there is rarely a direct upward route from the fire to the outside. In most instances, at least some horizontal ventilation is required.

Horizontal Ventilation. Horizontal ventilation is achieved by opening windward and leeward doors to create an airflow through the spaces in which the combustion products are collecting. Fresh air flowing in through a windward doorway moves the combustion products out through the leeward doorways (Fig. 10.4). The leeward doors should always be opened first. Portholes should also be opened; however, small portholes are not very effective in removing smoke and heat.

Combination Vertical and Horizontal Ventilation. When the fire is below deck, there may be some difficulty in moving smoke and heat out of the ship. In some instances, a combination of vertical and horizontal ventilation will work. A horizontal flow of air may sometimes be created over a hatch on the deck above the fire. This airflow can produce a venturi effect that pulls smoke and heat upward from the lower deck (Fig. 10.5). A properly placed portable fan will help move the air more rapidly. The doors to uninvolved areas

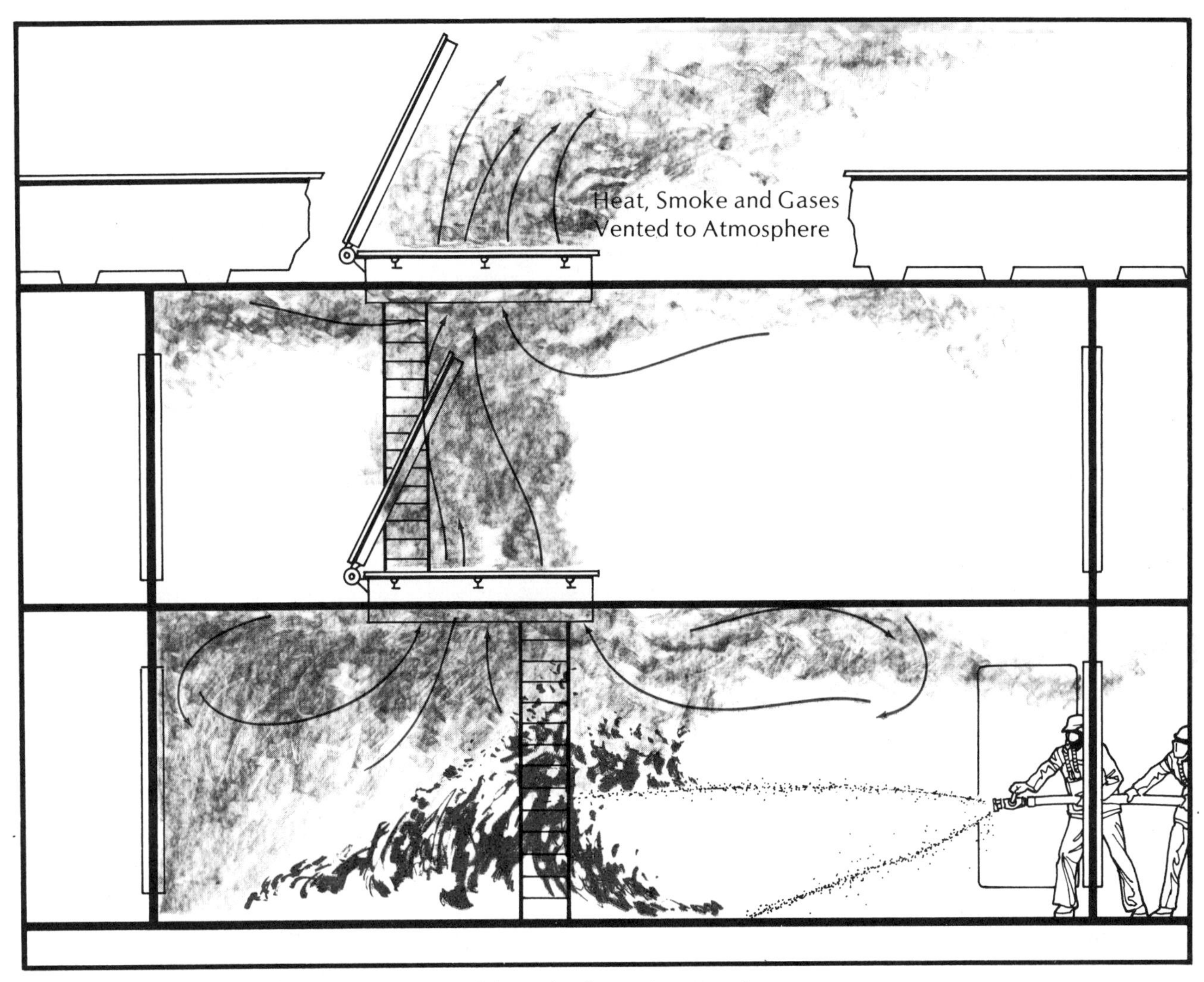

Figure 10.3. Vertical ventilation, directly upward from the fire to the atmosphere.

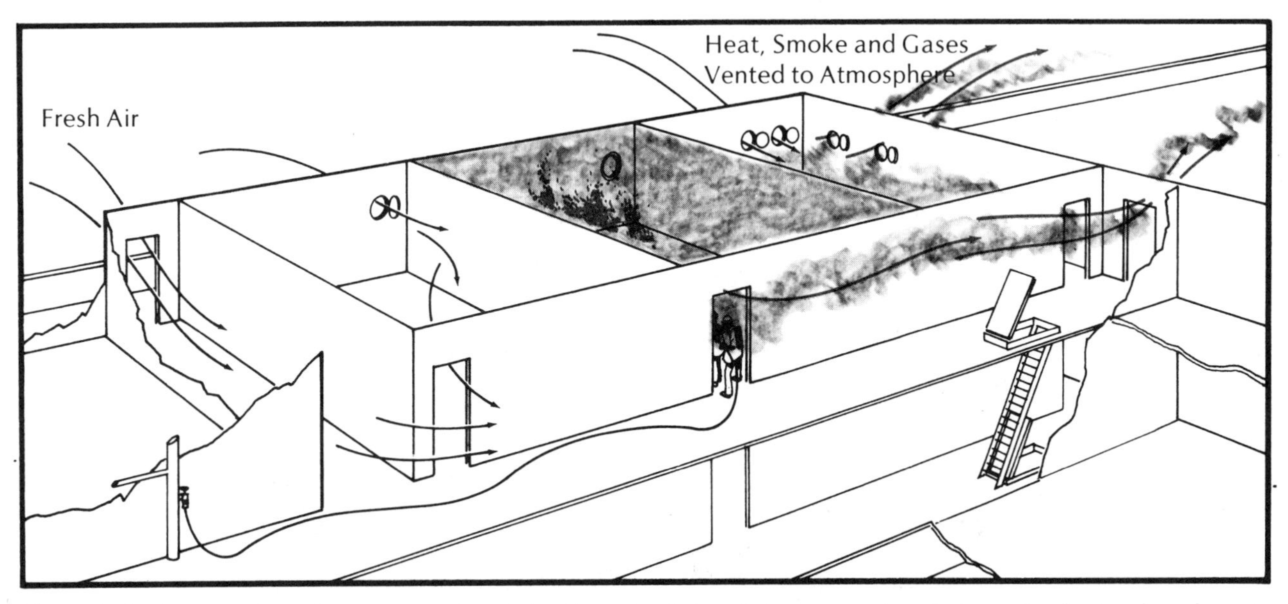

Figure 10.4. Horizontal ventilation. Fresh air entering through windward doorways and portholes pushes heat and smoke out leeward doorways and portholes.

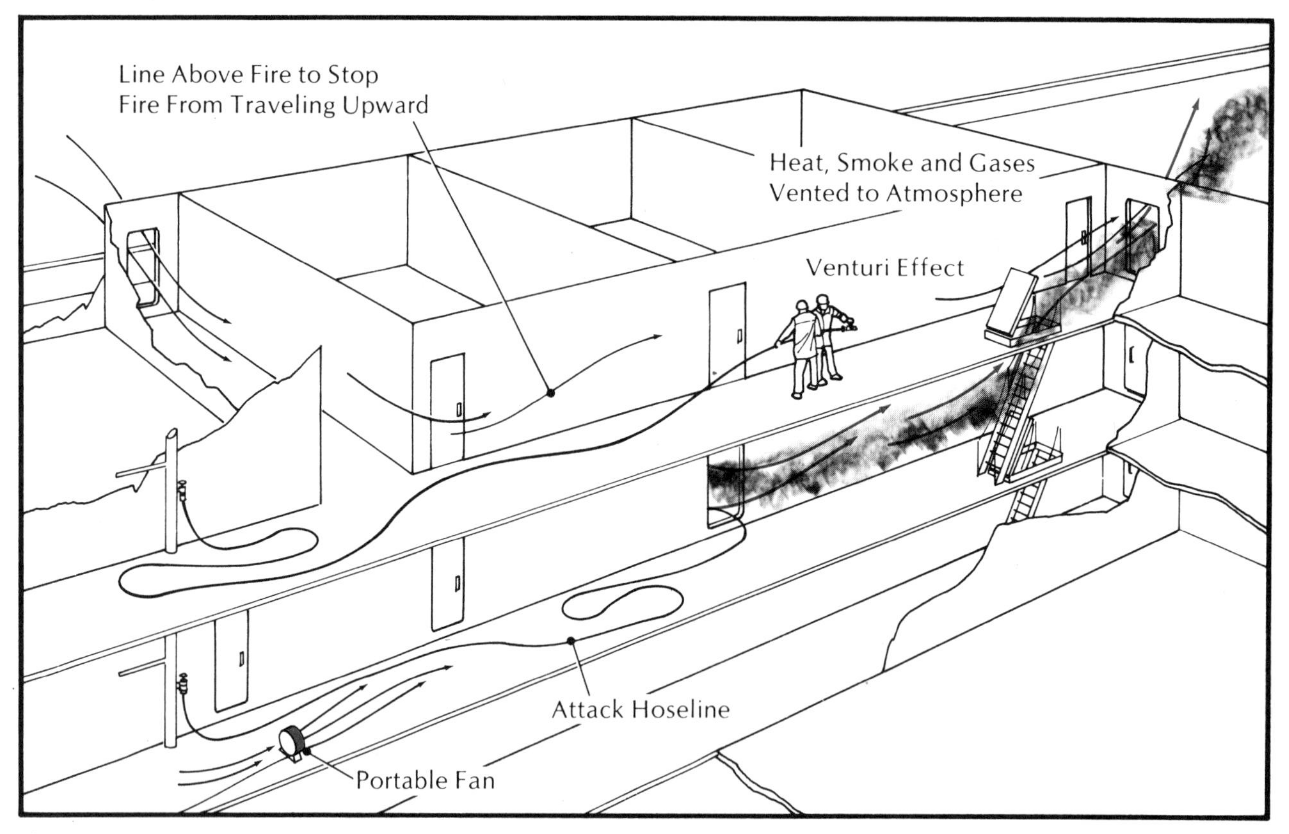

Figure 10.5. Combination venting. An airflow is created above the fire. It pulls combustion products up from the involved deck and out the doorway.

should be closed, to keep out contaminated air. These doors should remain closed until venting has been completed.

Mechanical Ventilation. Smoke-contaminated air can be moved out of compartments, along passageways and up through deck openings with properly positioned portable fans (red devils). The fans should be placed to push and pull the air in order to establish an airflow from the contaminated area to the outside. In some instances the ship's mechanical air intake system can be used in conjunction with portable fans. An alternative, if power fans are not available, is the use of a windsail. The windsail can be rigged to force clean air into the contaminated area while the ship is under way. The smoke then would exhaust through natural "down stream" opening.

Protecting Exposures

Protecting exposures means preventing the fire from extending beyond the space in which it originated. If this can be accomplished, the fire can usually be controlled and extinguished without extensive damage. To protect exposures, the fire must virtually be surrounded on six sides; firefighters with hoselines or portable extinguishers must be positioned to cover the flanks and the spaces above and below the fire. The officer in charge must also consider fire travel through the venting system. Crewmen must be dispatched to examine and protect openings in the system through which fire might enter other spaces.

Rescue

The rescue of trapped personnel is an extremely important aspect of every firefighting operation. Rescue may be the first step in the operation, or it may be delayed because of adverse circumstances. For example, suppose someone is trapped in a compartment that is located beyond the fire. If some firefighters can get past the fire while others control it, the rescue may be accomplished immediately. If the fire cannot be controlled easily, it may be best to attack and control the fire before attempting the rescue.

The decision as to when to attempt a rescue is a difficult one. If the rescue attempt is delayed, a direct attack with fog could push the fire into the area where personnel are trapped; an indirect attack could generate enough steam to scald them. On the other hand, a holding action may be feasible while an alternative route is used to make the rescue. The decision involves the twofold problem of protecting lives and protecting the vessel. This problem is not always solved by assuming that the lives of trapped persons are more important than the vessel. The vessel is, above all, the sanctuary of the crew. A delay in controlling the fire, due to imprudent rescue attempts, could result in an uncontrollable fire, loss of the vessel and forced abandonment. The fire situation could force a decision to attack the fire in a manner that might be detrimental to trapped personnel, but that would save the vessel and other crew members.

Overhaul

Overhaul is begun after the main body of fire is extinguished. It is actually a combination of two procedures, an examination and a cleanup operation.

Overhaul can be a dangerous procedure. Records show that land-based firefighters are injured more during overhaul than during any other operation. This is attributed to a letdown after the fire is controlled, leading to a degree of carelessness and a lack of regard for personal safety.

Examination and Extinguishment. The objectives of the examination are to find and extinguish hidden fire and hot embers and to determine whether the fire has extended to other parts of the ship. This is an important aspect of firefighting that should be conducted as seriously as the attack on the fire. Overhaul personnel should make use of four senses—hearing, sight, touch and smell. They should trace the length of all duct systems, look into them, and touch and smell them, to determine the extent the fire has traveled. They should inspect all overhead spaces, decks and bulkheads in the same manner. They must be thorough and especially watchful where wiring or piping penetrates through bulkheads or decks; fire can travel through the smallest crevice.

Any materials that might have been involved with fire, including mattresses, bales, crates and boxes should be pulled apart and examined. Materials that might reignite, especially bedding, baled cotton and bolts of fabric, should be removed from the fire area. They should be placed on a weather deck, with a charged hoseline manned and ready to extinguish any new fire.

Smoke-blackened seams and joints should be checked carefully. Areas that are charred, blistered or discolored by heat should be exposed until a clean area is found. If fire is discovered, the area should be wet down until it is completely extinguished.

Cleanup. At the same time debris should be cleaned up and free water should be removed. Any unsafe conditions should be corrected. For example, hanging lagging should be removed; boards with exposed nails should be picked up and placed in containers; hanging wires should be secured; and all debris should be removed, to make the fire area as safe as feasible.

Dewatering. Free water can impair the stability of a vessel. Every effort should be made to limit the accumulation of water in large compartments and cargo holds. These efforts should begin with the use of water patterns that allow maximum cooling with minimal quantities of water; preference should be given to fog sprays over solid streams. Only as much water as is absolutely necessary should be used.

As soon as water is used for extinguishment, unwatering procedures should be started. The lack of portable dewatering equipment on merchant ships may create a problem. If debris clogs the fixed piping system, it may be necessary to follow a rather complex backflooding procedure to clear the suction strainers.

Structural Weakness. Steel plating and support members can be weakened considerably by high temperatures. This weakening may not be apparent unless there is visible deformation. In all cases where structural weakness is suspected, a careful inspection should be made. Weakened members should be supported by shoring or strongbacks.

Fire Under Control

A fire may be considered to be under control when

1. The extinguishing agent is being applied to the seat of the fire; i.e., streams from initial lines (and backup lines if they were required) have been able to penetrate to the seat of the fire and have effectively begun to cool it down. At this point, men with shovels should be able to turn over burned material to expose hidden fire.
2. The main body of fire has been darkened. At this point, the fire cannot generate enough heat to involve nearby combustible materials.
3. All possible routes of fire extension have been examined or protected. This is, basically, a combination of the exposure protection and overhaul procedures discussed earlier.
4. A preliminary search for victims has been completed. The preliminary search should be conducted at the same time as the fire attack, ventilation and exposure protection

procedures, if possible. As soon as the fire is under control, a second and more comprehensive search should be undertaken. Areas that were charged with smoke and heat must be closely examined. Searchers must look in closets, under beds, behind furniture and drapes and under blankets. An unconscious person must be removed to fresh air immediately. If the person is not breathing, rescue breathing must be started immediately.

Fire Out

Before a fire can be declared completely out, the master of the vessel must be assured, by the on-scene leader, that certain essential steps have been taken. These include

1. A thorough examination of the immediate fire area, to ensure that
   a. All paths of extension have been examined and opened where necessary.
   b. Ventilation has been accomplished, and all smoke and combustion gases have been removed.
   c. The fire area is safe for men to enter without breathing apparatus. This can be verified by the use of a flame safety lamp or an oxygen indicator. (While an oxygen concentration of 16% will support life, it is wise to wait until a reading of 21% is obtained.)
2. A complete overhaul of all burned material.
3. The establishment of a rekindle watch. One crew member (more if the fire has been extensive) must be assigned to do nothing but check for reignition, and to sound the alarm if it occurs. A second crewman can be assigned to patrol the exposures and the paths of possible extension.
4. The replacement or restoration of firefighting equipment. Used hose should be replaced with dry hose. The used hose should then be cleaned, flushed, dried and rolled for storage. (This is especially important with unlined hose, which may be used anywhere aboard ship except in machinery spaces.) Nozzles should be cleaned and installed on the dry hoses.

   Portable extinguishers, whether partially or fully discharged, should be recharged or replaced.

   Breathing apparatus should be cleaned, facepieces sterilized and cylinders or cannisters replaced. The entire unit should be stowed, ready for the next emergency. Additional cylinders or cannisters should be ordered at the first opportunity.

   If the sprinkler system was activated, the sprinkler heads (automatic type) should be replaced, and the system restored to service. Activated detection systems that must be reset should likewise be restored to service.
5. A damage control check. A thorough examination should be initiated to determine if the vessel has been damaged by the fire. The high temperatures associated with fire can cause decks, bulkheads and other structural members of the ship to warp or become structurally unsound. When this occurs, temporary support should be provided by shoring. Any other repairs necessary to the well-being of the vessel should be undertaken immediately. Any necessary dewatering operations should be started.
6. A muster should be conducted to account for all ship's personnel.

Critique

Soon after the fire is out and the fire protection equipment has been restored to service, a critique should be held. The critique need not be a formal affair; in fact, a good time to hold it is while crew members are having a cup of coffee before going back to their normal duties.

The crew has just put out a fire on a ship at sea. This is quite an accomplishment, and they have every reason to be proud. However, while the details are still fresh in their minds, they should consider several questions: How could they have done better? More important, how could the fire have been prevented? If they had the same fire again tomorrow, would they fight it the same way? Could they have accomplished the same result with less physical punishment to firefighters? With less damage to the ship?

All this should be discussed, along with anything else pertaining to the fire. The officer in charge should encourage suggestions and recommendations, and write them down. Worthwhile ideas should then be made a part of the prefire plan.

FIRE SAFETY

Coast Guard regulations require that a fire party be organized and trained on every U.S. flag vessel. The fire party may be broken up into several teams with different duties. The leader of the fire party should be an experienced officer with the authority to administer fire prevention training programs and to direct firefighting operations.

Hose Team

One of the most important units within the fire party is the hose team. A hose team ideally should have four members to operate proficiently; the Coast Guard recommends at least two people for a 3.8-cm (1½-in.) hoseline, and three people for a 6.4-cm (2½-in.) hoseline.

The key member and leader of the hose team is the nozzleman, who controls the nozzle and directs the stream onto the fire. In many instances, the nozzleman must make decisions before an officer arrives on the scene. The nozzleman must have the training and discipline to advance the team close to the fire, to ensure that the water is directed into the seat of the fire. This is a responsible position, and it should be assigned to a crew member who has received training in firefighting at a maritime facility. The nozzleman should also be thoroughly familiar with the ship's design and construction features.

The backup man is positioned directly behind the nozzleman. He takes up the weight of the hose and absorbs some of the nozzle reaction, so that the nozzle can be manipulated without undue strain. To be able to maintain his position, he must work in unison with the nozzleman. The other hose team members are positioned along the hose to assist in maneuvering and advancing the nozzle.

It is a good idea to use engine room personnel to handle the hoselines assigned to protect engineering spaces, since they understand the machinery and are familiar with that part of the ship. In addition, these crewmen will more likely be in the vicinity of the fire when it occurs. By the same reasoning, the other crew members should be assigned to fire stations near their work stations, when possible.

Advancing the Hoseline. When an emergency occurs, the hose should be run out *before* the fire station hydrant is opened. Without water, the hose is light and easy to handle; it can be advanced quickly. Once the hose is charged with water, it becomes heavy and difficult to advance. Firefighters become tired from moving the additional weight of the water, especially if the hose must be manhandled up or down ladders and along narrow passageways. If they are wearing breathing apparatus, their labored breathing depletes the oxygen supply more rapidly than normally.

The hose should be run out as follows: The nozzleman and backup man pick up the first section of hose and advance toward the fire. The third team member picks up the center section and advances it. The fourth team member remains at the fire station to open the hydrant. When the nozzleman is in position, he asks for water. As the water fills the hose, the third and fourth team members should straighten out any kinks and check hose couplings for leaks. Leaky couplings should be tightened with a spanner. Upon calling for water, the nozzleman should open the nozzle slightly, to allow trapped air to escape. The nozzle should be closed when water begins to flow. The hoseline is then ready for use.

During drills, hose should be run out, and the nozzle should be positioned to attack a simulated fire. The training should be as realistic as possible. Hose teams should practice maneuvering the hose below decks, through passageways, and up and down accommodation ladders and narrow hatches.

Using the Hose Stream. The manner in which hose streams are applied depends on the fire situ-

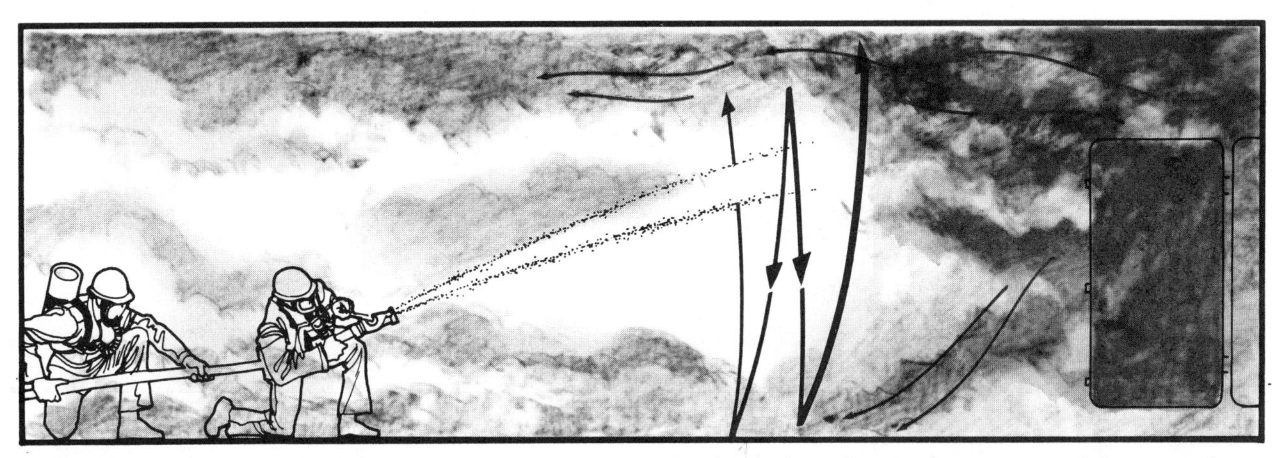

Figure 10.6. A fog stream is used to push flames, heat and smoke ahead of an advancing hose team. Firefighters must keep low to allow the heat to pass above their bodies.

ation. The nozzleman must know what type of stream to use, and how to use it, under different fire conditions.

Passageway–Compartment Fire. When flames have traveled out of a compartment and into a passageway, it is essential that the compartment be reached. The hose stream must be directed into the seat of the fire. The flames in the passageway must be knocked down before the nozzle can be positioned properly. This is best accomplished by advancing as close to the flames as possible and keeping low to the deck (Fig. 10.6). Then the nozzle should be opened to the fog position. The stream should be moved up and down so that the water bounces off the bulkhead and the overhead, and into the flames. This will push the heat and flames ahead of the nozzleman, who should continue to advance until he reaches his objective.

Steam will be produced when the stream hits the flames and hot gases. This and the smoke will make visibility very poor. A backup hoseline should be brought into position behind the first attack line as quickly as possible. The backup line can be used to protect the advancing hose team, or it can be directed onto the fire if a larger volume of water is required to gain control of the situation.

Fire in a passageway must never be attacked from opposite directions. If it is, one of the hoselines will push flames, heat and smoke directly at the other hose team (Fig. 10.7).

Hidden Compartment Fire. To attack a substantial fire behind a closed door, the charged hoseline should first be positioned outside the door. Then the door should be opened only enough to insert the nozzle. Using the door to protect his body, the nozzleman should sweep a fog stream around the compartment. Both the nozzleman and the backup man should crouch as low as possible, to allow the heat and steam to pass overhead (Fig. 10.8). After a few seconds, the door may be opened a bit more. If conditions permit, the team should enter the compartment and advance until they can hit the seat of the fire with a straight stream.

Other Fire Party Personnel

Other crewmen in the fire party are assigned to specific duties or teams. Several crewmen must be available to act as searchers. Under the cover of hoselines, they search for trapped personnel. Still other crew members are assigned to check for fire extension, to ventilate the fire area or to act as messengers if necessary.

Protective Clothing

If a fire has been burning for any length of time, it can reach temperatures exceeding 538°C (1000°F) and produce severe concentrations of smoke and noxious gas. Firefighters who are not sufficiently protected against these hazards cannot press their attack against the fire. They may have to retreat or be burned or overcome. If they become casualties, then they reduce an already limited firefighting force.

At a minimum, each member of the hose team should be equipped with a water resistant coat or jacket, rubber boots, a hard hat and work gloves. This clothing will help protect against heat, hot water and steam. The approved fireman's suit shown in Figure 10.9 will provide adequate protection and is recommended as suitable for shipboard firefighting operations.

Respiratory protection is best provided by self-contained breathing apparatus. Members of the hose team must be trained in the use of this equipment. They must know its limitations but have confidence in its ability to protect them in a hostile atmosphere. (*See* Chapter 15 for a discussion of breathing apparatus.)

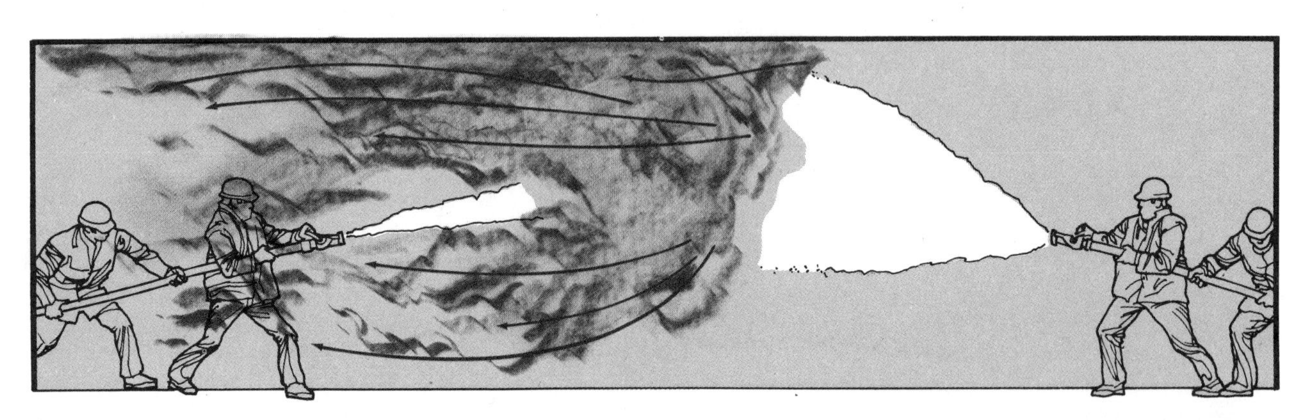

Figure 10.7. When two hose teams attack a fire from opposite sides, the team with the weaker stream is placed in jeopardy.

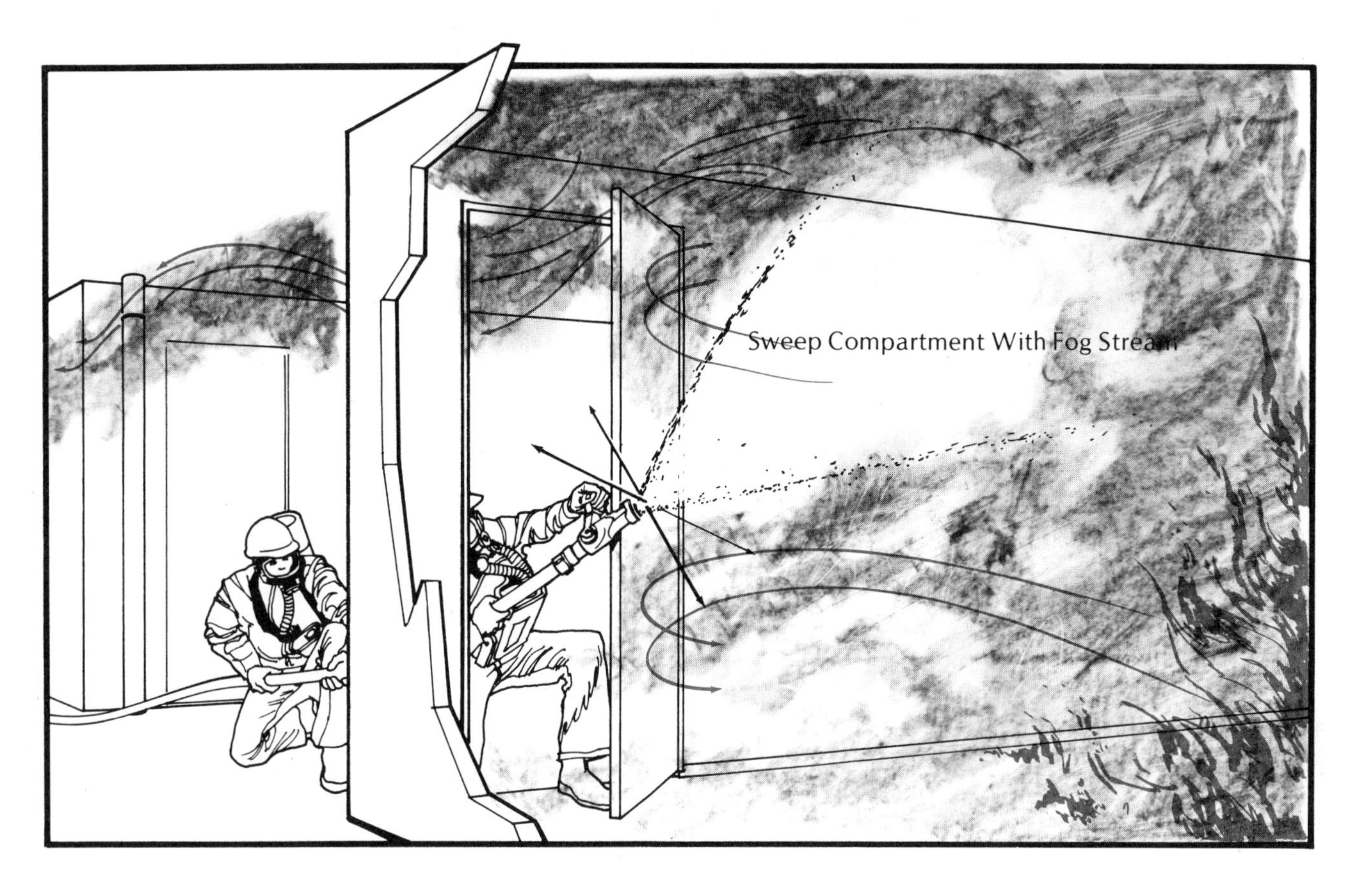

Figure 10.8. Fighting a fire in a closed compartment. The door is opened slightly and used as a shield. The fog stream is swept back and forth across the compartment. The hose team crouches low.

In some situations, the first hose team at the fire will not have time to don protective clothing or breathing apparatus. They may have to make the initial attack immediately to keep the fire from progressing beyond control. In such situations, they must use common sense. It would be poor judgment to abandon a firefighting position where they were not experiencing any difficulty. The position could be held for a short but essential time without protective gear. However, if they cannot knock down the fire, and heat and smoke are threatening their position, they should back away from the fire. The nozzleman should use a fog stream to block the heat. The team should continue to back away until they reach a position they can hold without undue hardship. Meanwhile, a backup team should be donning protective clothing and breathing apparatus. The backup team should relieve the men on the line as soon as possible.

FIGHTING SHIPBOARD FIRES

In this section, 15 different shipboard fire situations are described. The recommended procedures for fighting each fire are then detailed, from the alarm through overhaul.

Cabin Fire

The Fire. The burning material is in a wastebasket in a far corner of the cabin. The flames have spread to a desk and have ignited drapes at a porthole. The cabin door opens into the cabin from an inboard passageway.

Confining the Fire. Upon seeing smoke seeping from the door, a crewman sounds the alarm. The alarm is acknowledged by another crewman. The door is cool to the touch, indicating that the fire has not spread across the cabin. The crewman who discovered the fire opens the door, notes the fire situation and determines that no one is in the cabin. He then leaves the cabin and closes the door, being careful not to lock it. He has taken the first step in confining the fire.

Sizeup. The crewman's quick assessment of the fire situation reveals that *1)* rescue is not a problem and *2)* the burning materials are ordinary combustibles, best extinguished by water.

Attack. The fire station near the cabin includes a 9.5-liter (2½-gal) water extinguisher. The crewman activates the extinguisher outside the closed door. When he is sure the extinguisher is

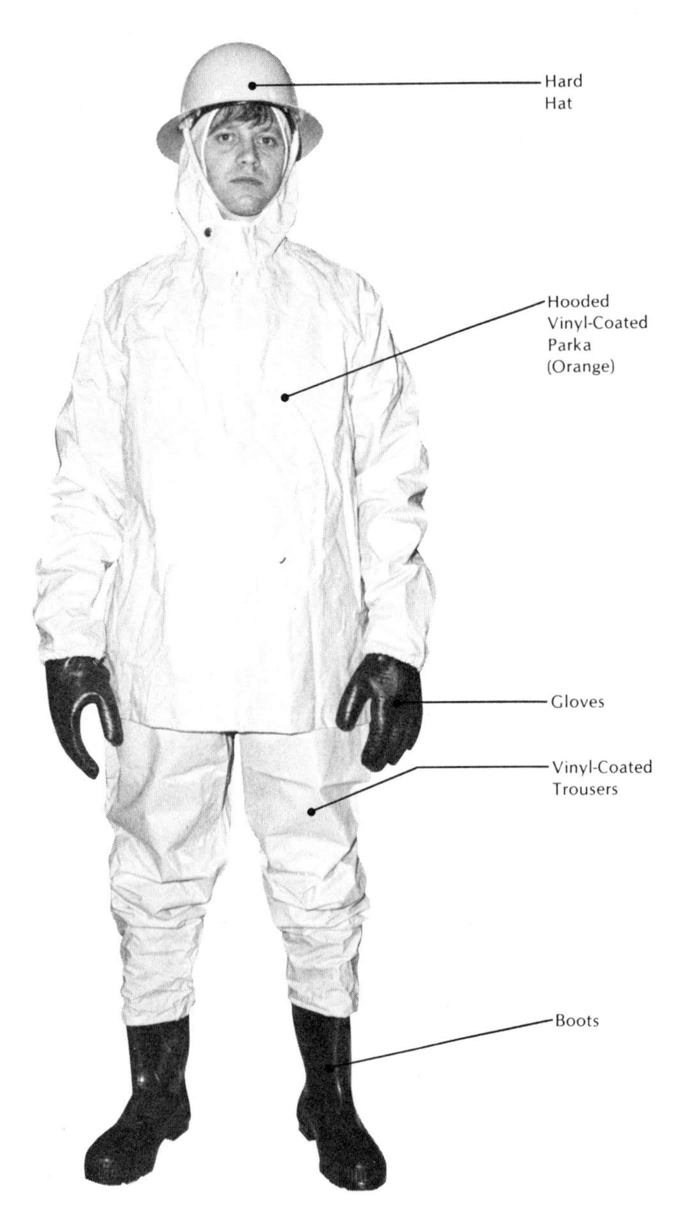

Figure 10.9. An outfit that meets U.S. Coast Guard requirements for protection against water and heat. (Courtesy C. J. Hendry Co.)

operating, he opens the door, directs the stream at the base of the flames first then moves the stream upward to hit the higher flames. By placing his finger over the tip of the nozzle, he develops a spray stream that may be a little more effective. Because he cannot completely extinguish the fire, he leaves the cabin, again closing the door.

Backup. The firefighting team assigned to the location runs out the hoseline and brings the nozzle into position outside the cabin door. Upon being notified that the fire was not completely extinguished, they charge the hoseline with water, open the door and direct water into the remaining flames. This is very important. Whenever possible, *the initial attack should be backed up with a secondary means of attack* (Fig. 10.10).

Protecting Exposures. During the attack, the officer in charge sends crewmen into adjacent spaces, around, above and below the fire area, to check for fire extension. This fire does not extend out of the cabin, because it was discovered early and extinguished properly.

Overhaul. The drapes and all other burned or charred materials are placed in buckets and thoroughly soaked with water. Other debris and the water are cleaned up and removed from the cabin. During overhaul, the cabins above, below and adjacent to the fire cabin are again carefully inspected for fire travel.

Engine Room Fire

The Fire. A bucket of oil spills on solid decking and ignites when it contacts a hot manifold. The liquid covers about a 0.93-m^2 (10-ft^2) deck. Flames cover the entire spill and are beginning to travel up a bulkhead.

Attack. The alarm is sounded. The initial attack is made with a portable extinguisher (dry chemical, CO_2 or Halon). The objective is to quickly knock down the flames (Fig. 10.11A).

Confining the Fire. The fire party immediately shuts down the venting systems and closes hatches and doors in the vicinity of the fire.

Backup. The engine room is equipped with a semiportable extinguishing system. Its hose is run out and used to continue the attack if necessary. In addition, hoselines from the water-main system are advanced into position to assist in the attack.

Dry chemical, CO_2 and Halon extinguishing agents have very little cooling power. It is highly probable that metals in direct contact with the fire will retain enough heat to reignite the oil spill. To ensure cooling and prevent reignition, water in the form of low velocity fog is directed onto the metal surfaces (Fig. 10.11B). This is done carefully to keep the oil from splashing and to prevent water from being directed into nearby electrical equipment.

Protecting Exposures. While the attack on the fire progresses, a second hoseline is run out from the fire main. This second line is advanced into the boiler room and positioned to cool areas directly exposed to the fire. The bulkhead is hot to the touch; it is cooled with water fog, for maxi-

Figure 10.10. Whenever possible, the initial attack should be backed up with another means of attack.

mum absorption of heat with minimum runoff. The stream is directed onto the bulkhead as long as steam is produced when the water hits the bulkhead. Other spaces adjacent to the fire area are protected similarly.

Ventilation. As the attack is made, doors and hatches in the upper parts of the engine room (preferably those opening directly to weather decks) are opened to vent the products of combustion. Once the fire is completely out and surrounding structural members have been cooled, mechanical venting is used to remove the combustion gases and draw in cool air. This is permissible, since the duct is vented directly to the outside. In most cases, operation of the exhaust fan at the top of the engine room on low speed would accomplish the necessary venting.

Bilge Fire

The Fire. Flammable liquid in the engine room bilge has ignited, creating a substantial body of fire that is increasing in size.

Sizeup. The location of the fire indicates a flammable-liquid fuel requiring CO_2, fog or foam as

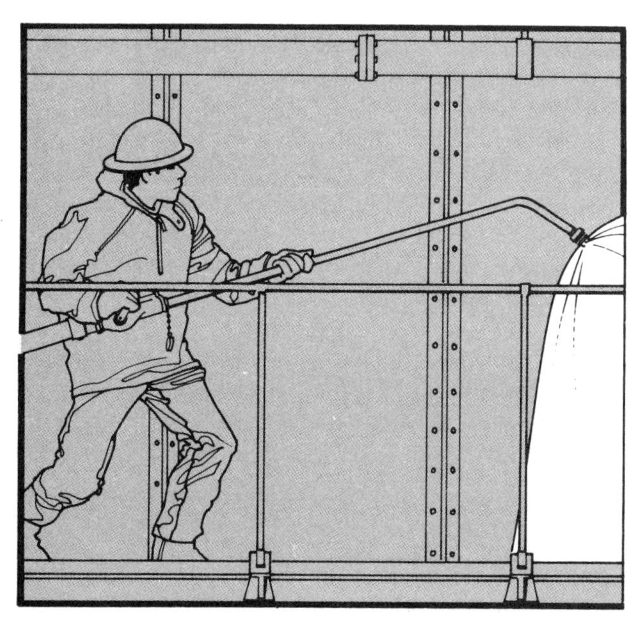

Figure 10.11. Engine room spill fire. **A.** Initial attack is made with a portable or semiportable dry chemical, CO_2 or Halon extinguisher. **Figure 10.11. B.** The attack is backed up with a hoseline used to cool metal surfaces to prevent reignition.

the extinguishing agent. Because the liquid fuel is free to spread as the vessel rolls and pitches, the crewman discovering the fire is careful to protect his flanks and rear. The fire is large enough so that it is beyond the capability of small portable extinguishers.

Attack. The alarm is sounded. The initial attack is made with the semiportable CO_2 extinguishing system. One crewman runs out the line while another activates the system. The agent is directed below the deck plates and into the bilge, as close to the fire as possible. To cover the entire width of the fire, the nozzle is swept back and forth. As the flames in one area are knocked down, the stream is directed onto a new area. The nozzleman is extremely careful to protect his flanks and rear as he advances the nozzle.

It is essential that all the flames be extinguished before the supply of CO_2 runs out. If manpower is available, portable CO_2 extinguishers can be used to supplement the primary attack and to protect the nozzleman.

Confining the Fire. During the attack, all doors and hatches to the engine room are closed. All engine room ventilators are shut down. This confines the fire and ensures that it will not travel outside the engine room. With the engine room sealed off, the crew is prepared to use the fixed CO_2 total-flooding system, should this become necessary.

Backup. The fire party runs out hoselines from the fire-main system. The nozzles are positioned to protect those fighting the fire, and the lines are charged with water. A carefully applied fog pattern will be used to cool hot metal surfaces reducing the possibility of reignition.

The CO_2 attack fails to extinguish the fire. The hoselines, already in position, are used to attack the fire with low-velocity fog applicators. The low-velocity fog dilutes the oxygen above the fire, knocks down flames, cools metal surfaces and generates steam, all of which contribute to extinguishment. The applicators can be directed through small openings and poked into confined areas. In addition, the fog applicator can be manipulated so that water is not sprayed on machinery unnecessarily.

Attacks with the semiportable CO_2 system and fog streams will control and extinguish a bilge fire. If they do not, a very serious condition exists: The ship's power plant is threatened. Then the fixed CO_2 extinguishing system must be used. The fixed system is regarded as a last resort. However, when it must be used, the decision to activate the system should not be delayed. It is better to make an early decision to use the CO_2 flooding system than to allow the fire to do extensive damage to the engine room. If flooding with CO_2 extinguishes the fire, then the engine room is lost only temporarily; it can be restored to operating order. On the other hand, a delay may allow the fire to damage machinery and electrical equipment beyond repair. The results then are a disabled engine room and loss of propulsion.

Fixed CO_2 System. Before the fixed CO_2 system is activated, all personnel are evacuated from the engine room. Saturation with CO_2 will reduce the oxygen content below the level required to sustain life. The engine room should have been sealed during the initial attack. If not, all openings should be closed, and ventilation systems shut down, at this time. The CO_2 system is activated on the order of the master or, when designated, the officer in charge of the fire party.

Protecting Exposures. The areas adjacent to the fire are continuously observed for fire extension and to ensure that CO_2 is not leaking from the engine room. The crewmen assigned to spaces fore and aft of the engine room wear breathing apparatus. This precaution is necessary because the spaces to be examined may be contaminated by smoke and/or CO_2. These crewmen are also equipped with handlights (flashlights) to improve visibility.

Reentry. The CO_2-saturated area is reentered with caution. Although there are no hard and fast rules concerning reentry, many factors must be considered. How hot was the fire? If oxygen is allowed to reach the fire area, will metal in that area be hot enough to cause reignition? Is it essential that the engine room be restored as fast as possible because of heavy seas? Or are the seas calm and without navigational hazards, so that entry may be delayed? The engine room should not be entered for at least an hour, primarily to allow the heat to dissipate. The injection of CO_2 into a sealed area will extinguish a flammable-liquid fire almost immediately. However, since CO_2 has no cooling effect, metal surfaces remain hot. It is a "damned if you do and damned if you don't" situation; the fire is out, but the threat of reignition makes the area dangerous.

After an hour, entry is attempted by a two-man search team dressed in protective clothing and using breathing apparatus with lifelines attached (Fig. 10.12). They enter through the highest access door into the engine room. If they find the heat excessive, they will leave the area immediately and wait at least another 15–30 minutes before their next attempt. When they are able to

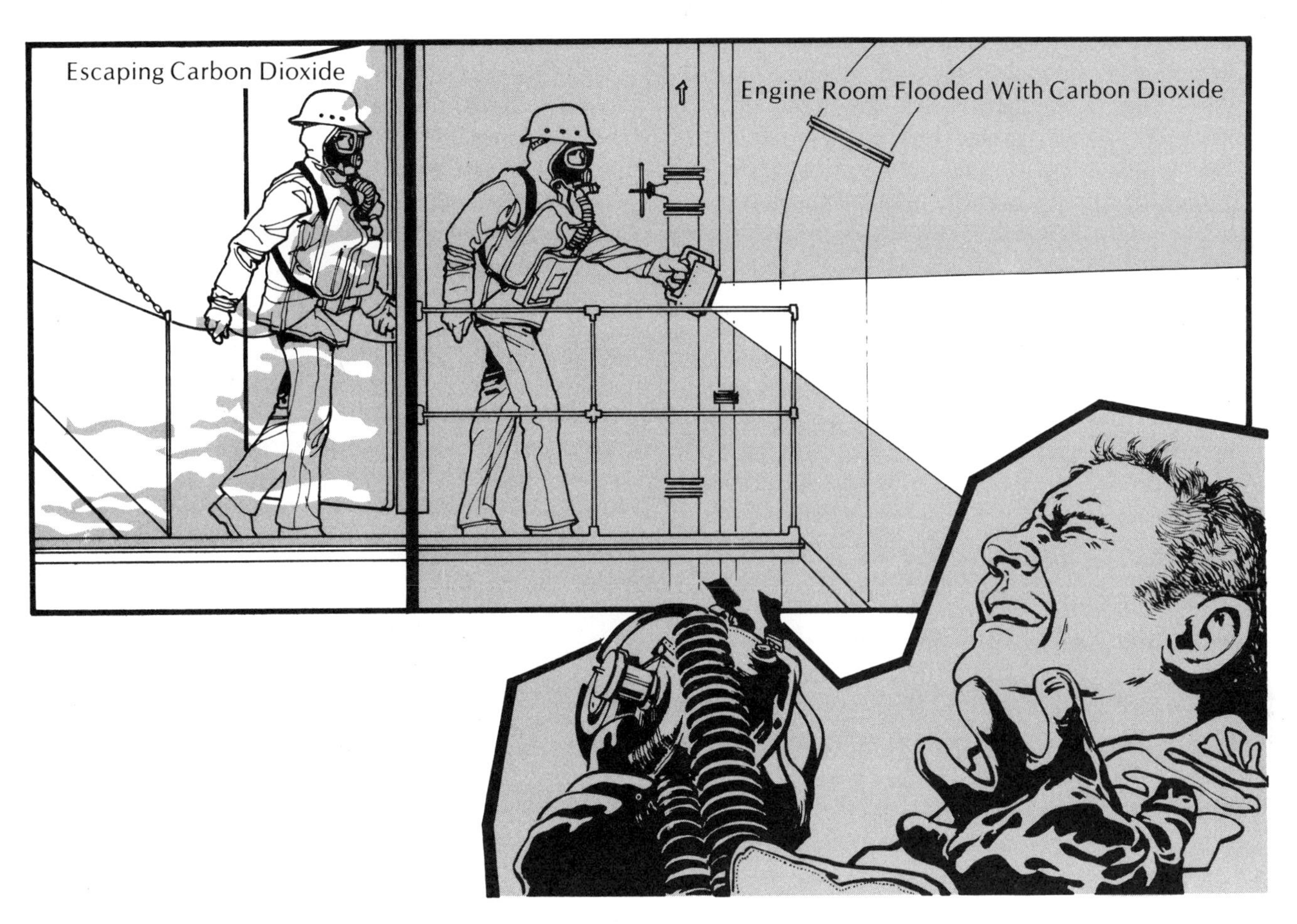

Figure 10.12. Reentering a space that has been flooded with CO_2. The breathing apparatus is tested before the space is entered. Protective clothing and a lifeline are used. Firefighters work in two-man teams. They move through the entryway quickly and close the door immediately to prevent CO_2 from escaping.

tolerate the temperature, the team proceeds to the lower level of the engine room. They use the hoselines from the unsuccessful fire attack to cool metal surfaces near the fire. *They do not remove the facepieces of their breathing apparatus, since the atmosphere will not support life.* Removal of the facepiece would result in almost instantaneous collapse and death. (Fig. 10.12). The team keeps track of the time they spend in the engine room to ensure that they leave before their air supplies run out. They are also timed by crewmen outside the fire area. After cooling the metal surfaces, the men leave the area the same way they entered. They remove their facepieces only after leaving the engine room and shutting the access door.

The crewmen selected for this task are thoroughly familiar with the engine room. They are so chosen because visibility may be restricted by smoke. Visibility will be further reduced by the steam generated when the cooling water hits the hot metal surfaces.

Entry at the highest level is recommended because CO_2 is heavier than air. Entry through a door that is level with the fire could allow excessive amounts of CO_2 to be lost when the door is opened. The high entry level also immediately exposes the crewmen to the highest temperatures they will encounter, since the heat rises to the upper parts of the engine room. If the men can tolerate the heat at the upper level, it will not present a problem as they move down. A disadvantage of high level entry is that it forces the crewman to climb down and then up ladders. It also makes rescue of the team more difficult if something should go wrong. As this discussion implies, the entire process is dangerous. Crewmen engaged in the operation must thoroughly understand their duties and the problems involved. They must be fully aware of all safety requirements.

Once the metal has been cooled down, the engine room is ventilated with the mechanical venting system. If there was any reason to keep an inert atmosphere in the lower portion of the engine room, then natural ventilation would be employed, and after a short time the upper area of the engine room would support life. Since we have cooled the potential source of ignition and want to regain control of the engine room, mechanical ventilation is indicated. After the space has been vented for a reasonable time, a portable gas detector is used to determine whether any fuel vapors remain. Since none are present, a flame safety lamp or portable oxygen meter is used to ensure that the engine room contains sufficient oxygen for breathing. The lamp burns in all parts of the engine room, so the crew is allowed to enter without breathing apparatus.

Overhaul. The oil remaining in the bilges is blanketed with a layer of foam to prevent reignition until it can be removed. The source of the oil leak is found and repaired.

Boatswain's Locker Fire

The Fire. A member of the deck gang notices smoke coming out of a deck hatch opening in the forward section of the vessel. He turns in the alarm.

Sizeup. This hatch leads to the boatswain's stores, other storage areas and the chain locker. The only access to the area is down a ladder and through a passageway. This section of the vessel is not protected by a fixed fire-extinguishing system. The smoke is heavy, and visibility is limited. Exactly what is burning cannot be determined, but the area is used primarily for the storage of class A materials.

Confining the Fire. The crewman closes the deck hatch to reduce the supply of oxygen to the fire.

Precautions. The fire party is dressed in full protective clothing and breathing apparatus, since heavy smoke and extreme heat can be expected.

Attack. Hoselines are advanced to the hatch and charged. The hatch cover is cooled with water and then opened. The nozzleman and backup man climb down the ladder with the nozzle, and crewman on deck assist in advancing the hose down the ladder. The nozzle is advanced toward the glow of the fire. The advancing firefighters keep low, for better visibility and a lower heat concentration. The nozzleman uses short bursts of water fog to reduce the heat. Heated gases and steam pass over the firefighters' heads and move out through the access hatch. When the nozzleman is in a position from which the stream can reach the fire, he uses a solid stream to penetrate into the seat of the fire. When the stream hits the fire, steam is generated and visibility is greatly reduced. He shuts off the nozzle to permit the steam and smoke to lift. When visibility is regained, he moves in and completes the extinguishment of the fire.

Backup. An additional hoseline is positioned at the hatch opening and charged. This line can be used to protect the initial attack team if they are forced to withdraw because of a burst hose or excessive heat.

Protecting Exposures. While the attack is under way, other crewmen examine the compartments adjacent to the fire. Combustible materials are moved away from bulkheads. Heated bulkheads are cooled with water in the form of fog.

Ventilation. The fire area is ventilated after the fire is out. Since the area is not served by a mechanical ventilation system, portable fans are employed.

Overhaul. Because of the quantity and nature of the materials stowed in a boatswain's locker, overhaul is extensive. All charred material is moved to weather decks and thoroughly soaked.

Paint Locker Fire

The Fire. Smoke is discovered issuing from a paint locker in the machinery spaces, through a partially opened door. This paint locker has a manual CO_2 system.

Sizeup. If there is not much smoke, the fire is in the early stages. Large volumes of dense smoke indicate that the fire has been burning for some time.

Attack 1 (Light Smoke). The alarm is sounded. The door is opened further, and the locker is visually examined to determine what is burning. A smoldering class A fire (rags, rope, paint brushes) is extinguished with a multipurpose portable dry chemical or water extinguisher.

Backup. A hoseline is run out and charged with water, to be used to wet down any smoldering material.

Attack 2 (Heavy Smoke). The door and vents are closed, and the manual CO_2 system is activated.

Backup. A hoseline is run out and charged with water, as a precautionary measure, for the protection of exposures and to cool off paint locker bulkheads.

Confining the Fire. The fire was isolated when the door was closed.

Protecting Exposures. All sides of the paint locker are continuously examined to ensure that the fire is not extending.

Ventilation. The locker is not opened or ventilated until all bulkheads, decks and overheads are cool to the touch. Then, with a charged hoseline in position, the door is opened and the compartment is allowed to ventilate long enough for a normal level of oxygen to return to a small space. Test for a safe oxygen level.

Overhaul. Fire-damaged material is removed from the locker. Expended CO_2 cylinders are replaced. The hoseline is kept in readiness until overhaul is complete and CO_2 protection is restored.

Galley Fire

The Fire. A fire involving the deep fryer on the galley stove has extended to the duct system.

Sizeup. The alarm is sounded. Sizeup indicates that a class B fuel is involved. There is an exposure problem, because the fire has entered the venting duct.

Attack. Using an appropriate portable extinguisher (CO_2, dry chemical or Halon), galley personnel first direct the extinguishing agent onto the burning oil. They are careful not to spatter the burning liquid. Then, with a sweeping motion, they direct the agent into the hood and duct area (Fig. 10.13). Since the range is fueled with gas, they shut off the fuel supply to the pilot lights and the burners. (The pilot light was extinguished in the attack.)

The firefighters do not shut off the exhaust fan in the duct over the range. They use it to pull the extinguishing agent into the duct and spread it through the duct. However, once the flames are knocked down, they turn the exhaust fan off. If CO_2 is used, the exhaust fan must be secured. After the flames are out, water spray must be used to cool metal surfaces.

Backup. A hoseline is run out to the galley door and charged with water. The hoseline is thus in position to support the attack or to cool down hot metal surfaces.

Confining the Fire. All ventilation to and from the galley is shut down.

Protecting Exposures. The galley exhaust duct system is examined, from beginning to end, for fire travel inside and outside the ducting. Every compartment through which it passes is checked. A hoseline is positioned where the duct vents to the outside, in case fire shows at that point.

After the visible flames are extinguished and the metal surfaces have been cooled, the duct power venting system is turned on. Since the firefighters suspect that there is fire in the duct, they remove the grease filters. With the fan in operation, they direct a fog stream up the duct. The fan helps pull the water into the duct. This cools the metal ducting and pushes heat and smoke out of the duct. The duct inspection plates are opened to check for fire.

Ventilation. After the fire is completely out, the smoke and heat are removed from the galley by mechanical ventilation.

Overhaul. All grease residues are cleaned off the stove, hood and ducting. The filters are cleaned or replaced. The range is restored to operation.

Fire in an Electrical Control Panel

The Fire. Smoke is discovered issuing from the rear of the main electrical control panel in the engine room. The alarm is sounded.

Sizeup. This is a class C fire. The rear of the panel is not readily accessible because of the sheet metal construction of the cabinet. The insulation on live electrical wires is burning. As the wires get hotter, their resistance increases* and additional heat is generated; this can extend the fire and cause irreparable damage to electrical components.

---

* The resistance of pure metals—such as silver, copper and aluminum—increases as the temperature increases. (*Electricity for Marine Engineers,* p. 68, prepared by the MEBA Training Fund.)

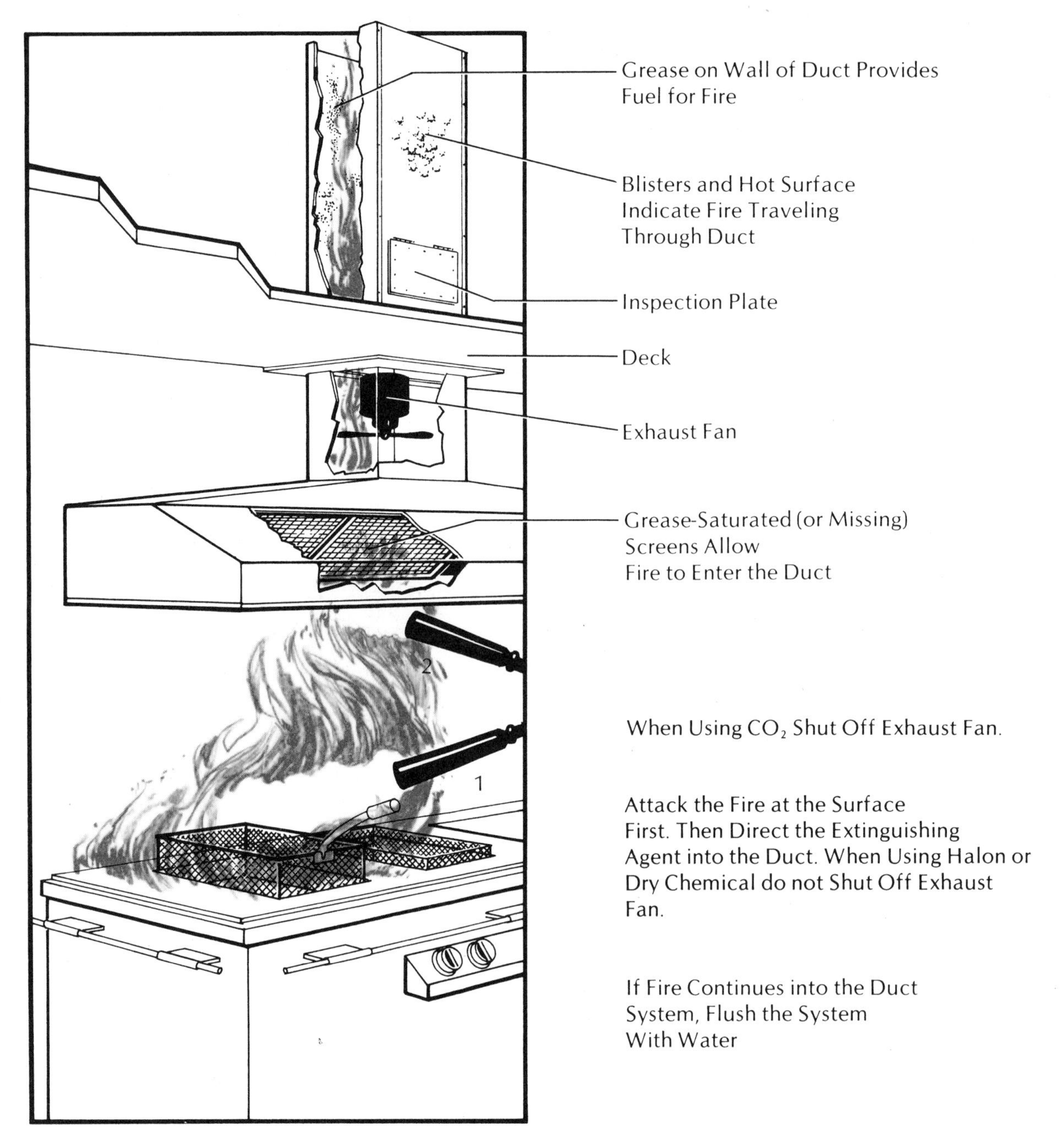

Figure 10.13. Fighting a galley range fire. 1. Attack the fire at the surface first, then 2. direct the extinguishing agent into the duct.

Attack. The fire party immediately attempts to deenergize the circuit or the equipment. This serves the double purpose of protecting firefighters from electrical shock and reducing the ability of the fire to extend.

The initial attack is made with a CO_2 or Halon extinguisher, either portable or semiportable (Fig. 10.14). There are several reasons for using these agents. First, the person charged with deenergizing the equipment may not have completed his assignment. Since CO_2 and Halon are not conductors of electricity, those attacking the fire are protected from electrical shock, provided they do not touch the equipment. Second, these agents do not leave a residue and will not damage delicate electrical components. Third, firefighters cannot easily get at the fire, because of the cabinet; the extinguishing agent must be applied through the ventilation slots. A gas will more readily flow through these small openings. If CO_2 or Halon is not available, a dry chemical extinguisher may be used. However, dry chemical leaves a residue that could damage electrical contacts and could be difficult to clean up.

Backup. The initial attack is backed up with additional portable or semiportable extinguishers approved for use in class C fires. The Coast Guard

Figure 10.14. Fire in an enclosed electrical cabinet should be attacked with CO_2 or Halon. The agent should be directed into the unit through the cabinet vents.

regards any fire involving electrical equipment as a class C fire, even if the equipment is deenergized.

Confining the Fire. The fire is isolated by deenergizing the equipment and knocking down the flames.

Protecting Exposures. All wiring and equipment near the fire is checked for fire extension. All equipment that is electrically connected to the involved panel is also examined. Electric cables are traced along their entire length, especially where they pass through decking and bulkheads. It is found that the fire was confined by quick deenergizing of the electrical equipment, extinguishment of the flames and natural cooling of hot components.

Ventilation. The burning insulation has given off irritating and toxic fumes. The ventilation system was shut down when the alarm was sounded. It is reactivated as soon as the fire is out. If this cannot be readily accomplished, the firefighter should be equipped with protective breathing apparatus.

Caution. If the machinery space is small and several semiportable CO_2 extinguishers are used, the oxygen content in the air may be reduced enough to make breathing difficult. Then protective breathing apparatus should be used during the attack, until ventilation has been accomplished. If the ventilation system cannot be reactivated, anyone entering the fire area should wear such breathing apparatus.

Overhaul. Overhaul procedures are carried out by engine department personnel. These crewmen are best able to open up fire-damaged equipment with minimum damage to that equipment. After overhaul, when the fire is completely out, the involved equipment is inspected and repaired. Then the power is restored.

Cargo Hold Fire on a Break-Bulk Cargo Ship

The Fire. The smoke detection system alarm indicates smoke in no. 2 hold, lower 'tween deck. The vessel is 3 days out of port and 4 days from her destination. There are no nearer ports. Fair weather is the forecast for the next 48 hours.

Sizeup. The ship's manifest and cargo stowage plan are consulted. They disclose the following information regarding no. 2 hold:

- The lower hold contains heavy machinery in wood crates.
- The lower 'tween deck contains baled rags, cartons of paper products and bags of resin. (The alarm system indicates this area as the fire location.)
- The upper 'tween deck contains cartons and crates of small automotive parts and rubber tires.

Smoke is observed coming out of the ventilators for no. 2 hold. The main deck over the forward port section of the hold is much warmer than the surrounding deck areas.

The sizeup points out several important considerations regarding cargo hold fires and how they should be fought. For a direct attack with hoselines, the cargo above the fire must be moved so that the lower hatches can be opened. This would take at least several hours; and time is important! Every second of delay allows the fire to gain in intensity. It could become uncontrollable while the cargo is being moved. Suppose CO_2 were injected into the fire hold while crewmen were attempting to reach it for a direct attack. Eventually, the cargo hatch would have to be opened for the hoselines. Some CO_2 would escape, air would enter the hold, and the smothering effect of the CO_2 would be destroyed.

In a direct attack, it would be difficult to bring water to bear on the seat of the fire. If the fire were below the top layer of cargo, then more cargo would have to be moved to reach the fire. Pouring large amounts of water onto the burning area would not ensure extinguishment. In addition, the runoff could cause a stability problem and could damage cargo.

For all these reasons, fire in a loaded cargo hold should be fought indirectly, using a carbon dioxide flooding system. The agent can be brought to bear on the fire rapidly. When properly used, CO_2 has exhibited a high success rate for controlling and, in many instances, totally extinguishing hold fires. The master must have patience and confidence that the fire can be contained and extinguished with CO_2.

Attack. An indirect attack is employed. The hold is first sealed off; the seal *will be maintained until the vessel reaches port,* where shoreside firefighting units are available. The following actions are taken (Fig. 10.15):

- All hatch covers are checked to ensure that they are securely dogged down.

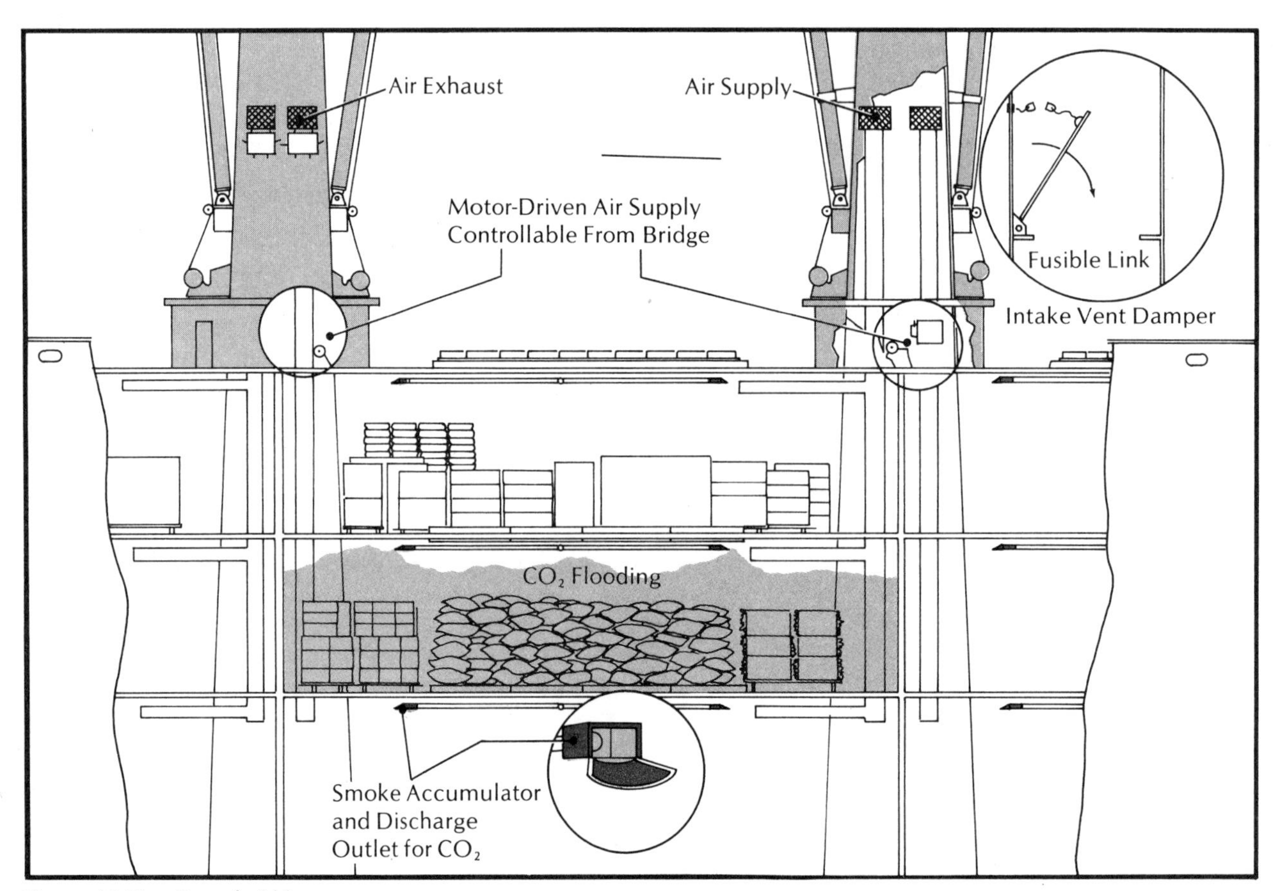

Figure 10.15. Cargo hold layout.

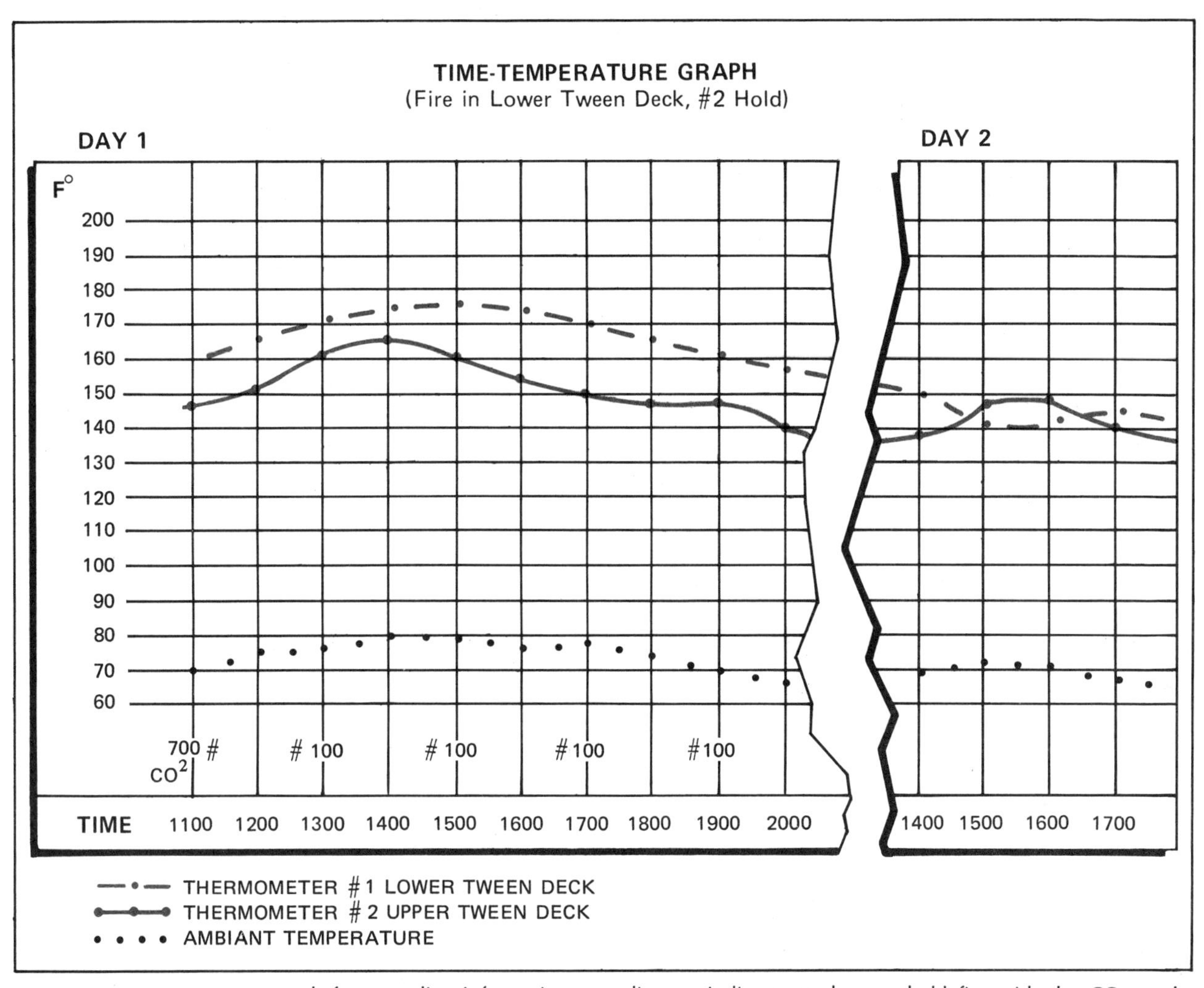

Figure 10.16. Temperature graph for recording information regarding an indirect attack on a hold fire with the CO_2 total-flooding system.

- One or more hoselines are run out on deck and charged. The lines will be used to cool hot spots on deck and (if necessary) on the exterior of the hull.
- Ventilator dampers are closed, and openings are covered with canvas that will be wetted down at frequent intervals.
- Instructions for the ship's CO_2 system are studied to ensure that the proper number of CO_2 cylinders are discharged into the right compartment.

As the CO_2 is discharged into the lower 'tween deck, crew members are stationed at vents, hatch covers and entrances to adjacent holds. They check for heat or smoke that is being pushed out by the CO_2. These signs indicate points where CO_2 could leak out or air could leak in. The leaks are sealed with duct seal or strong cargo tape. When the attack is started, the time, temperatures (readings from thermometers placed on hot spots on bulkheads of adjoining holds and/or compartments) and amount of CO_2 discharged are recorded on a line graph (Fig. 10.16). This information will be used to check the progress of the fire and determine how well it is being controlled. Temperatures will be entered on the graph hourly, and the time and amount of each additional CO_2 discharge will be recorded.

Patience is required when a fire in class A materials is being extinguished with CO_2. The process is slow. The oxygen content of the atmosphere in the hold must be reduced to 15% to extinguish flaming fire, and much lower to extinguish smoldering materials. The temptation to take a peek and see how things are going must be avoided; no opening must be made to release heat or smoke. If the hatches are opened, air is allowed to enter the hold and precious CO_2 will escape. This changes the atmosphere in the hold, permitting the fire to rekindle. Since the supply of CO_2 on a ship at sea is limited, none can be wasted. Crew members must be continually alert to detect any leaks from the involved cargo space.

Confining the Fire. By sealing the hold, the crew took the first step in confining the fire. The holds fore and aft of the fire are now checked. If any bulkheads are hot, combustibles located near or against them are moved, or the bulkheads are cooled with water fog (Fig. 10.17).

Exterior bulkheads (in this case the hull of the ship) are checked visually from the deck. Any blistering or discolored paint is investigated by a crewman in a boatswain's chair, lowered over the side (sea conditions permitting). Some of the hull plating is hot, indicating it is in contact with fire. A stream of water is allowed to flow down the side of the ship to cool these spots.

The deck over the fire is watched, especially where containers are stored over the hot spots. Hazardous cargo stored on deck near the fire is moved to a safer area. If this were not possible, the deck would have to be cooled constantly.

It is not possible to check the area directly under the fire, except by checking the bulkheads in the adjacent holds.

Protecting Exposures. The holds fore and aft of the fire were checked for hot spots before the CO_2 was discharged. However, the application of CO_2 was not delayed for this examination. Pyrometers are placed on the hottest spots on decks and bulkheads, at each level of the hold (Fig. 10.17C). Where hot spots are not present but the bulkhead is to be monitored for temperature, the pyrometers are placed midway between the deck and the overhead. The temperatures are read hourly, to monitor the progress of the fire. Pyrometers are secured to the bulkhead with cargo tape or duct seal if they are not of the magnetic type. Hoselines are positioned on deck, ready to be used in cooling bulkheads if required.

Once the CO_2 is discharged, leakage from the fire hold may make the adjacent holds untenable without breathing apparatus. The oxygen content of these holds is checked continually, to ensure the safety of crewmen performing monitoring tasks.

Ventilation. Absolutely no ventilation is attempted until the ship is in port and additional help is available. By the time the ship arrives in port, temperatures in the hold will have abated and will have remained fairly constant for a period of 24 hours. The hatch cover is then partially (and cautiously) opened. Charged hoselines are in position, and men with breathing apparatus are ready to advance the hoselines if some smoldering fire still exists.

Natural ventilation is employed to make the *upper* 'tween deck tenable. The inert atmosphere in the lower 'tween deck and lower hold is maintained.

Overhaul. After the upper 'tween deck has been declared safe for men by a marine chemist (or a qualified ship's officer if a marine chemist is not available), offloading is begun. Charred material is kept separate from other cargo, placed on a noncombustible deck and wet down if necessary. The upper 'tween deck is cleared of all material involved in the fire, and the area over the hatch covers is cleared. Then the lower 'tween deck is opened. Again men with breathing apparatus stand by, ready to advance charged hoselines if necessary.

When the hold is declared safe, the cargo involved in the fire is offloaded. Sufficient additional cargo is offloaded to ensure that there is no hidden fire. The hold is completely ventilated,

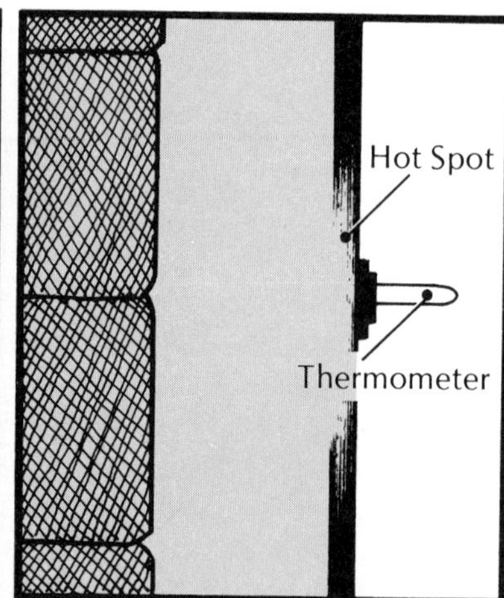

Figure 10.17. A. If possible, cargo should be moved away from hot bulkheads. **B.** Hot bulkheads may be cooled with water fog, applied until steam is no longer produced. **C.** Pyrometers are placed halfway between the deck and overhead.

and the CO_2 system is restored. The fire area is carefully examined for structural damage. Then the hold is reloaded with cargo that was not involved in the fire.

Tanker Fire

The Fire. During the transfer of gasoline from no. 3 center tank to a shoreside facility, a cargo hose ruptures at the flange on the ship. The gasoline spills on deck and immediately catches fire.

Sizeup. The cargo transfer was just started, so the cargo tanks are nearly full. The wind is light and blowing across the starboard quarter, toward shore. The scuppers have been plugged, and as yet no gasoline has entered the water. The alarm was given promptly, and all crew members are aware of the fire and have manned their firefighting stations.

Attack. The general alarm and the ship's whistle are sounded to alert shoreside personnel and nearby ships. Help is immediately requested from local fire departments.

At the same time, the cargo pumps are shut down to stop the flow of fuel to the fire. Although some fuel will continue to drain from the cargo hose and pipelines, the amount is small compared to what the pumps would feed into the fire.

Crewmen activate the nearest foam monitor between the fire and the deckhouse, on the windward side of the fire. The monitor has a greater reach and volume than the handlines. The operator directs the foam onto a nearby vertical surface, so it runs down and forms a blanket on the burning gasoline. An alternative method is to lob the foam onto the near edge of the fire and move the nozzle slowly from side to side. This also allows the foam to build up a continuous blanket. A foam handline is run out and used to blanket any area that the monitor cannot reach. A hoseline equipped with a fog nozzle is run out and charged to protect the crewmen on the foam monitor from excessive heat (Fig. 10.18). Firefighters are careful not to destroy the foam blanket by haphazardly directing water into the foam. An unbroken blanket of foam is maintained over the gasoline spill until all sources of ignition are eliminated.

If small open patches of fuel are allowed to continue burning, their heat will start to break down the foam blanket. The patches will grow in size, and eventually fire will again cover the entire surface of the spill. The same thing will happen if the fire is not completely extinguished before the supply of foam is depleted. Thus, the entire surface of the flammable liquid must be blanketed with foam, and the blanket must be maintained until the fire is out.

It may be difficult to maintain a full foam blanket where the foam can spread out in all directions, as on a weather deck. If the spill is large, a lot of foam will be needed; a shipboard foam system does not provide an inexhaustible supply. This problem can be solved, at least partially, by shutting off the fuel source as quickly as possible.

If the fire is extinguished by the foam, crewmen should be careful not to disturb the foam

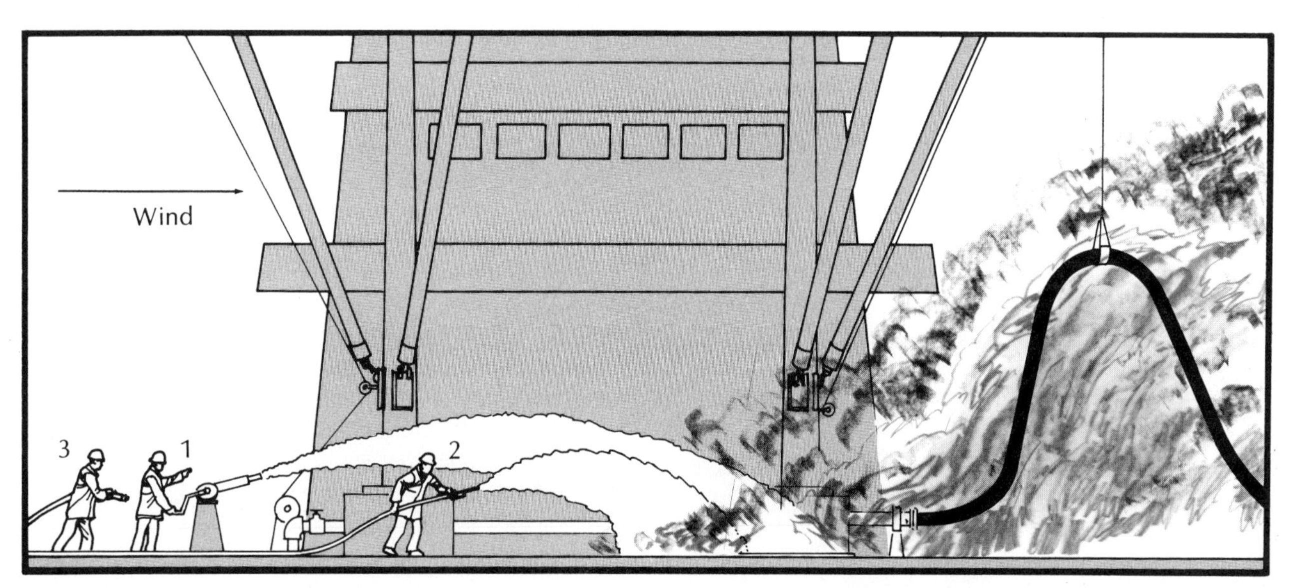

Figure 10.18. **1.** With the cargo pumps shut down, the spill fire is attacked from the windward side with a foam monitor. **2.** Foam handlines are used to cover areas that the monitor cannot reach. **3.** A water fog nozzle is used to protect personnel.

blanket by walking through it unnecessarily. Hoselines should be kept available for immediate use; deck plates and other structural metal will be hot, and reignition could occur. Any burning paint, hose or gaskets should be extinguished with a fog stream applied carefully so as not to break up the foam blanket.

If the foam attack is not successful, the fire can be controlled and extinguished with fog streams. Several hoselines should be positioned to sweep the burning area with fog. The attack should be started from the windward side; this allows the wind to carry the fog into the fire, providing greater reach. The wind also carries heat and smoke away from the firefighters. As the streams are swept from side to side, they should be kept parallel with the deck. The nozzlemen should advance slowly; they must not move into the fire too rapidly, or the flames may get around their flanks and behind them. They must not advance to the point where reflash could envelop them in flames. Water fog will knock down and push flames away, but it will not provide a smothering blanket. The possibility of reignition must be considered at all times.

Even after the fire is extinguished, the fog streams should be used to sweep the fire area and other hot surfaces, to cool them down. The water will flow over the side, so it can be used without restriction. It is better to apply too much water fog than to risk reignition. The water must be applied until three things are accomplished: *1)* the supply of fuel to the fire is shut off, *2) all* metal surfaces are cool to the touch and *3)* the flammable liquid is diluted or washed overboard.

If the foam and fog attacks both fail and the fire continues to increase in size, the safety of the crew must then become the prime consideration. The attack should be abandoned, and the ship evacuated. Firefighting operations should not be continued to the point at which crewmen are trapped and have no avenue of escape.

Confining the Fire. The fire is isolated by quickly covering it with a blanket of foam and promptly securing the ship's cargo pumps. As soon as the fire in the vicinity of the cargo valves is extinguished, crew members, protected by fog streams, close all valves and ullage openings. Openings in the deckhouse are closed and protected with hoselines. Ventilation intakes drawing air from the vicinity of the fire are shut down. Electrical equipment in the vicinity of the fire is deenergized.

Protecting Exposures. The ship and the shoreside complex are threatened by the fire. Other ships tied up at the same or adjacent docks are also jeopardized. These ships began preparations to get under way when the general alarm was sounded.

At the oil storage depot, water spray and foam systems were activated to cool tanks and piping. Hoselines with fog nozzles were advanced and charged for the same purpose. Additional foam equipment and foam concentrate were requested by the involved ship and were brought to the scene.

Overhaul. Once the fire is extinguished and the leak secured, the remaining spilled gasoline is cleaned up. The foam blanket is maintained until the gasoline is removed or recovered. Then, all cargo hoses are checked or replaced. Shoreside fixed and portable foam monitors are kept in position until all gasoline fumes have dissipated.

To allow an examination of the spill area underneath the pier, the tanker is warped to another position away from the cargo handling area. The foam supply is replenished without delay, and the ship's fire protection system is restored to duty. Then cargo offloading operations are resumed.

CONTAINER FIRES

A fire involving ordinary combustibles in a modern, well-built container will frequently extinguish itself by consuming all of the oxygen in the container. In the following examples we will assume that this did not happen.

Container Fire on Deck

The Fire. A 12.2-m (40-ft) aluminum container stowed on deck in the forward section of the ship is giving off smoke. The aluminum is discolored from heat.

Sizeup. The container is in the center of a three-tier stack and is surrounded on three sides by other containers. It was packed at a stuffing shed at the terminal and can be expected to contain a variety of materials, none reported to be hazardous.

Attack. The alarm is sounded. Crewmen check the container labels and the cargo manifest to determine the contents of the involved container and adjacent containers. At the same time, a hoseline is advanced to the involved container and charged. It is used to cool down the container with fog. The nozzleman stands back so he will not be scalded by the steam that is generated. A second hoseline is run out to the adjacent containers. Neither hose stream will reach the seat of the fire; however, they will help contain the fire while preparations are made for final extinguishment.

The cargo manifest indicates that water is the proper extinguishing agent. Now crewmen chisel to punch a hole about 2.54 cm (1 in.) in diameter near the top of one side of the container, close to the hottest area. A piercing applicator (Fig. 9.5) or the pike end of a fireaxe could also be used. A short applicator is attached to a combination nozzle, and the low-velocity head is removed. The applicator is inserted into the hole, and the nozzle is opened. The nozzleman floods the entire container, even though this may not always be necessary.

If the cargo in the container is very valuable and can be damaged by water, CO_2 or Halon can be introduced into the container through the opening. Six or more portable 6.8-kg (15-lb) CO_2 extinguishers should be used for the initial discharge. The opening should then be plugged, and additional CO_2 discharged hourly, until the fire is out.

Confining the Fire. The fire is confined as long as it does not extend from the container. Hoselines are used to cool the outside of the container and prevent the fire from burning through until the proper extinguishing agent has been applied to the inside of the container.

Protecting Exposures. The cargo in adjacent containers and the cargo in the hold below the main deck are exposed to the heat of the fire. A hoseline is used to protect the adjacent containers and to cool the deck. If the fire were intense and a container located right on the deck were involved, it would be advisable, as a precautionary measure, to inspect the compartment immediately below the main deck. The fire could extend downward by conduction.

Overhaul. The container is opened, and its contents are removed and examined. Any fire that is discovered is soaked with water. This step may be delayed until the container has been unloaded in port. However, the crew must ensure that the fire is completely out. This would require the use of additional water, directed into as well as on the outside of the container.

Container Fire in a Hold

The Fire. The smoke detection system indicates smoke in no. 4 hold, lower section, starboard side. The ship is at sea, and the nearest port is 3 days away.

Sizeup. The no. 4 hold is fully loaded with containers, and there are two tiers of containers on top of the hatch cover.

Attack. The alarm is sounded. The fire party opens the emergency escape hatch and notes a minimum of smoke and no heat. Visibility within the hold is fair (15 m (50-ft)). The on-scene leader declares that the hold can be entered. Two crew members, familiar with the stowage of containers on the ship, are equipped with breathing apparatus, lifelines and portable lights.

The crewmen enter the hold. When they reach the lower level, they still encounter light smoke and no heat so they decide to proceed. A hoseline is lowered to them, and crewmen on deck are ready to pay out additional hose as needed.

The two-man team locates the burning container. They first cool the outside of the container with water fog. Then they make an opening into the container, insert the applicator and discharge water into the container. The applicator is held in place until sufficient water has entered the container. Then the team examines the involved container and adjacent containers for hot spots. Since several hot spots are found, the involved container is flooded again with water; an additional hoseline is advanced into the hold, and water fog is used to cool the hot spots on adjacent containers. When additional examinations indicate that the fire is out, the hoselines are withdrawn, the hold is secured and the detection system is restored to service.

If the initial examination had disclosed a considerable amount of smoke and heat, with less than 15 m (50 ft) of visibility, the hold would have been sealed. It then would have been flooded with CO_2 as previously described.

The fire confinement, exposure protection and ventilation procedures are the same as for the cargo hold fire described earlier in this chapter.

Overhaul. No attempt is made to overhaul the containers until the ship is in port. At the dock, the containers are unloaded without the need for crewmen to enter the hold to attach the lifting mechanism. If the CO_2 total-flooding system had been activated, the inert atmosphere could be maintained during the unloading. Charged hoselines are positioned at dockside to fight any reignition that might occur when the containers are opened.

LNG Spill Involving a Leak

Liquefied natural gas † is a hydrocarbon fuel composed mostly of methane. It burns cleanly, with little or no visible smoke. The flame height is greater than that of other hydrocarbon fuels, and the radiant heat produced is much more intense. In the liquid state, LNG weighs about half as much as water; its liquefication temperature is approximately −162°C (−260°F). Its volume increases 600 times as it changes from a liquid at its boiling temperature to a gas at atmospheric pressure and 15.6°C (60°F). When the temperature of the vapor rises to approximately −112°C (−170°F), it weighs the same as air. It is transported in the liquid state for economic reasons.

LNG is colorless, odorless and severely damaging to the eyes and throat. The liquid causes frostbite on contact with the skin. It causes embrittlement fractures in ordinary steel but may be safely handled in stainless steel, certain nickel steel, certain copper alloy and aluminum containers. It is usually odorized by adding methyl mercaptan as an aid in detecting leaks of the vapor. The ambient vapor is not irritating to the eyes or throat.

Many safeguards are built into ships that transport LNG, to combat spills and fires. These vessels are equipped with deck water spray systems to control spills, prevent brittle fracture of the deck plating and facilitate fast warmup of the vapors to minimize the fire hazard from cloud drift. The spray system is also used to help prevent ignition. If ignition does occur, the spray system will provide a water curtain to protect vital areas, such as the bridge and gas control room, from the intense radiant heat of the fire. It also will cover most piping and tanks with a cooling barrier of water. The radiant heat could build up enough pressure in uncooled tanks and piping to cause them to rupture.

Every LNG ship is equipped with enough large dry chemical skid units (Fig. 10.19) to protect the entire weather deck area in case a fire occurs and extinguishment is desired. The dry chemical would be used to extinguish a small spill fire where the LNG spill could be controlled. If a large spill were to occur, as from a high energy impact (collision), fire would be almost a certainty. The spread of fire would be controlled with dry chemical, while the fire was allowed to burn itself out. The water spray system would be used to prevent other tanks from becoming involved.

† *See* Chapter 5 for a discussion of liquefied natural gas (LNG) and Chapter 9 for a discussion of the special fire protection systems installed on LNG tankers. Because LNG constitutes an almost unique spill or fire hazard, the material in those chapters will be reviewed and enlarged upon here.

Special protective clothing must be worn by personnel handling LNG spills. This clothing must consist of at least rubber gloves, a face shield and protective clothing. At least two self-contained breathing apparatus should be available on the vessel. Proximity- and entry-type suits for approaching an LNG fire, closing a valve or other necessary actions may also be provided. Such equipment should be readily available, and personnel should practice donning the gear quickly as preparation for emergency situations.

The Spill. During offloading, a section of piping develops a leak. A quantity of LNG spills onto the deck.

Sizeup. The spill is small, and the liquid is being vaporized rapidly by the continuous flow of water spray on the decking below the loading arm. (This is standard practice while LNG is being loaded or discharged.) The wind is blowing offshore, taking the vapors away from the vessel and the shore installation. There are no other vessels in the vicinity.

Confining the Spill. The source of the spill is isolated by means of control valves on each side of the leak. Ignition is prevented by actuating the water spray system. The crew and the shore installation are alerted to the emergency, and all pumping equipment and ventilation intakes are shut down.

Protecting Exposures. The steel deck and shell plating are protected by the water spray system. Personnel involved in controlling the leak don protective clothing. Other crew members run out and charge hoselines, and place water monitors at the ready. The dry chemical hose reels are advanced but not pressurized. (They will be pressurized only if the LNG ignites.)

Attack and Overhaul. With the source of the spill shut off, the water spray vaporizes the liquid and/or flushes it over the side. When all the LNG has been removed, the leak is repaired. Then all emergency equipment is secured, and normal offloading operations are resumed.

LNG Spill Involving Fire

Let us now suppose that fire occurs on the same vessel, at the same dock, with the wind blowing from the water and exposing the shore facility.

The Fire. During offloading, a pipe on the vessel ruptures. The subsequent spill catches fire.

Sizeup. The fire is small but spreading. A light wind is blowing the flames across the deck of

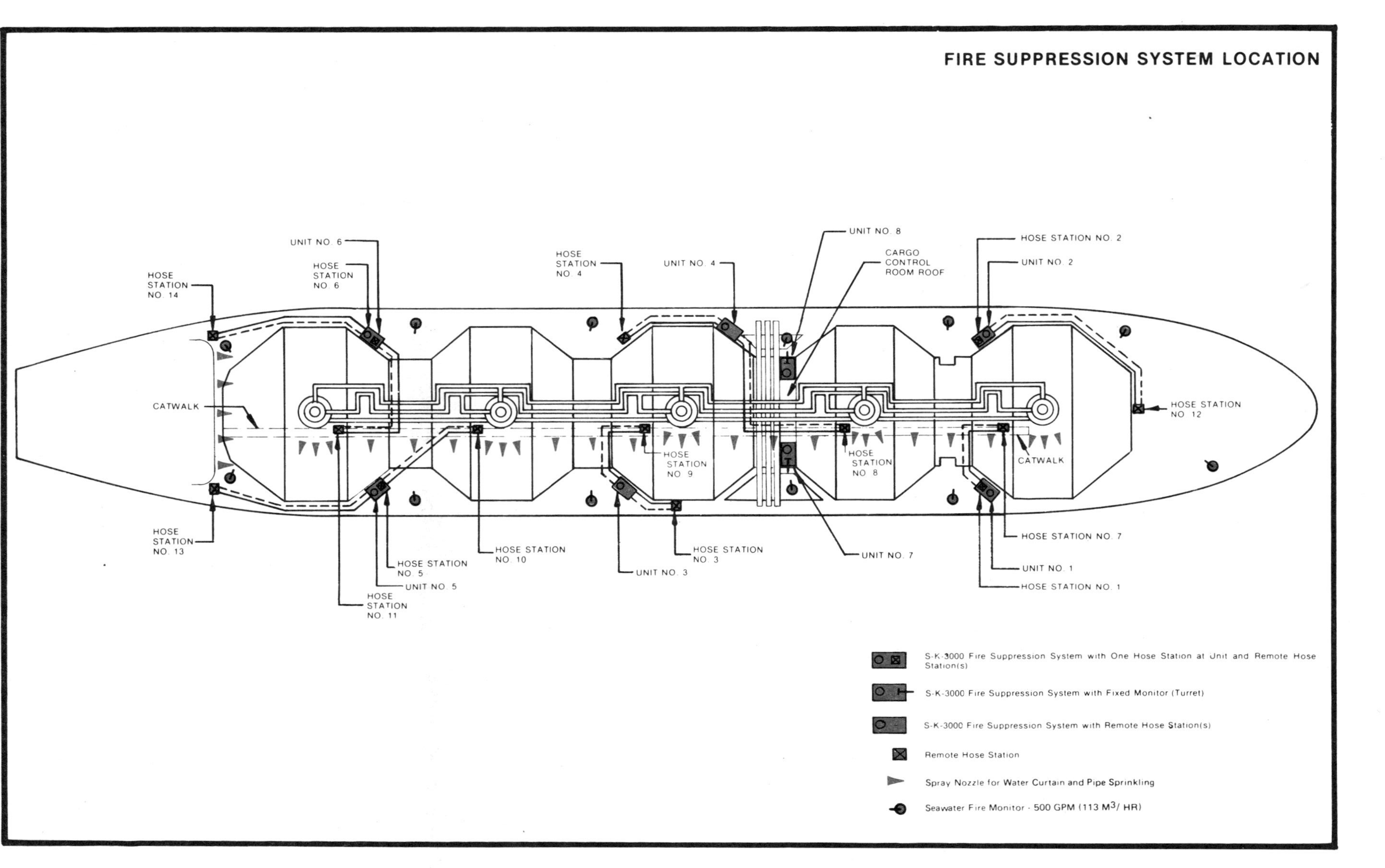

Figure 10.19. Typical dry chemical fire suppression system for LNG vessels. (Courtesy Ansul Company)

the ship toward the shore installation. There are other ships in the vicinity. Under present regulations, no other ships would be permitted at the same dock while LNG is loaded or unloaded.

Confining the Fire. The alarm is sounded. The cargo pumps are shut down, and the deck water spray system is activated. The crews of other ships at the pier and shore personnel are alerted to take emergency measures to protect their vessels and facilities. The proper valves are shut down to stop the leak and hoselines are run out and charged with water. Two teams using water fog advance along with the person designated to close the isolation valves, to protect him from the radiant heat. All deck openings are closed, and all ventilation intakes are secured.

Attack. The hose reels and the dry chemical system nearest the spill are run out and charged with dry chemical to the nozzle. Since extinguishment is desired, the dry chemical hoselines are used. To extinguish a spill on the loading arm, the turret monitor for the dry chemical unit on the dock side of the ship is used. It can be actuated locally or remotely, to discharge the agent from the turret in a fixed pattern.

Protecting Exposures. The deck water spray system protects the cargo tanks, piping and deckhouse. The protective spray system on the shore installation is also activated.

The ship's cargo hoses are disconnected. The master requests a tug to move the vessel. If the fire is extinguished, the request can be cancelled. However, it is important that the ship be ready to move as soon as possible.

Overhaul. Once the leak is isolated and the fire is extinguished, the remaining LNG is washed overboard. If the spill is very large and the wind is blowing toward the terminal facility or the vessel's housing, it may be desirable to let the fire burn under controlled conditions until all the LNG is consumed. However, a small spill would not normally present a hazard after extinguishment, if the water spray system is operating. Moreover, the longer the structural metal is exposed to high heat, the greater is the chance of structural damage. Thus, a small spill would usually be extinguished immediately.

LNG Spill due to a High Energy Impact (Collision)

As of this writing, there has not been a recorded high energy impact involving an LNG ship. However, given the planned construction of 100 LNG ships worldwide within the next decade, it is conceivable that such a collision could occur. We do not know what would actually happen during and after such a collision, so the following case is mostly speculation. In this hypothetical case, very little is said concerning the second vessel involved in the accident. The type of vessel, its cargo and the experience and training of its crew would, of course, affect firefighting efforts. For example, we assume below that the two vessels become locked together. The crew of the second vessel would actually do everything in their power to move away from the LNG vessel. Moreover, they would attempt to fight the fire on their ship with the firefighting systems and devices carried on board. However, given these limitations to our hypothetical situation, we can suggest what might happen.

The Fire. An LNG vessel is hit broadside with sufficient force to penetrate the double hull and LNG container. It is believed that the second vessel would have to be moving at a speed of at least 28–37 km/hr (15–20 knots). The two vessels remain locked together. The escaping LNG fills the void space surrounding the tank and, in contacting seawater, commences to boil very violently. It produces a gas overpressure in excess of the pressure that can safely be relieved through the void relief valves on spherical tank ships only and the hull opening. At the time of the collision, the LNG gas is ignited. Shortly thereafter, the pressure buildup results in further rupture of the hull, with fire spreading into the sea with the spilled LNG. The second vessel is totally engulfed in flames from the LNG, while the LNG vessel experiences fire only in the area of the collision.

Sizeup. Of the five LNG tanks on the vessel, only one tank is breached. Each tank contains 25,000 m^3 (approximately 150,000 barrels) of LNG. The wind is blowing across the LNG ship and down the length of the other vessel. The vessels are in open sea, with no land or other vessels in the vicinity. The alarm is sounded.

Confining the Fire. The deck water spray system is activated. Hoselines are run out and activated with high-velocity fog to supplement the fixed-spray system. Hose-reel lines are run out from the dry chemical units but are not pressurized at this time. Since the other vessel is engulfed in flaming vapors and must be assumed to be immobilized, the LNG vessel maintains its propulsion and steering systems in operation. They are needed to jockey the two vessels as necessary to alter the present relative wind position temporarily to effect a rescue, if possible, of the other crew. This action could also increase the intensity of heat and flame hitting the uninvolved LNG tanks, perhaps beyond the ability of the water spray system to protect these tanks. As a result, additional LNG tanks could become involved.

Protecting Exposures. Four LNG tanks are in danger of becoming involved. At present, they are being protected by the water spray system. Not much can be done to aid the other vessel, beyond rescue of its crew. The fire on the second vessel cannot be combated until the spilling LNG is completely consumed. The rate of release of LNG from the hull will determine how long it will take for the contents of the involved tank to be consumed.

Attack. The situation requires that efforts be directed toward controlling the spread of the fire, while the LNG from the involved tank is consumed. This is accomplished by *1)* effecting a position for rescue of the other crew, and then maneuvering to change the relative wind more toward the beam (to decrease the heat intensity on the uninvolved tanks; and *2)* continuing to cool adjacent structures. Additional cooling is provided by fire-main hoselines and deluge water flow from the fire-main monitors. No attempt is made to extinguish the fire.

At least several hours will be required for all the LNG in the involved tank to burn up. During this time, there is no relaxation of efforts to control the fire and keep it from spreading.

Overhaul. Eventually, the spill rate begins to decrease. When this becomes evident, the firefighters advance as many water fog applicators as possible to the fire. The fog streams are directed at the breach in the hull, so that flames will not flash into the double hull as the liquid flow diminishes. Further, they use the vessel's inert gas system to make the breached void inert (as well as the LNG tank as it is draining). This minimizes the danger of explosion when all the liquid has drained and only gas is left. They continue this cooling after the flames on both vessels have been extinguished, and until all metal structures are too cool to cause reflash.

As the two vessels move against each other, the friction could provide sufficient heat for reflash, so the contact area is carefully observed. Water is sprayed until all danger of reflash has passed. Maintaining the tank and void inert also aids in preventing reflash.

When it is considered safe to do so, the two vessels are separated. During this operation, the crews again carefully guard against explosion. The LNG vessel is able to maintain stability within a safe range; its structural integrity is not decreased by the lengthy burnout. These are, of course, predictions. They may be valid only if maneuvering and structural cooling procedures are accomplished as described above.

Comments on High Energy Impact without Fire. It is highly unlikely such an impact could occur without a fire. Since it is conceivable, we must consider the effect of the vapor cloud drifting toward an inhabited area. All present predictions of possible effects of vapor cloud drift are based on instantaneous release of the total content of the tank. This in itself is unrealistic, as the breach in the hull would not normally release all the LNG at once. If it did happen, a drifting vapor cloud of such magnitude could present a fire hazard to large inhabited areas. However, it is more likely that a small cloud would form and would dissipate within a few hundred yards of the vessel. Deliberate ignition of the vapor cloud to prevent a catastrophe might well be a self-sacrificing gesture (ignition stops the forward drift of the vapor cloud). It probably would not be necessary, though, given the present design of LNG vessels and the existing navigational safeguards.

Fire on a Passenger Ferry

The Fire. Fire in a passenger vehicle is filling the vehicular alley with smoke. Some people are leaving their cars, while others are blowing their horns to attract attention.

Sizeup. The car is on the main deck of a drive-on–drive-off passenger and vehicle ferry. The fire is in the engine compartment of the car. There are 7 cars in front of the involved car, and 10 cars behind. Three passengers in the involved car are able to leave it. There are two passenger decks above the main deck. The vehicular alley is open at either end, with two crossovers to adjacent vehicular alleys. There is a walkway running the entire length of the vehicular alley.

Attack. The alarm is sounded. The water spray, or manual sprinkler, system for the involved alley is activated. This suppresses the fire while hoselines are advanced into position. It also reduces visibility and tends to keep the passengers in their cars. One crewman immediately takes a portable extinguisher (CO_2, dry chemical or Halon) to the involved auto. He cannot readily open the hood, so he directs the stream into the engine compartment through the radiator grill and from below on either side of the engine. A hoseline is run out and charged. Under its protection, the passengers are removed from the car and taken to the forward part of the vessel (into the wind).

Confining the Fire. The cars to the rear are the most threatened, but neither they nor the burning car can be moved. The fire can best be prevented from spreading by quick extinguishment using portable extinguishers and hoselines.

Protecting Exposures. The water spray system and the hoselines provide sufficient protection for exposed vehicles. The passengers of exposed cars are taken from the vehicle alley until firefighting operations are completed.

Ventilation. The ferry's speed is reduced, and it is maneuvered broadside to the wind. This reduces the draft that would tend to accelerate the fire. Once the fire is out, the original course is resumed; the natural draft quickly removes the smoke and heat.

Overhaul. The engine compartment is overhauled by disconnecting the battery, ungrounded strap first. Any smoldering insulation is wet down. Hot spots are cooled with spray from a partially opened nozzle. Some gasoline has spilled, and it is flushed from the deck. Upon docking, if there is any suspicion of flammable vapors in the alley, the cars should not be driven off the ferry. Instead, they should be pushed or pulled off by the crew or shoreside help.

SUMMARY OF FIREFIGHTING TECHNIQUES

A comprehensive fire prevention program is of prime importance in the day-to-day operation of any vessel. However, there may be a time when, in spite of all the crew's precautions, they are faced with a hostile fire aboard ship. Weekly fire drills should have provided crewmen with the skills they need to control and extinguish the fire and conclude their voyage successfully.

At the first indication of fire, the alarm must be sounded. Then the fire must be located; sometimes this will be easy and at other times it may be quite difficult. However, the alarm will bring people to help locate the fire.

After the fire is located, the firefighters must determine what is burning to know what extinguishing agent or agents to employ. The extent of the fire and the best method of attack must also be determined. Whatever the method of initial attack may be, it should be backed up with a second, and more substantial, means of attack. For example, a small fire might be attacked initially with portable extinguishers. Then charged fire-main hoselines or a semiportable system would be advanced as backup if the initial attack fails to control the fire. An initial attack with a semiportable system would be backed up with charged hoselines; an initial attack with hoselines would be backed up with more or larger hoselines. However, water must be used prudently on a vessel, because of the stability problems that it can cause.

The fixed extinguishing system is, in most cases, a backup that should be used only as a last resort. This is especially so in the case of an engine room fire. Use of the fixed CO_2 system requires that the engine room be evacuated, and the loss of power and steering for a long period of time may lead to worse problems than the fire. However, when the last resort is the only remaining hope, it must be used. Fixed systems may be used for the initial attack on cargo hold fires because they are effective in such confined spaces and because it is not essential that anyone enter the holds during a voyage.

Fires directly involving hazardous materials generally produce dangerous fumes. The materials may also react violently to normal extinguishing agents. When a ship's fire party is in doubt as to the correct procedures for fighting a fire involving a hazardous substance, the Coast Guard may be consulted. The Coast Guard will provide information concerning the proper procedures and safety precautions. Here are three organizations that may be contacted:

- The National Response Center, located in Coast Guard headquarters in Washington, D.C. The tollfree telephone number is (800) 424–8802.
- The Chemical Transportation Emergency Center (CHEMTREC). The tollfree telephone number is (800) 424–9300.
- The Chemical Hazards Response Information System (CHRIS) of the Coast Guard, located in Washington, D.C. The telephone number is (202) 426–9568.

The fire must be confined to the space in which it originated. This may be accomplished by controlling the flow of air to and from the fire area; by cooling the adjacent bulkheads, deck and overhead; and by directing an extinguishing agent onto the fire to reduce its intensity or its ability to radiate heat to other combustibles. Trapped victims must be found and removed to safety or protected with hoselines or ventilation until the fire is under control. The exposures, the six sides of the box that contains the fire, must be checked and protected.

Then, finally, after the fire is out, the overhauling begins. The fire debris must be examined, bit by bit, to ensure that there are no smoldering embers. All paths of extension must be checked. If the fire has been extensive, a fire watch should be set. All firefighting equipment must then be placed back in service, and any structural damage must be rectified. The cause of the fire should be determined, as a first step in preventing the recurrence of the same type of fire.

BIBLIOGRAPHY

Fire Fighting Manual for Tank Vessels, C.G. 329 U.S. Coast Guard, Department of Transportation, Washington, D.C.

Marine Officers' Handbook, Edward A. Turpin and William A. MacEwen, Cornell Maritime Press, Inc., Cambridge, Maryland. 1965.

Damage Controlman, U.S. Navy Training Manual, U.S. Navy Training Publication Center, Washington, D.C. 1964.

Fire Fighting—Ship, Bureau of Ships Manual, Chapter 93, U.S. Navy, Washington, D.C.

Proceedings of the Marine Safety Council, CG-129, U.S. Coast Guard, Department of Transportation, Washington, D.C.

Fire Chief's Handbook, 4th ed., James Casey, Dun-Donnelley Publishing Corporation, New York, N.Y.

Protection of Tugboats, Towboats & Barges

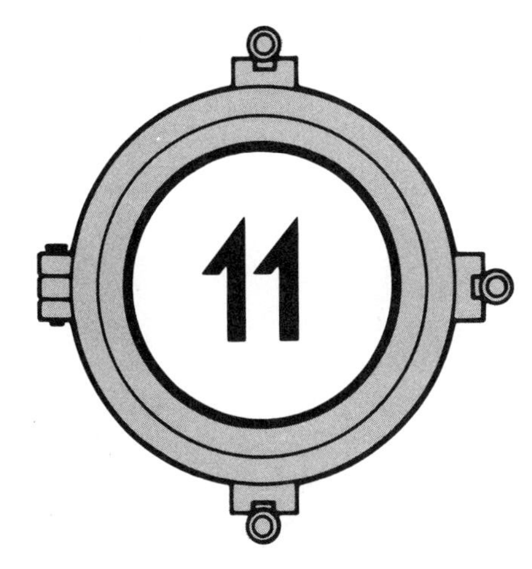

Millions of tons of cargo are transported on rivers, lakes, canals and intercoastal waterways each day. The vast majority is carried on unmanned barges without propulsion, pushed or pulled by tugboats or towboats (Figs. 11.1 and 11.2). Barge transportation has become international with the introduction of Lash and Seabee vessels. These ships carry barges and their contents to almost every corner of the world. More and more freight is being transported by ocean-going barges and tugboats, and both tugs and barges are increasing in size, carrying capacity and speed (Fig. 11.2).

Every conceivable raw material, commodity and manufactured product is transported by barge. Many of these materials are highly combustible; other types of hazardous cargo include explosives, radioactive materials, corrosives, irritants and poisonous materials. The remainder are mostly ordinary class A combustibles, which can easily become involved with fire. Yet with all these millions of tons of cargo, carried day in and day out in all weather conditions, there is a very low incidence of major accidents and fires.

SAFETY

Safety is a matter of vital and continuing concern to the barge and towing industry—concern for the welfare of the men and women employed aboard towboats; concern for the general public living and working along waterways; and concern for the safety and integrity of the cargo. Although the industry has a remarkable safety record, it continues to work toward improving that record. Industry representatives cooperate fully with the responsible agencies, particularly the U.S. Coast Guard. Towboat and barge personnel are trained in the practical aspects of safety and fire prevention. General safety and firefighting theory and tactics are an important part of the curriculum at industry training schools. Many barge and towboat owners conduct private safety training schools along with on-the-job training sessions; some arrange for crewmen to take courses at state fire training schools for "hands-on" learning. In addition, the American Waterways Operators, Inc., publishes basic safety and fire prevention literature, safety posters and checkoff forms for barge safety inspections (*see* Table 11.1 at end of this chapter).

The teaching of safety is essential if crewmen are to have the proper attitude toward safety; it is the crew who must make safety work, day after day. Some of the ways in which crewmen can help make tugboats, towboats and barges safer for all concerned are as follows:

1. Avoid exposure to dangerous situations whenever possible. Lost or broken equipment can be replaced—you cannot.
2. Do not unnecessarily expose yourself to the chance of falling overboard. Whenever possible, work where you are protected.
3. Don your life jacket and work vest properly; keep them fully fastened when there is any chance you may fall overboard.
4. Never stand when riding in a skiff or small boat.
5. Avoid going out on the tow alone after dark. If you must do so, advise the pilothouse watch before going out, and report your return.
6. Keep guardrails and lifelines in place and pulled up snugly. Do not hang or sit on lifelines.
7. Do not lean against lock walls, docks or other shore structures while you are on the boat or the barges.

Figure 11.1. Towboats and barges move millions of gallons of combustible petroleum products on rivers, lakes and waterways every day. (Courtesy American Waterways Operators, Inc.)

8. Always carry loads on your outboard side.
9. Use a pike pole to handle lines or wire beyond the edge of the boat or barge.
10. Observe "No Smoking" areas carefully. Smoking is never allowed in bed, in paint lockers, at oil or fueling docks or on the decks of petroleum or flammable-cargo vessels. Discard cigarette butts in the proper receptacles. Never throw burning material over the side.
11. Learn the locations of fire hydrants and portable extinguishers, and know how to use them. Report every fire immediately upon discovery.

Figure 11.2. Each year there is an increase in the amount of cargo moved along coastal waters by ocean-going tugboats and barges. (Courtesy American Waterways Operators, Inc.)

12. Except in an emergency, never run or jump on the job. Never engage in horseplay on the boat or tow.
13. Stay alert at all times. Watch for tripping hazards, open hatches and slick spots on deck.
14. Always close open manhole covers, or place guards around them if they must remain open.
15. Never walk on dry cargo barge hatch covers.
16. Keep your hands and feet away from places where they are liable to be crushed.
17. Always wear goggles or eyeshields when you are chipping, burning, grinding or scraping, or when your eyes are exposed to wind-blown dust or other irritants.
18. Do not wear loose and ragged clothing around rotating or moving machinery.
19. Make sure portable ladders are securely set. Watch out for cracked or broken rungs and rails.
20. Keep all areas of the deck (including gangways, walkways and outside stairways) free of oil, grease, debris, ice and other foreign substances.
21. Stand clear of all lines and cables under tension. Do not straddle lines that are being tightened.
22. Always place ratchets to tighten lines pulling inboard.

23. Do not make a line fast to a bitt or timberhead on which there is already a line.
24. Keep your fingers out from between bitts or timberheads and the wires or ropes being handled.
25. When you are feeding a line onto a bitt or timberhead, always work from the "dry" side. Stay clear of "working" lines.
26. Do not stand in the bight of a line at any time.
27. Never swim off a boat or barge.
28. Always lift loads by bending your knees. If the load is too heavy, get some help!
29. Know the safe and proper way to do your job. If in doubt, ask your supervisor.
30. Report all injuries immediately to the proper authority.
31. Always use your approved flashlight when you are on deck after dark.
32. Keep all portable gangways and walkways secured so they will not slip when used.
33. If you notice that any item of equipment is damaged, or discover any hazardous or dangerous condition, report it to your immediate supervisor.
34. Beware of hanging fenders from moving towboats as they come alongside stationary tows, docks, pilings or sea walls.

FIRE PROTECTION EQUIPMENT FOR TUGBOATS AND TOWBOATS

Tugboats and towboats are both used to move barges, in addition to their harbor duties. Owing to their construction, tugboats are better adapted to the towing of barges in open water, where they are subjected to heavy winds and waves. Ocean-going tugboats range in size up to 356 metric tons (350 gross tons). Their lengths range from 30.5–45.7 m (100–150 ft), and their engines from 1120–6710 kilowatts (1500–9000 hp). Tows are normally pulled, but in the newer integrated tug–barge configurations the barges are pushed (*see* Fig. 11.2).

Towboats are the power units that propel single barges or multiple-barge tows made up of 40 or more barges. These vessels are designed to work in the protected waters of rivers and canals. Special rudder arrangements, and one to four propellers (powered by individual diesel engines) in a Kort nozzle, provide the control necessary to navigate the restricted channels of rivers and canals. A tow of about 10 tank barges represents about 13,600,000 liters (3,000,000 gal) of petroleum products; this is a common towing assignment for such vessels.

Some of the most common sizes of towboats and tugboats are shown in Figure 11.3, and towboats up to 58 m (190 ft) in length, with 16.5-m (54-ft) beam, 2.6-m (8.6-ft) draft and up to 7.5 megawatts (10,000 hp) have recently been put in service. Modern tugboats and towboats are powered by diesel engines. They are both outfitted with the same basic types of fire protection equipment.

Automatic Fire Detection Systems

While not normally used, the fire detectors used on tugboats and towboats are almost always set up to activate an alarm rather than a fire extinguishing system. There are two reasons for not installing automatically operated extinguishing systems in engine rooms:

1. A system that automatically floods the engine room with an extinguishing agent can jeopardize the lives of personnel in that space.
2. The flooding of an engine room would cause the loss of propulsion. During a critical navigating maneuver, this could result in a serious accident.

Detectors are sometimes used to trigger firefighting systems in such spaces as paint lockers, lamp lockers and small storage rooms, as these spaces are not usually occupied. Wherever an automatic system is used, the proper warning devices should be installed, and warning signs posted (*see* Chapter 9).

On tugboats and towboats, automatic fire detection systems are used primarily in engine rooms. The detectors most often employed are pneumatic detectors and combination heat and smoke detectors.

Pneumatic Detectors. In the pneumatic detector, heat from a fire causes air within a diaphragm or hollow metal tube to expand. The resulting pressure is used to actuate switches that will turn on alarm bells, shut off power venting systems and close dampers. Pneumatic detectors are very reliable, because they have few moving parts. They should be tested at least quarterly to ensure proper operation.

Heat and Smoke Detectors. Some newer vessels are equipped with combination heat and smoke detectors. These devices are extremely sensitive, and they react faster than the pneumatic-type detectors. The detectors are placed in engine rooms and in living areas. In some systems, the

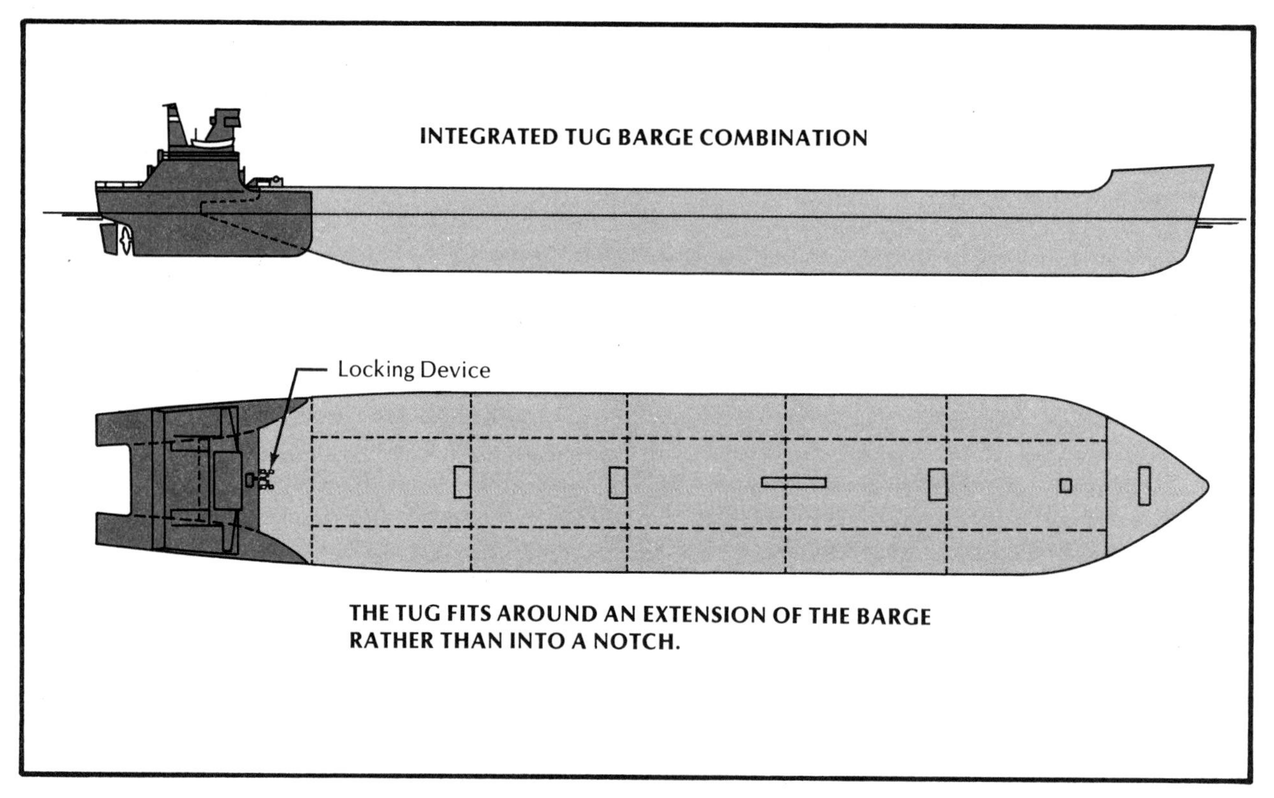

Figure 11.3. A modern integrated Tug Barge system.

detectors transmit an electrical signal to an annunciator when they sense fire or smoke. The annunciator display board indicates, with lights, the location of the detector that sent the alarm and the cause of the alarm. Smoke is indicated by an amber light, and flaming combustion by a red light. An audible alarm is sounded simultaneously. The detectors and circuits can be monitored. A breakdown in the system is indicated on the annunciator display board by a blue or white light and a buzzer that is distinctly different from the audible fire alarm.

Combination heat and smoke detectors are sometimes installed as individual self-contained units. When such a unit senses smoke or fire, it sounds an audible alarm and flashes a light. (*See* Chapter 6 for a detailed discussion of fire detectors and detection systems.)

Fixed Fire-Extinguishing Systems

Fire-Main System. The fire-main system is the basic firefighting system for tugboats and towboats. In most systems, 6.4- or 7.6-cm (2½- or 3-in.) piping carries water from the pumps to the fire stations. The water pumps have capacities ranging from 570–1900 lit/min (150–500 gal/min). Generally, two pumps are installed, with one in service and one as a backup pump. Because tugboats and towboats do not always have enough space to separate the pumps, they may both be located in the same general area.

The fire stations are usually located at the main deck level, on exterior bulkheads. Each fire station has a water outlet with a control valve and a 3.8-cm (1½-in.) connector. The connector is often fitted with a wye gate so that two 3.8-cm (1½-in.) hoselines can be connected into the fire station. A wide variety of nozzles are used since some towboats are not required to carry specific nozzles.

The most widely used nozzle is similar in operation to a garden hose nozzle; it produces a good fog stream and a straight stream. Plastic combination nozzles are also being used on some vessels. They can provide a solid stream and 30° and 60° fog streams. The ability to vary the width and type of stream is advantageous in firefighting operations. Straight streams have greater reach and penetration which allow the attack to be made from a distance. Fog streams have excellent heat absorption qualities, and their conical shape protects firefighters from the fire's heat when a close-in attack is necessary.

Some older vessels still carry smooth-bore nozzles that are both inefficient and dangerous. If the smooth-bore nozzle of a charged hoseline is dropped and becomes free, the nozzle will whip back and forth and can cause injury if it hits a crew member. If the nozzleman is forced to retreat and abandon the hose, he may not be able to shut off the water flow. There is a proposal to do away with this type of nozzle in the marine services.

Some towboats and tugboats do not have a fire-main system (*see* Chapter 9). Their only firefighting water supply is a pipe outlet with a connection for a deck washdown hose. While this setup is very ineffective, it can be used to extinguish fire *if the attack is made early, while the fire is small.* Firefighters would have to advance close to the fire, because the nozzle will probably produce a poor water stream. (*See* Chapter 10 for a description of the use of water streams to attack fire.)

Engine Rooms. Fixed carbon dioxide (CO_2) or Halon 1301 total-flooding systems are installed in the engine rooms of some tugboats and towboats. These systems are similar to those described in Chapter 9, but are smaller in scale. When fire is discovered, the system is activated from outside the engine room by pulling two release cables in the proper sequence. The cable pulls are usually located just outside the doorway(s) leading from the engine room to the main deck or passageway. Before the CO_2 is released into the engine room, a warning horn sounds. The engine room must be evacuated at that time, because the CO_2 will lower the oxygen content below the level necessary to sustain life. For effective extinguishment, openings to the engine room must be secured and made as airtight as possible. This is especially important on towboats, where there are large windows in the bulkheads and skylight vents in the overhead. (*See* Chapter 10 for a description of the use of the CO_2 total-flooding system in an engine room.)

Paint and Lamp Lockers. Small CO_2 or Halon 1301 flooding systems are often used to protect paint and lamp lockers and deck gear-storage spaces. These small systems may or may not be activated by fire detectors. Normally, they are activated manually. However, if a system can be operated automatically, a discharge warning horn or bell must be part of the system. If a space is protected by an automatic extinguishing system, its doors should be kept closed when it is unoccupied. An open door would allow the gas to flow out of the space if the system were activated.

If a locker is protected by a manual system, a separate detection system sounds the alarm when it senses fire. Then the space must be checked for occupants before all openings are closed to contain the extinguishing agent. An external means of closing vents and louvers and an excess pressure release device are required for each small compartment protected by gas extinguishing agents. The system is activated with the usual pull cables. (If the fire is small, crewmen should attempt to extinguish it by some means other than the gas flooding system, which will discharge 45.4 kg (100 lb) CO_2 or Halon into the locker.)

Semiportable Systems

Semiportable carbon dioxide extinguishing systems are used to fight fires in engine rooms. The usual system consists of one or two 22.7- or 45.4-kg (50- or 100-lb) cylinders, a length of hose and a horn-type discharge nozzle. The nozzle handle is long, and the nozzle control lever can be locked in the open or closed position. With the lever locked in the open position, the long horn can be used to discharge CO_2 into places that are difficult to reach.

Semiportable Halon systems are also used. This system usually consists of one or two cylinders of agent at a pressure of about 1310 kilopascals (190 psi) at normal temperatures. For faster release, as in the case of explosion suppression systems, the gas is pressurized with nitrogen to pressures as high as 6900 kilopascals (1000 psi). The agent is directed onto the fire with an on–off nozzle connected to the cylinders by a length of rubber hose. (*See* Chapter 8 for a detailed description of both CO_2 and Halon semiportable systems.)

Cautions. Small tugs and towboats have small engine rooms. The rapid discharge of several CO_2 cylinders into a confined space could lower the oxygen content to a dangerous level. The person using the extinguisher could pass out and be injured as he falls. Humans are also affected by exposure to small concentrations of CO_2, even though the oxygen content of the air may not be reduced to the danger level. If a fire persists and CO_2 must be used in a small area, crewmen must wear breathing apparatus, and work in two-man teams if possible.

Carbon dioxide extinguishers must not be used to purge fuel tanks. Recently an explosion occurred when a CO_2 extinguisher was used to purge a small fuel tank. It has been known for some time that the flow of CO_2 through the discharge nozzle produces static electricity. Usually this is not considered to be dangerous during firefighting operations, however, during the purging the discharge horn was close enough to the fuel tank rim to permit a static spark to jump from the horn to the tank, ignite the vaporized fuel and cause the explosion.

Portable Extinguishers

Tugboats and towboats are required to carry portable fire extinguishers capable of extinguishing class A, B and C fires. (*See* Chapter 8 for a full description of such extinguishers.)

FIGHTING TUGBOAT AND TOWBOAT FIRES

In this section, two fire situations and the recommended firefighting procedures are described from the alarm to overhaul. The terminology and the procedures detailed in the *Firefighting Procedures* section of Chapter 10 should be reviewed at this time.

Electrical Fire on a Harbor Tugboat

The Fire. The main generator is giving off dense smoke, and the insulation on the windings is starting to burn.

Sizeup. The vessel is under way with a tow. The generator is supplying electricity for the tug's lighting, radar and communication needs. The generator is located in the engine room on the port side aft.

Attack. The alarm is sounded. The circuit breaker is tripped to take the generator off the line. This should be done from the engine room, if possible; otherwise, the breaker on deck or in the pilothouse can be tripped. The diesel engine driving the generator is shut down to protect crew members attacking the fire. The auxiliary generator is started to provide electricity for the engine room lights, the navigation equipment and the general service pump.

One crewman advances a portable CO_2, dry chemical or Halon extinguisher as close as possible to the generator. He directs the stream into the generator windings.

Confining the Fire. The fire was confined by removing the electrical load and shutting down the generator drive engine. This prevents the production of additional heat and the overheating of wires that run to other areas.

Protecting Exposures. Crewmen remove whatever combustibles they can carry from the vicinity of the generator. Since the entire engine room is exposed, the best protection is to quickly knock down the flames and cool the generator. Once the generator is taken off the line and the driving mechanism secured, the fire can be controlled.

Ventilation. The burning insulation gives off large quantities of irritating smoke. Mechanical ventilation is used to clear the engine room after the fire is extinguished.

Overhaul. A great deal of heat remains in the copper generator components. This is absorbed by wet canvas and burlap, placed on the outside of the generator housing.

Fire in a Deck Gear-Storage Space

The Fire. Cordage, wooden tackle and burlap waste are burning on the deck inside the storage space. The fire is extending up the forward bulkhead. The alarm is sounded.

Sizeup. The storage space is on the main deck; it runs 7.62 m (25 ft) across the beam of the vessel and is about 4.6 m (15 ft) wide with a 2.4 m (8 ft) overhead. It is entered through port and starboard doors. A small vent is located in the overhead. In addition to deck gear, the storage space contains cartons of toilet tissue and paper towels and some lumber. The main body of the fire is located about 2.4 m (8 ft) from the starboard doorway. The space is not equipped with a CO_2 or Halon flooding system.

Attack. Since firefighters are able to enter the space, an attack is made with a portable extinguisher. A dry chemical extinguisher could quickly knock down the flames if properly applied. However, it might not extinguish the fire completely, because class A combustibles tend to smolder. A portable water extinguisher could do the job if the stream is properly applied, but an extensive fire may be beyond the extinguisher's capability.

Backup. Hoselines are positioned at the two doorways to back up the extinguisher. The hoselines are charged to the nozzle. The fire is so intense that the crewman with the extinguisher is forced to retreat. An attack must be made with a hose stream. The port-side hoseline will be used, so that the stream does not push the fire across the entire compartment.

The port-side line (line 1 in Fig. 11.4) is advanced to the space. The nozzleman uses a spray pattern and crouches low. He sweeps the overhead with a short burst as he enters, to cool off the hot gases. Then he advances into the space, directing short bursts of fog at the base of the flames. The starboard door is opened to allow the heat and smoke to vent outside. The starboard

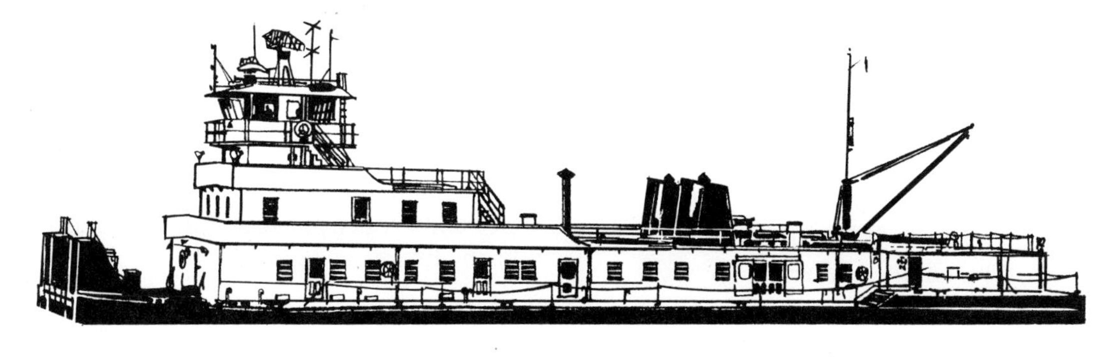

TOWBOATS	Length (Feet)	Breadth (Feet)	Draft (Feet)	Horsepower
	117	30	7.6	1000–2000
	142	34	8	2000–4000
	160	40	8.6	4000–6000

TUGBOATS	Length (Feet)	Breadth (Feet)	Draft (Feet)	Horsepower
	65–80	21–23	8	350–650
	90	24	10–11	800–1200
	95–105	25–30	12–14	1200–3500
	125–150	30–34	14–15	2000–4500

Figure 11.4. Standard dimensions and power of towboats and tugboats (including ocean-going tugs). (Courtesy American Waterways Operators, Inc.)

hoseline (line 2) is used to knock down any flame that is pushed out the door. It is not at any time directed into the door. Its function is only to prevent the extension of flames outside the door.

Protecting Exposures. All areas adjacent to the fire are checked. If fire has entered these areas this fire must also come under attack. If fire has not entered the area but bulkheads are hot, combustibles must be moved away from the bulkhead and the bulkhead cooled with a fog stream.

Overhaul. Once the fire is knocked down and darkened, the burning material is pulled apart and wet down. When it can be picked up, it is taken out on the deck and saturated with water. The spaces adjacent to the storage space are carefully inspected for fire extension. All materials that are stowed against or near the bulkheads are moved to allow a complete inspection.

FIRE PROTECTION FOR BARGES

At one time or another, barges carry almost every conceivable type of flammable cargo. A tow may consist of similar barges that are all carrying the same commodity, or a variety of different types of barges carrying different cargoes. In the latter case, the diversity of cargoes and storage methods can complicate firefighting operations.

Types of Barges

The hulls of most inland waterway barges are similar in length, width and draft because they all must be able to navigate the same waterways. The final configuration of a barge is, however, determined by the cargo it will carry and the size of the locking system.

Open-Hopper Barge. This type of barge is used primarily to move sand, gravel, rock, coal, logs, lumber and fertilizer. With slight modifications, it can be used to transport almost any solid commodity, in bulk or packaged. The hopper barge is usually a double-skinned, open-top box; the inner shell forms a long hopper or cargo hold (Fig. 11.5).

Covered Dry-Cargo Barge. This barge is similar to the open-hopper barge, but it is equipped with watertight covers for the entire cargo hold (Figs. 11.5 and 11.6). It is generally used to carry grain products, coffee, soybeans, paper and paper products, lumber and building materials, cement, iron and steel products, dry chemicals, aluminum and aluminum products, machinery and parts, rubber and rubber products, salt, soda ash, sugar and sometimes packaged goods.

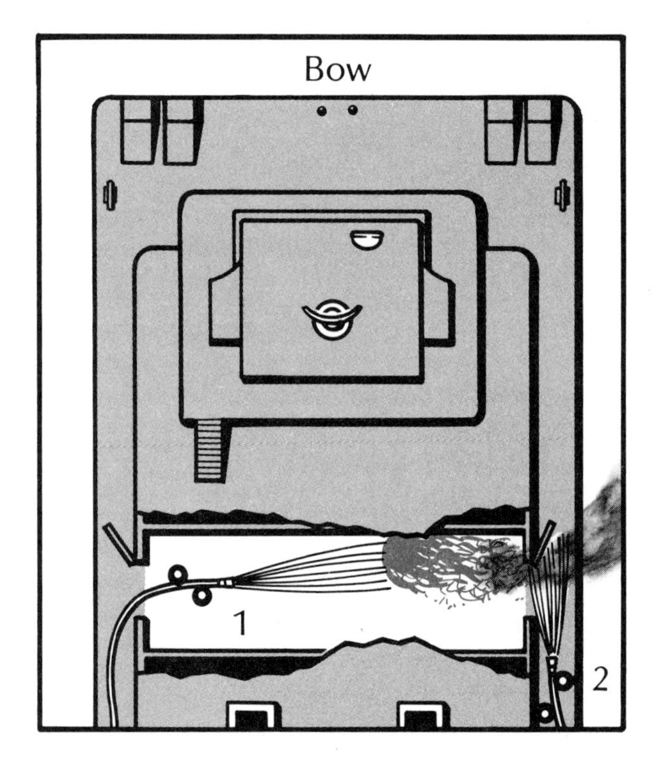

Figure 11.5. Location of the fire and placement and use of hoselines. Note that line 2 is directed so that its stream does not block the flow of combustion products out of the compartment.

Tank Barges. Three basic types of tank barges (Fig. 11.5) are used for the transportation of liquids. On *single-skin* tank barges, the bow and stern compartments are separated from the midship by transverse collision bulkheads. The entire midship shell of the vessel constitutes the cargo tank. For strength and stability, this huge tank is divided by bulkheads. The structural framing for the hull is inside the cargo tank.

Double-skin tank barges have, as the name implies, an inner and an outer shell. The inner shell forms the cargo tanks; the tanks are free of internal structural members and thus are easy to clean and to line. Double-skin barges are used to transport poisons and other hazardous liquids that require the protection of a void space between the outer and inner shells.

Barges with *independent cylindrical tanks* (Fig. 11.7) are used to transport liquids under pressure or liquids that are offloaded by pressure. In some cases, cylindrical tank barges are used to carry cargoes at or near atmospheric pressure when special tank lining or insulation is required. The barge itself is generally of the open-hopper type, with the tanks nested in the hopper. The tanks are then free to expand or contract, independently of the hull. For this reason, cylindrical tank barges are preferred for high temperature cargoes such as liquid sulphur and refrigerated cargoes such as anhydrous ammonia.

The three most common sizes of tank barges are shown in Figure 11.5. More than 3200 tank barges, with a total cargo capacity of over (6,300,000 tons) are in service today. The majority are used for the transportation of petroleum and petroleum products—approximately (214,-000,000 tons) annually.

Since 1946, movements of bulk chemicals by barge have been increasing steadily. Chemicals now comprise one of the most important liquid commodities transported by water. The U.S. Coast Guard lists approximately 400 chemicals that are transported or proposed for transport by barge. Some examples are anhydrous ammonia, which is transported under a pressure of 1724 kilopascals (250 psi) or refrigerated to −33°C (−28°F); liquefied sulphur, which is moved at 127–138°C (260–280°F); and liquefied methane,

OPEN HOPPER BARGE

Length (Feet)	Breadth (Feet)	Draft (Feet)	Capacity (Tons)
179	26	9	1000
195	35	9	1500
290	50	9	3000

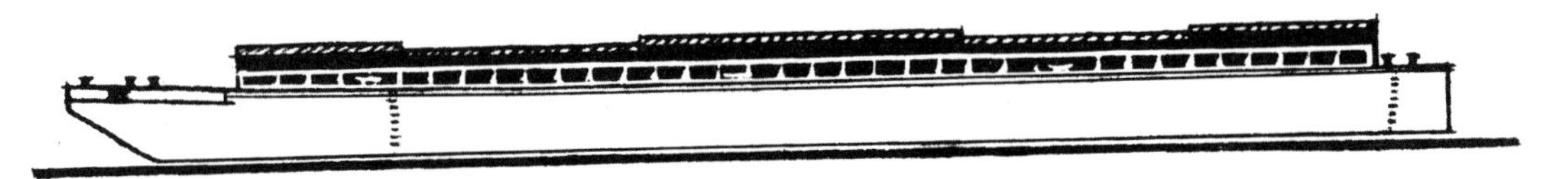

COVERED DRY CARGO BARGE

Length (Feet)	Breadth (Feet)	Draft (Feet)	Capacity (Tons)
175	26	9	1000
195	35	9	1500

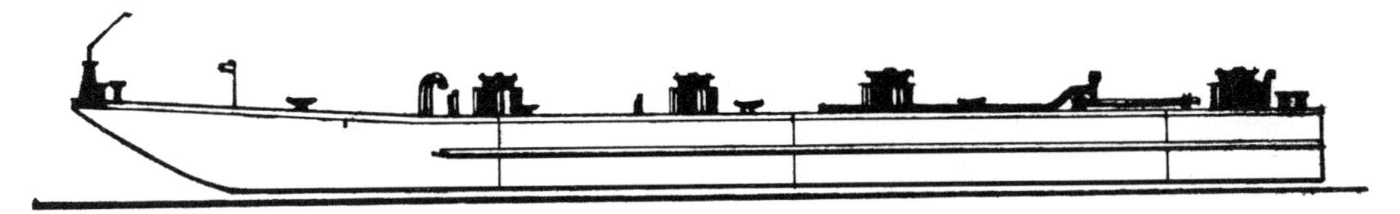

LIQUID CARGO TANK BARGE

Length (Feet)	Breadth (Feet)	Draft (Feet)	Capacity (Tons)	Capacity (Gallons)
175	26	9	1000	302,000
195	35	9	1500	454,000
290	50	9	3000	907,000

Figure 11.6. Three common barge configurations. (Courtesy American Waterways Operators, Inc.)

which is transported at −161°C (−258°F). Barge-mounted tanks are used to transport liquid hydrogen at −252°C (−423°F) and liquid oxygen at −183°C (−297°F).

Chemicals and chemical products now make up about 3% of the total cargo transported by barge. The number of chemicals and the volume carried are both expected to continue to increase.

Deck Barge. This is a hull box, generally with a heavily plated, well-supported deck. Deck barges usually carry machinery, vehicles and heavy equipment (Fig. 11.8).

Other Barges. Among the various other types of barges (Fig. 11.8) are rail *carfloats, scows* and barges of special construction. The latter include

Figure 11.7. Typical covered dry-cargo barge with rolling weathertight hold cover. (Courtesy American Waterways Operators, Inc.)

Figure 11.8. Giant independent cylindrical tanks are nested into open-hopper barges for transporting chemicals. (Courtesy American Waterways Operators, Inc.)

self-unloading barges for cement and grain, *derrick* and *crane* barges and those designed to carry special cargo such as the Saturn space vehicle.

Lash and *Seabee* barges are, essentially, inland waterway barges of the hopper type, with cargo hold hatch covers. They carry almost every type of material except bulk liquids. They are equipped with special fittings with which they are loaded into ocean-going vessels especially designed to carry barges.

Ocean-going barges are similar in configuration to inland barges. They are usually larger than inland barges, since they are not restricted by the need to navigate inland waterways.

Fire Protection

The fire protection equipment carried aboard barges is very limited. In almost every case, it consists only of two portable fire extinguishers. Barge owners try to conform to the Coast Guard regulation requiring that these extinguishers be provided and maintained. However, the barges are often left unattended at dockside, where they are subject to vandalism and theft. As a result, the portable extinguishers are often missing, and the barge is left without any fire protection.

On some large fuel barges, the small pump room is protected by a CO_2 flooding system. The system usually consists of two 22.7-kg (50-lb) cylinders, piping and discharge horns. The cylinders are secured to the outside pump room bulkhead. When fire is discovered, the system is activated manually.

Early discovery of a barge fire is another problem, since barges are unmanned and do not have fire detection systems. A fire that starts after a barge is loaded will probably not be discovered until it has reached an advanced stage. (*See* Chapters 1 and 2 where fire prevention and safety practices during loading are outlined.)

A fuel barge must be loaded and unloaded in strict compliance with all safety regulations. The barge must be manned and properly grounded, and all spark-producing tools must be removed. Smoking and the use of open lights are absolutely forbidden. All hatch covers should be closed. Open hatches allow volatile fumes to cascade over the coaming onto the deck (Fig. 11.9). If the fumes are ignited, fire could flash back into the tank, causing a massive fire and explosion. Soundings should be taken through the ullage ports, where the flame screen which must always be installed will help prevent fire from flashing back into the tank.

FIGHTING BARGE FIRES

There are two situations in which a tugboat or towboat crew should fight a barge fire:

1. When the fire is small, so that it can be fought with portable extinguishers or with a hoseline from the tow vessel
2. When the involved barge is so positioned in the tow that the fire is an immediate threat to the tow vessel

Small Barge Fire

A small fire involving barge cargo can usually be extinguished with portable appliances. The fire debris should be thoroughly overhauled. Any ma-

DECK BARGE	Length (Feet)	Breadth (Feet)	Draft (Feet)	Capacity (Tons)
	110	26	6	350
	130	30	7	900
	195	35	8	1200

CARFLOATS	Length (Feet)	Breadth (Feet)	Draft (Feet)	Capacity (Railroad Cars)
	257	40	10	10
	366	36	10	19

SCOWS	Length (Feet)	Breadth (Feet).	Draft (Feet)	Capacity (Tons)
	90	30	9	350
	120	38	11	1000
	130	40	12	1350

Figure 11.9. Three less common barge configurations. (Courtesy American Waterways Operators, Inc.)

terial or section of the barge in the vicinity of the fire should be carefully examined and wet down with water. This is vital, to ensure that no smoldering embers are left to reignite the fire. If the material that was burning can be carried easily, it should be taken to the towboat deck and wet down with a hoseline. As another precautionary measure, a crewman should be stationed at the site of the fire for several hours to watch for signs of active fire.

If there is any doubt about the fire being completely out, the barge should be dropped from the tow as soon as possible. If there is a choice, the barge should be towed to a terminal, where land-based firefighters can position their apparatus and equipment. The local fire department and the U.S. Coast Guard should be notified, so they can prepare to receive the barge.

Protecting the Tow Vessel

When flames from a burning barge jeopardize the tow vessel, that vessel must be protected; it is the only source of power for maneuvering the tow. Hoselines should be positioned to attack the main body of fire and protect the tow vessel itself. Two lines should be used to knock down the flames; a third line should be positioned to protect the other firefighting personnel with a fog pattern and to cool exposed surfaces on the tow vessel (Fig. 11.10).

If possible, the tow should be maneuvered so that the flames, heat and smoke are carried away from the tow vessel. If the hoselines cannot keep the fire from the tow vessel and maneuvering is restricted, the tow vessel should release the tow and then move to control it from another position. The tow must be controlled until it can be grounded. A tow that is burning out of control can be disastrous to other vessels and to shore installations.

Figure 11.10. Volatile fumes can be seen leaving the open fuel-barge hatch. Hatches must be closed during loading and offloading.

Extensive Barge Fire

If a barge fire cannot be extinguished by the tow vessel's crew, the barge should be grounded or secured to the shoreline if possible. This action will

1. Isolate the barge from the towboat. (If the burning barge can be separated from the other barges in the tow, the fire will be isolated and confined to one barge.)
2. Minimize the danger of the barge becoming a navigational hazard.
3. Make the barge accessible to land-based firefighters if it is grounded on or close to the shore.

When an involved barge is to be grounded, the tow operator should notify his company, the U.S. Coast Guard (captain of the port) and the local land-based fire department. The exact location of the burning barge and, if possible, the nature of the burning cargo should be provided. This information is very valuable to responding Coast Guard and fire department units as well as other emergency and environmental agencies.

The location should be given in as much detail as possible; e.g., "The barge is tied against the east bank of the river at the 98.7 mile marker. We can see a gravel road." On the basis of this information, the local fire department may be able to determine whether they can reach the site with fire apparatus. If the location cannot be reached by land, the fire department can notify the U.S. Coast Guard of that fact.

Knowledge of the fuel is important for two reasons. First, it indicates the extinguishing agent that must be used. Water could be drafted (pumped) at the site to fight a class A fire. But for a class B fire, foam would have to be carried to the site. Second, the fuel type indicates the need for special precautions; e.g., if the burning material were a chemical that produced toxic fumes, protective clothing and breathing apparatus would have to be provided for firefighters. If the burning barge were near a populated area, emergency evacuation procedures might have to be initiated; or, if the barge were carrying explosives, the area would be evacuated and other vessels would be warned to stay clear of the area.

Fires on Ocean-Going Barges

Ocean-going barges can become involved with fire, and shore line communities are rarely willing to allow a burning barge to be grounded on their beaches. The firefighting equipment carried by a large sea-going tug is only slightly more effective than that of a large towboat. The difference is simply the tug's ability to pump water at a greater volume and pressure.

Being in open water is of some help, since the tow can be maneuvered freely to take advantage of the wind. However, a rough sea can make maneuvering difficult and dangerous.

It is extremely important that the fire and the tug's position be reported immediately, whether the fire is on the tugboat itself or on a barge. If

Figure 11.11. When fire threatens the tow vessel, it must be protected. Lines 1 and 2 attack the fire directly; line 3 protects firefighters and the tow vessel. Quick beaching of the tow is imperative.

assistance is required, it will be on its way as soon as possible. Meanwhile, other vessels can be warned to keep clear if a burning barge or loose barges present a navigation hazard.

Tugboat in Pushing Position. Fire on a barge that is being pushed can endanger the tugboat. The fire is fairly close to the tug, and it has a path by which to travel to the tug. If a decision is made to fight the fire, the barge should be positioned so the wind carries the flames, heat and smoke away from the tug. If the fire is at the far end of the barge, away from the tug, then the tug should be positioned to bring the wind directly from the stern (Fig. 11.11). This should force the fire away from the tug and slow its travel along the barge. At least two handlines from the tug should be run out to attack the fire. If the tug is equipped with a deck monitor, it can be used in the attack.

When the fire is near the tug end of the barge, the wind should be brought across the tug's beam or from the stern quarter (Fig. 11.12). Handlines should be run out to attack and flank the main body of fire. The deck monitor can also be used, if available.

When there is no wind, the tow should first be brought to a stop. Then it should be moved slowly to create a slight airflow that causes the flames, heat and smoke to move in a direction favorable to firefighting and away from the tugboat (Fig. 11.13).

During the firefighting operation, the tug should be prepared for a stern tow. If the fire cannot be controlled, the tug should break away from the pushing position and take the barge on

Figure 11.12. When the fire is at the far end of the barge, the wind should be brought astern of the tug. The fire should be attacked with at least two hoselines.

Figure 11.13. When the barge fire is near the tug, the barge should be maneuvered so the wind pushes the fire away from the tug and toward the shortest paths of possible fire travel.

a towline astern. The barge must be kept under control with a towline so that it does not become a navigation hazard.

Towline Astern. Fire on a barge on a long towline astern does not threaten the tugboat. The fire is essentially isolated. It may consume the cargo on the barge, but it is not a threat to life. The tugboat crew should not attempt to attack the fire, for several reasons. Since it is important to keep the towing rig intact, the tug cannot maneuver freely. Even if the tugboat managed to reach the barge, the transfer of crewmen to the barge would be dangerous and it would be difficult to get hoselines into position. Because of the risks involved, the crew should not attack the fire. Instead, they should report the fire and their position and request assistance.

In Harbor. The fire and the tug's position should be reported immediately. If the tow is being pushed or is alongside, it should be maneuvered so that the fire does not endanger the tug. The burning barge may be released from these tow positions, but a towline should be maintained to keep the barge under control. Nearby tugs without tows can assist in this by approaching from windward and getting lines on the barge.

The tug should move out of the main channel. The burning barge may be grounded if there is shallow water nearby that can be reached without jeopardizing the tug, other vessels or shore installations. The barge may be grounded on an empty beach. However, if explosives, volatile fuels or toxic chemicals are involved, this may be unwise; the beaching could endanger nearby buildings and people.

The important thing is to control the burning barge—with a towline or by grounding or beaching—until a fireboat reaches the scene. If no fireboat is available and there is nowhere to ground the burning barge safely, the barge should be towed to open water and allowed to burn itself out.

Figure 11.14. If there is no wind, the tug must be maneuvered to create a favorable air movement.

Table 11.1. Standard Vessel Safety Inspection Checkoff Form.*

VESSEL ______________ *DATE* ______________ *INSPECTED BY* ______________

QUARTERS	*YES*	*NO*	*ACTION*
Floors in good condition—free of slippery spots, slippery rugs or other articles			
Stairtreads in good condition & adequately non-slip			
Handrails in place & secure			
Electrical equipment grounded			
Fan blades guarded			
Adequate number of ash trays			
"No Smoking" areas clearly defined			
First aid cabinet adequately stocked			
General housekeeping satisfactory			
GALLEY			
Floors in good condition—free of slippery spots, slippery rugs or other articles			
Stairtreads in good condition & adequately non-slip			
Handrails in place & secure			
Stove in good working order, free of grease & stove guard in use			
Electrical equipment grounded			
All utensils & related items safely stored			
Ice box alarm tested—or does door open outward			
Stores stacked safely—heavy items toward bottom			
General housekeeping satisfactory			
ENGINE ROOM			
Decks & steps clear of oil and grease			
Stairtreads in good condition & adequately non-slip			
Handrails in place & secure			
Bilges clean			
Equipment guards in place			
Tools in good repair & properly stored			
Power tools grounded			
Lifting gear functioning properly			
Goggles provided at grinder with instructions to use			
Container provided for rags			
Engine room supplies properly stored			
Escape hatches clear & accessible			
Any unnecessary tripping hazards			
General housekeeping satisfactory			
FIRE FIGHTING EQUIPMENT			
Extinguisher provided at each designated station			
Extinguishers tested annually (note date & recharge on each)			
Extinguishers weighed periodically (note date on each)			
Fixed CO_2 system inspected & tested (note date last tested)			
Fire hose & nozzle provided & connected at each station			
Fire hose in good condition			
Spanner wrench provided at each hydrant			
Fire axes in good repair & properly positioned			

*Courtesy, American Waterways Operators, Inc.

Table 11.1. Standard Vessel Safety Inspection Checkoff Form.* — continued

BARGE ______________ DATE ______________ INSPECTED BY ______________

	YES	NO	ACTION
Are warning signs in good condition & properly located? a. No Smoking b. No Visitors c. No Open Lights			
Are decks kept clean, oil-free & clear of hazards?			
Are void hatch/manhole covers securely dogged?			
Are tank hatches in good condition with good paint-free packing to ensure a gas-tight seal?			
Are butterworth plates & other deck openings in good condition with all bolts or camlocks in place and with a proper gasket?			
Are there sufficient ullage screens in good condition on board for each ullage opening?			
Are pressure/vacuum valves in good working order with flame arrestors in place & in good condition?			
Are adequate number of portable fire extinguishers aboard & in good condition & properly charged?			
Are fire hoses of sufficient number & length, properly stowed and in good condition?			
Are proper number & type of nozzles aboard, in correct location & in good condition?			
Are proper number of fire axes, suitably located and in good condition?			
Are cargo pump emergency stops in good operating condition? Are they clearly marked?			
Are machinery guards in place & in good condition?			
Are mooring lines in good condition & of sufficient length?			
Are proper gangways/access ladders aboard, of sufficient length, clean & in good condition?			
Are running lights, mooring lights & warning signals in good condition & properly displayed?			
Is electric wiring, including electrical portable cords, and speaker wires in good condition?			
Are sufficient portable fenders provided?			
Where fitted, are pump room ventilators in good operating order?			
Are pump engine spark arrestors properly maintained?			
Are storage spaces free of an accumulation of flammable material?			
Is absorbent material available to control minor deck spills?			
Are cargo pump glands & seals tight and in good condition?			
Are tank ladders in good condition?			
Are all handrails & stanchions in place and in good condition?			

FIRE FIGHTING EQUIPMENT	YES	NO	ACTION
Alarm bells & lights tested			
Fire stations properly marked & numbered			
Fire pumps functioning properly			
Fire station bill properly posted			
Note date of last fire drill			

DECK			
All decks & stairways free of slippery areas			
All ladders & stairway treads in good repair and adequately non-slip			
Handrails in place & secure			

Table 11.1. Standard Barge Safety Inspection Checkoff Form.* — continued

VESSEL ____________ *DATE* ____________ *INSPECTED BY* ____________

	YES	*NO*	*ACTION*
Lifelines in good condition & in place			
Manhole covers in good condition & in place			
Life boat in good condition			
Life boat boom operable & in good condition			
Life rings in good condition & in place			
Life ring lights operable & properly secured			
Equipment guards in place			
Stores stacked safely—heavy items toward bottom			
Any unnecessary tripping hazards			
If fresh air breathing apparatus is required: is it in good condition (note date last tested)			
Work vests & life jackets in good condition and in sufficient number			
Deck fittings in good condition			
Deck wires, ratchets & lines in good condition			
All navigation lights operable			
Gasoline storage in safe, open area			
General housekeeping satisfactory			

REMARKS:

BIBLIOGRAPHY

Big Load Afloat, American Waterways Operators, Inc., 1973. Washington, D.C.

Safety Manual, American Waterways Operators, Inc., 1977. Washington, D.C.

Fire Protection Handbook, 14th edition, National Fire Protection Association. 1976. Boston, Mass.

Waterfront Fires, Robert E. Beattery, National Fire Protection Association. 1975. Boston, Mass.

Protection of Offshore Drilling Rigs & Production Platforms

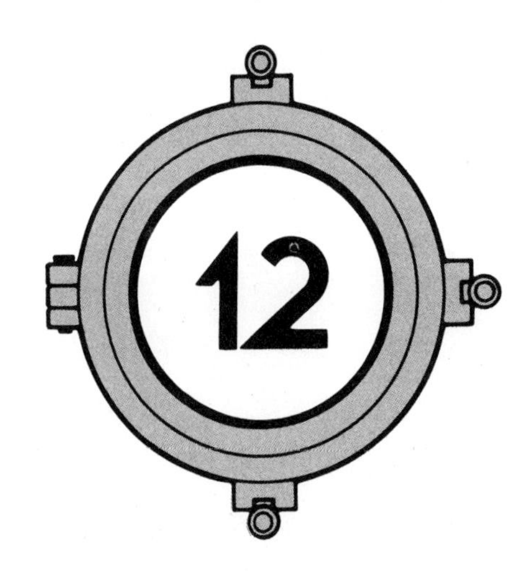

Offshore oil-drilling rigs and production rigs have varied configurations. The size and shape of each unit and the components that make it up depend on its function, the site at which it will be used and how it will be transported to that site. Offshore units are generally divided into two broad classes, mobile drilling units and fixed drilling or production platforms.

Mobile units are rigs that can be towed or can move under their own power from one location to another. When a self-propelled mobile unit is under way, the U.S. Coast Guard considers it to be subject to all maritime regulations, including navigational and fire protection regulations. Thus, the discussions and descriptions contained in the first 10 chapters of this book apply to all self-propelled mobile units.

Fixed units are rigs that are permanently secured to the seabed (Fig. 12.1), e.g., artificial islands, fixed structures and mobile units that are resting on the seabed. Thus, an offshore platform whose jacket (steel base) is secured to the seabed is a fixed unit. Fixed units are governed by (and must conform with) the *Rules and Regulations for Artificial Islands and Fixed Structures on the Outer Continental Shelf* (U.S. Coast Guard publication CG320) and regulations contained in *Oil, Gas, and Sulphur Leases in the Outer Continental Shelf, Gulf of Mexico Area* (U.S. Department of the Interior).

SAFETY AND FIRE PREVENTION

A manned offshore unit is the workplace, home and recreational area for its crew. The machinery spaces, processing and support equipment and the living, recreation and galley spaces are compacted into the smallest, most effective area possible. This is a potentially dangerous situation, since the crew is always close to hazardous drilling or production operations. Yet a safe environment must be maintained for and by the people who work on the unit. Every member of an offshore crew must think safety, work safely and remain constantly conscious of the hazards of his environment.

Safety

Safety is more than simply being careful. It includes knowing what is unsafe and how to avoid the careless actions and inactions that can make an area unsafe. Poor safety practices result from a lack of safety knowledge; carelessness results from a disregard for that knowledge. Both can lead to disaster.

A crewman who operates a piece of welding equipment without training may not know the safety rules. A person who smokes in a "no smoking" area may be acting in direct violation of safety rules he knows and understands. A supervisor who sees that a valve is leaking but does nothing to ensure that it is repaired may only be accused of inaction. However, the fires and injuries that can result from these practices will *not* be affected by any subtle differences in intent. Safety requires the full and continuous participation of every offshore worker.

Fire Prevention

Fire prevention on a drilling or production unit requires a twofold effort. The unit is subject to both the hazards of a ship and the hazards of similar land-based installations. As on a ship, careless smoking, hot work (burning and welding), and improper maintenance and electrical malfunctions are the most common causes of fire.

Safety rules and regulations, common sense and complete cooperation in fire prevention pro-

Figure 12.1. A fixed offshore unit of recent design. The platform is supported by a steel base fastened to the seabed.

Figure 12.2. Heat and sparks from welding and burning operations could cause a fire if proper precautions are not taken.

grams are the crew's main defenses against the outbreak of fire (*see also* Chapters 1, 2 and 4). A continuing education program and visual reminders will help maintain an awareness of the need for extreme care at all times.

The main causes of fire can be eliminated. Workers can refrain from smoking in bed and in restricted areas. They can ensure that all safety regulations (including an inspection of the area and providing a fire watch) are followed before and during burning and welding operations (Fig. 12.2). These operations should conform to Coast Guard regulations and the recommendations set forth in the *Manual of Safe Practices in Offshore Operations.* The offshore unit and its machinery can be maintained properly and carefully. All maintenance work should conform to the requirements of the owner and recognized industrial organizations such as the American Association of Oilwell Drilling Contractors (AAODC), the American Institute of Electrical Engineers (AIEE), the American Petroleum Institute (API) and the American Society of Mechanical Engineers (ASME). These nonprofit organizations can provide up-to-date technical data and recommended safe practices for the operation and maintenance of all mechanical and electronic equipment.

Oil Spills*

As Fire and Safety Hazards. It is hardly necessary to remind anyone in the petroleum industry that any oil outside a pipeline or other equipment designed to contain, process or use oil is a fire hazard and can result in serious injury, loss of life and/or extensive property damage. It must also be recognized that leaking or spilled oil on floors, decks, ladders, stairways or walkways presents slipping and falling hazards that can lead to serious injuries or fatalities.

Prevention of Spills. Pollution prevention demands individual effort. Pollution may occur through accident or the malfunction of equipment, but it often occurs because of poor housekeeping or failure to follow good operating practices. Here are some guidelines that will help prevent injuries and fires:

1. *Good housekeeping*—keep it clean. As previously mentioned, loose oil is a fire and safety hazard. Oil on decks in open sumps, in buckets, dripping from loose connections—any oil outside its proper container—is a fire hazard and can easily be ignited by sparks from various sources. Whenever any oil is spilled, it should be cleaned up immediately; oil on walkways can cause serious, and sometimes fatal, accidents. Oily rags should not be left lying

* The discussion on oil spills is adapted from the *Manual of Safe Practices in Offshore Operations,* 2nd revision. Offshore Operators Committee, 1972.

around or be allowed to accumulate anywhere but in a suitable container.

2. *Connections.* All pipeline and hose connections should be made in accordance with the best oil field practice. A leak in either a pipeline or a hose should be repaired immediately. Otherwise, the pipeline or hose should be taken out of service until it can be repaired or replaced. Open-ended lines should be closed by blind flanges or bull plugs to prevent accidental discharge; if this is not practical, drip pans and sumps should be provided.
3. *Drip Pans.* Drip pans or their equivalents should be placed under any equipment from which pollutants may escape into the surrounding water. This equipment should include, but should not be limited to, pumps, prime movers, broken connections and sampling valves. Permanent drip pans must be piped to sumps that are protected against the occurrence of fire and pollution.
4. *Sumps.* Adequate sumps and drainage systems must be installed wherever there is a possible source of pollution. Drip pans, bleed-off lines, gauge columns and such devices should be piped to sumps. Sumps should be designed to accommodate normal drainage. They should be located as far as practical from any source of ignition. Sumps should be covered, so that no spark can fall into or ignite the oil they contain. Satisfactory means must be provided to empty the sumps to prevent overflow.

Emergency Remote Shutoffs

The emergency remote shutoff system is, in essence, a fire and spill prevention system. Remote shutoff stations are located throughout the offshore rig. If a production pipeline is ruptured allowing the production to escape, the lever at any remote station may be pulled and automatically closes a set of valves, shutting down the production flow. At the same time, the production pumps are deenergized and a second set of valves is opened to allow vapors to bleed into the atmosphere.

FIRE DETECTION SYSTEMS

Since fire is a continuous threat to offshore units and to those who man them, most units have automatic fire detection systems. The detection devices in these systems are usually pneumatic tube detectors, heat and smoke detectors and combustible-vapor detectors. In most cases, the detectors are wired to sound the alarm when they are actuated; they are also set up to activate automatic fire extinguishing systems in unattended spaces. Manual fire alarm systems are also installed on most offshore rigs, and fire can be reported via the telephone and intercom systems.

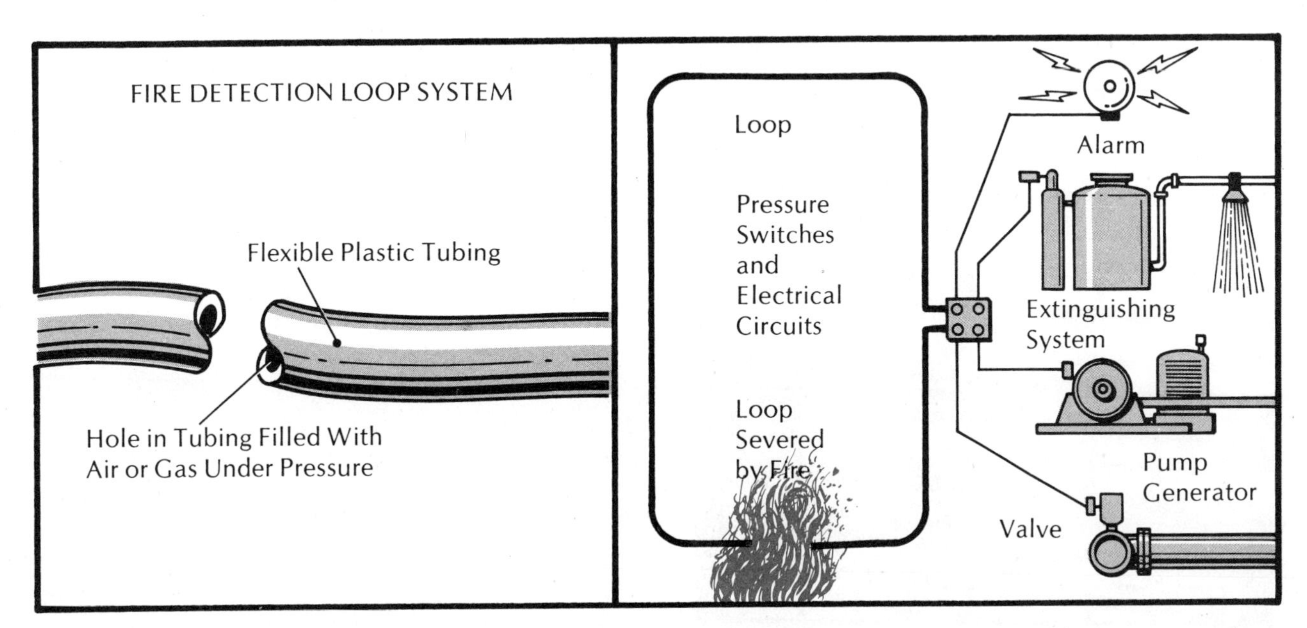

Figure 12.3. Fire detection loop system. **A.** Plastic tubing whose hollow core is filled with gas or air under pressure. **B.** When the tubing is severed, the pressure is lost. The reduced pressure allows switches to activate emergency equipment, including alarms, extinguishing systems, generators and product control valves.

Fire Line Automatic System (Pneumatic Tube Fire Detector)

The fire line automatic system is used to detect fire in open spaces and to activate alarms and/or firefighting equipment automatically. A fire line system is a flexible plastic or metal tubing that is strung around the outside of the entire structure, to form a loop. The tubing has a hollow core, which is filled with gas or air under pressure (206.84 x 10^3 pascals (30 psi)). The ends of the tubing are connected to pressure switches that can activate alarms and other devices electrically (Fig. 12.3).

How the System Works. When any portion of the tubing is burned through by fire, the air or gas inside the hose is released. The pressure in the tubing decreases, allowing pressure switch contacts to move and close electric circuits. These circuits may be set up to activate the fire alarm system, automatically shut down valves in pipelines, shut off remote emergency valves to incoming product, deenergize product pumps and energize fire-main pumps.

The fire line system is simple in design and very reliable. Its primary drawback is its vulnerability to false alarms. It will activate the equipment that it controls whenever the plastic hose is severed; e.g., when it is accidentally cut or damaged by abrasion.

Several fire line loop systems may be installed on an offshore unit. The additional loops would be used to protect specific areas; e.g., if the platform has several deck levels, a separate loop may be used for each deck. The pressures within the loops can be supervised individually at a central console (Fig. 12.4).

A separate fire line loop can be used at the well head location. The well head loop can be set up to activate the fire pumps and open a deluge valve, allowing a large volume of water to be pumped to the well head area and discharged through water spray nozzles (Fig. 12.5). (Well head protection is discussed in more detail later in this chapter.)

Figure 12.5. A spray nozzle primed to discharge a large volume of water into the well head area in case of fire.

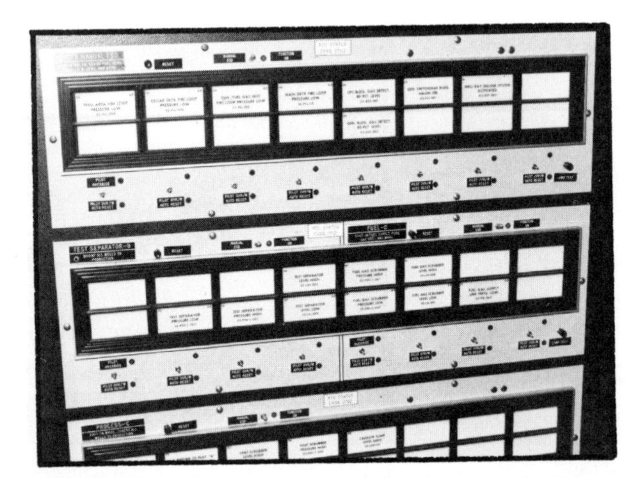

Figure 12.4. A central console is used to monitor all the fire loop systems installed on a rig.

Other areas that may be protected by individual fire line loops are compressor and generator rooms. The loops in these rooms are made up of metal tubing rather than plastic tubing, and use fusible metal plugs. When fire occurs, the plugs melt, allowing the air or gas to escape and the pressure in the tubing to drop. In addition to sounding the alarm, the loss of pressure can be used to shut down machinery and equipment in the protected space.

Heat and Smoke Detection System

Heat and smoke detectors are not used on all manned units. When they are used, they are installed primarily in living spaces, spaces housing electronic gear and storage areas. They cannot be used in outside areas, where winds might carry away the heat and smoke of a fire.

In living spaces, heat and smoke detectors are normally used only to actuate an alarm when they sense fire. In an electronic gear room, they may be used to activate an automatic Halon or

CO_2 flooding system, in addition to sounding the alarm. In the latter case, the detection system can also be wired to cut off the power, shut down exhaust fans and close ventilation openings.

Combustible-Gas Detection System

Combustible-gas detectors are used extensively on drilling and production units. These sensors are placed in such areas as switch gear, compressor and generator rooms, as well as in living spaces (Fig. 12.6) and galleys. They can also be installed to protect product pipelines, manifolds and well heads. Briefly, the combustible-gas detection system sounds the alarm when it senses the buildup of dangerous concentrations of flammable vapors. (*See* Chapter 6 for a more detailed description of the system.)

Manual Fire Alarm System

Whether or not fire detection systems are installed on an offshore rig, it is imperative that the crew be watchful for the occurrence of fire. Many times, an alert crewman discovers fire before even the most sophisticated fire detector is actuated. In spaces that are not protected by such devices, the crewman *is* the fire detector. When fire is discovered, the alarm should be sounded immediately, and firefighting procedures begun. Every

Figure 12.6. A combustible gas detector installed in living space. The detector senses the presence of flammable vapors in the surrounding air.

Figure 12.7. Emergency alarm box. The Fire and Abandon alarms can be sounded by pushing the proper buttons.

new crew member should be shown the locations of emergency alarms and taught how to use them.

The fire alarm system most commonly used on offshore units is an electrical system powered by the generator. Batteries serve as the emergency power source. Fire alarm boxes (Fig. 12.7) are located on all levels of the unit, in open and enclosed spaces. Each alarm box has three buttons—yellow (or orange), red and black.

The yellow (or orange) button is pushed to sound the fire alarm. This activates a siren that produces a warble, or double-pitch sound. At the sound of the alarm, each crewman should report directly to his fire emergency station, as given in the station bill.

The red button activates a steady pitched siren that alerts the crew to prepare to abandon the unit. Crew members must proceed to the positions assigned them in the station bill for this purpose.

The black button shuts off the emergency signal. It is used only in the event of a false alarm or accidental activation of an alarm.

Telephones are usually located near alarm boxes; they can be used to communicate the details of the emergency to the person in charge or the fire control team. A public address system can also be used to communicate with most parts of the rig.

FIREFIGHTING SYSTEMS AND EQUIPMENT

In general, the fire protection systems installed on self-propelled mobile units are similar to those installed on ocean-going vessels. The propulsion plant space may be protected by a CO_2 flooding system and a semiportable system. A fire-main system must be installed, and it must conform in all respects to Coast Guard regulations governing similar systems on ships. (*See* Chapter 9.)

The systems installed on fixed offshore units tend more to automatic detection and extinguishment than the systems that protect their land-based counterparts. The reason is obvious: There is no land-based fire department to respond to an alarm from an offshore rig. The crew must fight the fire alone, using the equipment installed on the rig.

Actually, as compared to land-based installations, the typical manned drilling or production unit is well protected against fire. Most of the firefighting equipment described in Chapters 8–10 is available on most rigs. If this equipment is used properly, the crew should be able to control most fires. Many oil and gas companies conduct comprehensive training in both fire safety and fire control. Personnel assigned to offshore units are generally required to demonstrate a knowledge of firefighting equipment and an ability to use the equipment against various types of fire.

Fire-Main System (Fixed Units)

The fire-main system is a system of piping that carries water from the pumps to fire stations located throughout the unit. In warm climates, the piping is filled with water. In cold climates, the piping is either dry or filled with a mixture of water and antifreeze. Although a single-main system may be used, most units have loop-type systems. (*See* Chapter 9.) Two diesel-powered pumps (one as the primary supply, and the second as backup) supply seawater to the system at sufficient volumes and pressures to produce good hose streams for firefighting.

Fire stations are located so that the hose stream from one station will overlap the hose stream of the adjacent station for complete coverage of a fire at any point on the unit. At each fire station, a 152–305 m (50–100 ft) length of 3.8– 6.4 cm (1½- or 2½-in.) hose, with a nozzle, is connected to the pipe outlet. A valve at the outlet controls the flow of water to the hose; during use, the valve should be fully opened.

Adjustable fog nozzles are used on most fixed units. (*Note:* These nozzles are not currently permitted by USCG regulations. A regulation has been proposed, but to date, nothing has been finalized.) This type of nozzle can deliver a solid stream or either of two fog patterns. When the control lever is pushed all the way forward, the nozzle is closed (Fig. 12.8A). When the lever is pulled back, water is discharged from the nozzle; the further back the lever is pulled, the greater the water volume (Fig. 12.8B). The water stream pattern is selected by rotating the nozzle barrel; a straight stream is obtained by rotating the barrel clockwise and fog patterns are obtained by rotating the barrel counterclockwise to the 30° or 90° setting (Fig. 12.8C).

The fog cone discharged from the nozzle is a fine spray that can knock down a sizable fire when properly applied. Although the cone is hollow, it can absorb large amounts of heat, which makes it ideal for cooling down hot surfaces. A wide fog cone is also excellent protection for firefighters approaching an extremely hot fire. In open areas, the fog pattern can be used continuously during the approach to the fire. However, in enclosed passageways, a continuous fog stream will push the fire ahead into unburned areas of the structure. In such a case, the fog stream should be applied intermittently, in short bursts. (*See* Chapter 10.)

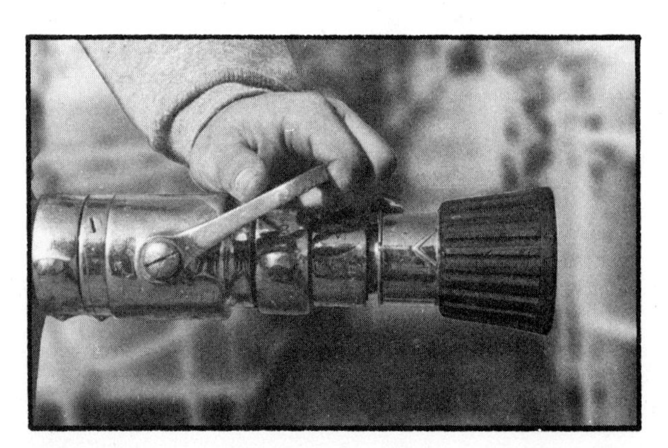

Figure 12.8. The adjustable fog nozzle can deliver a solid stream or two different fog patterns. **A.** The fully closed position. **B.** The fully open position with the stream selector set for a solid stream.

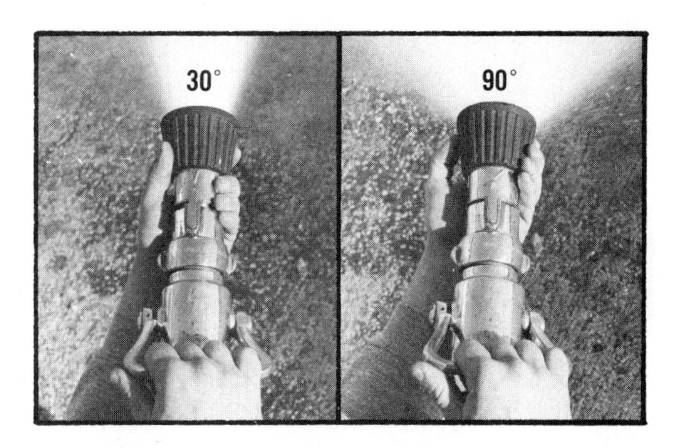

Figure 12.8C. The adjustable fog nozzle will deliver a 30° or 90° fog pattern as well as a solid stream.

Foam Stations. Foam stations (Fig. 12.9) are sometimes located at fire-main stations, so that either water or foam may be applied from the station. If foam is to be used, the fog nozzle is removed from the hose and replaced with an aspirating-type foam nozzle. (*See* Chapter 8 for a discussion of the operation of this nozzle.) The water control valve is opened; then the eductor valve is rotated to the foam position so that foam concentrate is picked up and mixed with water in the hoseline.

Monitor Nozzles. Monitor nozzles are placed in fire-main systems to produce heavy streams of water for fire attack or for cooling exposures. A monitor nozzle can be operated easily by one crewman and can produce a 1892–7570 lpm (500–2000 gpm) fog stream. A stream of that size can knock down and extinguish an extensive fire and can also be used to cool metal supports and deck surfaces during a fire, to prevent weakening and buckling of the structure. Monitor nozzles can attack and control a large fire from a relatively safe distance, especially from the windward side of the fire. They can also be used to protect firefighters who are advancing to the fire with handlines. On units with expanded-metal decks, the use of large volumes of water is not dangerous, since the water runs off rapidly and does not cause a weight problem.

A distinct advantage is gained when monitor nozzles are positioned so they can reach the helicopter pad (Fig. 12.10). In the event of an accident involving a helicopter fire on the pad, heavy streams (preferably fog streams) can be directed from one or more monitors to blanket the entire landing pad in water. The water will either extinguish the flames or keep them under control until hoselines or dry chemical extinguishers can be advanced for a close-in attack.

Figure 12.9. Some fire stations have a foam concentrate storage tank as an integral part of the fire-main system. By opening the eductor valve the hoseline is supplied with a foam solution.

Carbon Dioxide and Halon 1301 Flooding Systems

Carbon dioxide (CO_2) and Halon 1301, total flooding systems are generally installed in generator and compressor rooms and spaces housing electronic equipment. These systems are also occasionally used to protect storage spaces on platform structures. Either system consists of a group of cylinders containing the extinguishing agent, a manifold, piping, valves and discharge nozzles. The system may be set up for manual activation or automatic activation by fire detectors.

Figure 12.10. Monitor nozzles, positioned to reach the helicopter pad, can apply large volumes of water from a safe distance.

When the system is activated (either manually or automatically), the extinguishing agent is first discharged into a time delay. This delays the discharge of the agent into the protected space for about 20 seconds. During this period, a warning alarm sounds. All personnel must evacuate the space immediately, closing the door tightly as they leave. In most systems, exhaust fans and dampers are closed automatically by pressure switches that are actuated by the agent when it is discharged. The reasons for evacuating and totally closing the protected space prior to the discharge of the agent have been discussed in earlier chapters.

Foam Systems

Oil storage tanks, including the "gun barrel" tank, can be protected by any of several types of foam systems. The tank construction usually determines which system affords the best protection. Fixed roof storage tanks (the tanks used on offshore units) can be protected by either subsurface foam injection systems or tankside foam chamber systems.

Subsurface Foam Injection System. In the subsurface system, mechanical foam is produced by a high-back-pressure foam maker located at a distance from the tank site. The foam is forced through piping into the bottom of the tank. It bubbles up to the surface of the stored liquid product, where it forms a floating vaportight blanket (Fig. 12.11). The foam blanket extinguishes the fire.

Subsurface systems can only be used to protect tanks containing petroleum products. Polar solvents and water-miscible fuels will break down the foam, destroying its effectiveness.

Foam Chamber System. This system makes use of one or more foam-dispensing chambers. The chambers are installed on top of the shell of the tank, near the roof joint. Two types of chambers are used. One type deflects the foam onto the inside wall of the tank, so it cascades down onto the surface of the burning liquid. The other type, called a *Moeller chamber,* directs the foam onto the liquid surface via a flexible tube.

Piping connects the foam chambers to a foam house at a remote location; the chambers are normally empty. When a fire is to be extinguished, foam solution is produced at the foam house and pumped to the foam chambers. The foam is aerated in the chambers to form a mechanical foam and is then dispensed onto the burning fuel.

Fires involving polar solvents must be extinguished with a foam that is compatible with these liquids. The foam manufacturer's recommendations regarding the type of foam concentrate, its strength (3% or 6%) and the rate of injection should be followed carefully.

Fighting Storage Tank Fires. If fire develops in a storage tank, the foam system should be used for the initial attack. Monitor or handline water streams should be directed onto the tank shell as soon as possible. Cooling the tank shell helps prevent the temperature of the liquid product from increasing rapidly. It also cools the metal at the liquid surface; this helps the foam blanket form a tight seal. If the metal in this area is excessively hot, the water in the foam will boil off and the foam blanket will break down.

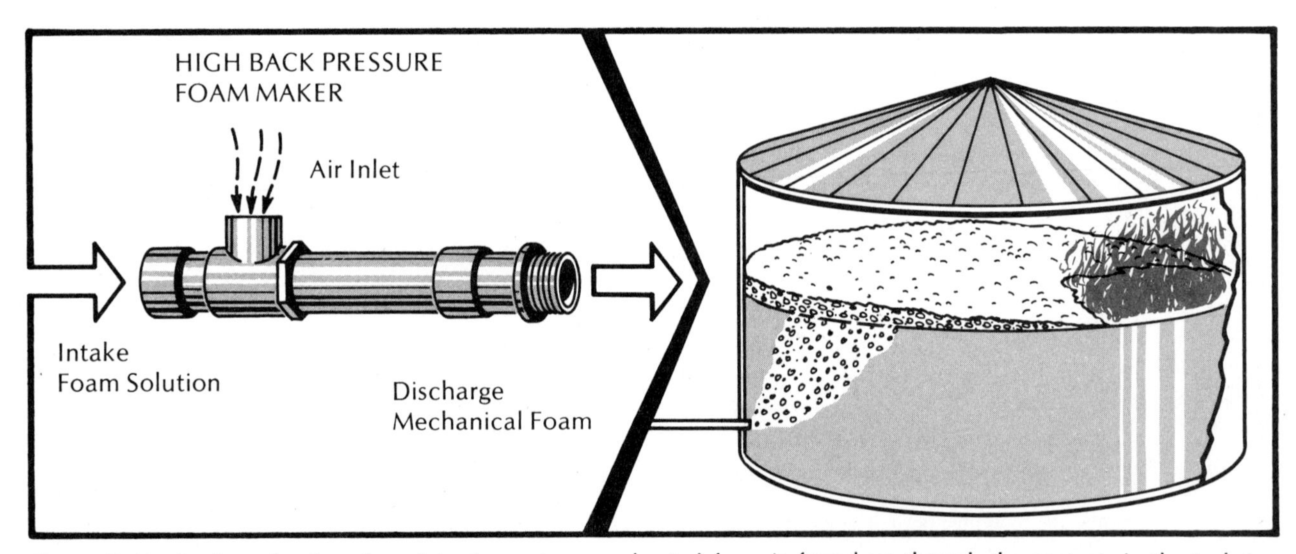

Figure 12.11. In the subsurface foam injection system, mechanical foam is forced up through the contents in the tank to form a floating blanket that smothers the fire.

Water Spray System

Water spray systems are used on some offshore units to protect the well head and adjacent areas. Each system consists of a piping arrangement, valves, spray discharge heads and water pumps. For manual operation, the water flow control valve must be opened, and the water pumps started. These are time-consuming operations that allow the fire to increase in size and intensity. Automatic operation is thus desirable; it can be attained by using a fire loop detection system to activate the water spray system.

When fire is detected by an automatic system, the fire pumps are activated and a deluge valve is simultaneously opened. (A deluge valve is a fast opening device that allows the full volume of water to flow without delay.) Water pumped into the system is discharged through spray heads in a pattern that provides an umbrella of water over the area. Exposures are protected from radiant heat by water that is directed onto walls and/or equipment to keep their surfaces cool.

If a fire involves product that is escaping from a ruptured pipeline or a leaky connection, the spray system should not be shut down until the flow of product is stopped. Even if the spray system extinguishes the fire, it should be allowed to operate as long as product is moving to the site of the fire. This serves two purposes: It prevents the product from reigniting, and it cools metal structures in the fire area.

Automatic Sprinkler System

The living, recreation, office and galley spaces on a fixed offshore unit can be protected by an automatic sprinkler system. The system would consist of piping, fusible-link sprinkler heads, valves, pumps and a pressure tank.

When fire causes one or more sprinkler heads to open, the initial water supply comes from the pressure tank. As the water flows out of the tank, alarms can be actuated by the water itself or by the pressure drop in the tank. The same action is used to start the fire pump, which supplies additional water to the system. Most fires can be brought under control by the water discharged from one or two sprinkler heads. The system should not be turned off until the fire is definitely out or until a hoseline can be positioned to handle any remaining fire. After a system has been used, it should be restored to operation according to the manufacturer's directions.

Sprinkler systems are very reliable if they are well maintained. When a system fails to operate properly, it is usually because the wrong valves are closed. To avoid this, all valves should be marked with their operating positions (such as "Keep open at all times") and, if necessary, sealed in the proper position.

SPECIAL FIREFIGHTING PROBLEMS

Offshore units are unique in their structural design, and they present special firefighting problems. In land-based structures and on ships, the floor or decking is generally solid; this structural feature prevents or slows the vertical travel of fire. Most platforms have expanded-metal supports and decks, often four or five levels high. This open decking not only allows fire to travel upward, but it also allows fire to drop downward.

The problem can be a very serious one. For example, suppose a fire originates in some oil drums stored on a middle deck level. The fire will travel upward by convection and radiation and extend to structures and materials above the fire area. At the same time, burning liquid can drop through the decking and ignite combustibles on the deck below. Moreover, expanded metal is highly susceptible to heat, and the decks may expand, warp out of shape and lose their strength.

Such a fire must be attacked rapidly at its point of origin. However, the crew should also position hoselines above and below the fire deck, to extinguish fire that might extend to those areas and to keep the metal decking cool. If monitor nozzles are available, a "blitz" attack can be made. In a blitz attack, firefighters hit the fire with everything at their disposal. Finesse is not important in this situation; getting the fire out is the important goal. Water damage need not be considered, since the large volumes of water will simply plunge into the sea.

Helicopter Fires

Serious fires can originate on the helicopter pad, which is often placed directly above living and office spaces. The deck of the pad is solid metal or wood, and access is usually by a single staircase. A crash, with a resulting fire, could cause burning fuel to spill off the pad. The fuel would run down the side walls of the structure, carrying fire downward for several decks. Such a fire would test the knowledge and experience of even the best-trained firefighters.

The fire should be attacked at each burning level, preferably with fog streams. The streams should be directed first at deck level and then worked upward, with a sweeping motion, across the flames. Dry chemical can also be used to knock down the flames quickly, but water streams are necessary to cool hot surfaces (to prevent reflash). A combined dry chemical and water attack would be effective, since the two agents are compatible. The fire on the pad must be attacked and extinguished, since it is the source of the fire and the spilling fuel. As long as the pad fire continues to burn, any spilled fuel will cause problems as it runs downward.

If anyone is trapped in the helicopter or isolated on the landing pad, the attack at the pad level becomes imperative. If a rescue is possible, the initial attack on the pad fire can be made with portable dry chemical or AFFF (Aqueous Film-Forming Foam) extinguishers. The dry chemical or foam can be used to knock down flames and open a path to the aircraft. The firefighters should advance at a steady pace, with the wind at their backs, directing the agent with a sweeping motion. They must be wary of reflash, and not move in so fast that the fire burns behind them. The use of dry chemical alone to fight the fire is an emergency measure; handlines should be advanced and placed in operation as quickly as possible. The cooling effect of water is essential to extinguish the fire and to minimize reflash.

A good standard procedure is to station personnel with portable dry chemical or AFFF extinguishers at the ready for all landings and takeoffs. An immediate attack on a fire resulting from a crash could save lives and keep the fire from increasing in size and intensity. Helicopter and landing deck operations should conform to the recommendations published in the *Manual of Safe Practices in Offshore Operations*.

Fires in Living Spaces

Fires in living spaces on an offshore unit should be attacked basically as described in Chapter 10. The only difference would be in the confinement and overhaul procedures. There are "dead spaces" in walls and partitions and between ceilings and roofs on offshore units. These spaces provide a channel for fire travel not usually found on ships. They must be checked carefully for fire extension, since they can allow fire to persist undetected for some time. Dead spaces must be opened fully during overhaul if they show any signs of heat or fire.

Well Head Fires

Some well head fires are readily controlled by the crew, especially when down hole safety devices function properly. When safety devices fail, the fire may be beyond the capability of platform crews to extinguish. Destructive explosions, usually damaging to the structure, often accompany the outbreak of the fire. Fire detection and fire extinguishing systems may be disrupted; for example, automatic valve-closing devices may fail to work, or the fire main may be damaged beyond use. When the situation threatens the crew's safety, abandonment procedures should be initiated without delay.

Even when the firefighting systems survive the explosions, they may be incapable of extinguishing a well head fire. The escaping fuel results in an extremely intense fire, like a gigantic blowtorch. If the flames are directed horizontally toward the platform, the situation is untenable. However, if the flames are directed upward, heat will be convected up and away from the platform. Even then, the radiant heat is intense enough to endanger all exposures. If a decision is made to stay on the platform, the exposures must be protected by cooling with water streams. The water should be applied directly onto the surfaces of the exposures. Heavy stream monitor nozzles should be used to protect exposures wherever possible. The water spray system at the well head (if there is one) should be used to keep that area cool. If the flames are extinguished but the flow of fuel is not stopped, additional explosions could occur as the fuel reignites.

There is no easy way to determine when to abandon an offshore rig, and when the crew can safely stay to fight a fire. Many factors must be considered. However, one rule should remain foremost during the decision-making process: *Protect the lives of the crew*. If the situation is judged to be dangerous to personnel, then the unit should be abandoned. In any case, it may be wise to evacuate all personnel who are not needed to combat the fire.

Support Vessels

The supply and rig tender vessels that service offshore units carry portable fire extinguishers, firemain systems, CO_2 flooding and semiportable systems and dry chemical semiportable systems. Most of these systems are manually operated, although newer vessels may have automatic extinguishing systems activated by detection devices. Fire detectors may be found on some rig tender vessels, primarily in paint and lamp lockers. They are wired to sound the alarm when fire occurs and, in some instances, to activate a 22.7 or 45 kg (50 or 100 lb) CO_2 cylinder to flood the locker. (Automatic fire extinguishing systems are not installed in rig tender engine rooms for the reasons given earlier in this book—the need to evacuate the space and the loss of propulsion. For these same reasons, the manual CO_2 flooding system is used only as a last resort. *See* Chapters 6, 8 and

9 where all these fire protection systems have been described.)

Dry Chemical Semiportable Extinguisher. Rig tender vessels often carry combustible products and liquids on their open deck spaces, aft of the bridge. Among these combustible materials are drilling mud, lubricating oils for machinery and a wide array of supplies. On some rig tenders, a 226.8 kg (500-lb) dry chemical semiportable extinguisher is located on the main deck, near the cargo. This extinguisher is used as the primary attack unit for fires involving the deck cargo.

Before the dry chemical extinguisher is activated, the hose should be run out to its full length, with the nozzle closed. Then the nozzle should be positioned to attack the fire. The unit may be activated by pulling the manual release lever, allowing nitrogen to flow from its cylinder into the dry chemical tank. The nitrogen pressurizes the dry chemical, but a "burst-disk" delays the release of the chemical to give it time to become fluidized. When the tank pressure reaches 1379 x 10^3 pascals (200 psi) the disk bursts, and both nitrogen and dry chemical are released into the hoseline. (Nitrogen is an inert gas that contributes to the extinguishment of the fire.)

The attack should be made from windward, if possible. The nozzle should be directed with a wrist-flicking motion, to sweep the agent across the flames. It should be kept parallel to the deck or pointed slightly downward. If the nozzle is pointed upward, the dry chemical will block the nozzleman's vision; in addition, some of the agent will pass over the flames and do no good. Smaller portable extinguishers can be directed onto the flames on either side of the main attack path.

Dry chemical can knock down a fire quickly, which makes it an ideal initial attack agent. However, the dry chemical attack must be backed up with a secondary means of extinguishment. The compatibility of dry chemical and water makes hoselines the obvious choice as a backup. (Although foam is usually recommended for class B fires, water, especially when applied in a fog pattern with sufficient volume, has the capacity to extinguish sizable flammable liquid fires.) If the burning material and the deck are not cooled with hoselines, a reflash may occur. Since the fire is on an open deck with drainage into the sea through scuppers, large amounts of water can be applied from several hoselines without affecting the vessel's stability.

BIBLIOGRAPHY

Fire Protection Handbook. 14th ed. NFPA, Boston, 1976

Manual of Safe Practices in Offshore Operations. 2nd revision. Offshore Operators Committee, 1972

Bryan, JL: Fire Suppression and Detection Systems. Los Angeles, Glencoe Press, 1974

Rules and Regulations for Artificial Islands and Fixed Structures on the Outer Continental Shelf (No. CG320)

Oil, Gas and Sulphur Leases in the Outer Continental Shelf, Gulf of Mexico (U.S. Department of the Interior Publication)

Fire Safety

Part III

The four chapters in this final part deal with the safety of ship's personnel during and after emergencies. Chapter 13 covers the organization and training of personnel to handle emergencies in an orderly and efficient manner. This is an extremely important first step toward minimizing the hazards of any emergency situation and ensuring the safety of those involved. Chapter 14 is a fairly complete discussion of first aid techniques that may be applied aboard ship. Chapters 15 and 16 cover the personal safety equipment and safety devices carried on U.S. flag vessels. The equipment and devices described in those chapters have been mentioned in several earlier chapters.

Organization & Training of Personnel for Emergencies

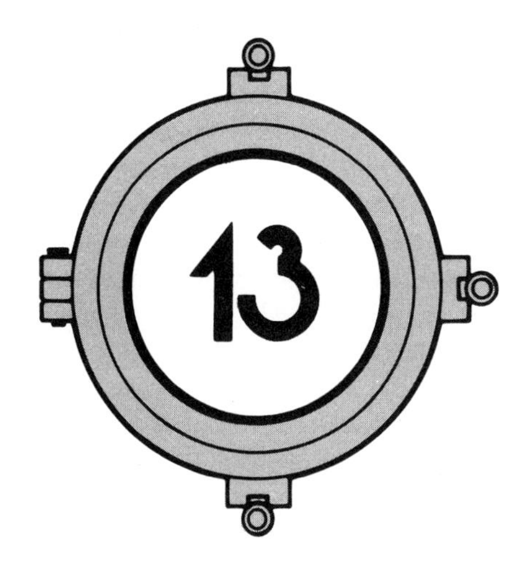

Every land-based emergency service, such as a fire department or police department, is carefully organized to accomplish its goals. These goals are accomplished when certain people properly perform certain assigned tasks. Thus, the emergency service must first be organized in terms of personnel; responsibilities and duties must be clearly set forth and a chain of command, with a single, ultimately responsible chief officer, must be established. Then the service must be organized in terms of tasks; personnel must be trained to know what to do in an emergency, when to do it and how to use the necessary equipment.

Aboard a ship, the crew constitutes the emergency services—in particular, the fire department. There already is a normal chain of command on every vessel—from the master through his officers to their departments. This chain of command does not change during emergency situations; the master and his department heads remain responsible for the efforts of the crew. The station bill lists the duties of crew members during emergencies; drills and training sessions are held to ensure that the crew will be capable of performing these duties properly if the need should arise. The equipment needed to carry out these duties is carried on board. Thus, all the elements of an emergency service are available. The effectiveness of the crew as a firefighting force depends only on how well these elements are assembled.

ORGANIZATION OF PERSONNEL

The organization of personnel aboard ship resembles in many ways the organization of a large industrial plant. In an industrial plant the general manager is the top executive; he carries out the policies of the owners—the corporation. However, he and the corporation are subject to federal, state and local laws affecting the plant and its employees. The plant manager is assisted by his subordinates and is responsible for their performance. These subordinates are the heads of such departments as personnel, production, transportation, engineering and security.

Aboard ship, the master is the top "executive." Like the plant manager, the master follows the instructions of the owners. He, too, is subject to applicable laws. The laws that govern the master's operations are the maritime laws as set forth in the laws Governing Marine Inspection and U.S. Coast Guard rulings and regulations. Since the maritime laws and regulations are designed to provide safety at sea for passengers, crew and ship, they are quite strict; inevitably the master is charged with the responsibility for violations.

Aboard ship, the master's authority is second only to God's (an old mariner once said that is so only because God has seniority). The master's authority is derived from government regulations and the instructions of the ship owners. However, the ship owners can in no way authorize a ship's master to act contrary to any federal regulations.

The responsibilities of the master are tremendous. Under ancient but still valid laws of the sea, he is responsible for practically every action taken aboard his ship, by himself or by his subordinates. While he may delegate his authority, he cannot, in any way, relieve himself of responsibility for the acts of those whom he authorized to act.

The master is not, however, alone as he paces the bridge of his ship. As assistants, he has the chief mate and other deck officers and the chief engineer and other engineering officers. These people demonstrated their competence and ability in difficult and searching examinations, before they were granted licenses. In addition, each officer was required to serve in various subordinate ranks before sitting for the Coast Guard license examination. Although a master must have proven qualifications, he may depend on competent assistants to help him in carrying out his many duties. In this regard, the law holds each licensed or documented seaman responsible for his actions in carrying out his duties.

Like an industrial plant, a ship is organized into various departments, each under control of a department head—the chief mate, chief engineer and chief steward (Fig. 13.1). The chief mate is second to the master in the chain of command. Aside from being responsible for carrying out the orders of the master, he is usually in charge of safety, lifesaving and firefighting equipment and the training of the crew. He coordinates the work of his department and the lifesaving and firefighting drills. However, instructional sessions and training drills are planned with the master, who is responsible for training in the use of firefighting, lifesaving and other emergency equipment. The master should place in the log an entry indicating that he has reviewed and approved the training plans. Another entry should be made to note that the actual drills have been completed.

During the planning of drills and training sessions, the chief mate should consult with the chief engineer, especially when the engine room is selected as the location of a fire drill. The chief engineer is jointly responsible, with the master, for training the crew in the use of all emergency equipment.

THE STATION BILL

The station bill is a muster list required by federal regulations. It lists the special duties and duty station of each member of the crew during emergencies, and the signals for these emergencies (Fig. 13.2). In one column are listed the duty station and duty assignment of each member of the crew during a fire or other emergency situation; in another column are listed the boat station and duty assignment of each crew member during an abandon ship procedure.

Normally, the master draws up the station bill when he takes command of a vessel. The station bill is then used for all voyages of that vessel under his command. The master makes an introductory statement and signs the station bill before sailing. He then ensures that copies are posted in conspicuous locations in the vessel, particularly in the crew quarters.

Locator Numbers

The makeup of a ship's crew changes somewhat with each voyage, but the emergency duties and

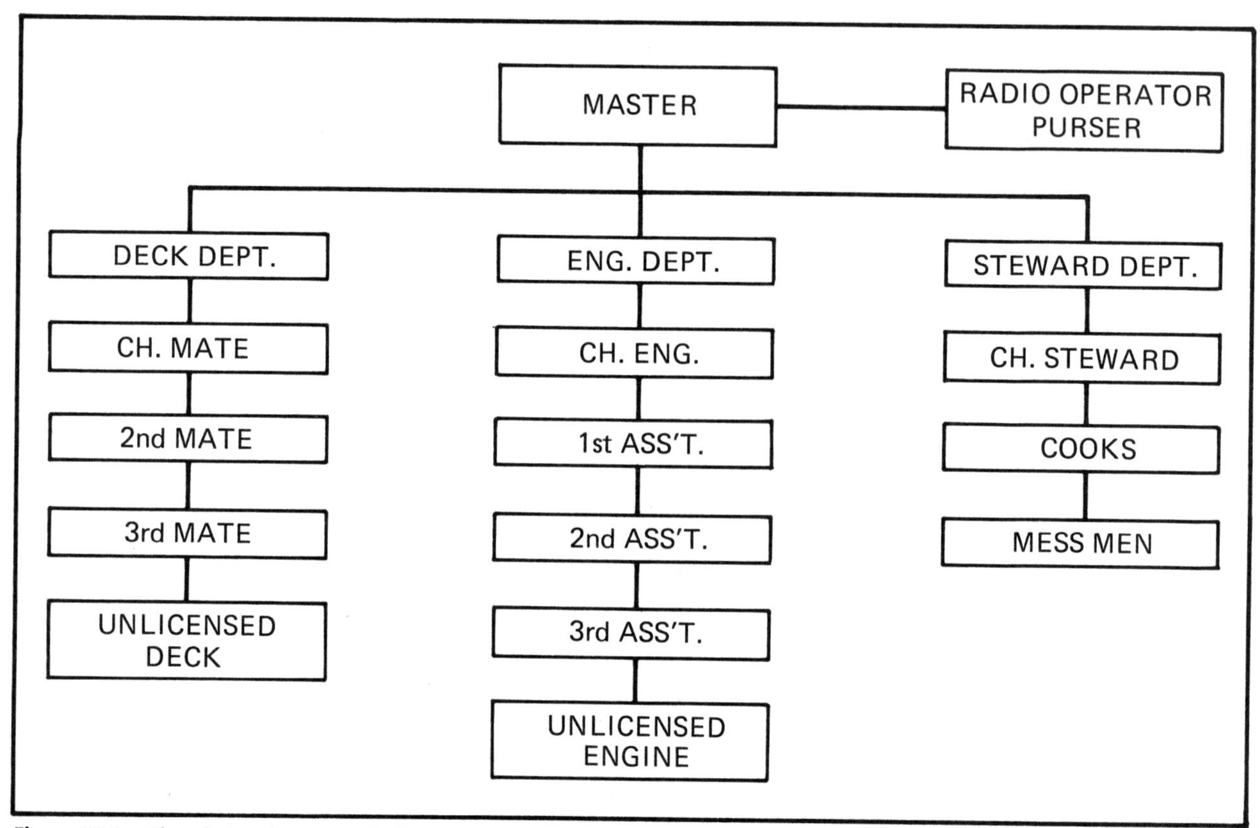

Figure 13.1. The chain of command aboard ship.

CG 848 (Rev. 10-50)
U. S. COAST GUARD

UNITED STATES COAST GUARD

SPECIMEN OF A STANDARD STATION BILL PREPARED FOR FREIGHT AND TANK SHIPS CARRYING PERSONS IN ADDITION TO CREW

STATION BILL

SIGNALS

.... (Name of ship) .... (Name of company)

FIRE AND EMERGENCY—Continuous blast on ship's whistle and general alarm bells for a period of at least 10 seconds.
ABANDON SHIP—More than 6 short blasts and 1 long blast on the whistle and the same signal on the general alarm bells.
MAN OVERBOARD—Hail, and pass the word "MAN OVERBOARD" to the bridge.
DISMISSAL—From FIRE AND EMERGENCY stations, 3 short blasts on the whistle and 3 short rings on the general alarm bells.

WHERE WHISTLE SIGNALS ARE USED FOR HANDLING BOATS

Lower boats—1 short blast on whistle
Stop lowering boats—2 short blasts on whistle
Dismissal from boat stations—3 short blasts on whistle

INSTRUCTIONS

1. Entire crew shall familiarize themselves with the location and duties of their emergency stations immediately upon reporting on board.
2. Each crew member shall be provided with an individual supplementary station bill card which must show in detail the special duties to perform.
3. Entire crew shall be instructed in the performance of their special duties and crew on watch will remain on watch on signal for emergency drill.
4. Every person participating in the abandon-ship drill will be required to wear a life preserver and entire boat crew shall assist in removing covers and swinging out boats.
5. Emergency Squad will assemble with equipment at scene of action immediately upon the emergency signal.
6. Stewards' department will assemble and direct passengers, properly dressed and wearing life preservers, to embarkation stations.
7. Person discovering FIRE shall immediately notify the bridge and fight the fire with available equipment.
8. Immediately upon the FIRE AND EMERGENCY signal, fire pumps to be started, all watertight doors, ports. and air shafts to be closed, and all fans and blowers stopped. Fire hose to be led out in the affected area as directed.
9. Upon hearing the signal, "MAN OVERBOARD," throw life ring buoys overboard, stop engines and send lookout aloft. Emergency boat crew consisting of all seamen shall immediately clear lee boat for launching.
10. During periods of low visability, all watertight doors and ports below the bulkhead deck shall be closed, subject to the Master's orders.

No.	RATING	FIRE AND EMERGENCY STATIONS	No.		ABANDON SHIP—BOAT STATIONS
		DECK DEPARTMENT			
A.	Master	On the bridge. In command, all operations.	A.	Lifeboat No. 1	In command. On bridge in charge all operations.
1.	Chief Mate	At scene of emergency. In charge.	1.	Lifeboat No. 2	In command. In charge launching lifeboats amidship.
2.	2d Mate	On the bridge. Relieve the watch.	2.	Lifeboat No. 3	In command. On the bridge. Relieve the watch.
3.	3d Mate	Prepare all lifeboats for launching. In charge.	3.	Lifeboat No. 4	In command. In charge launching lifeboats aft.
4.	Radio Operator	Radio room. At instruments.	4.	Lifeboat No. 1	Attend Master's orders and instructions.
5.	Boatswain	Emergency squad. Provide life lines.	5.	Lifeboat No. 1	2d in command. Attend forward gripes and falls.
6.	Able Seaman	[illegible] squad. Relieve the wheel.	6.	Lifeboat No. 2	2d in command. Attend forward gripes and falls.
[illegible]		Provide extra length of hose and spanner.	7.	[illegible]	[illegible]
	[illegible]	Emerg[illegible]		[illegible]boat No. 3	[illegible] and attend boat [illegible]
	Able Seaman	Assist 3d Mate pre[illegible]			
15.	Ordinary Seaman	Bridge. Act as messenger.	15.	Lifeboat No. 4	Release boat chocks and secure drain cap.
16.	Ordinary Seaman	Emergency squad. Act as messenger.	16.	Lifeboat No. 1	Lead out and attend boat painter.
17.	Ordinary Seaman	Assist 3d Mate prepare lifeboats for launching.	17.	Lifeboat No. 2	Lead out and attend boat painter.
		ENGINE DEPARTMENT			
18.	Chief Engineer	In charge of Engine Department.	18.	Lifeboat No. 1	Assist in general operations.
19.	1st Assistant	Engine room. In charge.	19.	Lifeboat No. 2	Assist in general operations.
20.	2d Assistant	In charge of fire room and steam smothering apparatus.	20.	Lifeboat No. 3	Assist in general operations.
21.	3d Assistant	Attend main steam smothering line.	21.	Lifeboat No. 4	Assist in general operations.
22.	Jr. Engineer	Attend CO² or foam smothering system.	22.	Lifeboat No. 1	Turn out forward davit and assist at forward falls.
23.	Jr. Engineer	Attend CO² or foam smothering system.	23.	Lifeboat No. 2	Turn out forward davit and assist at forward falls.
24.	Jr. Engineer	Engine room. At fire pumps.	24.	Lifeboat No. 3	Turn out forward davit and assist at forward falls.
25.	Pumpman	Assist 2d Assistant Engineer in fire room.	25.	Lifeboat No. 1	Release after gripes and attend after falls
26.	2d Pumpman	Emergency squad. Assist with [illegible]	26.	[illegible] No. 4	Lead out and attend boat painter.
27	[illegible]	[illegible] main [illegible]			
	[illegible]	[illegible] Engineer.	[illegible]	Lifeboat No. [illegible]	[illegible] and
[illegible]	Fireman	Fire room. Assist [illegible] Assistant Engineer.	43.	Lifeboat No. 3	Turn out davits and assist at fa[illegible]
44.	Storekeeper	Emergency squad. Provide inhalator.	44.	Lifeboat No. 2	Turn out davits and assist at falls.
45.	Wiper	Engine room. Act as messenger.	45.	Lifeboat No. 1	Turn out after davit.
46.	Wiper	Assist 3d Officer prepare lifeboats for launching.	46.	Lifeboat No. 1	Stand by life ring buoy, ready for use.
47.	Wiper	Emergency squad. Assist with fresh air mask.	47.	Lifeboat No. 2	Stand by life ring buoy, ready for use.
48.	Wiper	Emergency squad. Assist with inhalator.	48.	Lifeboat No. 1	Turn out davits and assist at falls.
		STEWARDS' DEPARTMENT			
49.	Chief Steward	Arouse, warn, and direct passengers. In charge.	49.	Lifeboat No. 1	Arouse, warn, and direct passengers. In charge.
50.	Chief Cook	Secure galley.	50.	Lifeboat No. 3	Lead out and attend boat painter.
51.	2d Cook	Assist Chief Cook secure galley.	51.	Lifeboat No. 4	Lead out and attend boat painter.
52.	Mess[illegible]	Close all ports and door[illegible]	[illegible]	Lifeboat N[illegible]	[illegible] davit.
		[illegible] all [illegible]			[illegible] Arouse, wa[illegible]
[illegible]	Utilityman	[illegible] aft.	56.	Lifeboat No. 4	Turn out after davit.
57.	Utilityman	Assist 3d Mate prepare lifeboats for launching.	57.	Lifeboat No. 3	Stand by life ring buoy, ready for use.
58.	Galleyman	Assist 3d Mate prepare lifeboats for launching.	58.	Lifeboat No. 4	Stand by life ring buoy, ready for use.

NOTE.—For additional information see notice entitled STATION BILLS, DRILLS AND REPORTS OF MASTERS, Form 809 A.

....
Master.

This specimen station bill has been prepared for freight and tank ships that carry a crew of 35 to 58 persons, which vessels are equipped with 4 lifeboats. In view of the various types of fire fighting and lifesaving equipment on board vessels of this class, this specimen is to be used only as a guide in making up suitable station bills in compliance with the regulations. Copies of this specimen may be obtained from the office of the Officer in Charge, Marine Inspection, U. S. Coast Guard.

U. S. GOVERNMENT PRINTING OFFICE 16—34270-3

Figure 13.2. Typical ship's station bill.

station assignments remain the same. Thus, station bill assignments are not made by name, but by *locator number*. As a crewman signs aboard a ship, he is given a number; his emergency duties and station are then the ones that are listed on the station bill for that number. Some steamship lines call the locator number the *articles number*. At least one line refers to it as *bunk number,* feeling perhaps that a crewman might forget his articles number but would remember his bunk number. Whatever it may be called, this number identifies the duties and station of each officer and crewman in emergency situations.

Emergency Stations and Duties

Thus the particular station to which an officer or crew member reports and the duties he is expected to perform depend on both his locator number and the type of emergency that must be dealt with. The loud ringing of general alarm bells and the sounding of the whistle indicate whether he is to report to his fire and emergency station or his abandon ship station.

The emergency duties assigned to a particular crewman should, whenever possible, be similar to the normal work activity of that person. For instance, stewards department personnel should be assigned to assist passengers; deck department personnel should be assigned to run out hose and lifeboats; and the engineering department should be assigned to run out hoses in the machinery space, with which they are most familiar.

Signals

Fire and Emergency Stations. The fire signal is a continuous blast of the ship's whistle for not less than 10 seconds supplemented by the continuous ringing of the general alarm bells for not less than 10 seconds (Fig. 13.3). The NFPA recommends that a vessel in port and not under way sound five prolonged blasts on the whistle or siren to alert other ships and shore authorities that there is fire aboard. However, the proper fire signal is prescribed by the authorities of each port. When fire is discovered aboard a ship that is in port, it is imperative that the local fire department be summoned. Ship-to-shore radio, telephone or a pierside fire alarm box may be used for this purpose.

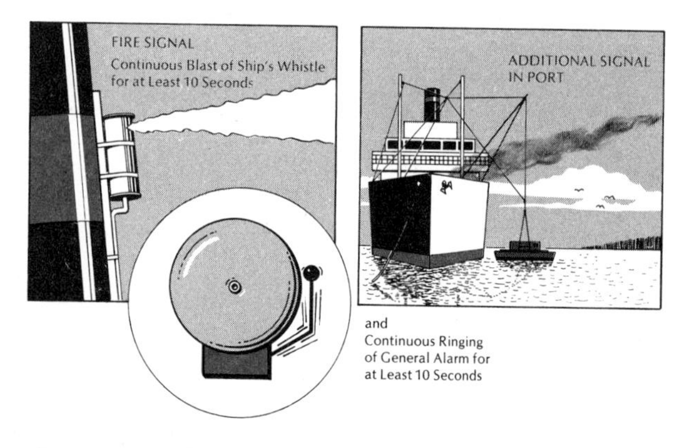

Figure 13.3. Ship's fire signal.

Ship's officers should be aware of the fire alarm procedures at each of the ports visited by their ship. When a ship is approaching port with fire aboard, the port's fire department and the U.S. Coast Guard should be notified by radio of that fact as contained in 33 CFR 124.16. The Coast Guard might want to give the ship special mooring instructions and the fire department can have specialized firefighting equipment ready, if necessary.

Dismissal from fire and emergency stations is signaled by the sounding of the general alarm three times supplemented by three short blasts on the whistle.

Boat Stations and Abandon Ship. The fire and emergency signal is sounded by the officer of the deck when fire is discovered. However, the sounding of the boat stations and abandon ship signal should be authorized only by the master (or by his replacement in case of illness or injury). The abandon ship signal is more than six short blasts and one long blast on the ship's whistle and on the general alarm bell. The master designates which boats are to be used. His instructions are communicated to officers and crew either by loudspeaker or by passing the word, or both.

The sounding of the boat stations and abandon ship signal does not authorize the lowering of boats. On the sounding of the alarm, those assigned to boat stations (by the station bill) must move quickly to those stations and await further instructions. One short blast on the whistle is the signal to lower away. Two short blasts on the whistle is the signal to stop lowering the boats. Dismissal from boat stations is signaled by three short blasts on the whistle. The U.S. Coast Guard *Manual for Lifeboatmen, Able Seamen and Qualified Members of the Engine Department* (CG175) contains detailed information on abandon ship procedures and lifeboat operation.

A decision to abandon ship because of fire should not be made hastily, even when the fire is severe. Vigorous and intelligent firefighting, maneuvering of the ship to take advantage of the wind until help arrives and the use of CO_2 flooding systems are alternatives to abandon ship procedures. In its *Fire Fighting Manual for Tanker-*

men (CG329), the U.S. Coast Guard cautions against recklessly abandoning a burning vessel. The manual points out that more lives have been lost by launching the boats too soon and by crewmen going over the side in panic than by remaining on board. It continues with the advice to fight the fire *always,* even if only as a rear guard action to increase the chances of survival.

Man Overboard. The man overboard signal is a hail and shout by the person who saw the victim fall: "Man overboard, port (or starboard) side!" The word should be quickly passed to the bridge. The person who witnessed the fall should try to keep the victim in sight while heaving life rings overboard in his general direction. Standard procedures for recovering a person overboard are described in the *Merchant Marine Officers' Handbook,* Knight's *Modern Seamanship* and *Manual for Lifeboatmen, Able Seamen and Qualified Members of the Engine Department* (CG175).

EMERGENCY SQUAD

An emergency squad is a group of crewmen selected by the master for their special training to deal with emergencies. The chief mate (assisted by the boatswain) is normally in command of the emergency squad. The rest of the squad should be made up of crewmen trained in the use of fire, emergency and rescue equipment. Candidates for the emergency squad would be crew members who are highly knowledgeable in emergency procedures and have earned certificates for their proficiency.

A mustering location for the emergency squad should be included in the station bill. The mustering location could be on either wing of the bridge, at a designated position on the main deck or wherever the master feels would be best. However, the chosen location should be one that the members of the squad can reach promptly—i.e., in less than 2 minutes. On larger vessels with more than one emergency squad, there should be a mustering location for each squad.

Mustering Signals

A special signal should be designated by the master to assemble the emergency squad. This signal should be one that will not be confused with the general alarm and navigational signals. Coded signals may be used to summon the emergency squad without alarming passengers. Coded signals also permit the use of a terminal number to designate one of several possible mustering locations. For instance, suppose the mustering signal is the sounding of the numbers 2–2–2. Then a fourth number, from 1–5, could indicate in which of five possible locations the emergency squad is to muster. (The use of more than five location numbers is not recommended, because of the increased possibility of a miscount.) Of course it is important that all squad members know the locations indicated by the terminal number. These should be posted with the station bill, in the areas where squad members usually work and on cards attached to their berths.

Signals of all sorts can be misinterpreted, and today most ships have loudspeaker systems. The use of these systems to muster the emergency squad avoids the possibility of misinterpreted signals. Even when the squad is summoned verbally, a code can be used to avoid alarming passengers. All that is required is a simple, easily recognized (and somewhat bland) name for the emergency squad, such as the "ready team" or "squad fire." Then an announcement like "The ready team will assemble on the port side of hatch number 3" would mean that the emergency squad should report promptly to that location with their equipment.

Training

An emergency squad is a team. A team is a group of people brought together to accomplish a common goal. The word *team* brings to mind the words *coordination, cooperation* and *training.* Training is absolutely essential, since without it there can be little coordination or cooperation. Training consists basically of two parts, which must come in the following order:

1. A teaching–learning process, in which the necessary knowledge is communicated to the trainee
2. Practice and demonstration of the necessary skills, using the proper equipment.

As an example, fire drills are practice and demonstration sessions. They must come after crewmen have learned what to do; otherwise, they can serve no purpose except to reinforce bad habits.

Under an able and understanding leader, proper training will gradually produce coordination and cooperation among members of the emergency squad. After several practice sessions they will indeed be operating as a team.

As mentioned earlier in this chapter, the master is responsible for all ship's functions, including those he assigns to subordinates. Thus, although the master assigns the training of the emergency squad (and the rest of the crew, for that matter) to his chief mate, he should review and approve the plans for proposed lessons and drills. These sessions are made more meaningful when the master personally observes them and then discusses them with the persons in charge.

The members of the emergency squad should attend periodic instructional sessions dealing with the variety of emergencies that could occur aboard ship. At each session, a problem could be presented, solutions discussed until a satisfactory one is found and the necessary tools and equipment handled for familiarity. Then the regularly scheduled fire drills would be demonstrations of efficiency rather than training sessions.

CREW FIREFIGHTING TRAINING

The emergency squad may be called upon to deal with many emergencies, such as collision, man overboard and a lost or damaged rudder; but when the fire signal is sounded, all hands are involved. The station bill lists an assigned task and station for each member of the crew. Therefore, all crew members should receive some training in firefighting.

All crewmen should receive instruction in how to transmit a fire alarm (Fig. 13.4). The type of alarm equipment carried and its locations will vary from ship to ship and may not be familiar to new members of the crew. The ship's interior phone system, its use for reporting fires and plans for its use to maintain communications during emergencies should be explained. Ships in port are usually in the greatest danger of fire, for a number of reasons (*see* Chapter 1 and 2). For the ships safety, every crew member must know how to summon land-based fire departments when in port. Since many crew members may be ashore, manpower is at its lowest and there may be no power available for the ship's pumps.

Everyone aboard should know how and when to use each type of fire extinguisher. Crewmen who are assigned to hoselines or to activate fixed firefighting systems require additional training. For example, a crewman assigned to a fire station should know how to couple hose, attach and operate a nozzle with spray and solid streams and use fog applicators.

The Coast Guard recently expanded its licensing and certification examinations to include more questions on firefighting and emergency equipment. Furthermore, the witnessing of fire drills conducted by the crew and the testing of firefighting equipment are vital parts of the vessel inspection procedure. Therefore, Coast Guard marine inspectors will require that crew members demonstrate their knowledge of the proper operation of the firefighting systems installed on their ship (Fig. 13.5). These are additional reasons for ensuring that every crewman is fully trained in fire fighting procedures, as far as they affect his assigned duties.

A master can consider himself fortunate if his crew includes hands who have received formal hands-on firefighting training. However, regardless of the background of the crew, shipboard

Figure 13.4. Crewmen must know how to **(A)** operate the ship's manual fire alarm system, **(B)** report a fire via the ship's telephone and **(C)** summon the land-based fire department when the ship is in port.

Figure 13.5. Knowledge of a firefighting system includes how to operate it, how it works, where it is used and when to use it.

training is necessary for both the emergency squad and the crew as a whole. Since the chief mate is usually in charge of the crew member assigned to handle an emergency situation, he is usually the training leader or instructor. However, when someone with experience in firefighting or teaching is available, that person might be assigned to direct firefighting instructional sessions. Then the chief mate and the instructor should together draw up lesson plans for the instructional sessions and prefire plans for combating fires in the various parts of the ship.

The Four-Step Instructional Method

There are, of course, a number of ways to familiarize students with firefighting equipment and teach the required skills. The four-step method outlined below has been used successfully on a number of ships. The instructor works through all four steps whenever a new topic is discussed or a new piece of equipment is operated. The *key points* are the basic steps in an operation. A half dozen or so key points, in the proper order, may have to be learned for the operation to be successful. Failure to perform one step properly might ruin the operation.

Step 1: Preparation. Find out how much the crew member (trainee) knows about the subject or the equipment under discussion. Arouse his interest. Encourage discussion.

Step 2: Presentation. Explain, illustrate and demonstrate the operation. Use the most effective types of instructional materials to discuss each part of the operation. Emphasize and illustrate the *key points.*

Step 3: Confirmation. Have the crew member actually handle the tool or equipment. Have him explain its operation. Make sure he repeats the *key points.*

Step 4: Demonstration. Have the crew member demonstrate what he has learned. He should use the tool, operate the equipment or (if the lesson is not concerned with manual operation) describe it aloud or in writing. The training session is not over until the crew member has demonstrated that he has learned what the lesson was supposed to teach (Fig. 13.6).

In teaching or in monitoring a crew member's work, the instructor must correct errors as they occur. This criticism must not be of a personal nature; there should be no "bawling out." Criticism should be directed at the work, not at the worker.

Guidelines for Course Planning

Even the most experienced teacher should prepare carefully for each lesson. The amount of preparation will depend on the subject matter and the type of lesson. However, the preparation should include the following:

1. The instructor acquires a list of the crew members who are to attend the session, by name and rating. He must have assurance that these crew members will not be called away for routine chores during the training session. He should set up the training session to last no more than 1 hour; training sessions become boring after an hour, and crew members end up learning little.
2. The instructor lists the information to be presented and picks out the key points. (*Safety precautions* are always key points.)
3. The instructor assembles everything he will need for the session. He selects the training area, perhaps on deck or in a cabin; if the lesson involves a particular space on the ship, he might have the session there. (If he wishes to use the engine room, he should first secure the permission of the chief engineer. Chief engineers, as a rule, do not welcome basket parties or conventions within their domain.) He then assembles the tools or equipment to be discussed at the session, a blackboard and other appropriate teaching aids. Finally, he ensures that pencils and notebooks will be available to all crew members attending the session.
4. Prior to the actual training session the instructor reviews the material he intends to present.

Figure 13.6. The four steps in the suggested instructional technique. The instructor performs the first two; seamen perform the last two under the direction of the instructor.

Sample Lesson (Oxygen Breathing Apparatus)

The following sample lesson illustrates the four-step instructional method and the pre-session preparation required of the instructor. Although the lesson deals specifically with breathing apparatus, the applicability of the method to other topics should be obvious. Prior to the session, the instructor assembles the following materials:

- One oxygen breathing apparatus, complete with canister and carrying case
- One lifeline
- Copies (one for each crew member) of a diagram of the operating cycle, showing how exhaled breath reaches the chemicals and produces oxygen
- At least one copy of this book. (Chapter 15 contains the information that the instructor will discuss.)

Step 1: Preparation. The instructor tells the crew members that he intends to teach the safe operation of the oxygen breathing apparatus. (For simplicity, in this lesson the apparatus will be referred to as the *OBA*.) He spells out the performance objectives of the lesson, and how he intends to measure them.

The instructor then asks whether any of the students have any knowledge of the OBA, what it is, what its purpose is and where it might be used. He encourages discussion by asking if anyone has had experience with the OBA in school, aboard this ship or on a previous ship. Throughout, he tries to arouse interest and get the crew members talking.

Step 2: Presentation. The instructor shows the OBA to the crew members, describes its construction and explains its operation. He uses a diagram to show the airflow through the device, and dons the OBA while explaining the procedure. He notes that a copy of the complete operating instructions is located on the inside cover of the carrying case, then emphasizes the key points:

1. The OBA must be donned properly. The fitting of the facepiece is critical—neither smoke nor gases can be permitted to enter.
2. A lifeline must always be attached to anyone entering a smoke-filled or oxygen-deficient compartment.
3. A canister must be inserted into the mask after its protective cap is removed.
4. The timer must be set and when its alarm sounds the crew member *must* leave the contaminated atmosphere.
5. The used canister must be removed and disposed of properly. A canister is good for one use only.
6. The OBA must never be used in a compartment that may contain flammable or combustible gas.

Step 3: Confirmation. Each crew member is allowed to examine the OBA and canister. Each is asked to repeat the key points. The instructor encourages seamen to ask questions and take part in the discussion of the device.

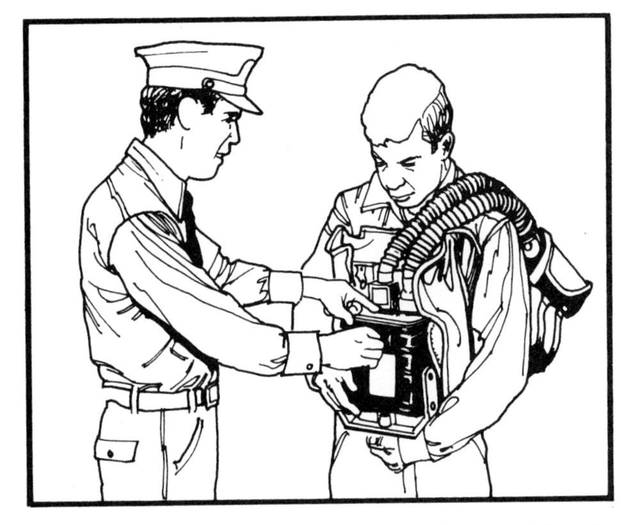

Figure 13.7. Instruction and practice in the use of breathing apparatus help develop confidence in the equipment.

Step 4: Demonstration. Each crew member is required to don the OBA, starting with removal of the device from its container. He then performs each of the steps recommended by the manufacturer of the OBA. The instructor corrects any errors as they are made (Fig. 13.7).

BIBLIOGRAPHY

Faria LE: Protective Breathing Apparatus. Bowie, Md, Robert J. Brady Co., 1975

Fire Department, City of New York. Training Bulletins.

Noel JV, Capt. USN: Knight's Modern Seamanship. 13th ed. Princeton, Van Nostrand, 1960

National Fire Protection Association. National Fire Codes. Standard No. 311: Ship Fire Signal. Boston, NFPA, 1977

Turpin EA, Mac Ewen WA: Merchant Marine Officers' Handbook. Cambridge, Md, Cornell Maritime Press, 1965

United States Coast Guard. Fire Fighting Manual For Tank Vessels. CG-329. Washington, DC, GPO, 1974

United States Coast Guard. Manual For Lifeboatmen, Able Seamen and Qualified Members of the Engine Department. CG-175. Washington, DC, GPO, 1973

United States Coast Guard. Manual For the Safe Handling of Flammable and Combustible Liquids and Other Hazardous Products. CG-174. Washington, DC, GPO, 1976

United States Coast Guard. Proceedings of the Marine Safety Council. Vol. 33. 10:177, 1976

Emergency Medical Care*

The medical emergencies that arise in firefighting situations are not limited to burns. They may range from simple skin scratches to life-threatening problems. The fire itself is, of course, a source of thermal burns. Inhaling smoke from the fire can poison the victim, but all the types of injuries normally associated with any accident situation can occur during firefighting, owing to the restricted work space, the rolling of the vessel, poor footing in water-soaked compartments and poor visibility due to smoke. In addition, smoke may cause respiratory arrest, and firefighters under strain may have heart attacks. Both require immediate action on the part of the rescuer. Even drowning in water-filled compartments is a possibility.

The rescuer must protect the patient from additional harm, correct life-threatening conditions, treat minor injuries and keep the patient stable until medical help can be reached. The rescuer's role includes:

- Removing the patient (victim) from any situation threatening his life or the lives of rescuers
- Correcting life-threatening problems and immobilizing injured parts before transporting the patient
- Transporting the patient in a way that minimizes further damage to injured parts
- Administering essential life support while the patient is being transported
- Observing and protecting the patient until a medical staff can take over
- Administering care as indicated or instructed.

* The material in this chapter has been adapted from Grant H, Murray R: Emergency Care. 2d ed. Bowie, Md, Robert J. Brady Co, 1978.

United States Coast Guard regulations ensure the presence of qualified rescuers aboard each vessel. The Coast Guard requires that every applicant for an original license as deck or engine officer aboard a U.S. merchant vessel possess a first aid certificate issued by the U.S. Public Health Service or a certificate of satisfactory completion of the American National Red Cross course in standard first aid and personal safety. Further, the applicant must have a currently valid card certifying that he has satisfactorily completed a course in cardiopulmonary resuscitation (CPR). These cards are issued by the American National Red Cross and the American Heart Association.

TREATMENT OF SHIPBOARD INJURIES

The officer on watch should administer first aid in the case of a life-threatening injury aboard ship. Otherwise, this officer should send the injured person to the ship's medical officer for treatment. The officer on watch should determine as nearly as possible the cause of the accident or the reason for the injury. He should enter the particulars in the watch log, along with the names of any witnesses. Additionally, an entry must be made in the ship's records of every injury reported to him, the patient's signs and symptoms, and the treatment administered.

Most steamship companies provide their ships with forms for reporting accidents that result in injuries to crew members. These forms must be made out by the officer investigating the accident. If an injury results in loss of life or in an incapacitation for more than 72 hours, the master must notify the nearest Marine Inspection Office of the U.S. Coast Guard. This notice must be followed by a report in writing and in person to the officer in charge of marine inspection at the port in which the casualty occurred or the port of first arrival.

Every shipboard injury should be investigated, at least with respect to its cause and possible corrective actions. The investigating officer should take statements from witnesses as part of this investigation. The master and the ship owner should ensure that any corrective action indicated by the investigation be taken promptly.

Emergency Care Supplies

First aid kits and emergency care supplies* should be carried on every vessel at all times. The ship's medicine chest should include at least the following items:

1. Splints for Immobilizing Fractures, i.e., padded boards of 4-ply wood, 7.6 cm (3 in.) wide, in lengths of 38, 91.6 and 137.2 cm (15, 36 and 54 in.), cardboard, plastic, wire-ladder, canvas-slotted, lace-on and inflatable splints.
   a. Triangular bandages for fractures of the shoulder and upper arm
   b. Short and long spine boards and accessories for safe removal of victims and immobilization of spinal injuries
2. Wound dressings
   a. Sterile gauze pads in conventional sizes for covering wounds
   b. Soft roller bandages 15.2 cm (6 in.) wide and 4.57 m (5 yd) long, for the application of large dressings, for securing pressure dressings to control hemorrhage, and for securing traction splints or coaptation splints (to join the ends of broken bones)
   c. Sterile nonporous dressings for closing sucking wounds of the chest (either plastic wrap or aluminum foil)
   d. Universal dressings, approximately 25.4 × 91.4 cm (10 × 36 in.), folded to 25.4 × 22.9 cm (10 × 9 in.), for covering large wounds including burns, and for compression, padding of splints, or application as a cervical collar
   e. Adhesive tape in widths of 2.5, 5.1 and 7.6 cm (1, 2 and 3 in.)
   f. Large safety pins
   g. Bandage shears

* For additional information regarding emergency care supplies, see Chapter 6 of "The Ships Medicine Chest and Medical Aid at Sea," HEW (H.S.A.) 78-20-24, U.S. Government Printing Office, Washington, D.C.

3. Sterile saline (for burns)
4. Supplies for acute poisoning:
   a. Activated charcoal
   b. Syrup of ipecac
5. Potable water for eye and skin irrigation
6. Oropharyngeal airways
7. Bag-mask resuscitators
8. Blood pressure monitoring apparatus.

DETERMINING THE EXTENT OF INJURY OR ILLNESS

Before the rescuer can begin emergency care, he must rapidly but effectively examine the patient to determine the seriousness of the illness or the extent of the patient's injuries. Many poorly trained attendants base emergency care only on the obvious injuries. This approach can be quite dangerous for the patient: an obvious injury may be relatively minor and pose no real threat to life, while a hidden and undetected injury may result in the patient's death.

The information available to the rescuer consists of *1)* what the patient tells him; *2)* what the crew and other witnesses tell him; *3)* what he is able to observe about the patient's obvious injuries; and *4)* what he is able to observe about how the injury was produced. Information from these sources is combined with a thorough check of the patient. The rescuer then may judge the extent of the injuries and prepare to administer the emergency care.

Classifying Injuries

Although accidents occur in many ways and for many reasons, each type of accident commonly produces certain "standard" injuries. For example, a firearms accident is generally expected to produce a soft-tissue injury, while broken bones usually result from falls. One seldom thinks of a fire as producing anything but a burn, and the only injury usually connected with poisoning is damage to internal organs. These are, however, only the obvious injuries. They often result from the accident, but they may not be the only types of injuries to have occurred.

Many secondary injuries, partially or completely concealed, may result from any accident if it is serious enough. For the rescuer to recognize all the patient's problems, he must have a complete understanding of the types of injuries, both obvious and hidden, that may be produced in an accident. Table 14.1 lists these injuries. The rescuer must realize that several injuries may be produced in any accident. He must look to the

Table 14.1. Types of Accidents and the Injuries They Produce

TYPES OF ACCIDENTS	THE EXPECTED INJURY (USUALLY OBVIOUS)	OTHER POSSIBLE INJURIES (NOT NECESSARILY OBVIOUS)
MOTOR VEHICLES	1,2,3,4	
FALLS	2,3	4
FIRES, EXPLOSIONS	1	2,3,4
SWIMMING AND BOATING	4 (DROWNING)	1,2,3
FIREARMS	1	2, 4
POISONING BY SOLIDS, LIQUIDS, GASES	4	1
MACHINERY	1,2,3,4	
ELECTRIC SHOCK	4 (CARDIAC ARREST)	1,2,3
	1 SOFT TISSUE INJURIES 2 FRACTURES 3 DISLOCATIONS 4 INTERNAL INJURIES	

mechanism of injury for clues as to the extent of physical damage. Thus, for example, a rescuer must never assume burns to be the only firefighting injury.

Diagnostic Signs and Their Significance

Diagnostic signs are a set of indicators that the rescuer should use in evaluating the patient's condition. With training and practice crewmen can use these signs to determine how best to provide emergency care. The basic diagnostic signs are

- Respiration
- Pulse
- Blood pressure
- Skin temperature
- Skin color
- Pupils of the eyes
- State of consciousness
- Ability to move
- Reaction to pain.

Each sign or combination of signs indicates something about the patient and what should be done to help him. The signs can be observed quickly, with minimal equipment.

Respiration. The normal adult breathing rate is about 12 to 15 breaths per minute. Both the rate and the depth of breathing are important. To determine the patient's breathing rate and depth, look, listen and feel for air exchange. Look for movement of the chest, and listen and feel for air exchange at the mouth and nose. Table 14.2 lists several respiration observations and the conditions they indicate.

Pulse. The pulse is an indication of heart action. The normal pulse rate in adults is 60 to 80 beats per minute. Pulse readings are generally taken at the wrist. However, this may be difficult in an emergency situation where there is a great deal of movement, or where shock has resulted in an extremely weak pulse.

Table 14.2. Interpretation of Respiratory Observations

Diagnostic Sign	Observation	Indication
Respiration	None	Respiratory arrest
	Deep, gasping, labored	Airway obstruction, heart failure
	Bright red, frothy blood with each exhalation	Lung damage

Table 14.3. Pulse Observations and Indications

Diagnostic Sign	Observation	Indication
Pulse	Absent	Cardiac arrest, death
	Rapid, bounding	Fright, hypertension
	Rapid, weak	Shock

To determine the pulse rate, place the fingers (not the thumb) over the carotoid artery in the neck or the femoral artery in the groin. Both these arteries are quite large, and both lie close to the surface. If no pulse can be detected at these points, listen to the patient's heart by placing your ear directly on the patient's chest or by using a stethoscope. Table 14.3 lists the major pulse observations and indications.

Blood Pressure. Blood pressure is the pressure that circulating blood exerts against the walls of the arteries. There are actually two different blood pressures, *systolic* and *diastolic*. Systolic pressure is the pressure exerted while the heart is contracted (when blood is being pumped through the arteries). Diastolic pressure is the pressure exerted while the heart is relaxed (when blood is returning to the heart). Both blood pressures are measured by a device called a sphygmomanometer, which is used in conjunction with a stethoscope.

The rescuer should take blood pressure readings as soon as he can after checking for and correcting any life-threatening emergencies. He should record the pressures and the time when they are first taken. If at all possible, he should continue taking blood pressure readings until he turns the patient over to the medical officer, physician or onshore rescue squad. Such a record is of help to the physician in determining the proper treatment.

Blood pressures are read in millimeters of mercury (mm Hg). Although blood pressure levels vary with age and sex, there is a useful rule of thumb: Normal systolic pressure for men is 100 plus the age of the patient; their normal diastolic pressure is 65 to 90. For women, both pressures are usually 8 to 10 mm lower than those of the man.

To measure blood pressure, secure the cuff of the sphygmomanometer around either arm of the patient, just above the elbow. Follow the directions on the cuff for the proper placement of the pressure diaphragm over the artery. Find the brachial artery by palpating the arm in front of the elbow.

Close off the valve on the bulb. Inflate the cuff with the rubber bulb until the needle of the dial stops moving with the pulse. (This is usually a point between 150 and 200 on the dial.)

Place the stethoscope diaphragm over the artery in front of the elbow, and slowly release air from the bulb by opening the valve. The point on the dial at which the first sounds of a pulse are heard through the stethoscope is the *systolic* pressure.

Continue to release air from the bulb slowly, while listening through the stethoscope. The point on the dial at which the pulse sound begins to fade and disappear is the *diastolic* pressure. Record the pressures by writing the systolic pressure over the diastolic pressure, as in 140/70 (Table 14.4).

Table 14.4. Blood Pressure Observation and Indication

Diagnostic Sign	Observation	Indication
Blood Pressure	Marked drop	Shock

Skin Temperature. Because the skin regulates the body temperature, changes in skin temperature indicate changes occurring within the body.

To determine the patient's skin temperature, feel his skin surface at several locations with your hand. Use the back of your hand, since it is more sensitive to temperature changes than the roughened fingers (Table 14.5).

Table 14.5. Skin Temperature Observations and Indications

Diagnostic Sign	Observation	Indication
Skin Temperature	Hot, dry	Excessive body heat (as in heat stroke), high fever
	Cool, clammy	Shock
	Cold, moist	Body is losing heat
	Cool, dry	Exposure to cold

Skin Color. Skin color is determined mainly by the blood circulating in blood vessels just below the skin. Thus, changes in color reflect an increase or decrease in the blood flow, or changes in the blood chemistry. However, darkly pigmented skin will obscure color changes.

Carefully examine the patient's face and hands for areas of abnormal skin color. Note whether the skin appears red, white or blue (Table 14.6).

Table 14.6. Skin Color Observations and Indications

Diagnostic Sign	Observation	Indication
Skin Color	Red skin	High blood pressure, carbon monoxide poisoning, heart attack
	White skin	Shock, heart attack, fright
	Blue skin	Asphyxia, anoxia, heart attack, poisoning

Pupils of the Eyes. The pupils of the eyes are good indicators of the condition of the heart and central nervous system. When the body is in a normal state, the pupils are the same size, and they are responsive to light. Changes and variations in the size of one or both pupils are important signs for the rescuer, especially in determining whether or not the patient is in cardiac arrest. In examining the patient's pupils, the rescuer should always consider the possibility that the patient wears contact lenses or has a glass eye.

Examine the pupils by gently sliding back the upper lids. Note whether the pupils are dilated (wide) or constricted (narrow). Examine both pupils, since some medical problems cause the pupils to be unequal in size. If the pupils are dilated, check their response to stimuli by flashing a light across them. In death, the pupils will not respond to light (Table 14.7).

Level of Consciousness. The normal, healthy person is alert, oriented, and able to respond to vocal and physical stimuli. A patient who is alert at first and then becomes unconscious may have suffered damage to the brain.

Carefully note the patient's level of consciousness when you first see him. Record any changes

Table 14.7. Pupil Observations and Indications

Diagnostic Sign	Observation	Indication
Pupils of the Eyes	Dilated	Unconsciousness, cardiac arrest
	Constricted	Disorder affecting the central nervous system, drug use
	Unequal	Head injury, stroke

Table 14.8. Levels of Consciousness

Diagnostic Sign	Observation	Indication
State of Consciousness	Brief unconsciousness	Simple fainting
	Confusion	Alcohol use, mental condition, slight blow to the head
	Stupor	Severe blow to the head
	Deep coma	Severe brain damage, poisoning

Table 14.9. Paralysis Observations and Indications

Diagnostic Sign	Observation	Indication
Paralysis or Loss of Sensation	Lower extremities	Injury to spinal cord in the lower back
	Upper extremities	Injury to spinal cord in the neck
	Limited use of extremities	Pressure on spinal cord
	Paralysis limited to one side	Stroke, head injury with brain damage

in consciousness, and relay this information to the physician (Table 14.8).

Paralysis or Loss of Sensation. When a conscious patient is unable to move his limbs voluntarily, or if they do not move when stimulated, the patient is said to be paralyzed. Paralysis may be caused by certain medical disorders (such as stroke) or by injury to the spinal cord. A patient suffering from paralysis does not feel or respond to pain in the affected parts. With some injuries paralysis is not complete, and the patient may have limited use of his extremities. In these cases, the limbs feel numb or there is a tingling sensation. It is important for the rescuer to remember that paralysis and loss of feeling are signs of probable injury to the spinal cord (Table 14.9). *The patient should not be moved until he is rigidly immobilized, since to do so might worsen the spinal injury.*

To determine if there is any paralysis, first ask the patient whether he has any feeling in his arms or legs; then ask him to move them. Do not move his limbs for him; see if he can do it by himself.

Reaction to Pain. Pain is a normal reaction to injury and a good indication of the location of an injury. However, certain injuries and medical disorders may interrupt this normal reaction. Ask the patient where he feels pain or discomfort. This information, along with observation of the patient and knowledge of the mechanism of injury, can indicate the type of injury (Table 14.10).

Table 14.10. Reaction to Pain: Observations and Indications

Diagnostic Sign	Observation	Indication
Reaction to Pain	General pain present at injury sites	Injuries to the body, but probably no damage to the spinal cord
	Local pain in the extremities	Fracture, occluded artery to extremity
	No pain, but obvious signs of injury	Spinal cord damage, hysteria, violent shock, excessive drug or alcohol use

EVALUATING THE ACCIDENT VICTIM

The rescuer must be able to *1)* rapidly evaluate the seriousness of obvious injuries, and *2)* analyze all other information to determine whether or not the patient has other, less obvious injuries. Here again, the rescuer must understand accidents and the injuries they can produce, the mechanisms of injury and the diagnostic signs and their significance.

One other tool is available to the rescuer—an actual survey of the patient. By combining the results of this survey with the other available information, the rescuer can analyze the patient's total condition accurately. The survey is divided into two parts. The *primary* survey is a search for immediate life-threatening problems. The *secondary* survey is an evaluation of other injuries, which do not pose a threat to life.

The Primary Survey

While several conditions can be considered life-threatening, two require immediate attention: respiratory arrest and severe bleeding. The need for immediate action in both these cases is obvious. Respiratory arrest sets off a vicious chain of events leading to cardiac arrest and then to death. Severe and uncontrolled loss of blood leads to an irreversible state of shock and again to death. In both instances, death will occur in a very few minutes if no attempt is made to help the patient. Thus the rescuer should begin the primary survey as soon as he reaches the patient. No diagnostic equipment is required for the survey; if one or more of the life-threatening conditions are found, the rescuer can start basic life-saving measures without delay.

Throughout the primary survey, the rescuer should be especially careful not to move the patient around any more than is absolutely necessary to support life. Unnecessary movement or rough handling might worsen undetected fractures or spinal injuries.

Check for Adequate Breathing. First, establish an open airway. Then look for the chest movements associated with the breathing process. At the same time, listen and feel for the exchange of air at the patient's mouth and nose.

If there are no signs of breathing, begin artificial ventilation *immediately* by the mouth-to-mouth or mouth-to-nose method. Do *not* leave the patient to get a resuscitator or other device. *Every second is critical to the patient.* If the patient does not start to breathe after three to five ventilations, go immediately to the next step in the primary survey.

Check for a Pulse. Check for heart action by feeling for the cartoid pulse in the patient's neck. The cartoid artery is a large vessel that lies close to the surface; it is easy to find in an emergency situation. If a pulse is present, continue artificial ventilation until the patient starts to breathe again.

If there is no pulse, immediately start cardiopulmonary resuscitation (CPR). Once again, *do not hesitate;* the patient's condition is very critical. Every second that the brain is without oxygenated blood, the chances for recovery decrease sharply.

At this point, the rescuer may decide to transfer the patient to ship's hospital, owing to his grave condition. CPR should be continued without interruption while the patient is being transported. However, if resuscitation efforts are immediately successful, the rescuer can go on to the next step in the primary survey.

Check for Severe Bleeding. Examine bleeding injuries carefully to determine whether they are actually as severe as they may appear. Many bleeding wounds that seem serious may be trivial. Control serious bleeding by direct pressure or by finger pressure on a pressure point. Use a tourniquet only as a last resort, when all other attempts to control the bleeding have failed.

Check for Other Obvious Injuries. At this point, life-support measures should have stabilized the patient; in most cases, the emergency will be over. Of course, there may still be problems that could later pose a threat to life, but they will not be as pressing as respiratory arrest and severe bleeding. Attention should now be directed to the other obvious injuries, in the order of their importance. Chest or abdominal wounds should be sealed, lesser bleeding wounds dressed, open fractures immobilized, and burns covered. The watchword is still "careful handling," so that unseen injuries are not worsened. When the obvious injuries are treated, the secondary survey begins.

The Secondary Survey

The purpose of the secondary survey is to find the additional unseen injuries that can often be worsened by mishandling. Examples are the closed fracture that is converted to an open fracture when the patient is moved to a litter, and the spinal injury that causes damage to the spinal cord when the patient is helped to his feet. Actually, the secondary survey is a head-to-toe examination during which the rescuer checks very carefully for specific injuries. It is conducted in the following manner.

Check for Scalp Lacerations and Contusions. Look for blood in the hair. If blood is present, separate the hair strands gently to determine the extent of the bleeding. Be very careful not to move the head while checking for scalp wounds, in case the neck has been injured. To check the part of the scalp that is hidden as the patient lies on his back, first place your fingers behind his neck. Then slide them upward toward the top of his head. This action develops a little traction, which is helpful if there is a neck injury.

Check the Skull for Depressions. Gently feel for depressions and protruding bone fragments. Again, be very careful not to move the patient's head any more than absolutely necessary.

Check the Ears and Nose for Fluid and Blood. Look in the ears and nose for blood or clear, waterlike fluid. The presence of either or both of these liquids indicates a possible skull fracture and damage to the brain. Blood, of course, comes from the lacerated brain tissue. The clear fluid is the cerebrospinal fluid that surrounds the brain and cushions it from shock. Blood in the nose alone, however, may mean only that the nasal tissue has been damaged.

Check the Neck for Fractures. Look and feel gently for deformities or bony protrusions in the neck. Normally the neck is symmetrical (even on both sides). However, sharp movement from side to side can separate the bony structures of the spinal column in the neck. In this case, you will notice that the head is in an abnormal position. If so, do not continue any further with this check. Immediately stabilize the patient's head with a cervical collar, rolled towels or a similar restraint. If the patient is conscious, tell him not to move his head, even slightly. A further check will provide information as to whether or not the spinal cord is damaged.

Check the Chest for Movement on Both Sides and for Fractures. From a position at the head of the patient, look to see if the chest is rising and falling in the normal manner. If the sides are not rising and falling together (one side may not be rising at all), there may be rib and lung damage. Gently feel the chest cage for broken ribs. Besides the depressions that are felt easily, a grating feeling may be caused by the movement of broken rib ends against each other.

Check the Abdomen for Spasms and Tenderness. Gently press against the abdomen. A "rocklike" abdomen or spasms indicate internal bleeding or a condition in which the contents of the internal organs have spilled into the abdominal cavity.

Check the Pelvic Area for Fractures. Look for swelling and discoloration, which are signs of a closed fracture. Feel for lumps and tenderness. Look for anything abnormal, for example the leg twisted too far to the side. Ask the patient if he has any intense pain in a particular area.

Check for Paralysis of the Extremities. Paralysis is a sign of spinal-cord damage. As a general rule, if there is no paralysis in the arms, but the legs are paralyzed, the back is broken. Otherwise, the spinal cord is intact. There are four ways to test a conscious patient for spinal-cord injury.

First, ask the patient if he has any sensation in his arms and legs. If he is able to feel the touch of your hand on his arms and legs, he probably does not have any spinal-cord damage. However, if he complains of numbness or a tingling sensation in his arms and legs, you should immediately suspect spinal-cord damage. In either case, carry out the rest of your survey.

Next, have the patient move both feet. If he can do so, it is a good indication that there is no spinal-cord damage. To be sure, ask him to raise his legs slightly, one at a time (only, of course, if he has no leg fractures). If he cannot, you must assume that he has suffered injury somewhere along the spinal cord.

To locate the general area of the injury, ask the patient to wiggle his fingers. If he can, have him raise his arms one at a time (again, only if no fractures are present). Then ask him to grip your hand as though he were going to shake it. If the patient cannot do this, his spinal cord is probably injured in the area of the neck. Lack of feeling and movement in the legs indicates spinal-cord damage in the lower back.

When paralysis points to some type of spinal-cord injury, immobilize the patient's entire body immediately. Use a long spine board, an orthopedic stretcher or some other long, rigid device. Remember that this is a very dangerous situation; any wrong movement might result in permanent paralysis or death.

Naturally, an unconscious patient cannot respond in the tests just described. However, you can check the condition of the spinal cord by pricking the skin of the hands and the soles of the feet (or the skin of the ankles above shoes) with a sharp object such as a pin. If there is no cord damage, the muscles will react and the arm or leg will jump. If the cord is damaged, there will be no reaction. As in the case of the conscious patient, a lack of reaction in the arms and hands indicates damage to the spinal cord in the neck. A reaction in the arms but not in the legs and feet indicates damage in the lower back.

Check the Buttocks for Fractures or Wounds. In many accident cases the buttocks go unchecked, even though they may have suffered serious injury. Feel carefully for irregularities in the body structure. Check for bleeding wounds that might not be obvious if the patient is lying on his back. If the check for paralysis has indicated possible spinal-cord damage, check the buttocks with as little movement of the body as possible. Otherwise, you may shift the patient slightly to allow a closer check.

TRIAGE

Triage is the sorting of accident victims according to the severity of their injuries. The reason for triage is simple: If patients are selected for treatment at random, those with minor injuries may be treated before those who have life-threatening problems. Some rescuers, when confronted with several accident victims, make the mistake of automatically caring first for the one who screams the loudest. However, the loud patient may have only minor cuts, while the quiet patient may be seriously injured or dying due to respiratory arrest, internal bleeding or deep shock. Another common error in a multiple patient situation is treating first the injuries that *appear* to be the most serious. A patient whose head is completely covered with blood looks grotesque and seems very seriously injured. In fact, he may have nothing more than a small, superficial cut on the scalp. On the other hand, a patient who appears to have only a slight chest wound may really have a punctured lung, which could cause him to bleed to death internally.

Accident victims should be sorted into three groups and treated according to their injuries: *1)* Those with high priority injuries; *2)* those with second priority injuries, and *3)* those with low priority injuries.

High Priority Injuries

- Airway and breathing difficulties
- Cardiac arrest
- Uncontrolled bleeding
- Severe head injuries
- Open chest or abdominal wounds
- Severe medical problems, such as poisoning or heart attacks
- Severe shock.

Second Priority Injuries

- Burns
- Major multiple fractures
- Back injuries with or without spinal-cord damage.

Low Priority Injuries

- Minor fractures
- Other minor injuries
- Obviously mortal wounds in which death appears reasonably certain
- Obvious death.

HEAD, NECK AND SPINE INJURIES

Serious head, neck and spine injuries can result during shipboard firefighting operations, especially as compartments become soaked with water. The risk of such injuries is increased if firefighters are not wearing protective headgear. If the vessel is rolling, firefighters may be thrown against bulkheads, equipment and cargo; this type of accident often causes head and neck injuries. In addition, one firefighter can injure another by the improper use of firefighting equipment. As the intensity of the fire situation increases, the chance for such injuries increases dramatically.

Signs of Skull Fracture

Many skull fractures can be diagnosed only with X rays. However, there are several important signs you can look for if a head injury is suspected but there are no obvious wounds:

- A deformity of the skull must be considered to be the result of a fracture until it is proved otherwise.
- Blood or a clear, waterlike fluid in the ears and nose is a good sign of a skull fracture.
- Discoloration of the soft tissues under the eyes may be present.
- Unequal pupils are an important sign of brain damage.

Evaluating a Patient for Brain Injuries

If the mechanism of an accident is sufficient to cause a skull injury, it is probably also sufficient to cause an injury to the brain. Several factors must be considered in determining whether the patient has suffered brain damage.

State of Consciousness. If the patient was unconscious immediately after the accident but then regained consciousness, he probably suffered only a brain concussion. Further damage to the brain is indicated if the patient gradually lost consciousness, or if he regained and then lost consciousness again. A blood clot may be causing pressure on the brain.

Awareness of Surroundings. Pressure on certain brain centers due to the injury may interrupt their function, causing disorientation, amnesia, or other similar reactions.

Condition of Pupils. Normally the pupils of the eyes are equal in size, and they constrict when exposed to bright light. Their unequal size or failure to react to light indicates that the brain is not functioning properly. If one pupil remains large when exposed to light while the other pupil constricts, damage to one side of the brain is indicated. A small flashlight (penlight) can be used to test the patient's pupils.

Emergency Care for Injuries to the Skull and Brain

In the case of a suspected or known skull or brain injury, the rescuer should proceed as follows.

- Maintain an open airway.
- Check for and stabilize associated neck injuries.
- Do not attempt to control drainage.
- Cover open wounds, but use little pressure.

- Do not remove impaled objects.
- Transport the patient without delay, but very carefully, to minimize movement and avoid bumping the head.
- Administer 100% oxygen during transportation (qualified personnel only).

The maintenance of an open airway is of primary importance for all injuries. It is doubly important in the case of the patient with a head injury, since this injury involves loss of the oxygen-carrying blood where it is needed most. Since the accident that caused the brain injury may also have produced a neck injury, the usual method of establishing an airway cannot be attempted. Tilting the head back too far might cause the death of a patient who has suffered a broken neck. The proper method is described below.

Neck Injury

A patient who has a head injury must always be suspected of having a neck injury as well. If the patient is unconscious, he should be treated as though he actually had a broken neck.

Surveying the Patient for Spinal Damage

The survey may be limited by the position in which the patient is first found. If he is found on his side, for example, the work of the rescuer is uncomplicated. The rescuer can closely examine the spine for deformity, lacerations and contusions, tenderness and other physical signs of injury. When the extent of the injury has been determined, the patient can be rolled very carefully into the proper position for immobilization and transportation. However, if the patient is found lying on his back, it is impossible to see many of the signs of injury. Moreover, it would be very dangerous to move him just for the purpose of visual examination. Instead, he should be examined for paralysis as described earlier in the section, under *Check for Paralysis of the Extremities.*

The rescuer may not find any obvious signs of spinal-cord injury but still suspect that there is one. If there is any question at all, the patient should be immobilized and transported as though he had a known spinal fracture, especially if he is unconscious. The following signs and symptoms are associated with spinal injuries.

Pain and Tenderness. The conscious patient feels pain and is able to point out the injury site. If the patient is on his side or stomach, run your fingers gently over the area of the suspected injury. When the fingers are directly over the injured area, the patient usually will complain of an increase in pain. If the patient attempts to move the injured area, the pain may also increase sharply. If there is no pain upon movement, there is probably no fracture or dislocation. (Observe this sign only if you have the opportunity to do so. Do not ask the patient to move merely to determine if there is pain. This action may worsen the injury.) If the patient is unconscious, the rescuer must rely on other means of determining the extent of the injury.

Deformity. In some cases the spine may appear crooked and bent out of shape. Of course, this sign usually may be observed only if the patient is on his side or stomach. If the patient is in the proper position, and heavy clothing does not conceal the injury site, run your fingers gently up and down the spine, feeling for bony protrusions. Do this carefully, so that the spine is not twisted further. Deformity is usually a very reliable sign of spinal injury; in the unconscious patient it may be the only reliable sign available. Remember, however, that there may be a spinal injury without obvious deformity.

Cuts and Bruises. Cuts and bruises on the face and neck are commonly associated with spinal fracture or dislocation. A patient who has bruises over his shoulders, or lower back and abdomen, may also have injuries to the spine. These signs may be observed in both conscious and unconscious patients. Again, bear in mind that there may be a spinal injury without these signs.

Paralysis. Probably the most reliable sign of spinal injury is paralysis of the extremities. To determine the presence and extent of paralysis, the rescuer should perform the tests listed under *Check for Paralysis of the Extremities,* or refer to Table 14.11.

Table 14.11.

A SUMMARY OF OBSERVATIONS AND CONCLUSIONS

	Observations	Conclusions
Legs	Can feel touch Can wiggle toes Can raise legs	Patient may not have a cord injury
Arms	Can feel touch Can wiggle fingers Can raise arms	
Legs	Cannot feel touch Cannot wiggle toes Cannot raise legs	Patient probably has an injury to the cord below the neck.
Arms	Can feel touch Can wiggle fingers Can raise arms	
Legs	Cannot feel touch Cannot wiggle toes Cannot raise legs	Patient probably has an injury to the cord in the area of the neck.
Arms	Cannot feel touch Cannot wiggle fingers Cannot raise arms	

Emergency Care for Injuries to the Neck and Spine

In the case of a known or suspected neck or spine injury, the rescuer should proceed as follows.

- Apply and maintain traction on the head.
- Restore the airway and ensure adequate breathing.
- Control serious bleeding by direct pressure.
- Make a complete body survey.
- Immobilize the patient before moving.
- Take sufficient time.
- Administer 100% oxygen during transportation (qualified personnel only).

Apply and Maintain Traction. Regardless of the position in which the patient is found, one rescuer should immediately station himself at the head, where he can apply traction. He should grasp the patient's head with his fingers under the chin, and exert a steady pull upward and slightly to the rear. The upward pull helps to keep any broken bone fragments from overriding and severing or doing further damage to the spinal cord. The slight backward pull helps to open and maintain the patient's airway. The rescuer applying traction must remain in this position until the patient is completely immobilized or until it has been determined that he has no spinal injuries.

A cervical collar should be applied at the same time that the head is placed in the traction position. The collar ensures that the patient's head does not fall forward or roll to the side if the rescuer must remove his hands for any reason. If a cervical collar is not available, a substitute collar can be fashioned from a bath towel, a folded multitrauma dressing or a folded blanket. When applying a substitute collar, the rescuer should ensure that the patient's head is not moved any more than is absolutely necessary.

Restore the Airway and Ensure Adequate Breathing. The usual method of tilting the head as far back as possible to establish the airway may worsen a spinal injury. Instead, when traction has been applied, resuscitation measures can be started if necessary. First attempts should be directed toward ventilating the patient's lungs. With one rescuer holding the patient's head in the traction position, another should try to ventilate the patient by the mouth-to-mouth technique or with a positive-pressure device such as a bag-mask resuscitator. Even though the head is not tilted in the best airway position, ventilation may be successful.

If the airway is not opened sufficiently to allow ventilation, do not attempt to increase the opening by tilting the head further back. Instead, use either the chin-lift method or the jaw-lift method of creating an airway. Both methods can be performed by the rescuer who is maintaining traction, and they will actually aid in the traction efforts.

If the patient is unconscious, an oropharyngeal airway such as the S tube can be inserted. This will ensure a proper exchange of air and minimize the problem of the relaxed tongue blocking the throat. Great care must be taken while inserting the airway, to avoid undue neck movement. The airway can be left in place during transportation, and it will be useful if the patient must be resuscitated. It should be remembered, however, that when the patient regains consciousness the airway must be removed immediately. Left in place, it will cause the patient to gag and vomit, creating additional problems and placing an added strain on injured parts. The throat of an unconscious patient should be suctioned periodically, to remove collected fluids.

Control Serious Bleeding by Direct Pressure. When breathing problems have been corrected, attention should be directed to any seriously bleeding wounds. Pressure dressings should be placed on these wounds, but with as little movement of the body as possible. If the bleeding cannot be stopped by pressure dressings, one rescuer may have to maintain manual pressure on the wound while the patient is being immobilized. No attempt should be made to dress less serious wounds before immobilization is completed.

Make a Complete Body Survey. While one rescuer is holding the patient's head in the traction position, another rescuer should make a complete body survey using the method described earlier. If the survey indicates that there is no cord damage anywhere along the length of the spine, attention can be given to other injuries, and the normal treatment sequence can be followed. On the other hand, if the survey shows spinal-cord damage, the patient must be rigidly immobilized before anything else is done.

Immobilize the Patient Before Moving. The method of immobilization depends upon the circumstances of the accident. If the patient is lying on the deck and is easily accessible, there is little problem in preparing him for transportation. An orthopedic stretcher or a full backboard is the most desirable device for immobilization. The patient can be immobilized on either of these stretchers with a minimum of body movement. More detailed information on the removal of victims with neck and spine injuries is given later in this chapter.

RESPIRATION PROBLEMS AND RESUSCITATION

If for some reason the air supply to the lungs is restricted or stopped, the brain does not get enough oxygen to survive. Then the brain signals that regulate heart and lung activity slow down and stop. As the actions of the brain, heart and lungs cease, so does life itself. Thus, rescuers must act promptly when they find that a patient's respiration is blocked or stopped.

Airway Obstructions

The most obvious airway obstruction is an accumulation of foreign matter in the mouth, throat or windpipe. Vomit, blood, phlegm, and foreign objects that cannot be coughed up or swallowed tend to create dangerous obstructions.

A less obvious but equally dangerous airway obstruction results from unconsciousness. During unconsciousness, regardless of the cause, the muscles that control the lower jaw and tongue relax. This usually leads to an obstruction of the throat when the patient's neck is bent forward. The bending of the neck causes the lower jaw to sag. Since the tongue is attached to the lower jaw, it drops against the back of the throat and over the voice box, blocking the airway.

Recognition of Airway Obstruction

A rule of thumb that may be used to survey a patient for airway obstruction is to tilt the patient's head backward and

- *Look* for breathing movements.
- *Listen* for airflow at the mouth and nose.
- *Feel* for air exchange.

The rescuer should *not* assume that a patient is breathing adequately unless he can hear and feel an exchange of air through the mouth and nose, and see that the chest is rising and falling. For this, he should place his ear close enough to the patient's mouth and nose to hear and feel the exchange. In cases of complete obstruction, there will be no detectable movement of air. Cases of partial obstruction are easier to detect and may be identified by listening. Noisy breathing is a sign of partial obstruction of the air passages. "Snoring" usually indicates air-passage obstruction by the tongue, as in the case of a bent neck. "Crowing" indicates spasms of the larynx (voice box). A "gurgling" sound indicates foreign matter in the windpipe. Under no circumstances should a "noisy" breathing condition go untreated.

Cyanosis

A dependable sign that the brain is getting too little oxygen is cyanosis. This condition is characterized by a noticeable blue or gray color of the tongue, lips, nail beds and skin. In blacks and other patients with dark complexions, the blue or gray color may be noted at the tongue and nail beds only.

Treatment of Airway Obstructions

The head should be kept tilted back throughout all the following steps. If any step opens the airway, it is not necessary to go any further. But it is necessary to ensure that the patient continues to breathe properly.

Step 1: Quickly Clean Out the Patient's Mouth. Using your finger, quickly sweep the patient's mouth clear of foreign objects, broken teeth or dentures, sand or dirt, and so forth.

Step 2: Tilt the Patient's Head Back. Place the patient on his back with his face up. Tilt the patient's head backward as far as possible, so that the front of the neck is stretched tightly (Fig. 14.1). If necessary, elevate the patient's shoulders with a blanket roll, to keep the head tilted back. Never put a pillow, rolled blanket or other object under the patient's head. This will defeat the purpose of the head tilt by bending the neck forward and perhaps blocking the airway even more.

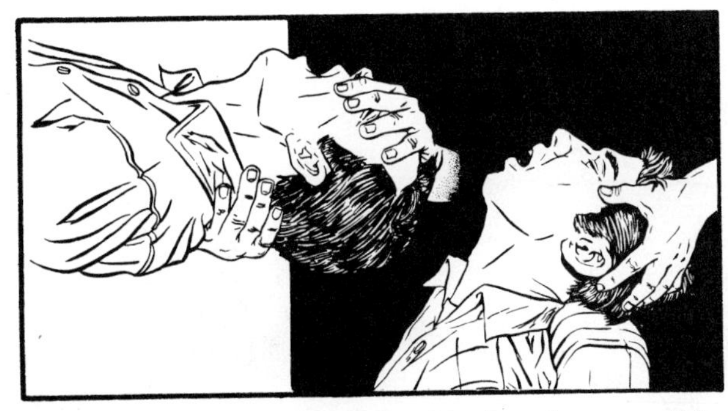

Figure 14.1. Tilt the patient's head back as far as possible to open the airway.

If the airway opens and the patient starts to breathe alone when the head is tilted back, go no further. Otherwise go on to step 3.

Step 3: Force Air into the Lungs. If the head tilt does not open the airway, try to force two or three good-size breaths quickly into the patient's lungs through the mouth, while holding the nostrils pinched shut (Fig. 14.2). This forced ventilation may be enough to start spontaneous respiration, or it may dislodge a partial obstruction that has been restricting breathing.

Watch the patient's chest for movement indicating that your breaths are reaching his lungs. If the patient's chest rises and falls with two or three quick breaths, the airway is unobstructed. If forcing air into the patient's mouth does not open the airway, it is necessary to go on to the following steps.

Step 4: Lift the Jaw. If both the head tilt and the forced ventilation fail to get air into the patient's lungs, it may be necessary to increase the stretch of the neck to get the tongue out of the way. A jaw-lift method can be used.

To pull the tongue as far forward as possible, insert your thumb between the patient's teeth, with your fingers under his chin. Pull his jaw forward, allowing the lower teeth to be positioned higher than the cutting edges of the upper teeth (Fig. 14.3). Take care not to hold or depress the tongue during this procedure.

If it is not possible to insert your thumb in the patient's mouth because of clenched teeth, try the two-hand jaw lift. Grasp the angles of the patient's jaw. With both hands just below the earlobes, lift the jaw forcibly upward so that the lower teeth are in front of the upper teeth. Be sure that you do not flex the head

Figure 14.2. Force several good-size breaths into the patient's lungs while pinching his nostrils closed.

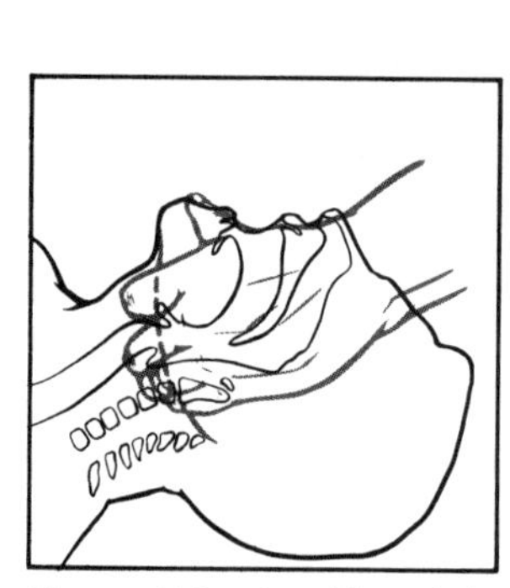

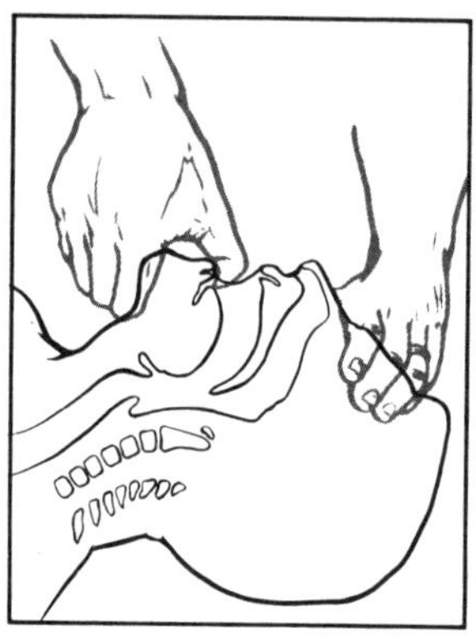

Figure 14.3. Jaw lift and chin lift.

forward when you attempt to pull the jaw forward.

The tongue is now in the extreme forward position, and it is unlikely to be blocking the air passage. Another quick breath into the patient's mouth will determine whether or not the airway is clear. If it is clear, artificial ventilation may be carried out. If the airway is still not open, go on to step 5.

Step 5: Clear the Airway. When steps 2–4 all fail, an object is probably lodged so deeply in the patient's throat that the quick sweep of the mouth in step 1 failed to reach it. Try to reach the object with your extended index finger. If this fails, attempt to dislodge the object by concussion.

Turn the patient on his side, and administer a few sharp slaps to his back between the shoulders. Once again, sweep your fingers inside the patient's mouth to see if the object has been dislodged.

If sharp slaps between the shoulders fail to clear the obstruction, you can attempt the abdominal thrust shown in Figure 14.4. However, if the patient is in desperate condition, do

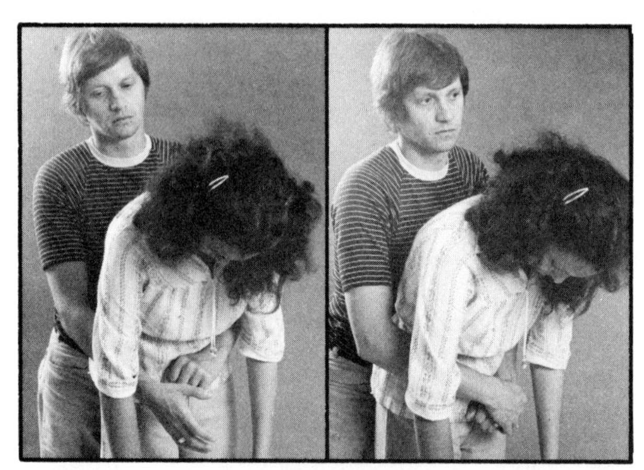

Figure 14.4. Abdominal thrust (Heimlich maneuver). A. Positioning the hands. B. Applying the force. C. The maneuver for a patient in the prone or supine position.

not waste time trying to clear foreign matter from the airway. Forcing air into his lungs is more important, and this often succeeds despite some blockage. Speed is of the essence. If it is obvious that efforts to open the airway will not be immediately successful, radio for medical assistance and arrange for transport of the patient to a medical facility without delay. Surgical procedures will most likely be needed to save his life. Meanwhile, repeat the five-step procedure.

Mouth-to-Mouth Resuscitation

This technique has been proved both experimentally and clinically to be the most effective means of artificially ventilating a nonbreathing patient.

With the air passage maintained by maximum extension of the head (as described in the preceding section), pinch the patient's nose shut with your thumb and forefinger. This will prevent air from escaping when you blow into the patient's mouth. Take a deep breath, open your mouth wide, and place it over the mouth of the patient, making a tight seal (Fig. 14.5). Quickly blow your full breath into the patient's mouth until you can feel the resistance of the expanding lungs and can see the chest rise. Remove your mouth from the patient's mouth, and allow him to exhale.

Repeat this breathing cycle every 5 seconds, or about 12 times per minute. Each breath should provide at least 100 cc (about 2 pints) of air. This is twice the amount of air in a normal breath. Expired air contains about 16% oxygen and from 4% to 5% carbon dioxide. Thus, the double-size breaths ensure adequate oxygenation of the blood and removal of the carbon dioxide from the patient's lungs.

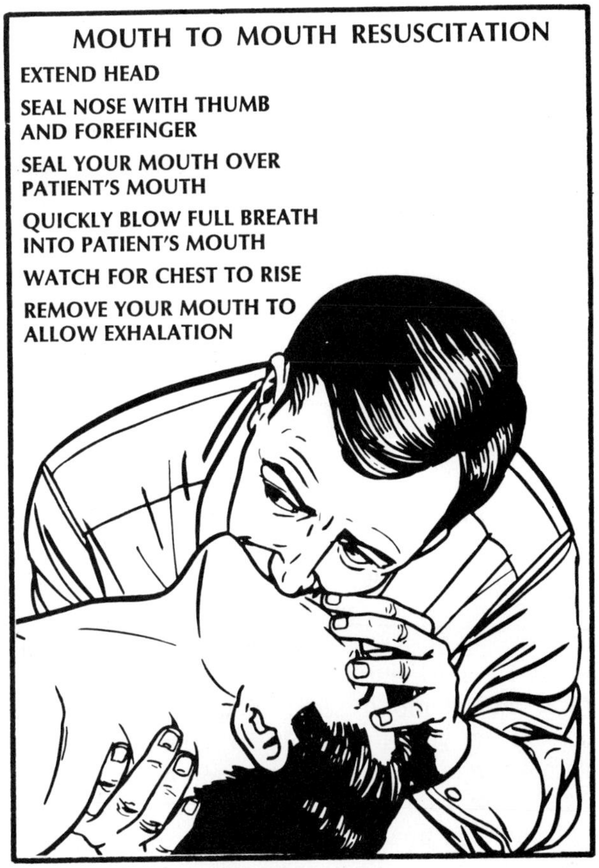

Experience has shown that the three most common errors committed by rescuers while performing mouth-to-mouth resuscitation are:

1. Inadequate extension of the patient's head, so that the airway is not properly established
2. Failure to open the patient's mouth wide enough
3. Forgetting to seal the patient's mouth and nose.

If the patient's stomach becomes distended (bulged) with air from the inflations, turn his head to the side. Compress his stomach gently to expel the air. The bulging results when air slips past the epiglottis into the esophagus and stomach.

After resuscitation has been started, it should be continued until the patient is transported to a hospital, until he starts to breathe spontaneously, or until a physician pronounces him dead.

MOUTH TO MOUTH RESUSCITATION
EXTEND HEAD
SEAL NOSE WITH THUMB AND FOREFINGER
SEAL YOUR MOUTH OVER PATIENT'S MOUTH
QUICKLY BLOW FULL BREATH INTO PATIENT'S MOUTH
WATCH FOR CHEST TO RISE
REMOVE YOUR MOUTH TO ALLOW EXHALATION

Figure 14.5. Mouth-to-mouth resuscitation is the most effective way to ventilate a patient.

Mouth-to-Nose Resuscitation

This alternative method may be used if the patient has serious injuries to the lower jaw, or if he has a severely receding chin due to the lack of natural teeth or dentures.

The mouth-to-nose resuscitation technique is essentially the same as the mouth-to-mouth technique. Clamp the patient's jaw shut with your fingers, and cover his nose with your mouth. Blow your full breath into his nose. After each breath, allow the patient's mouth to open, to provide quick and effective exhalation.

Oropharyngeal Airways

Oropharyngeal airways are curved breathing tubes. They are inserted in the patient's mouth to hold the base of the tongue forward so it does not block the air passage. The rescuer cannot depend completely upon this type of device, however; the head must be tilted backward to provide the maximum opening.

There are two basic guidelines that determine whether or not an airway should be used:

1. If the patient is conscious and breathing normally, an airway should *not* be inserted because it will cause him to vomit.
2. If the patient is unconscious with breathing obstructed, an airway should be inserted if breathing remains obstructed after head tilt and artificial ventilation are attempted.

If the patient reacts by swallowing, retching or coughing after an artificial airway is in place, the airway must be removed quickly. Otherwise, it may make him vomit, increasing the likelihood of airway obstruction. Artificial airways should be employed only when the rescuer is trained in their use.

To insert an oropharyngeal airway, proceed as follows: Use one hand, with the thumb and index finger crossed, to pry the patient's teeth apart and hold his mouth open. With your other hand, insert the airway between his teeth. The curve should be backward at first, and then turned to the proper position as you insert it deeper (Fig. 14.6). This twisting maneuver prevents the tongue from being pushed further back into the throat, since the airway must be inserted over the tongue. If you have difficulty with the tongue, hold it forward with your index finger.

If the jaws are too firmly clenched for the maneuver described above, try to wedge them apart as follows: Insert your index finger between the patient's cheek and teeth, forcing the tip of your finger behind his teeth. If you have difficulty getting the teeth apart or inserting the airway, do not keep forcing. Instead, hold the head back and use mouth-to-mouth or mouth-to-nose resuscitation.

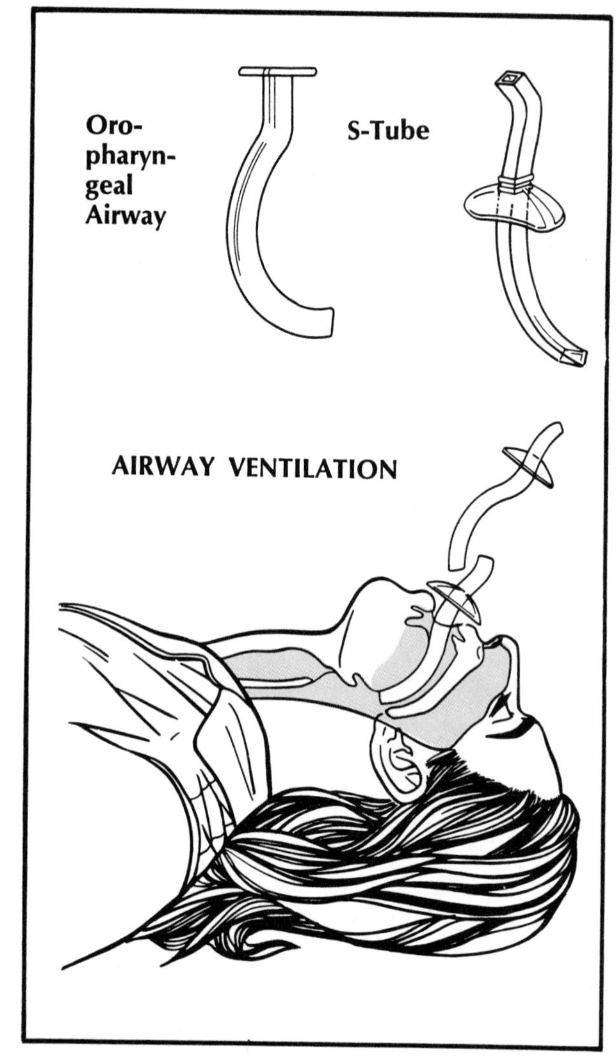

Figure 14.6. Inserting the oropharyngeal or S-tube airway.

Mouth-to-Airway Ventilation

The mouth-to-airway unit (commonly called the S tube) and various similar devices have been introduced to overcome objections to direct mouth-to-mouth contact. These tools should be employed only if the rescuer is trained in their use, and only if they are immediately available.

To use a mouth-to-airway device, proceed as follows: Tilt the patient's head back, and insert the airway in the manner described for oropharyngeal airways (Fig. 14.6). When using the S tube, make sure that the cupped flange is positioned properly. Prevent air leakage by pinching the patient's nose closed and pressing the flange firmly over his mouth. Hold his chin up so that the front of the neck is stretched. Then follow all of the other steps required for the mouth-to-mouth technique, such as rate and size of breaths.

Bag-Mask Resuscitators

Another valuable tool for artificial ventilation is the bag-mask resuscitator. This device consists of a facepiece fitted to a self-inflating bag. A special valve arrangement allows the bag to refill and the patient to exhale without removal of the unit from his face. A common problem with this device is failure of the operator to hold the facepiece firmly enough against the patient's face. The result is then a poor seal.

The bag-mask resuscitator should be used as follows: Hold the facepiece over the patient's face, and clamp it securely in place with one hand. Press your thumb over the rim of the mask, with your index finger over the chin part (Fig. 14.7). Use your third, fourth and fifth fingers to pull the chin upward and backward. Take a firm grip, but never poke your fingers into the patient's neck. Never push the mask down on the patient's chin, as this may bend the neck and obstruct the air passage.

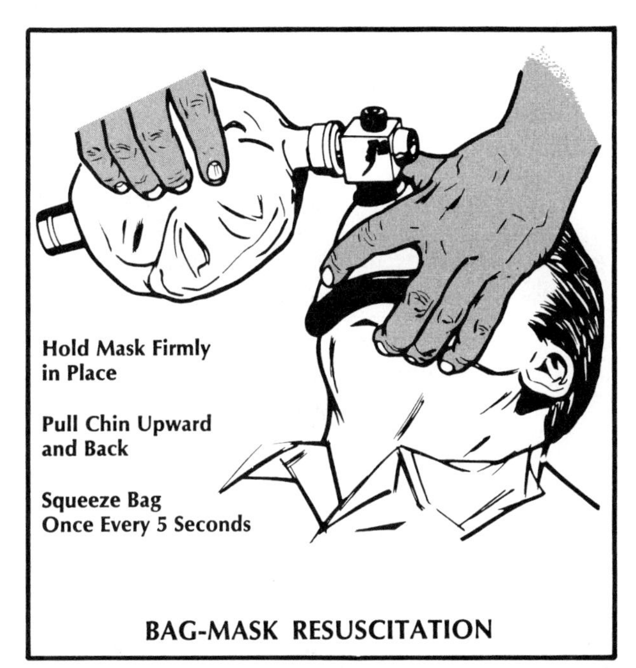

Figure 14.7. Proper positioning and use of the bag-mask resuscitator.

While holding the mask with one hand, squeeze the bag with your other hand about once every 5 seconds. The bag should be *squeezed* until the chest rises, and then *released* to allow exhalation. If you hear leakage, hold the mask more tightly and squeeze the bag more forcefully.

When the bag is released, the air inlet at the tail of the bag opens to allow it to refill. The valve at the mask prevents the patient from exhaling back into the bag. The bag should be released quickly to allow prompt valve action.

Be especially watchful for signs of vomiting. If the patient starts to vomit, discontinue the bag-mask operation immediately. Continuing the operation will force vomitus into the patient's windpipe, causing him to draw the fluids into the lungs, or creating a massive obstruction.

Mechanical Resuscitators

While many emergency care units carry mechanical pressure-cycled resuscitators as part of their equipment, the use of these devices is not recommended. Effective artificial ventilation depends on the volume of air introduced into the lungs, not the pressure that delivers it. Because the lungs are very elastic, the back pressure increases as they fill with air, and more and more pressure is required to deliver the proper amount of air. Moreover, if there is a partial obstruction of the airway, or if the patient's lungs have lost some of their elasticity, even more pressure is required to inflate them properly.

The problem with pressure-cycled resuscitators is that they inflate the lungs to a certain set pressure and then allow exhalation. But the machines have no way of knowing whether or not the proper amount of oxygen has been delivered. That is, they cycle from inflation to exhalation when they sense a certain back pressure, regardless of the amount of oxygen delivered. Often, they deliver an insufficient amount of oxygen, which is of little value to the patient.

Some mechanical resuscitators are equipped with override valves that deliver a constant flow of oxygen as long as a special valve is held open. The flow of oxygen bypasses the pressure-sensing device in the regulator. The operator is able to continue inflating the patient's lungs until the rising chest indicates that the proper amount of oxygen has been delivered.

Rescuers should remember that valuable time must be spent in obtaining a mechanical resuscitator and setting it up for operation. And the nonbreathing patient does not have much time. In addition, the mechanical resuscitator requires constant attention and, like any machine, may fail when it is needed most. It is best not to rely on a purely mechanical device to save a life.

CARDIOPULMONARY RESUSCITATION

Cardiopulmonary resuscitation (CPR) is another name for heart–lung resuscitation, or a combined effort to restore breathing and circulation artificially. Artificial circulation is produced when the chest is compressed by 3.8 to 5.1 cm (1½ to 2 inches), which squeezes the heart between the sternum and the spine. When the heart is squeezed in this fashion, blood is forced into the pulmonary circuit to the lungs (where it is oxygenated) and into the systemic circuit (through which it travels to all parts of the body). When the pressure is released (Fig. 14.8), the elastic chest wall causes the sternum to spring outward to its original position. The release of pressure on the heart results in a sucking action that draws blood into the heart from the veins and the lungs.

The blood is kept in constant motion as long as the heart is squeezed and released by the external chest compressions. The result is quite close to the circulation that is produced by a normally operating heart.

Signs of Cardiac Arrest

A heart attack can occur at any time. Under extreme physical and emotional stress, the risk of heart attack is much greater. Those responsible for emergency care should be aware that heart attacks are a common medical problem in firefighting situations. Both the physical exertion required of firefighters and the lack of oxygen due to smoke add to the probability of cardiac arrest.

You can determine rapidly and easily whether or not cardiac arrest has occurred by checking for three signs, all of which must be present if heart action has stopped.

No Respiration. Check for breathing as described above: *Look* for movement of the chest. *Listen* and *feel* for air exchange at the mouth and nose. If there are signs of breathing, there is no possibility of cardiac arrest. If there are no signs of breathing, there may be cardiac arrest. At any rate, the patient will need some sort of artificial ventilation as the minimum treatment.

No Pulse. Check for heart action by feeling for either the cartoid pulse in the neck or the femoral pulse in the groin. The cartoid and femoral arteries are quite large, and normally their pulses are easily felt if there is sufficient heart action to circulate blood. If a pulse can be felt, the patient is not in cardiac arrest and may need only to have his breathing restored or supported. However, if a pulse cannot be found, cardiac arrest is indicated. To be sure, check for the third sign.

Dilated Pupils of the Eyes. Check the pupils of the patient's eyes to see if they are dilated. Constricted (narrow) pupils indicate that there is blood circulation. Dilated (quite large) pupils indicate that blood circulation has stopped. Within 45 to 60 seconds after circulation to the brain ceases, the pupils begin to dilate. Within another minute the dilation is complete. Thus, if the pupils appear dilated, no oxygenated blood is being circulated to the brain. This sign, coupled with the first two, calls for immediate reestablishment of breathing and heart action by cardiopulmonary resuscitation.

The CPR Technique

The sequence of operations required in cardiopulmonary resuscitation is best remembered as the ABC technique.

- *A* stands for *airway:* Ensure that the patient has a clear airway, that the head is in the proper position, and that the throat is free of foreign objects.
- *B* stands for *breathe:* Inflate the patient's lungs immediately with four quick, good-size breaths, using the mouth-to-mouth technique, or mouth-to-nose if necessary. This provides the lungs with a high concentration of oxygen that is immediately available for circulation to the brain. If the establishment of a clear airway and the four quick ventilations do not start spontaneous breathing, check for the other two signs of cardiac arrest (no pulse, dilated pupils). Then go on to the next step.
- *C* stands for *circulate:* For this, the rescuer must
    1. Situate the patient properly.
    2. Locate the pressure point.
    3. Place his hands properly.
    4. Apply pressure.
    5. Interpose ventilations.

Any deviation from the proper procedure for placing the hands may result in damage to the ribs and underlying organs.

Locate the Pressure Point. Run one hand along the lower rib cage to a point in the center of the chest, where a flexible point called the *xiphoid process* is located. (See Figs. 14.8 and 14.9.) Measure three fiinger widths up (toward the neck) from the tip of the xiphoid. To do this, lay three fingers of one hand flat on the patient's chest, with the fingers touching each other and the lowest finger on the xiphoid tip.

Place Your Hands. Place the heel of your hand on the chest so that it is touching the third "measuring" finger. This hand should now be about three finger widths from the xiphoid tip. Its fingers should be pointing approximately across the chest. Remove the "measuring" hand, and place it on top of the other hand (Fig. 14.9).

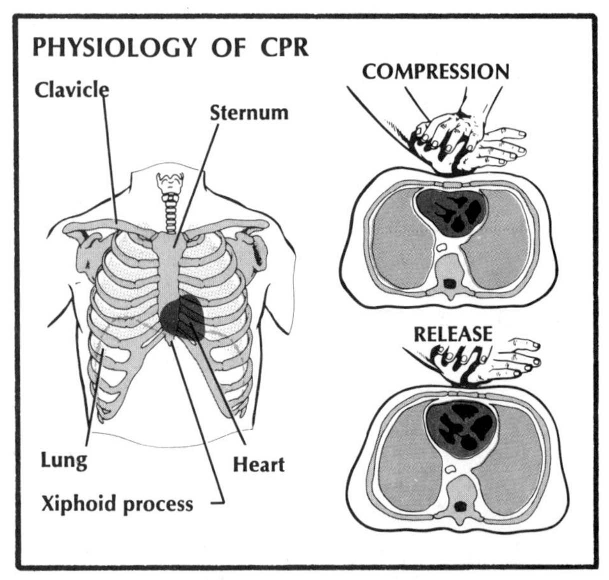

Figure 14.8. Cardiopulmonary resuscitation. The compression and release of the heart cause blood to be pumped to the lungs and throughout the body.

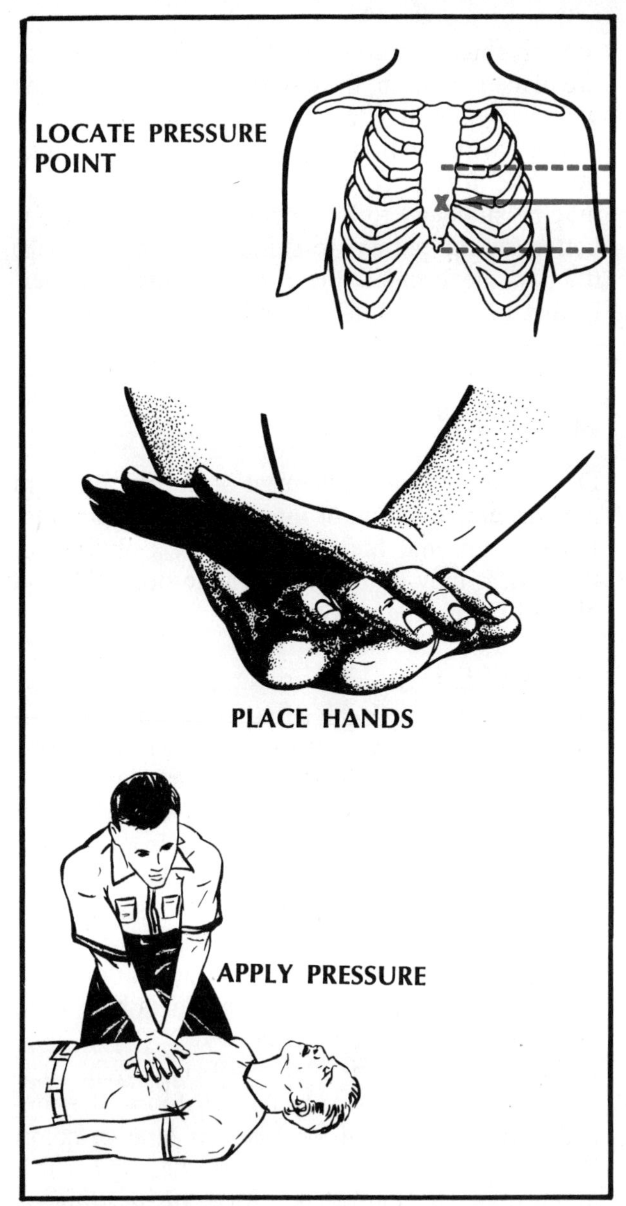

Figure 14.9. Finding the pressure point, placing the hands and applying the pressure in CPR.

Apply Pressure. With your shoulders directly above the victim's chest, press straight down, compressing the chest by 3.8 to 5.1 cm (1½ to 2 inches). Your elbows must not bend or flex. Pivot at the hips, making use of the weight of your head and shoulders to obtain the proper compression. A poor technique could result in fatigue and limit the time you are able to perform CPR.

Ventilate. Chest compressions without ventilation are of little value to the patient, since the only air exchange is that resulting from the chest movements caused by the compressions. This limited exchange is not sufficient to oxygenate the blood. (The proper ventilation procedures are given below.)

CPR with One Rescuer

When it is necessary to perform cardiopulmonary resuscitation without assistance, the following steps should be taken:

1. Place the patient on his back on a hard surface, preferably the deck.
2. As soon as you determine that the patient is in cardiac arrest (by checking the three signs), establish the airway. Ventilate the patient with four double-size breaths, without pausing between breaths.
3. Shift to the patient's side, and compress his chest 15 times, at a rate of 80 compressions per minute. Be sure that your hands are always in the proper position.
4. After the 15th compression, quickly pivot back to the patient's head. Inflate his lungs two times, again without any pause between ventilations.
5. Return to the chest, and again compress the chest 15 times, at the rate of 80 compressions per minute. Then perform two ventilations.
6. Continue the cycles of 15 chest compressions and 2 ventilations without interruption until the patient shows signs of recovery, or until you are relieved by competent medical personnel.

CPR with Two Rescuers

The most effective cardiopulmonary resuscitation can be accomplished by two rescuers working together. The two-man method is more effective and far less tiring than the one-man technique. The following steps should be taken:

1. Place the patient on his back on a hard surface.
2. Determine the patient's condition by checking for the three signs of cardiac arrest.
3. The first rescuer now positions himself at the patient's head. He establishes an open airway and quickly ventilates the patient with four double-size breaths.
4. The second rescuer positions himself at the patient's side. He starts manual chest compressions at the rate of one every second (60 per minute) as soon as the patient is ventilated.

5. After every five chest compressions, the first rescuer ventilates the patient with one quick, double-size breath. The rescuers must closely coordinate their efforts, so that the first rescuer will not attempt to ventilate the patient just as the second rescuer is compressing the chest.
6. Continue the cycles of one breath and five chest compressions without interruption.

If it is necessary to change rescuers during the resuscitation efforts, the switch should be made with as little interruption as possible. An effective means is to position the incoming rescuers on the opposite side of the patient. Thus, they can take over as soon as the outgoing rescuers tire.

Determining the Effectiveness of CPR Efforts

If the patient can be successfully resuscitated (that is, if biological death has not occurred), the effectiveness of CPR efforts can be measured by certain changes in the patient's condition. If the efforts are successful, the following changes *must* occur.

1. The pupils *must* constrict.
2. The patient's color *must* improve.
3. A pulse *must* be felt at the carotid artery with each heart compression.

The rescuer should realize that CPR has its limitations. It is doubtful that CPR will be of any help if there have been extreme crushing injuries to the chest, if large internal arteries have been cut open, if the skull has been crushed, or if the heart has ruptured.

The longer CPR is carried out without a positive response from the patient, the smaller his chances for survival. Rescuers (applying two-man CPR whenever possible) should not attempt more than 1 hour of continuous CPR. After 1 hour, there is no hope for the patient to recover. Rescuers should not feel guilty if they stop CPR after an hour; nothing more can be done for the patient.

Possible Complications of CPR

Damage to the rib cage and the underlying organs can be caused by improper placement of the rescuer's hands during chest compression. When the hands are placed too far to the right, the ribs may be fractured. They can then lacerate the lungs, and possibly the heart muscle itself. When the hands are placed too low on the sternum (breastbone), the bony xiphoid may be depressed too far, thus lacerating the liver. When the hands are placed too high, the collarbone may be broken where it joins the sternum.

Even with the hands placed in the proper position, the force required to compress the chest may be sufficient to break ribs. However, it is far better for the patient to suffer a few broken ribs than to die because the rescuer refused to perform CPR through a fear of inflicting injury.

BLEEDING

Depending on the mechanism of injury, a patient may develop external or internal bleeding or both. Because internal bleeding is the less obvious of the two, it may be the more dangerous. We shall discuss them separately.

Types of External Bleeding

Arterial bleeding is bleeding from an artery. It is characterized by a flow of bright red blood leaving the wound in distinct spurts. At times, the flow may be alarmingly heavy. Arterial bleeding is not likely to clot unless it is from a small artery or unless the flow of blood is slight. Arteries have a built-in defense mechanism: If they are cut through completely, they tend to constrict and seal themselves off. However, an artery that is not cut through, but is torn or has a hole in it, will continue to bleed freely.

Venous bleeding is bleeding from a vein. It is characterized by a steady flow of blood that appears to be dark maroon or even blue in color. Bleeding from a vein may also be heavy, but it is much easier to control than arterial bleeding. A real danger associated with venous bleeding in injuries to the neck is a condition known as *air embolism*. The blood in the larger veins is drawn to the heart by a sucking action that develops as the heart contracts and relaxes. This action may draw an air bubble into the open vein. If the air bubble is large enough, it can reduce the ability of the heart to pump properly. The heart may even fail completely. Even though venous bleeding may not appear heavy, it should be controlled quickly and effectively.

Capillary bleeding is bleeding from capillaries. It is characterized by the slow oozing of blood, usually from minor wounds such as scraped knees. Since the bleeding is from the smallest vessels, it can be controlled easily. However, because of the large amount of skin surface involved, the threat of contamination may be more serious than the blood loss.

Methods of Controlling External Bleeding

The most effective method of controlling external bleeding is direct pressure on the wound. In the case of an arm or leg wound not involving a fracture, direct pressure and elevation of the limb are recommended. There are few instances of bleeding where the flow of blood cannnot be controlled by this quick and efficient means.

Direct Pressure. When the bleeding is relatively mild, apply pressure to the wound with a sterile dressing, clean cloth or handkerchief (Fig. 14.10). Firm pressure on the wound for 10–30 minutes will, in most cases, stop the bleeding. To allow the patient some movement while the bleeding is being controlled, bind the dressing in place with a bandage.

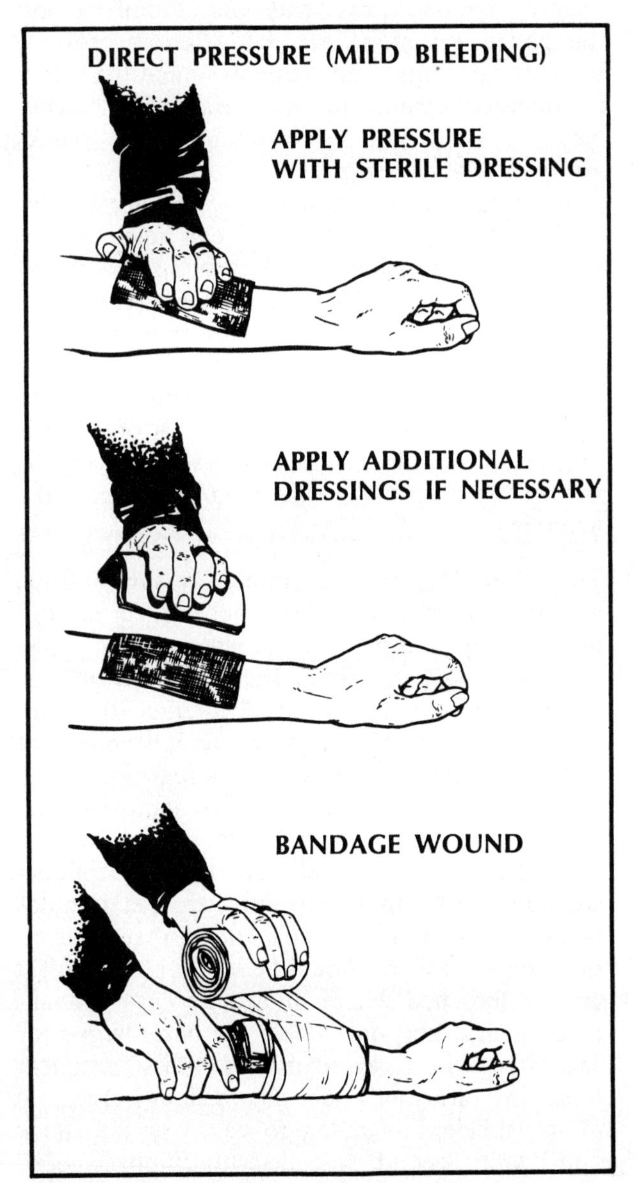

Figure 14.10. Controlling mild external bleeding with pressure. Once the dressing is in place, it should not be removed.

Do not attempt to replace the dressing once it is held in place, even if it becomes blood-soaked. Replacing a dressing releases the pressure on the cut blood vessels, interferes with normal coagulation and increases the likelihood of contamination. Instead of replacing the dressing, add another one on top of the soaked dressings, and hold them all in place. Continue this procedure until the patient is delivered to a medical facility.

If the bleeding is heavy, place your hand directly on the wound and exert firm pressure. If a cloth or handkerchief is immediately available, use it. But don't waste time trying to find a cloth or dressing; the patient's blood loss may have reached the critical point. If the bleeding continues, insert your fingers directly into the wound and attempt to compress the artery. Either squeeze it between your fingers or press it against a bony portion of the body.

Pressure Points. If the bleeding continues in spite of efforts to control it by direct pressure, the rescuer can apply finger pressure at a *pressure point.* The six major arterial pressure points are shown in Figure 14.11:

- The *brachial artery* controls bleeding from the arm.
- The *femoral artery* controls bleeding from the leg.
- The *carotid artery* controls bleeding from the neck.
- The *temporal artery* controls bleeding from the scalp.
- The *facial artery* controls bleeding from the face.
- The *subclavian artery* controls bleeding from the chest wall or armpit.

For Bleeding from the Arm. Apply pressure to a point over the brachial artery. To find the brachial artery, hold the patient's arm out at a right angle to his body, with the palm facing up. Between the elbow and the armpit, you will find a groove created by the large biceps muscle and the bone (Fig. 14.12). With your hand cradling the upper arm, press your fingers firmly into this groove. This will compress the brachial artery against the underlying bone. If the pressure is applied properly, you will not be able to feel a pulse at the patient's wrist.

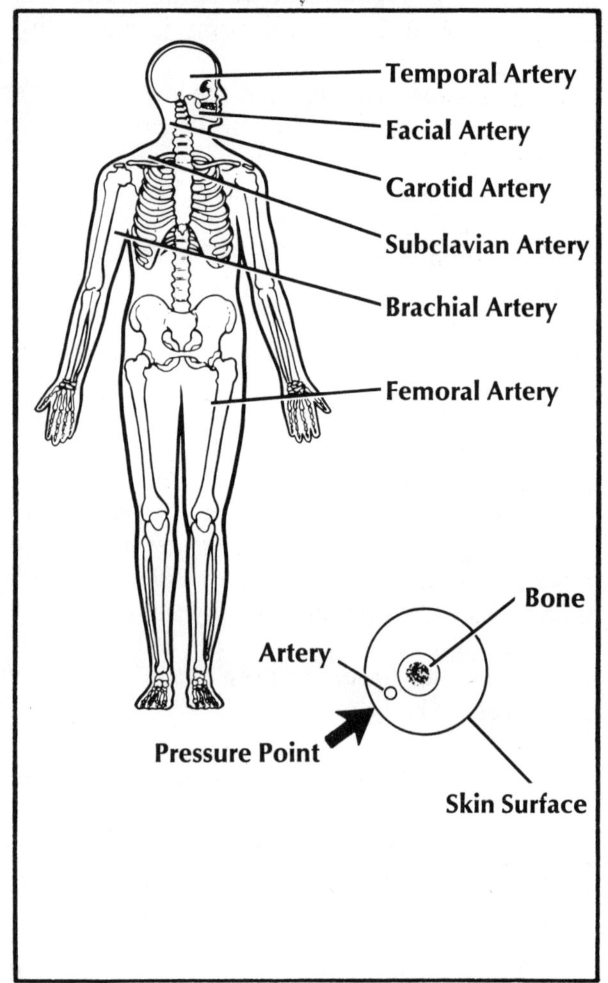

Figure 14.11. The six major pressure points. The inset shows how pressure at the pressure point squeezes the artery against a nearby bone to stop the flow of blood.

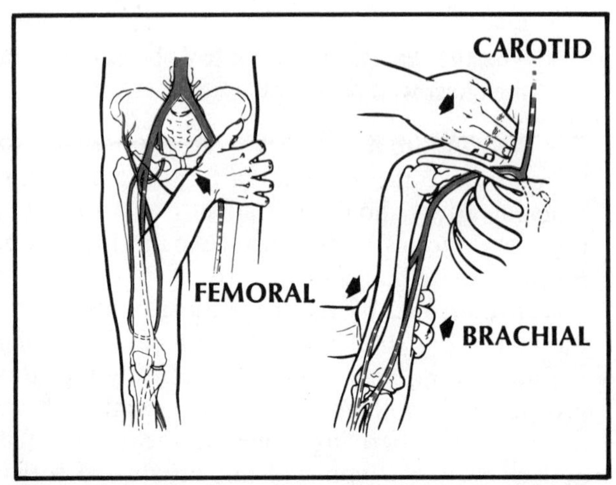

Figure 14.12. The femoral (leg), brachial (arm) and carotid (neck) pressure points.

For Bleeding from the Leg. Apply pressure to a point over the femoral artery. Locate the femoral artery on the inside of the groin, just below where the thigh joins the torso (Fig. 14.12). You will be able to feel pulsations at this point. Place the heel of your hand over the pressure point. Exert pressure downward toward the bone until it is obvious that the bleeding has been controlled. If the patient is very muscular or obese, you must exert considerable force to compress the artery.

For Bleeding from the Neck. Locate the windpipe at the midline of the neck. Slide your fingers around to the bleeding side of the neck and feel for the pulsations of the large artery. Place your fingers over the artery, with your thumb behind the patient's neck (Fig. 14.12). Exert pressure between your fingers and thumb so that the artery is squeezed against the vertebrae of the neck. Never apply pressure to both carotid (neck) arteries at the same time.

In most cases where major vessels are not involved, bleeding from the neck can be controlled by placing a dressing over the wound and applying direct pressure. However, when the pressure point must be used, extreme care must be taken to avoid producing unconsciousness by restricting the flow of blood to the brain. In addition, some patients may faint quite readily when pressure is exerted on a little bundle of nerve tissue in the neck. Care must also be taken not to squeeze the windpipe.

Tourniquet Pressure. If direct pressure and the use of pressure points do not effectively control the bleeding, a *tourniquet* should be used. However, a tourniquet should be considered only as a last resort. It must be used intelligently, with certain precautions and a full understanding of its function. If used improperly, a tourniquet may prove more harmful than effective, increasing the danger to the part to which it is applied.

In the past it was believed that a tourniquet should never be applied for more than 20 minutes at a time. However, it is now known that a tourniquet, once applied, should be left in place until it can be loosened where immediate care is available, as in a hospital. There are two reasons for this change in accepted practice. First, frequent loosening of a tourniquet may dislodge clots and allow sufficient bleeding to cause severe shock and death. Second, so-called "tourniquet shock" is now recognized as a very real danger to patients to whom a tourniquet has been applied. This type of shock is thought to be caused by harmful substances released by the injured tissues. These substances are held back by the tourniquet and then released into the general circulation when the tourniquet is loosened. Unless this shock is controlled, loosening of the tourniquet may prove fatal. Studies have shown that leaving the tourniquet in place causes more limbs to be lost, but more lives to be saved.

To apply a tourniquet, follow these steps (Fig. 14.13):

1. Select a place for the tourniquet between the heart and the wound—as close to the wound as possible, but not right at the edge of the wound.
2. Place a pad made from a dressing or a folded handkerchief over the main supplying artery. The pad will add to the pressure on the main artery and make the tourniquet more effective.
3. Place the constricting band around the patient's limb and the pad. If a commercial tourniquet is used, pull the loose end of the band through the buckle or friction catch, and draw it up tightly. If a cravat or other piece of material is used, knot the material. In the knot, insert a stick, rod or similar device that can be used to tighten the tourniquet.
4. Tighten the tourniquet just enough to control the bleeding. If it is too loose, it will be of no value. If it is unnecessarily tight, it will cause further damage to the limb.
5. Attach to the patient a notation indicating that a tourniquet has been applied. If a tag is not available, mark the patient's forehead with a pen or even blood. The marking will alert medical personnel that a tourniquet is in place; this fact might go unnoticed if the patient is covered with a blanket or if medical facilities are extremely busy.

APPLYING A TOURNIQUET

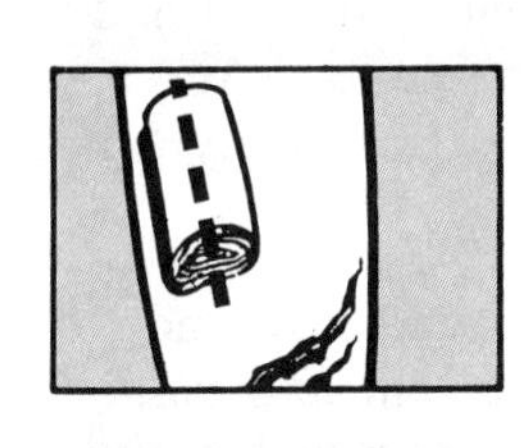

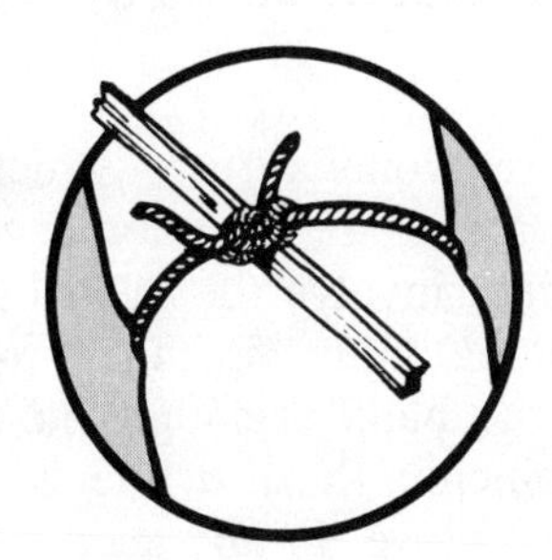

Place Pad Over

and Insert a Device

Tighten Only Enough
to Stop Bleeding

Use of material
that is too thin
may injure
blood vessels
and underlying
tissue

Figure 14.13. The steps in applying a tourniquet.

Signs and Symptoms of Internal Bleeding

Internal bleeding may be suspected when the mechanism of injury indicates internal damage and classic signs of shock are present, but there is no obvious injury. The signs of shock are:

- Rapid and weak pulse
- Pale, moist and cold skin
- Shallow and rapid breathing
- Thirst
- A weak and helpless feeling
- Shaking and trembling
- Dilated pupils.

In addition to these signs, the patient with internal bleeding may cough up bright red blood, or vomit blood that has the appearance of coffee grounds. The latter is an indication of bleeding in the abdominal organs; the abdomen will also become very stiff or develop muscle spasms. To estimate blood loss from closed wounds, figure an approximate 10% blood loss for each area of badly bruised tissue the size of a man's fist.

The problem of internal bleeding should not be taken lightly. Any patient exhibiting the signs described above should be considered a high priority patient. If the patient is unconscious and the mechanism of injury indicates that internal bleeding could have been produced, he should be treated accordingly.

Controlling Internal Bleeding

Patients with suspected internal bleeding should be treated in the following general manner.

1. Treat the patient for shock.
2. Expect the patient to vomit, and give him nothing by mouth.
3. If the bleeding is in an extremity, apply pressure to the injury site with a pressure dressing (a snug bandage over a bulky pad).
4. Transport a patient with abdominal or chest-cavity injuries immediately, but safely, to a medical facility. Such injuries represent a true emergency.
5. Administer oxygen (qualified personnel only).

WOUNDS

Wounds are injuries to the soft tissues of the body; they are classified as closed or open. Closed wounds, as the name implies, are injuries in which the skin surface is not broken and there is no external bleeding. Open wounds are injuries where the skin is torn and the underlying tissues are exposed. Bleeding may vary from slight to heavy.

Closed Wounds

The injury resulting from the impact of a blunt object is called a *bruise* or *contusion*. Although the skin is not penetrated, there may be a great deal of crushed tissue beneath the skin. Some bleeding always occurs at the time of the injury, and frequently for a few hours thereafter. Swelling generally develops 24 to 48 hours after the injury. A blood clot almost always forms at the injury site; the blood seeps into the surrounding tissues, causing a bluish discoloration, the "black and blue" mark.

Small contusions generally do not require emergency care unless they are associated with more serious problems such as internal injuries or fractures. A pressure dressing will reduce the bleeding and assist the natural healing processes.

Open Wounds

There are several categories of open wounds.

Abrasion. An abrasion is the least serious type of open wound. It is a scratching of the skin surface in which not all the layers of the skin are penetrated. A small amount of bleeding may result from an abrasion, but rarely more than a few drops. A great deal of dirt may be ground into the wound, so the possibility of contamination should be considered.

Incision. An incision is a wound that is made by a sharp object such as a knife or razor blade. The cut edges of the skin and tissue are smooth because of the sharpness of the object inflicting the injury. Obviously, if such a wound is deep, large blood vessels and nerves may be severed. Because the blood vessels are cut cleanly, incisions bleed freely. The bleeding from long and deep incisions is often quite difficult to control.

Laceration. A laceration results from the snagging and tearing of tissue, leaving a jagged wound that bleeds freely. It is usually impossible to see what important structures have been damaged by looking at the outside of the wound, since the jagged edges of the wound tend to fall together and obscure the depth. If important vessels have been torn, there is considerable bleeding, although usually less than from an incision. This is because the blood vessels are stretched and torn in a laceration. The cut ends curl and fold, which aids in rapid clot formation. An example of a laceration is the wound caused by a jagged piece of metal.

Puncture Wound. A puncture wound results from the disruption of the skin and tissue by a sharp, pointed object, such as a nail, ice pick or splinter. There is usually no severe external bleeding. However, in more serious puncture wounds, internal bleeding may be quite heavy.

Puncture wounds are classed as either penetrations or perforations. A *penetration* is a shallow or deep wound that damages tissue and blood vessels; it may result from a wide metal strip or a long, pointed shard of glass. A *perforation* is a deep puncture wound, such as a through-and-through gunshot wound that passes through nerves, bones and organs and causes great internal damage. A perforation differs from a penetration in that it results in an exit wound as well as an entrance wound.

Avulsion. Avulsions are wounds from which large flaps of skin and tissue are torn loose or pulled off. Avulsions might involve the eyeballs, ears or fingers. A common injury is the glove avulsion, caused when the hand is caught in a roller or other type of pinching hazard; the skin is stripped off much like a glove. In any accident that results in an avulsion, the rescuer should make every effort to preserve the avulsed part and transport it with the patient. It may be possible to restore the part with surgical techniques or at least to use the skin for grafts. The emergency treatment for avulsions requires application of large, bulky pressure dressings.

Traumatic Amputation. A traumatic amputation involving a finger, hand, arm or leg generally occurs when the extremity is torn off in an accident. Jagged skin and bone edges characterize the wound, and there may or may not be massive bleeding. As for other external bleeding, the most effective method of control is to use a snug pressure dressing over the stump. A tourniquet is rarely necessary.

Crushing Injury. Crushing injuries are caused when the extremities of the body are caught in some mechanical device. Open fractures are common in such accidents. There is usually a surface laceration of the bursting type, with extensive damage to the underlying tissues. Large, bulky dressings are required for emergency care. In cases where a limb has been severed by an extremely heavy crushing force, there is usually very little bleeding. The crushing action tends to close off the bleeding vessels as they are severed.

Emergency Care for Open Wounds

Emergency care for open wounds is directed toward stopping the bleeding and keeping the wounds clean:

1. Control the bleeding with direct pressure, the use of pressure points or, as a last resort, a tourniquet.
2. Prevent contamination of the wound by applying a *sterile* dressing.
3. Immobilize and elevate the injured part in the event of serious bleeding, if this will not worsen other injuries.

Dressing and Bandaging Materials

While the two terms are often confused, dressings and bandages are two separate items of supply. Dressings are applied to the wound to control bleeding and prevent contamination. Bandages are used to hold the dressings in place. A dressing should be sterile, but bandages need not be sterile.

Usually, a variety of dressings are carried as emergency care supplies. Separately wrapped sterile gauze pads, 10.2 cm^2 (4 $inches^2$), are the most common dressings. Large bulky dressings such as multitrauma and combination dressings are valuable where bulk is required for heavy bleeding, or where large areas must be covered. These dressings are especially useful for stabilizing impaled objects. Sanitary napkins are well suited for emergency care work because of their absorbent properties. While they are not usually sterile, they may be obtained separately wrapped, thus ensuring a clean surface at all times.

Do not apply the bandage too tightly, as this may restrict the blood supply to the affected part, resulting in grave complications. Do not apply the bandage too loosely (the most common error in bandaging), or it will not hold the dressing in place. The bandage must be applied rather snugly, since it stretches after a short time, especially when the patient can move the bandaged part. In bandaging extremities, leave the fingers and toes exposed wherever possible, so that color changes may be noted. Pain, pale skin, numbness and tingling are signs that a bandage is too tight.

Impaled Objects

Occasionally the rescuer will be confronted with a wound from which a piece of glass, a knife, a stick, or some other pointed object is protruding. When dealing with such an injury,

- Do not remove the object.
- Use a bulky dressing to stabilize the object.
- Transport the patient to a medical facility very carefully.

The removal of an impaled object may cause severe bleeding by releasing the pressure on the severed blood vessels; or, it may cause further damage to nerves and muscles. Occasionally, however, a portion of the object will have to be removed to allow transportation of the patient. Clothing may also have to be removed from the area of the injury, to make the wound more accessible. No attempt should be made to lift the clothing over the wound in the usual manner; instead, the clothing should be carefully cut away from the injury (Fig. 14.14). The rescuer must be extremely careful not to move the object while removing the clothing.

Bleeding around an impaled object may be controlled first by hand pressure if the bleeding is profuse. While one rescuer is applying pressure with his hand, another should be preparing dressings for the wound. The wound may be dressed in either of two ways. One method is to place several layers of a bulky dressing over the injury site, so that the edges of the dressings butt up against the object from both sides (see Figure 14.14). Another method is to cut a hole in a bulky dressing, slightly larger than the object, and then to pass the dressing very carefully over the object. In both cases, the bulky dressings keep the object from moving and exert direct pressure on

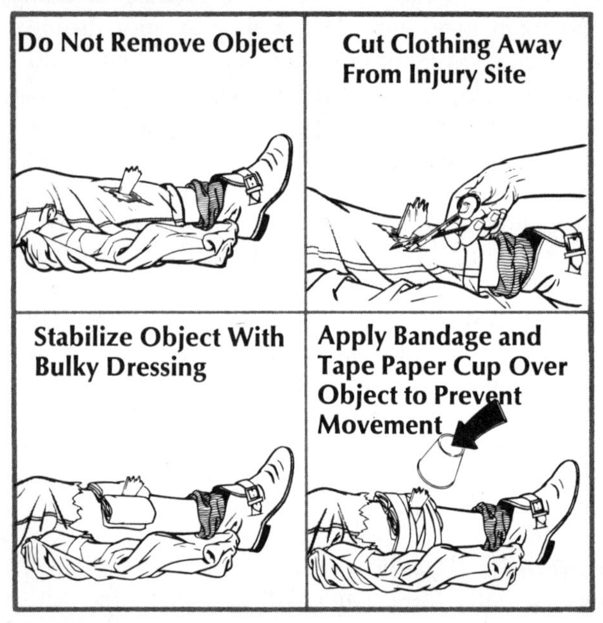

Figure 14.14. Emergency care for an open wound with an impaled object.

the bleeding vessels. Self-adhering bandaging material is well suited for use with this type of dressing. A paper cup taped over the object may help prevent accidental movement.

Objects Impaled in the Cheek. A foreign object impaled in the cheek presents a dangerous situation. It is the only case in which a foreign object should be removed from a wound. The cheek wall is fairly thin, and bleeding into the mouth and throat may be heavy. This bleeding cannot be controlled by pressure on the outside of the wound.

Carefully examine the wound. With your fingers, probe inside the patient's mouth to see if the object has passed through the cheek wall. If the object has come through the wall, carefully remove it by pulling it out toward the direction from which it entered the cheek. If the object will not come loose easily, leave it in place and pack compresses around it. In either case, be sure to place the head in a good position for drainage of blood.

When the object has been removed, pack the inside of the patient's cheek (between the cheek wall and the teeth) to prevent additional bleeding. This packing will not present too much of an obstruction if the patient has to vomit. Dress the outside of the wound in the usual manner, with a pressure dressing and bandage.

Objects Impaled in the Eye. Large foreign objects impaled in the eye should be removed only by a physician. Until medical assistance is available, such objects should be protected from accidental movement or removal. The following method is suggested.

Make a thick dressing of several layers of sterile gauze pads or multitrauma dressings. Cut a hole in the center of the dressing approximately the size of the eye opening. Carefully pass the prepared dressing over the impaled object, and position it so it is centered over the injury site (Fig. 14.15). This pad will serve as a cushion for a rigid shield that will be used to protect the eye.

Next, select a cup or cone of sufficient size to fit over the impaled object without the object touching the sides or top of the cup. A disposable

Figure 14.15. Emergency care for an eye wound with an impaled object.

drinking cup or a styrofoam hot-beverage cup usually works well. Position the cup over the pad, and fasten it carefully in place with soft, self-adhering roller bandage. This rigid protection will shield the object against accidental movement or inadvertent removal. It will also call attention to the fact that the patient has suffered a serious eye injury.

After the eye is protected against further injury by bandaging, the good eye should be securely bandaged also. This will reduce the movement of the injured eye that may be caused by constant movement of the good eye. (When one eye is attracted to a light source, the second eye moves with the first.)

SHOCK

Shock is the failure of the cardiovascular system to provide sufficient blood circulation to every part of the body. It can be caused in several different ways.

Types and Causes of Shock

Hemorrhagic Shock. This is caused by blood loss. The blood volume may be reduced by *1)* external bleeding, *2)* internal bleeding, or *3)* loss of plasma (the liquid part of the blood), as in the case of burned or crushed tissues.

Respiratory Shock. This is caused by insufficient oxygen in the blood. Respiratory shock results from an inability to fill the lungs completely, and is often seen in cases of severe smoke poisoning. Breathing may be impaired for other reasons as well:

- An open sucking chest wound, ribs broken away from the sternum, fractures to individual ribs and collapsed lungs can interfere with normal lung operation.
- An airway obstruction can prevent a sufficient amount of air from reaching the lungs.
- Spinal-cord damage can paralyze the muscles of the chest wall, causing the patient to breathe with his diaphragm alone.

The rescuer should be aware that respiratory shock, unlike the other types of shock, is not caused by impairment of circulation. At the outset the heart is operating normally, with the proper amount of blood. The blood vessels are constantly adjusting to keep the system full. However, the oxygen supply available for exchange in the lungs is not normal, and consequently the blood is not properly oxygenated. Inadequate air exchange in the lungs can produce shock as quickly as blood loss.

Neurogenic Shock. This is caused by loss of control of the nervous system. When the spinal cord is damaged in an accident, nerve pathways between the brain and the muscles are interrupted at the point of injury. As a result, the muscles controlled by the damaged nerves are paralyzed. These include the muscles in the walls of the blood vessels. The blood vessels can no longer change size in response to signals from the nervous system. They remain wide open, so a greater amount of blood is required to fill the vessels. Since the cardiovascular system contains only enough blood to fill the vessels in the normal state, circulation is impaired and shock develops quickly.

Psychogenic Shock. This is commonly known as fainting. Simple fainting is a reaction of the nervous system to such stimuli as fear, bad news, the sight of blood or a minor injury, rapid temperature changes and overexertion. Unless other problems are present, fainting is usually self-correcting. As soon as the head is lowered, blood circulates to the brain and normal functions are restored. Fainting can often be prevented if the head is lowered before loss of consciousness (by sitting down and placing the head between the knees). There are times when bystanders may tease the patient who has fainted. The rescuer should clear the area whenever possible to protect the patient from such abuse.

Cardiogenic Shock. This type is caused by inadequate functioning of the heart. Proper blood circulation depends on efficient and continuous heart operation. However, certain diseases and disorders weaken the heart muscle and cause it to operate at a reduced output. When the heart can no longer develop the pressure required to move blood to all parts of the body, circulation is impaired and shock results.

Signs and Symptoms of Shock

The signs of shock were listed briefly in an earlier section. In more detail, they are as follows:

- The eyes are dull and lackluster, a sign of poor circulation.
- The pupils are dilated, another reliable sign of reduced circulation.
- The face is pale and may be cyanotic. Cyanosis is an important sign of oxygen deficiency, in this case caused by reduced circulation.
- Respiration is shallow, possibly irregular or labored. The vital centers that regulate

breathing are slowing down, as are all life processes.

- The pulse is rapid and weak. The heart is working faster to make up for the reduced blood pressure and volume.
- The skin is cold and clammy. Blood has stopped circulating actively in the extremities and is collecting in the vital organs.
- There may be nausea, collapse, vomiting, anxiety and thirst.

Emergency Care for Shock

1. *Ensure adequate breathing.* If the patient is breathing, maintain an adequate airway by properly positioning the head. If the patient is not breathing, establish an airway and restore breathing through some means of pulmonary resuscitation. If both respiration and circulation have stopped, start CPR.
2. *Control bleeding.* If the patient has bleeding injuries, use direct pressure, pressure points or a tourniquet as required.
3. *Administer oxygen.* An oxygen deficiency will result from the reduced circulation. Administer 100% oxygen to the patient to compensate for this loss (qualified personnel only).
4. *Elevate the lower extremities.* Since blood flow to the heart and brain may have been diminished, circulation can be improved by raising the legs slightly. It is not recommended that the entire body be tilted down at the head, since then the abdominal organs may press against the diaphragm and interfere with breathing. Exceptions to the rule of raising the legs are cases of head and chest injuries, where it is desirable to lower the pressure in the injured parts. In these cases, the upper part of the body should be elevated slightly. Whenever there is any doubt as to the best position, the patient may be placed perfectly flat without adverse results.
5. *Avoid rough handling.* Handle the patient as gently and as little as possible. Body motion tends to worsen shock.
6. *Prevent loss of body heat.* Keep the patient warm, but guard against overheating, which can worsen shock. Place a blanket under the patient, as well as over him, to prevent loss of heat into the deck.
7. *Keep the patient lying down.* This avoids taxing the circulatory system at a time when it should be at rest. However, some patients, such as those with heart disorders, will have to be transported in a semisitting position.
8. *Give nothing by mouth.*

Anaphylactic Shock

Anaphylactic shock (or anaphylactic reaction) deserves special emphasis since it is a condition that should be considered a true emergency. Anaphylactic shock occurs when a person contacts or ingests something to which he is extremely allergic. The anaphylactic reaction may occur within a few seconds after exposure to an allergic substance. Thus, prompt recognition and treatment of the problem are vitally important.

The signs and symptoms of anaphylactic shock are:

- Itching or burning skin, especially about the chest and face (the skin may also be flushed)
- Hives over large areas of the body
- Swelling of the face and tongue
- Cyanosis visible at the lips
- Tightening or pain in the chest, wheezing, difficulty in breathing
- Weak or imperceptible pulse, dizziness, faintness or even coma.

Anaphylactic shock is a true emergency because it requires the injection of medication to combat the allergic reaction. Initial emergency care efforts should be directed toward life support.

Transport the patient to a hospital immediately and notify the hospital by radio. If the information is available, notify the hospital of the substance that caused the reaction and the means of contact (inhalation, injection or ingestion). Provide life-support measures as required, including pulmonary and cardiopulmonary resuscitation. Administer oxygen (qualified personnel only), and treat for shock.

BURNS

Many physiological effects result from exposure to heat. These include dehydration, heat exhaustion, blockage of the respiratory tract (edema) and burns.

Burns are damage to the skin (and underlying tissue) caused by high temperatures. The damage almost always arises from heat generated within the body by metabolic processes, rather than from heat directed at the body from outside. That is, the high temperature of the surroundings usually interferes with the elimination of heat produced within the body, thereby causing a rise in body temperature. There are relatively few instances in which the temperature of the surroundings is severe enough to cause a transfer of heat to the body.

The Skin

There are three layers of skin: the epidermis, the dermis, and the subcutaneous layer. The *epidermis* is made up of cells that are unusual because they can endure wear and tear; they are waterproof, and they contain sensory nerve endings. The *dermis* is the second layer of skin, made up of dense connective tissue that gives the skin its strength and elasticity. In this layer are hair follicles, sebaceous glands, and sweat glands. The *subcutaneous* layer is a layer of fatty tissue.

Classes of Burns

Burns may be classified according to cause and depth (or degree). The six major classes of burns by cause are:

- Thermal burns (caused by heat, including both flame and radiated heat)
- Chemical burns
- Electrical burns
- Cryogenic burns (caused by cold)
- Nuclear radiation burns
- Light burns (eye injuries).

Relative to depth, some physicians classify burns as either *partial* or *full* thickness. However, burns are usually classified by degree.

First-degree burns are burns involving the outer layer of skin (Fig. 14.16). A first-degree burn is a superficial injury, characterized by reddening of the skin. The reddening may be quite intense. A sunburn or mild scald is an example of a first-degree burn. While it may be quite painful, it will not cause scarring and will heal on its own

Second-degree burns are burns involving a partial thickness of the skin. A second-degree burn is characterized by deep reddening and blistering, caused by the injury of deeper layers of the skin and the capillaries found there. Plasma seeps into the tissues, raising the top layers of skin to form a blister. A second-degree burn, while deeper than a first-degree burn, does not injure the tissues to such an extent that they cannot heal themselves when treated with reasonable care. This is an important point that is not always recognized. Burns that are entirely second degree cause little scarring and do not require skin grafting. Sometimes, owing to the large body surface involved, they make the patient very ill and present a serious problem.

According to the NFPA, the following combinations of temperature and exposure produce second-degree burns of equal intensity:

Time of Exposure	*Temperature*	
1 sec or less	1093–1649°C	(2000–3000°F)
3 sec	371–482°C	(700–900°F)
15 sec	100°C	(212°F)
20–60 sec	71–82°C	(160–180°F)

Inhalation of air at these temperatures is said to produce burns of the respiratory tract with slightly longer exposure.

Third-degree burns are burns involving all layers of the skin, and sometimes underlying fat, muscle and even bone. These burns are also called full-thickness burns. A third-degree burn involves the entire thickness of the skin, with or without charring. Such a burn can never heal by itself, but requires the best surgical care. Lack of proper care may cause the patient to suffer months and even years of infection, disability and scarring.

Extensive third-degree burns can be extremely difficult surgical problems, requiring skin grafting of the involved areas at the earliest possible moment. Without grafting, the only way the wound can heal itself is by contracture, or drawing undamaged skin together to cover the damaged areas; the part that has been destroyed cannot be replaced except by dense scar formation. Third-degree burns may be the least painful of all types because of the extensive damage to nerve endings in the skin.

Determining the Severity of Burns

The amount of skin surface involved in a burn can be calculated quickly by using the "Rule of Nines." Each of the following areas represents 9% of the body surface: the head and neck; each arm; the chest; the upper back; the abdomen; the lower back and buttocks; the front of each leg; the back of each leg. The genital region is regarded as 1% of the body surface.

When the degree of the burn and the amount of body surface involved have been determined, the injury can be classified as to severity.

Critical (severe) burns are:

- Second-degree burns covering more than 30% of the body surface
- Third-degree burns covering more than 10% of the body surface

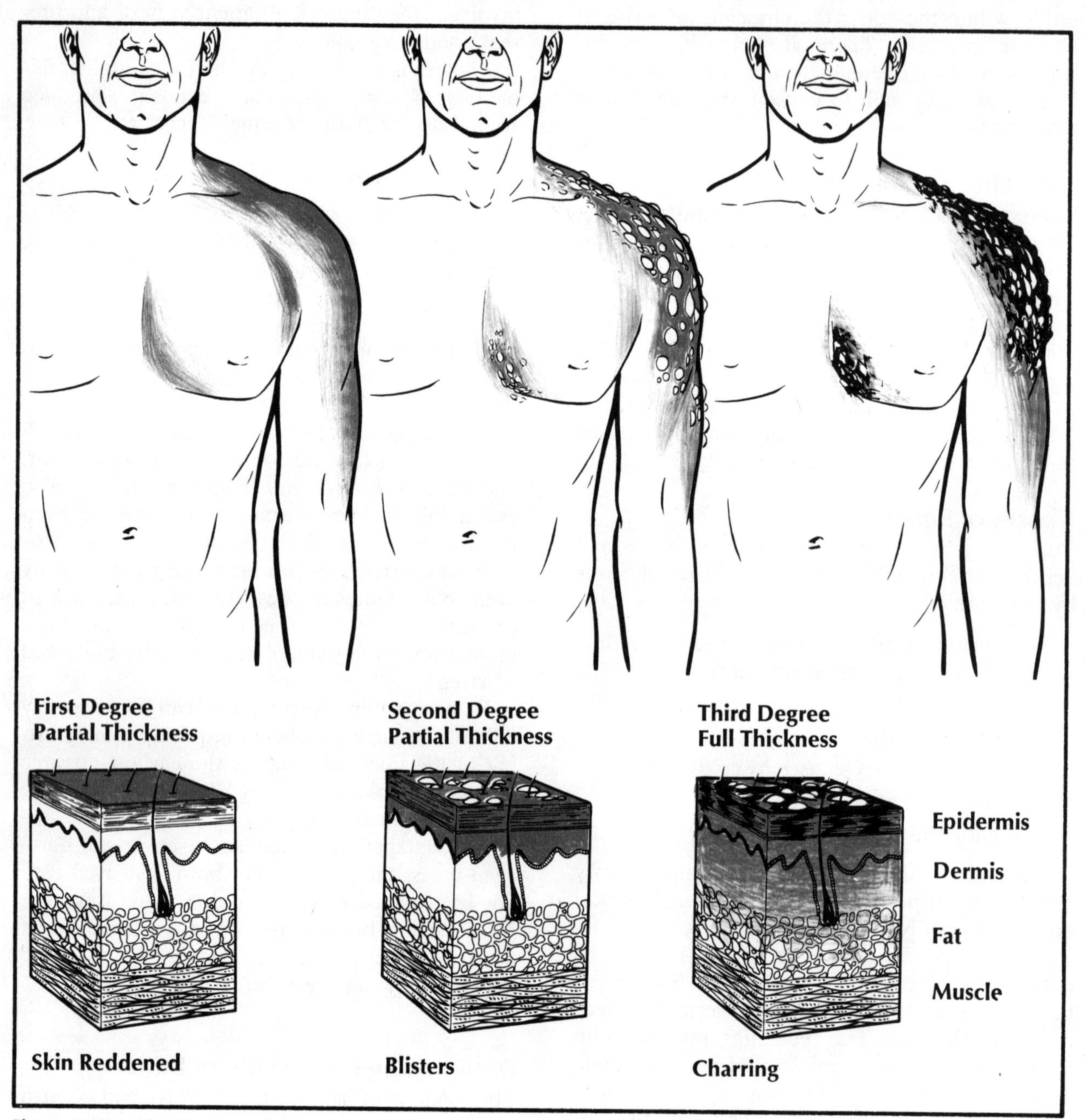

Figure 14.16. Classification of burns according to severity.

- Third-degree burns involving the critical areas of the hands, face or feet
- Burns complicated by respiratory tract injury, major soft-tissue injury, and fractures.

Moderate burns are:

- Second-degree burns covering 15% to 30% of the body surface
- Third-degree burns covering less than 10% of the body surface, including the hands, face and feet.

Minor burns are:

- Second-degree burns covering less than 15% of the body surface
- Third-degree burns covering less than 2% of the body surface
- First-degree burns covering less than 20% of the body surface, excluding the hands, face and feet.

Emergency Care Supplies

To give the burn patient the best possible care during treatment on the scene and transportation to a medical facility, the following equipment should be carried in the ship's medicine chest:

- Oropharyngeal airways of assorted sizes
- A bag-mask resuscitator with facepieces of various sizes
- Universal body dressing (a bed sheet is acceptable and is actually preferred because it can cover large areas, is economical and is easy to maintain)
- Universal extremity dressings (a sterile or clean terrycloth or turkish towel is preferred)
- At least 1000 cc of normal saline solution, including infusion tubes (without needles) for continuous treatment of chemical burns of the eyes and irrigation of the eyes when the eyelids are destroyed
- Sterile dressings in assorted sizes, for the head, face and neck, and between the fingers and toes
- Small suction cups to remove contact lenses from the eyes
- Bandage scissors
- Thermal blankets to maintain body heat.

Emergency Care of Burns

Burns are not treated, but rather cared for until medical attention can be obtained. Hospitals and physicians have particular courses of treatment for burns of different types. Any first aid treatment involving the application of ointments or sprays may make the physician's task more difficult. In any case, sprays or ointments that are applied by rescuers must be cleaned from the patient's body when he arrives at a medical facility. This is a long, tedious and often painful process.

Thermal Burns. General emergency care for patients with thermal burns is as follows: First make sure that the patient can breathe. Establish and maintain an open airway, and provide pulmonary resuscitation as required.

Check the patient for other injuries. A patient who has been burned may have fractures and lacerations caused by an explosion or the attempt to escape from the fire area. Lacerations and fractures must be treated as if no burns were present.

Cover the patient with a sterile or clean dressing (Fig. 14.17). Coverings such as blankets or other materials with a rough texture should not be used, because they can contaminate the wound. When the hands and feet are involved, dressings should be applied between the fingers and toes to minimize skin separation. Use sterile gauze

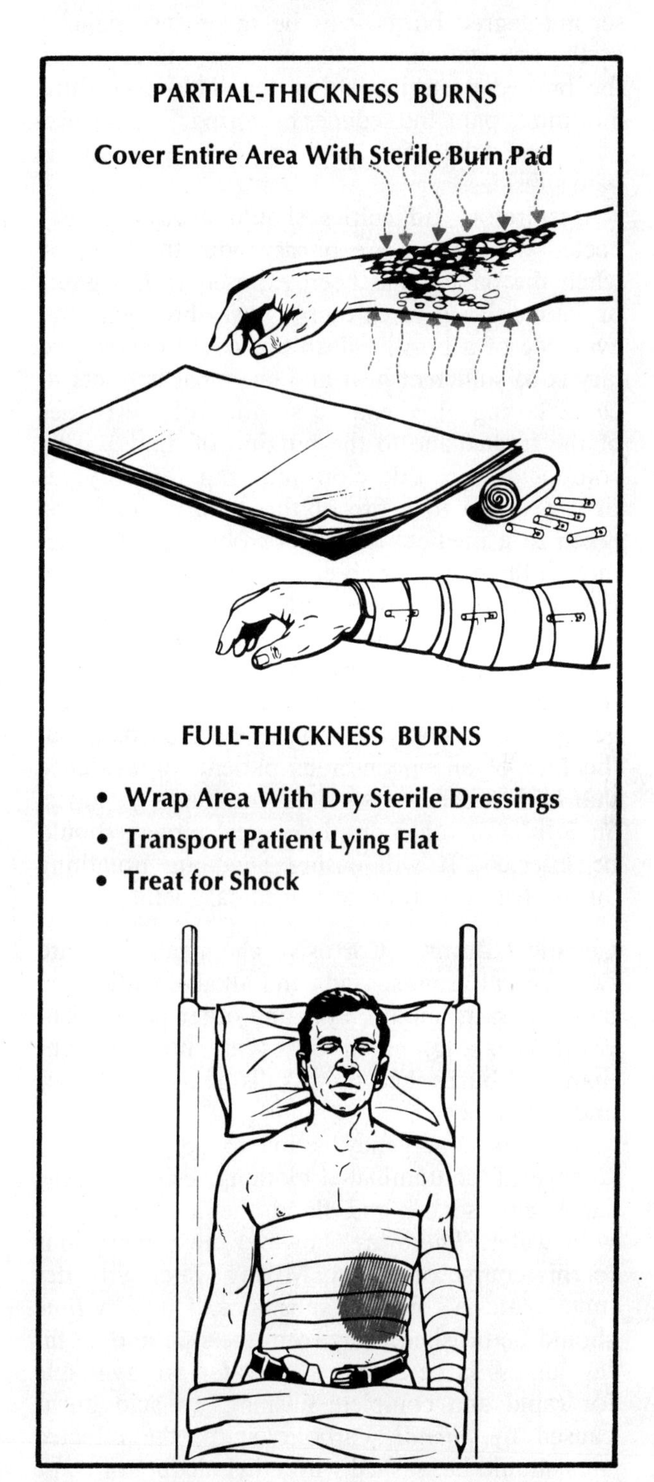

Figure 14.17. Emergency care for thermal burns.

pads for this purpose, and moisten them if sterile water is available. Moistened pads should also be used to cover burned eye areas.

Treat the patient for shock, and maintain his body heat. Do not delay transportation to a medical facility, but accomplish it smoothly and carefully.

Minor burns, such as first-degree and small second-degree burns, can be made less painful by the application of cold towels or immersion of the burned parts in cold water. This procedure minimizes pain and reduces blistering. The burned part should be kept in the cold water until all pain subsides.

Respiratory difficulties should always be expected when there are burns about the face, or when the patient has been exposed to hot gases or smoke. Even when there is very little external evidence of a burn, a flash fire could expose the larynx to sufficient heat and chemical irritants to cause laryngeal edema (a swelling of the tissues of the larynx due to the buildup of fluids). This complication can develop quite rapidly. Any patient exposed to a fire of this nature should be taken to a medical facility for observation, even though there may be no external signs of damage. Huskiness of the voice, mild respiratory distress or slight cyanosis should alert the rescuer to the problem. In some cases, the condition develops so quickly that the only relief possible is an emergency tracheostomy. When there are burns about the face of an unconscious patient, or evidence that he has been exposed to hot fumes or gases, an S tube or other oropharyngeal airway should be inserted. It will ensure adequate breathing during transportation to a medical facility.

Chemical Burns. Corrosive chemicals fall into two general groups: acids and alkalis. Either can burn the skin, mouth, eyes and other parts. Generally speaking, alkali burns are more serious than acid burns, because alkalis penetrate deeper and burn longer.

The first emergency action to be taken is to remove all contaminated clothing, especially the shoes and socks, and flood the affected areas with water. There are, however, two exceptions to this course of action. Mixing water with dry lime creates a corrosive substance. Thus *dry lime* should be brushed away from the skin and clothing, unless large amounts of water are available for rapid and complete flushing. In acid burns caused by *phenol* (carbolic acid), the affected area should be washed with ethyl alcohol or some other alcohol, since phenol is not soluble in water. Then the burn may be washed with water.

It is difficult to specify exactly how long a burned area should be irrigated with water. The water should be allowed to flow over the area long enough to ensure that all the chemical is flushed from the skin. The aftercare for chemical burns is similar to that for thermal burns. The area should be covered with a burn sheet, and the patient transported to a medical facility.

The rescuer should be careful not to get any of the hazardous chemicals on his own skin and clothing, and especially in his eyes. He should quickly remove any contaminated clothing as soon as his responsibilities to the patient are completed.

Electrical Burns. (*Caution: See* Techniques for Rescue and Short-Distance Transport, following in this chapter.) Electrical burns may be more serious than they first appear, since they often involve deep layers of skin, muscles and even internal organs. Basic emergency care for an electrical burn is to cover the site with a clean (preferably sterile) cloth and transport the patient to a medical facility. It is important to look for a second burn, which may have been caused by the path of the current through the body. The rescuer should also remember that in electrical accidents, the shock is likely to affect the patient's heart and lung action. In most cases the electrocuted patient will require CPR if he is to live.

Cryogenic and Nuclear Radiation Burns. These burns are discussed in the section on Environmental Emergencies, later in this chapter.

Special Care for the Eyes (Light Burns). Immediate emergency care is extremely important for chemical burns of the eyes. If acid or alkali burns are not treated immediately, irreparable damage may occur. The only emergency treatment possible is to dilute the chemical by flushing the eyes with large amounts of water. Sterile water is preferred, but if this is not available, ordinary running tap water should be used. (When chemicals are splashed into the eyes of a patient wearing contact lenses, the lenses should be removed immediately. Otherwise, they will prevent the water from getting to the corneal portions of the eye.)

Hold the patient's head, face up, under the running water. Have him hold his eyes open so that the globe and the undersides of the lids may be thoroughly irrigated. Tilt the head slightly to allow washing action from the nasal corner of the eye to the outside corner.

If running water is not available, have the patient hold his face down in a basin of water. Ask him to blink his eyes continually so that the necessary washing action is provided. Because it is natural to close the eyes when they are irritated, the patient may find it difficult to keep his eyes open during the irrigation. You may have to assist him by applying slight traction to the lids while the eyes are being flushed.

In port, irrigation of the eyes may be carried out during transportation to the hospital, using the fluids available in the ambulance. This will mean that the patient can be given medical treatment much sooner. At sea, the eyes should be irrigated immediately and repeatedly until a physician can be contacted by radio. In some cases, irrigation of the eyes may have to be continued all the way into port.

When a person suffers burns of the face from a fire, his eyes usually close rapidly due to the heat, thus protecting the globes. However, the eyelids remain exposed, and they may be burned along with the rest of the face. Since the treatment of burned eyelids requires specialized techniques, it is best to transport the patient without further examination of the eye. The eyelids should be covered with loose dressings during transportation to the nearest medical facility.

Light injuries are generally very painful. Some of the pain can be relieved by covering the eyes with dark patches.

If a patient is unconscious and his eyes remain open, the corneas may dry out, and ulcers will form. This condition will cause blindness, even though there are no other injuries to the eyes. Protect the eyes of an unconscious patient by maintaining their natural moisture. Close the lids and keep them closed, using tape if necessary. Normal tearing action will keep the surfaces of the globes moist. Be careful not to let the tape touch the globes.

FRACTURES AND INJURIES TO THE BONES AND JOINTS

There are three types of bone and joint injuries: fractures of bones, dislocations of bones, and sprains (injuries to ligaments). *Strains,* which are not injuries to the bones or joints, are often confused with sprains. Strains are injuries to muscles, caused by overexertion. The muscle fibers are stretched and sometimes partially torn. In most cases intense pain is the only sign of a strain.

Causes and Types of Fractures

In a fracture that is caused by *direct violence,* the bone is broken at the point of contact with an object. In a fracture caused by *indirect violence,* the bone is broken at a point other than the point of contact. The force that caused the break was transmitted along the bone from the point of impact. For example, a person who falls and lands on his hands or feet may suffer a broken arm or leg. A blow to the knees may fracture a hip. *Severe twisting forces* may cause fractures. For instance, a foot may be caught and twisted with sufficient force to fracture one of the leg bones. *Powerful muscular contractions* may cause pieces of bone to be pulled away. In addition, disease or aging may weaken bones sufficiently so that only a small force is needed to cause a break.

Fractures are divided into two basic categories, depending on whether a soft-tissue injury accompanies the fracture. An *open fracture* is associated with an open wound that extends between the fracture and the skin surface (Fig. 14.18). The soft-tissue injury may result from the tearing action of the broken bones, or from the object or force that caused the break. A *closed fracture* is a fracture without an accompanying soft-tissue injury. The injury must be determined by observing certain signs and symptoms. This injury is commonly called a "simple" fracture.

Signs and Symptoms of Fractures

Fractures are not always indicated by visible outward signs. Thus, the rescuer must be able to recognize other reliable signs before deciding on a course of treatment. However, whenever the mechanism of an accident is such that a fracture could exist, the rescuer should assume that it does exist.

Exposed bone ends are, of course, the surest sign of a fracture. Also, a *severe open wound* may have been caused by a force strong enough to fracture the bone directly under the wound at the same time.

Deformity is always a good sign of a fracture or dislocation. The rescuer should compare the suspected part with the unbroken similar part on

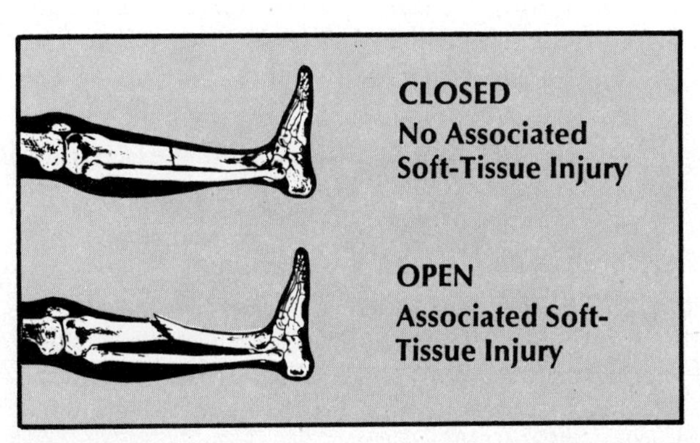

Figure 14.18. The two basic types of bone fractures.

the opposite side of the body. This helps in detecting differences in size or shape. Any depression of the skull should lead the rescuer to suspect a fractured skull; any depression of the rib cage is usually a sign of fractured ribs.

The patient's information is usually accurate. Sometimes the victim of a fracture has heard the bone snap or felt it break. *Pain and tenderness,* along with the patient's information, are usually reliable signs of a fracture. The site of a closed fracture can be found by gently pressing along the line of the bone; this is a helpful indicator when other more obvious signs are not present.

Grating is a sensation that can be felt by the rescuer when the broken ends rub together. However, this sign should not be sought intentionally, as it often increases discomfort and adds to the tissue damage.

Loss of use (disability) is a good sign of a fracture. The patient will not be able to use his injured arm or walk on his fractured leg. However, he should not be asked to try this only to determine whether a break has occurred. The patient often guards the injured part. In the case of a broken arm, he usually tries to hold it in the most comfortable position.

Dislocations

A dislocation is the shifting of a bone end that forms part of a joint, with injury to the surrounding ligaments and soft tissues. The joints most often affected by dislocations are the shoulders, elbows, fingers, hips and ankles. The signs of a dislocation are generally the same as those for a fracture. They always include

- Pain in the joint
- Deformity at the joint
- Loss of movement and pain when the patient attempts to move the joint.

A fracture or a dislocation may cause damage to nerves and blood vessels. In a dislocation, the bone end may be shifted a considerable distance from the joint. As it shifts, the bone end may move some nerves and blood vessels, or pinch others against other bones, resulting in paralysis or blood deficiency in the affected part. Numbness or paralysis below the dislocation site indicates a pinched or cut nerve. Loss of a pulse or coldness in the extremity is evidence of a pinched or severed blood vessel. If there is an indication that a blood vessel has been affected by the dislocation, immediate medical attention should be sought; the limb could be irreparably damaged by the reduced blood supply.

Sprains

Sprains are injuries in which ligaments are torn, usually by a forced motion beyond the normal range of the joint. Ankle sprains, for example, are caused when the body weight is thrown against a turned ankle. The areas of the body most commonly affected by sprains are the ankles and the knees.

Severe sprains often exhibit signs and symptoms similar to those of fractures and dislocations; sprains are sometimes mistaken for those more serious injuries. A dislocation almost always results in a deformity at the joint, while a sprain causes no such deformity. Other signs of sprains are:

- Pain during movement
- Swelling
- Discoloration.

Because dislocations and sprains exhibit the same basic signs as fractures, and because the rescuer does not have the formal training or the equipment necessary to make an exact diagnosis, he should treat all injuries to bones and joints as if they are fractures.

Emergency Care for Injuries to the Bones and Joints

Most fractures, especially those of the open type, appear gruesome and extremely dangerous, but they are rarely a threat to life. Unhurried and effective action by the rescuer may mean the difference between quick, complete recovery or a long, painful period of hospitalization and rehabilitation. No matter how short the distance to a medical facility, all fractures should be splinted before the patient is transported since the patient may not be treated immediately upon arrival at the medical facility.

In the case of a bone or joint injury, the rescuer should first care for the patient as a whole:

1. Ensure that the patient has an open airway and that he is breathing normally.
2. Stop any bleeding, and dress all wounds. In the case of an open fracture, the wound is dressed before the fracture is splinted. Bleeding wounds associated with open fractures may be controlled by direct pressure or, in extreme cases, by a tourniquet. Pressure-point control alone is not recommended, unless it is used only until a pressure dressing is applied.
3. Prevent shock.

When the patient's condition is stable, the rescuer should care for the bone or joint injury:

- Straighten any severely angulated fracture that can be straightened safely.
- Do not attempt to push back any bone ends.
- Immobilize the extremity before moving the patient.
- Immobilize the joints above and below the fracture.
- Immobilize dislocated joints, but do not attempt to reduce or straighten any dislocation.
- Apply slight traction during the splintering process.
- Splint firmly, but do not splint tightly enough to interfere with circulation.
- Suspect an injury to the spine in any accident that could cause such an injury, as well as obvious fractures or dislocations elsewhere.

Straightening Angulated Fractures. Slightly angulated fractures of the extremities do not usually present a problem. They can be immobilized in place with little trouble. On the other hand, severely angulated fractures pose a serious problem for both the rescuer and the patient. The angulation may make transportation to a medical facility quite difficult. The severe angle of the limb may be pinching or even cutting nerves and blood vessels at the injury site. Thus the rescuer should attempt to straighten all severely angulated fractures of the upper and lower extremities, with one exception: *Do not attempt to straighten fractures of the shoulders, elbows, wrists or knees.* Because major nerves and great blood vessels pass close to these joints, attempts to straighten fractures may actually increase the possibility of permanent damage.

The idea of straightening a severely angulated fracture may be quite distasteful because of a fear of causing additional pain. It is true that pain may be increased during the straightening procedure. However, it will be temporary, and it should decrease considerably after the splint is applied. It may be far more painful for the patient if he is transported with the limb in a severely angulated position. For angulated fractures other than those at the shoulders, elbows, wrists or knees, the rescuer should proceed as follows.

Cut or tear away the clothing that lies over the fracture site. If it is necessary to tear the clothing, do so very carefully to avoid moving the limb. Work slowly and deliberately. Make every effort to ensure that the broken ends of bone are not forced through the skin.

Grasp the extremity gently but firmly (Fig. 14.19). One hand should be directly below the break, and the other further down the limb for support. Have another rescuer provide countertraction by holding the patient firmly in place, especially the part of his body closest to the fracture site.

Apply traction steadily and smoothly. If any firm resistance is felt, do not attempt to correct the angulation forcibly. When the limb has been straightened, maintain traction on the extremity until the splinting device has been applied.

Do not attempt to straighten a dislocation. Movement of the displaced bones may damage nerves and blood vessels that lie close to the joints.

The Reasons for Splinting. When nature's supporting structure (the bone) is broken, some substitute support must be provided to prevent further injury and shock. When properly applied, a splint should:

1. Reduce the possibility that a closed fracture will become an open one.
2. Minimize the damage to nerves, muscles and blood vessels that might otherwise be caused by the broken bone ends.
3. Prevent the bone ends from churning around in the injured tissues and causing more bleeding.
4. Lessen the pain that is normally associated with the movement of broken bone ends.

Types of Splints. Any material or appliance that can be used to immobilize a fracture or dislocation is a splint. There are many types of commercially made splints, such as wooden splints, scored cardboard splints, molded aluminum splints, soft-wire splints and inflatable plastic splints.

The lack of a commercially made or specially prepared splint should not keep the rescuer from immobilizing a fracture or dislocation. A piece of wire or a tongue depressor inside a bandage may be sufficient to immobilize a fractured finger. An injured leg may be immobilized by bandaging it to the good leg, or by binding it in a pillow or blanket roll. A cane, umbrella or similar object may be used to splint a broken arm. Rolled-up newspapers also make good splints. A ladder may be used as a stretcher for transporting a patient with an injured back or spinal injury.

Splints are of two basic types: rigid and traction. Backboards, notched boards, molded splints,

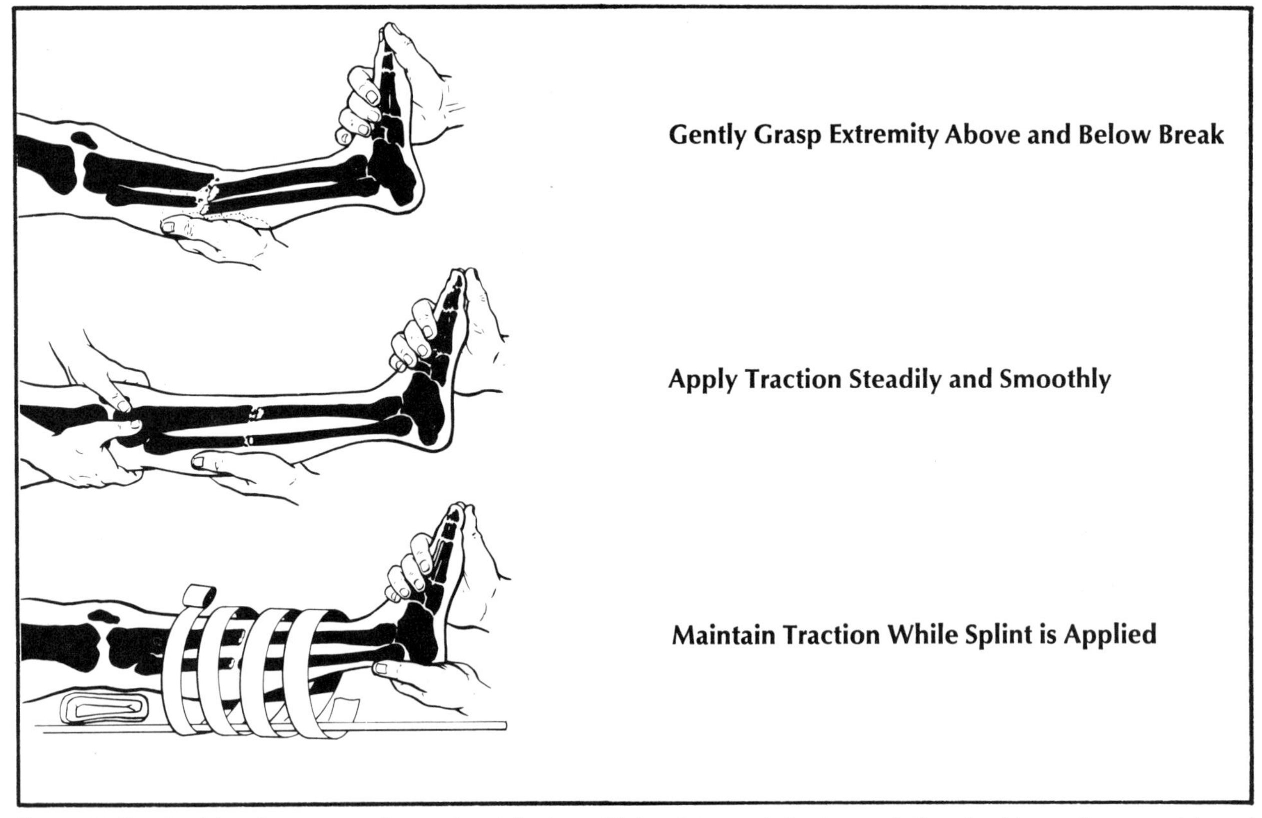

Figure 14.19. Straightening a severely angulated fracture. Dislocations and fractures of the shoulders, elbows, wrists and knees should not be straightened.

cardboard splints and inflatable splints are all rigid splints. A rigid splint, whatever its construction, must be long enough so that it can be secured well above and below the fracture site, to immobilize the entire bone.

Applying a Rigid Splint. To apply a rigid splint, grasp the affected limb gently but firmly, with one hand above the fracture and the other below it (Fig. 14.20). Apply slight traction by moving your hands apart. Have another rescuer place a padded splint under, above or alongside the limb. (There should be enough padding to ensure even contact and pressure between the limb and the splint, and to protect all bony prominences.)

Wrap the limb and splint with bandaging materials so that the two are held firmly together. Self-conforming, self-adhering roller bandage is especially well suited for this purpose. Make sure the bandaging material is not so tight that it affects circulation. Leave the fingers or toes of the splinted extremity uncovered, so that circulation can be checked constantly.

It is important to remember that rigid splints are effective only if they are *long enough* to immobilize the entire fractured bone; if they are *padded sufficiently;* and if they are *secured firmly* to an uninjured part.

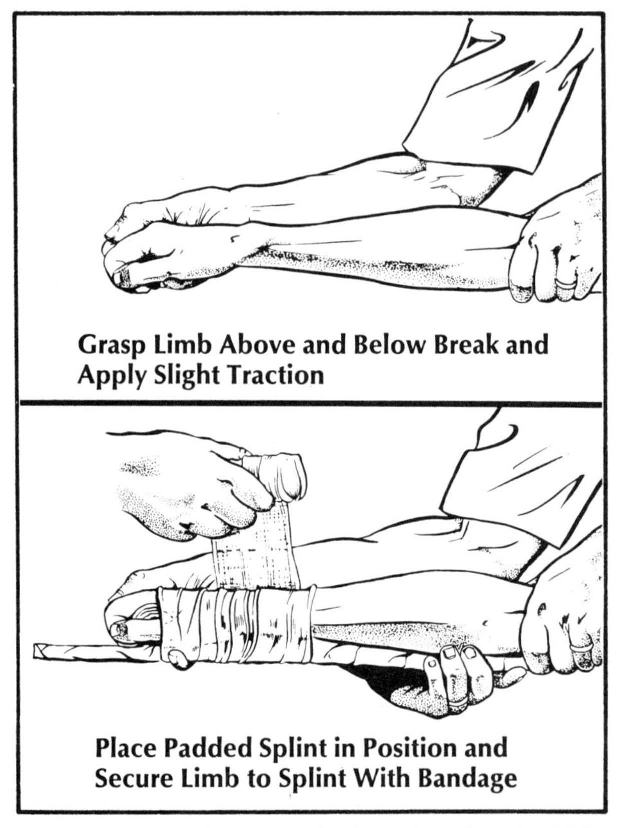

Figure 14.20. Applying a rigid splint. The splint should be long enough to be secured well above and below the fracture.

Inflatable splints are effective in immobilizing fractures of the lower leg or the forearm. They are of little value for fractures of the humerus (upper arm), or the femur (upper leg), since they do not extend past the upper joint in either case.

To apply an inflatable splint, gather the splint on your arm, so that the bottom edge of the splint is above your wrist (Fig. 14.21). With the hand of that arm, grasp the hand or foot of the affected extremity. Have another rescuer grasp the injured extremity above the fracture site. Apply gentle traction between your hand and the hand of the other rescuer.

While you maintain the traction, have the other rescuer slide the air splint over your hand and onto the patient's limb. See that it is properly positioned and free from wrinkles. While you continue to maintain traction, have the other rescuer inflate the splint by mouth. It should be inflated to the point where your thumb will make a slight dent when you press it against the splint.

When an inflatable splint is applied in cold weather and the patient is moved to a warmer area, the air in the splint will expand. This may cause too much pressure in the splint and on the injured part. It may be necessary to deflate the splint until the proper pressure is reached.

If the splint is of the zipper type, it is necessary to lay the limb in the unzippered splint and then to zip it up and inflate it. It will not be possible to maintain traction on the injured limb during the operation. Again, the splint should be inflated only by mouth, and only to the point at which you can make an indentation in the splint with your thumb.

- **Gather Splint on Your Own Arm**
- **Grasp Patient's Hand While Second EMT Grasps Limb Above Fracture**
- **Apply Traction**

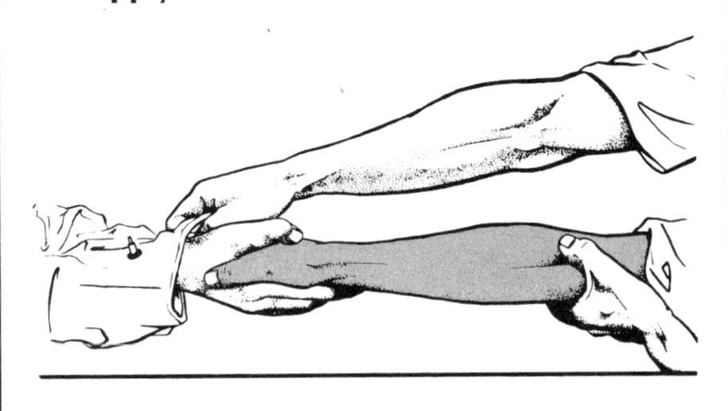

- **Slide Splint Onto Limb**
- **Inflate Splint by Mouth**

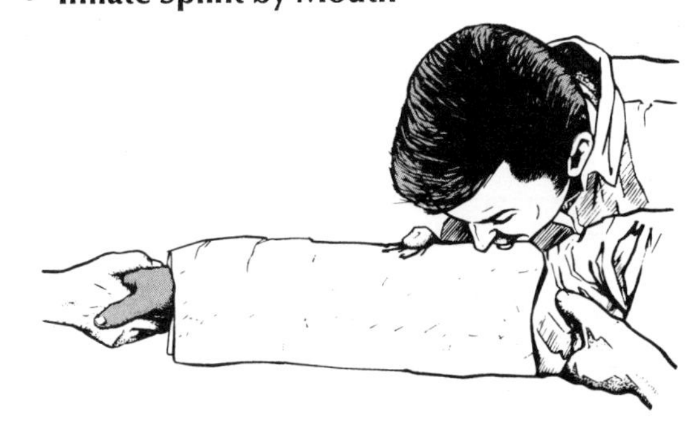

Figure 14.21. Applying an inflatable splint. The splint should be inflated only to the point where it can be pressed in easily with a thumb.

Other Immobilization Techniques. An arm sling made from a triangular bandage or other soft material is valuable for use with splints of the upper extremities. The sling serves several purposes: It helps immobilize the injured limb; it helps ease pain by taking some of the weight off the injured part; and it supports and further protects the limb during transportation.

To apply an arm sling, first place the splinted limb in a comfortable position. Then fashion the arm sling by placing the long edge of the triangular bandage along the patient's side opposite the injury (Fig. 14.22). Bring the bottom end of the bandage up over the forearm, and tie the two ends together. Make sure that the knot is not directly behind the patient's neck. Pin or tie the pointed end of the sling so that it forms a cradle at the elbow.

Another immobilization device is the sling and swathe. It is most effective when the patient has a fractured collarbone. To apply a sling and swathe, first apply an arm sling in the manner just described. Make a long swathe from 15.2-cm (6-inch) bandaging material. Circle the body and the arm sling with the swathe, and draw it up snugly to hold the injured part firmly to the body (Fig. 14.23).

Emergency Care for Injuries to the Upper Extremities

Injury to the Clavicle (Collarbone). The patient with a fractured clavicle typically sits or stands with the shoulder of the injured side bent forward. He generally has his elbow bent, with his forearm placed across his chest and supported by the other hand. The patient complains of pain in and around the shoulder. Any movement of the

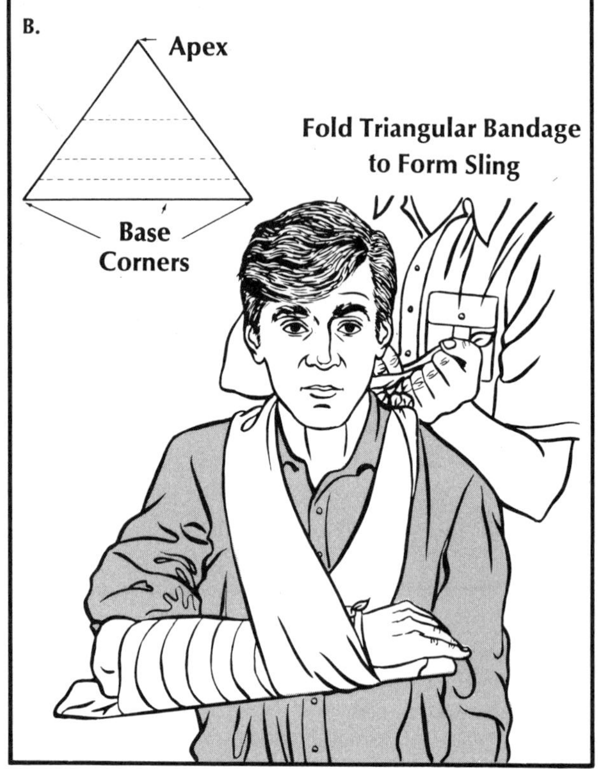

Figure 14.22. Applying an arm sling. **A.** Standard sling. **B.** Alternative sling.

shoulder or the arm on the injured side is painful. There may be swelling or an obvious lump in the area of the injury.

Have the patient fold the arm of the injured side across his chest in a comfortable position.

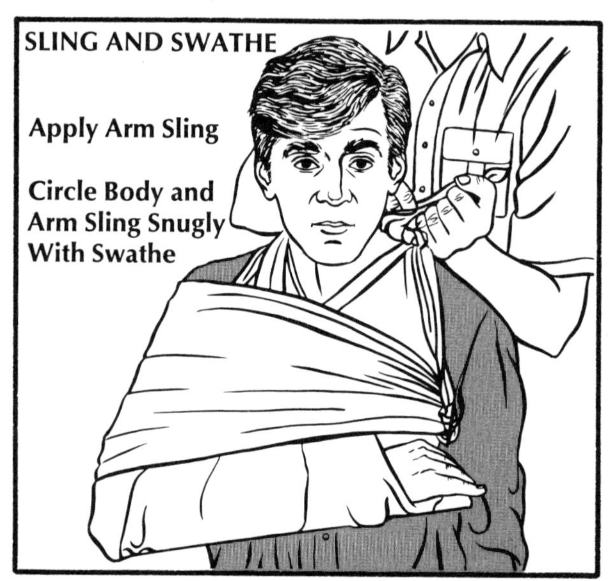

Figure 14.23. Applying a sling and swathe.

Place the arm in a sling, and hold the sling against the patient's body with a snug swathe.

Injury to the Humerus (Upper Arm). When the humerus is fractured, swelling and deformity may not be as evident as in other types of fractures. The patient complains of pain, especially when he moves the arm. The arm is tender when it is touched gently in the area of the fracture.

First correct any severe angulation. Immobilize the arm by securing it to a short board splint with two cravats or roller bandages, one just above the elbow and one just below the armpit (Fig. 14.24). Place the arm in a sling that supports only the wrist. The weight of the forearm will thus provide slight traction on the arm. If a short board is not available, use a sling and swathe to immobilize a fractured humerus.

Injury to the Elbow. Because bone movement could damage nerves and blood vessels, a frac-

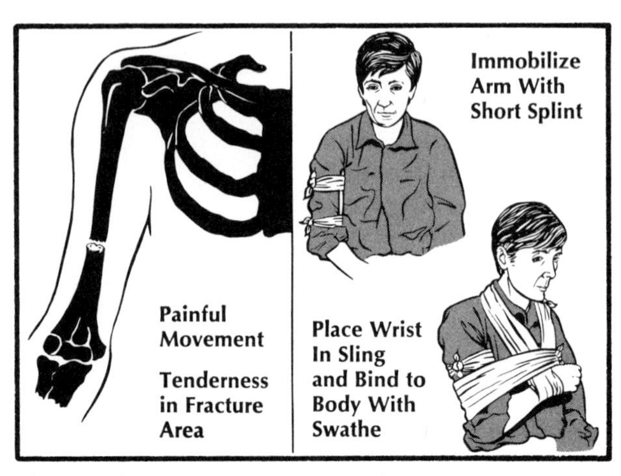

Figure 14.24. Emergency care for a fractured humerus.

ture of the elbow must be immobilized in the position in which it is found. It should not be twisted, straightened or bent in any direction.

If the extremity is found in a straight-out position, immobilize the limb in a well-padded splint that extends from the armpit to the fingertips (Fig. 14.25). If a splint of this length is not available, a rolled blanket may be used quite effectively. If the extremity is found in a bent position, immobilize the limb in the bent position with a wire ladder splint, a padded board bandaged to both the arm and forearm, or a sling and swathe (Fig. 14.25).

Injury to the Forearm and Wrist. When there is no angulation, splint the forearm in a well-padded rigid splint that includes both the elbow and the hand (Fig. 14.26). Place a rolled bandage or similar material under the palm to maintain the natural position of the hand. Secure the splinted arm in a sling.

When there is severe angulation of the bones of the forearm, straighten the angulation carefully with manual traction. Then splint the extremity in the manner just described. Inflatable splints are especially well suited for immobilizing fractures of the forearm.

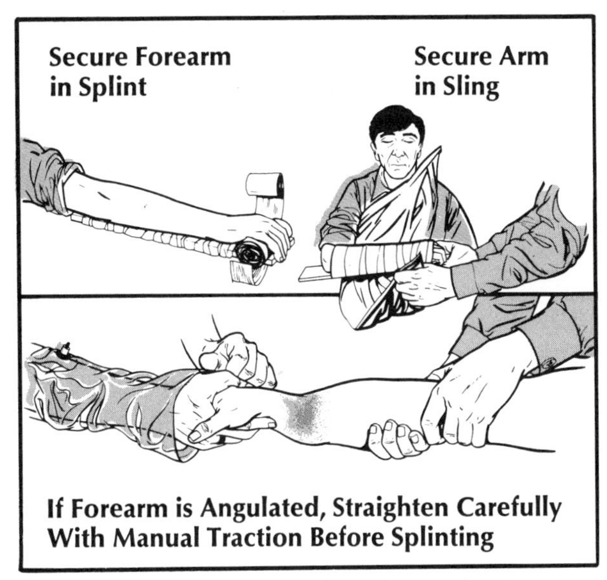

Figure 14.26. Emergency care for a forearm fracture. Note the rolled bandage between the splint and the patient's palm.

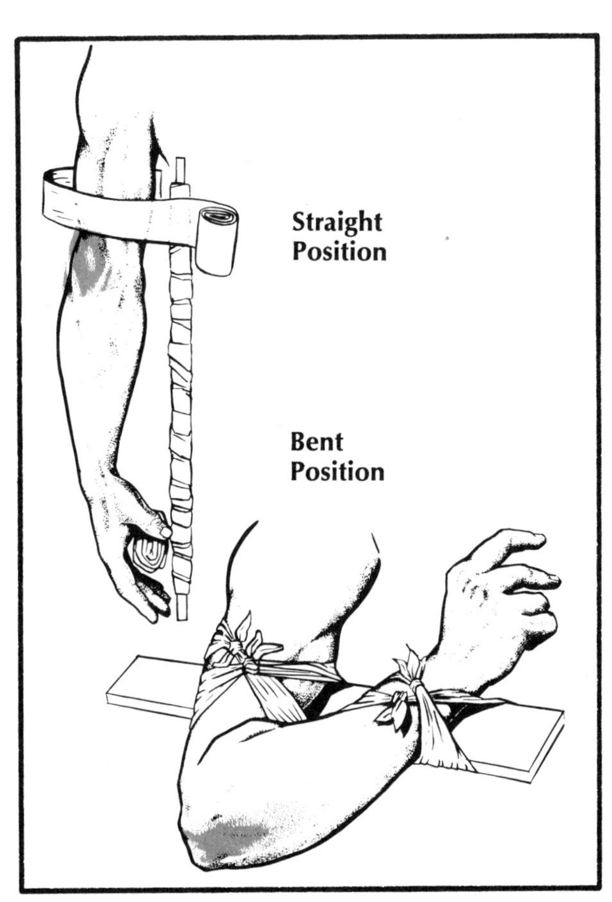

Figure 14.25. An injured elbow should be immobilized in the position in which it is found.

Injury to the Hand. A patient with a fractured hand experiences acute pain and tenderness. The joints of the injured bone appear much larger than the other knuckles.

The hand should be secured to a board splint that extends from beyond the fingertips to above the wrist. The hand itself should be bandaged so that it is maintained in the position of function. The arm should be placed in a sling to help lessen the pain. A fractured finger can be effectively splinted with a padded tongue depressor. The end of the wooden blade should extend well back into the palm to minimize movement at the joints.

Emergency Care for Injuries to the Lower Extremities

Injury to the Hip. A fractured hip may be immobilized with a long board splint, such as the long fracture board carried on ambulances. Use a well-padded board that reaches from the patient's armpit to his ankle (Fig. 14.27). Using cravats, tie this board to the ankle, lower leg, thigh, trunk and chest, so that the entire leg, pelvis and spine are immobilized.

A full backboard may also be used to immobilize a fractured hip. Slide the backboard very carefully under the patient, moving him as little as possible (Fig. 14.27). Place a blanket between the patient's legs, and bandage the legs together from the thighs to the ankles. Using long straps,

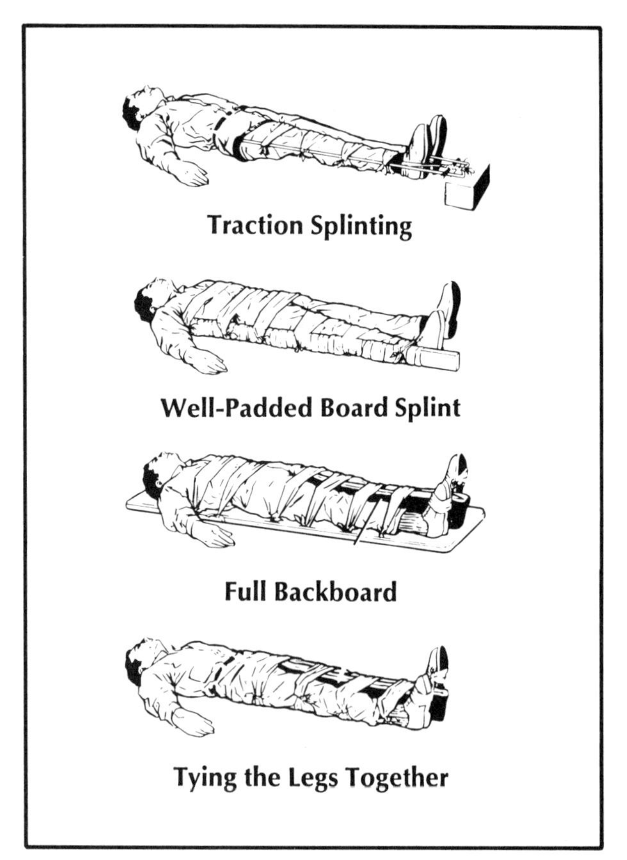

Figure 14.27. Methods of immobilizing a fractured hip.

secure the patient to the backboard from head to toe to ensure proper immobilization.

Dislocations of the hip are characterized by obvious deformity about the hip joint, and by the patient's resistance to attempts to correct the deformity. The leg is usually bent to some extent and turned inward.

To care for a dislocation of the hip, slide a long spine board very carefully under the patient. Move him as little as possible. Keep the leg on the injured side bent, using padding such as pillows to maintain it in the position in which it was found. Immobilize the patient on the board by securing him with straps.

Injury to the Femur (Upper Leg). Swelling may not be evident in the case of a fractured femur, because there is so much soft tissue in the thigh. However, pain and tenderness almost always accompany this injury. The patient may be in shock, owing to the amount of blood that has been lost into the tissues surrounding the fracture. The pain that comes with any movement of the injured extremity may contribute to the shock. There may be either severe angulation or relatively little deformity, depending on what caused the injury.

Severe angulation should be corrected by steady traction. Continue the traction by applying a half-ring splint, if available. If not, immobilize the injured extremity with a long board splint or a backboard, as in the case of a fractured hip.

Injury to the Knee. A fractured knee is painful, tender and swollen. If it is bent, the patient will not be able to straighten it. It may be possible to feel the gap between fragments of bone. The knee should be immobilized in the position in which it is found, to prevent further damage to nerves and blood vessels.

Improvise a means to immobilize the knee with available materials. Use a padded board splint to hold the leg in the position in which it was found. Make sure that all spaces between the splint and the leg are well padded.

A pillow can be used effectively to immobilize a fractured knee. Mold the pillow around the knee, in the position in which it is found. Use cravat bandages or belts to secure the pillow to the extremity. In moving the patient, take care to shift the injured limb as little as possible.

Injury to the Lower Leg. In a fracture of the lower leg, the usual pain and swelling are present, even if there is no deformity. Severe angulation should be corrected as previously described. Inflatable splints are especially well suited for immobilizing the lower leg.

To apply an inflatable splint to the lower leg, slide the splint over your arm until the lower end clears your wrist. Apply traction at the ankle and foot while another rescuer holds and supports the extremity above the fracture. Continue to maintain the traction while the other rescuer slides the splint over your hand and onto the patient's leg (Fig. 14.28). Make sure that the splint

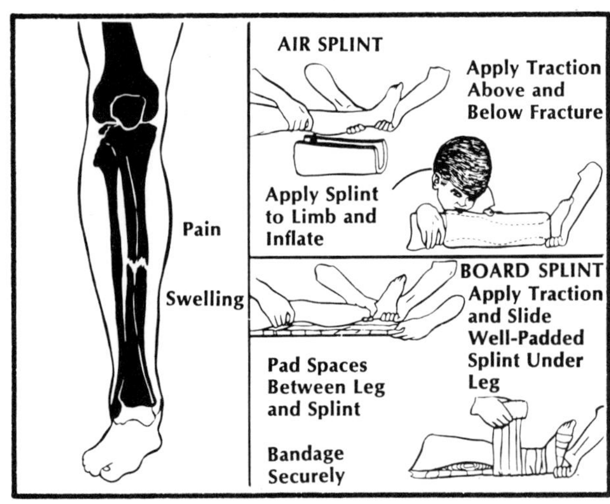

Figure 14.28. Emergency care for a fracture of the lower leg.

is wrinkle-free and that it covers the fracture site. Inflate the splint to the proper pressure.

If a board splint is to be used, take the toes of the injured leg in one hand, and the heel in the other. Pull gently to apply traction while another rescuer applies a well-padded board splint to the underside of the injured leg. Make sure that all spaces between the leg and the splint are well padded. Bandage the splint securely from the knee to the ankle.

Effective splints for the lower leg may also be made from pillows and blanket rolls.

Injury to the Ankle and Foot. It is often difficult to distinguish between a fracture and a sprain of the foot or ankle. Local swelling and pain characterize both types of injury. If it is impossible to tell the difference, splint the injured part.

A simple pillow splint is a quick and effective means of immobilizing an injured ankle or foot. The pillow should be molded carefully around the foot, and the edges secured with pins or cravats.

ENVIRONMENTAL EMERGENCIES

Environmental emergencies are injuries caused by the patient's surroundings. In firefighting situations, rescuers may have to deal with emergencies caused by exposure to heat and cold, radiation and air-borne poisons, as well as drowning.

Emergencies Caused by Heat

Normally, the body produces heat at a certain rate. If this heat can leave the body as it is formed, there is no change in body temperature. If heat leaves the body too rapidly, the body cools down. Heat is then produced at a greater rate, to bring the body temperature back to normal.

If heat leaves the body too slowly, the body temperature rises. As a result, the person is said to have a fever. The excess heat speeds up certain body processes, and additional heat is produced. Then, the body must eliminate not only the normal heat, but also the additional heat.

Heat produced within the body is brought to the surface mainly by the bloodstream. It escapes to the cooler surroundings by conduction and radiation. If air movement or a breeze strikes the body, additional heat is lost by convection. However, when the temperature of the surrounding air becomes equal to or rises above the body temperature, all the heat must leave by vaporizing moisture (sweat) from the skin. As the air becomes more humid (contains more moisture), the vaporization rate slows down. Thus, on a very humid, hot day, when the temperature is about the same as the body temperature, and there is little or no breeze, too much heat may be retained within the body. On such a day, or during several such days (a heat wave), medical emergencies due to heat are likely to occur.

Emergencies caused by heat are classified as heat cramps, heat exhaustion or heat stroke.

Heat Cramps. Heat cramps usually affect people who work in hot environments and perspire a great deal. Loss of salt from the body causes very painful cramps of the leg and abdominal muscles. Heat cramps may also result from drinking ice water or other drinks too quickly or in too large a quantity. The signs of heat cramps are

- Muscle cramps in the legs and abdomen
- Pain accompanying the cramps
- Faintness
- Profuse perspiration.

To provide emergency care for heat cramps, remove the patient to a cool place. Give him sips of salted drinking water (5 ml salt per liter (1 tsp salt per qt)). Apply manual pressure to the cramped muscle. Transport the patient to a medical facility if there is any indication of a more serious problem.

Heat Exhaustion. Heat exhaustion also occurs in individuals working in hot environments; it may be associated with heat cramps. It is brought about by the pooling of blood in the vessels of the skin. The heat is transported from the interior of the body to the surface by the blood. Blood vessels in the skin become dilated, and a large amount of blood collects in the skin. In addition, blood collects in the lower extremities when the patient is in an upright position. These two effects may lead to the inadequate return of blood to the heart, and eventually to physical collapse. Heat exhaustion can be prevented if the crew takes adequate water and salt tablets. Loose-fitting garments that allow cooling by evaporation can also aid in preventing heat exhaustion. However, firefighters should not shed their protective gear.

The signs of heat exhaustion are

- Weak pulse
- Rapid and usually shallow breathing
- General weakness
- Pale, clammy skin
- Profuse perspiration
- Dizziness
- Unconsciousness

- The appearance of having fainted (the patient responds to the treatment for fainting).

To provide emergency care for heat exhaustion, remove the patient to a cool place and remove as much clothing as possible. Have the patient drink cool water in which some salt has been dissolved, if he is conscious. If possible, fan the patient continually to remove heat by convection, but do not allow chilling or overcooling. Treat the patient for shock, and transport him to a medical facility if there is any indication of a more serious problem.

Heat Stroke. Heat stroke is a severe disturbance of the heat-regulating mechanism, leading to high fever and collapse. Sometimes this condition results in convulsions, unconsciousness and even death. Direct exposure to the sun, poor circulation, poor physical condition, and advanced age bear directly on the tendency toward heat stroke. It is a serious threat to life and carries a 20% mortality rate. Alcoholics are extremely susceptible.

The symptoms of heat stroke are

- Sudden onset
- Dry, hot and flushed skin
- Dilated pupils
- Early loss of consciousness
- Full and fast pulse
- The breathing is deep at first, but later shallow and even almost absent
- Twitching muscles, growing into convulsions
- Body temperatures reaching 40.5–41°C (105–106°F) or higher.

The rescuer must realize that heat stroke is a true emergency. Transportation to a medical facility should not be delayed. Remove the patient to a cool environment if possible, and remove as much of his clothing as possible. Make sure there is an open airway. Reduce the patient's body temperature promptly by dousing his body with water, or preferably by wrapping him in a wet sheet. If cold packs are available, place them under the arms, around the neck, at the ankles, and any place where blood vessels that lie close to the skin can be cooled. Protect the patient from injury during convulsions, especially from tongue biting.

Hyperthermia. In all cases of high body temperature (hyperthermia), rescuers should avoid extensive cold-pack treatment unless they are trained to administer it.

Radio contact with a physician and constant monitoring of the body temperature are critical when cold packs are used. Never allow ice or ice-filled objects to come into direct contact with the patient. The patient should be covered, and the cold packs placed around him. These cold packs should in turn be covered to help maintain their low temperature. The administering of cold packs should be a slow process. If the temperature is lowered too rapidly, the patient may go into shock. (This is why radio contact with a physician is strongly recommended.)

Emergencies Caused by Cold

General Cooling (Hypothermia). General cooling of the body is a true emergency. The patient should be transported to a medical facility as soon as possible.

Replace any wet clothing with dry clothing, and warm the patient. Since the body cannot generate adequate body heat, it is necessary to provide heat externally. All body surfaces should be warmed. To accomplish this, a hot water bottle, heating pads and the like may be useful, but no artificial heat should be placed next to the bare skin. If the patient is inside a warm place and is conscious, a hot bath will be most helpful. Hot liquids (again, only if the patient is conscious) will also speed the warming process.

Carefully monitor the patient's respiration and heartbeat. For a severely cooled patient, pulmonary or cardiopulmonary resuscitation may be required in the event of cardiac arrest.

Local Cooling. Local cooling injuries, affecting particular parts of the body, fall into two categories: frostbite and freezing. The parts most commonly affected by frostbite are the ears, nose, hands and feet. The symptoms of frostbite are progressive. First, the exposed skin reddens. Then, as exposure continues, the skin takes on a gray or white blotchy appearance, especially at the earlobes, cheeks and the tip of the nose. The exposed skin surfaces become numb, owing to reduced circulation. If the freezing process continues, all sensation is lost and the skin becomes dead white.

Gradually rewarm the frozen part by immersing it in warm water (specifically 39.4–41.7°C (103–107°F)). Make sure the affected part does not touch the container. In addition to rewarming, make every effort to protect the frozen area from further damage. Very gently remove anything that may cause constriction, such as boots, socks or gloves. Obviously, if the feet are involved the patient should not be allowed to walk. Carefully and thoroughly dry the warmed area to prevent recooling by evaporation.

The general condition and comfort of the patient are improved by hot, stimulating fluids such as tea and coffee. Coffee is especially good, since it both stimulates and helps to dilate the blood vessels. In many cases, as the part begins to thaw the pain is severe enough to require drugs for relief.

After thawing is complete, raise and lower the part rhythmically to stimulate the return of circulation. Avoid pressure on any part of the frostbitten area. Cover the affected part with a dry, sterile dressing. Do not allow the patient to smoke, as tobacco constricts the blood vessels and restricts circulation.

Deep frostbite, or freezing, is much more serious than frostbite. Like general cooling of the body, it should be considered a true emergency. Deep freezing of body tissues is characterized by a waxy, white appearance; the skin surface is quite hard and unyielding. It is likely that subcutaneous tissues are injured, and may actually be destroyed.

Arrange to transport the patient to a medical facility without delay. Keep the affected parts dry. Provide external body heat if possible, and provide pulmonary assistance or CPR as required.

Emergencies Caused by Poisoning

Poisons can enter the system in four ways:

- Ingestion (by mouth)
- Inhalation (by nose)
- Absorption (through the skin)
- Injection (into the body tissues or bloodstream).

In firefighting situations, it is rare to find a case of poisoning by ingestion or injection. Proper protective gear will eliminate these dangers. As an added safety measure, those at the fire scene should avoid gloved or bare hand contact with their mouths. Care should be taken to avoid ingesting any foods that were stored open in the fire area. Food containers and utensils should be cleaned before use.

Inhaled Poisons. Inhaled poisons may produce respiratory symptoms such as shortness of breath, coughing and cyanosis. The patient may pass into cardiac arrest if the respiratory problems are not corrected. The patient should be removed from the poisonous atmosphere and carried into fresh air. If he is breathing, his lungs should be flushed with oxygen (qualified personnel only). If he is not breathing, pulmonary resuscitation or CPR should be applied. A patient may show a temporary recovery from toxic gas poisoning and then go into respiratory arrest when left unattended. Anyone exposed to a toxic atmosphere should be kept under close observation.

Usually there is no indication of carbon monoxide poisoning until the patient collapses. The gas is odorless, tasteless and colorless, so the danger is not recognized until the patient passes out. He may have headaches and dizziness, but these are usually attributed to other causes and thus overlooked. There is only one sign of carbon monoxide poisoning that is usually reliable and unmistakable: The skin takes on a cherry-red color that is unlike any other symptom of illness. Since this color change may not be obvious with patients having dark complexions, assume the toxic gas is carbon monoxide and treat as follows.

Remove the patient to fresh air, and start resuscitation immediately. If possible, use a bag-mask resuscitator, so that the oxygen is administered as effectively as possible. If the patient is breathing spontaneously, a mechanical inhalator may be used. Transport the patient to a medical facility as soon as possible.

Absorbed Poisons. Absorbed poisons may cause irritation of the skin and mucous membranes and inflammation of the eyes.

To care for a patient who has absorbed poison through the skin, remove the contaminated clothing, including shoes, watches and rings. Flood the contaminated surface with water for at least 15 minutes. Do not use medication on the skin unless ordered to do so by a physician. If the poison has contacted the eyes, flush them with large amounts of water. Observe the patient closely for signs of shock, and be alert for changes in respiration and circulation.

Emergencies Caused by Explosions

An explosion is a very rapid release of energy. The magnitude of an explosion depends on several factors, including the type of explosive agent, the space in which the agent is detonated, and the degree of confinement of the explosion. The damage done by an explosion results from a shock wave that is generated by the release of energy. As the wave extends outward in all directions, two types of pressure are generated almost simultaneously. An *overpressure* (an increase over normal atmospheric pressure) surrounds each object as the shock wave hits it, tending to crush it inward. At about the same time, *dynamic pressure* (like a strong wind) strikes the object and tends to push it over and tear it apart. Any loose debris is picked up and propelled outward by the shock wave.

As the shock wave passes, the pressure decreases slightly (to below normal), and the airflow is reversed. This *suction* phase may cause further damage, although considerably less than that resulting from the shock wave.

Within the area of the blast, certain injuries may result from the shock wave itself. These include ruptured eardrums, ruptured internal organs, internal bleeding, and contusions of the lungs caused by the rapid changes in pressure. The lung injuries may cause pulmonary edema and hemorrhage. The resulting fluid congestion may decrease the amount of oxygen available for transfer to the blood, causing anoxia.

As explosive material detonates, a great deal of heat is generated. Although this heat is rapidly dissipated, people close to the point of detonation may be burned. The severity of burns, like other blast-related injuries, depends a great deal on the distance of the victim from the explosion. Unprotected skin areas, such as the face and hands, are especially vulnerable.

Since the shock wave causes loose materials and debris to be propelled outward, people may be injured by flying objects. These objects may cause abrasions, contusions and lacerations. If the objects are traveling at sufficient speed, they may also cause fractures or penetrate the extremities and vital organs. Heavy falling objects may cause typical crushing injuries, including severely bleeding wounds and fractures.

Rescuers should be prepared to deal with multiple and widely varied injuries in each patient. They should ensure an adequate airway, support respiration as required, control external bleeding, and splint fractures. Patients should be transported to a medical facility without delay, and the probability of severe internal injuries must be considered.

Drowning Emergencies

In drownings, the type of water entering the lungs is an important factor. In fresh-water drowning, the water in the lungs is absorbed into the bloodstream through the capillary walls. Two things happen: The blood vessels swell and in some cases burst, and the blood chemistry is thrown badly out of balance. The chemical imbalance is so great, in fact, that the heart goes into ventricular fibrillation. This is probably the principal cause of death in fresh-water drownings.

When a victim drowns in salt water, the process is reversed. Salt water is more concentrated than blood, so fluid is drawn from the blood into the lungs, causing pulmonary edema, or saturation of the lung tissues. As much as one-quarter of the blood volume may be lost into the lungs in a salt-water drowning; the victim may actually drown in his own fluids.

In all drowning cases, resuscitation measures must be started within a very few minutes if the patient is to survive. No effort should be made to drain water from the lungs; getting air into the patient without delay is of prime importance. Mouth-to-mouth resuscitation should be started even before the patient is removed from the water, if at all possible. If there are signs of cardiac arrest, CPR should be initiated when the patient can be placed on a firm surface. As soon as it is available, oxygen should be administered under positive pressure by a qualified professional.

Even if the patient has apparently recovered at the scene, he should be transported to a medical facility without delay. Delayed deaths after apparent recovery are common. They may be caused by pulmonary edema (fluid in the lungs) or other complications. In some cases, involved medical procedures are required to save the patient.

Special techniques for water rescue are discussed in the next section.

Emergencies Caused by Atomic Radiation

Radiation is a general term describing the transmission of energy. It takes several forms, including light, heat and sound. *Ionizing radiation,* the type to be discussed here, is dangerous because it cannot be seen, felt or heard. A person subjected to harmful ionizing radiation may be unaware of his exposure until instruments detect the radiation or until symptoms appear some time later.

There are three types of ionizing radiation. Alpha rays do little damage; they can be stopped by minimal shielding, such as clothing or even newspaper. Beta rays are more dangerous, but they still may be stopped by heavy clothing. Gamma rays are extremely dangerous. Gamma radiation can pass through clothing and *completely through the body,* inflicting great damage to body cells. However, the danger of alpha and beta rays should not be underestimated. They can enter the body via inhalation, consumption of contaminated food or through open wounds. Once radioactive particles are in the body, they continue to inflict cell damage until they are removed or until they decay.

Since ionizing radiation cannot be seen, felt or heard, some sort of detection instrument must be used to measure it. A Geiger counter is the device most commonly used, although ionization chambers and other devices may also be employed. The rate of radiation is measured in roentgens per hour.

If rescue is required, it should be performed quickly. Rescuers should wear protective clothing and breathing apparatus. The heavy clothing will shield them from alpha and beta radiation, and the breathing apparatus will prevent inhalation of radioactive particles. The victim should be approached away from the direction of smoke and air movement, as far as is possible. Radioactive particles may be carried in dust or in smoke. In a fire, smoke-borne radioactive particles will be a problem. Everyone should be removed from the path of the smoke.

The patient should be removed immediately, even if speedy evacuation violates the rules of good emergency care. Rescuers should remove their protective clothing and store it on the scene in a safe place. A radiological monitoring team can then evaluate it and decontaminate it.

Emergency Care for Patients Exposed to Radiation. There are four types of patients in radiation accidents.

1. The patient who has received external radiation over all or part of his body. Even if this patient has received a lethal dose, he presents no hazard to the rescuer, other patients or the environment.
2. The patient who has suffered internal contamination through inhalation or ingestion. This patient is not a hazard to the rescuer, other patients or the environment. The rescuer should clean away minor amounts of contaminated material deposited on the body surface during air-borne exposure. Then the patient should be treated for chemical poisoning, such as lead poisoning.
3. The patient who has suffered external contamination of the body surface and/or clothing by liquids or dirt particles, with problems similar to vermin infestation. Here, there is a potential hazard, and surgical isolation techniques must be used to protect rescuers. Cleansing measures must be used to protect other patients and the environment.
4. The patient with an open wound. When external contamination is complicated by a wound, care must be taken to avoid the cross-contamination of surrounding surfaces from the wound, and vice versa. The wound and surrounding surfaces should be cleaned separately, and sealed off when clean.

The general rules for handling radiation accident victims are as follows: Give lifesaving emergency assistance if it is needed. Determine whether physical injuries or open wounds are involved. Cover any open wounds with clean dressings, held in place with bandages; do not use adhesive tape. Place the victim on a stretcher if he is not already on one. Cover the stretcher, including the pillow, with an open blanket. Wrap the patient in the blanket, to limit the spread of contamination. If possible, obtain pertinent information, including rough radiological measurements, from those in attendance.

Decontamination. Rescuers should follow strict decontamination procedures after an exposure to radioactive materials, whatever the source.

Remove and save all clothing for evaluation by a radiological monitoring team. Do not burn the clothing, since that could release contaminated particles into the air in the form of radioactive smoke. Shower immediately, paying close attention to your hair, body orifies and body-fold areas. Decontaminate all emergency equipment under the supervision of the decontamination team.

TECHNIQUES FOR RESCUE AND SHORT-DISTANCE TRANSPORT

In most emergency situations, the victim must be removed from the accident site before he can be given complete emergency care. A victim overcome by carbon monoxide must be quickly moved to fresh air, a victim whose leg is pinned under wreckage must be disentangled and moved away from danger before his leg can be examined and splinted, and so on.

The actions required of rescuers in removing a victim will, of course, depend on the circumstances. However, these actions should be performed in the proper sequence. Although removal is discussed last in this chapter, it is the rescuer's first duty to his patient.

Rendering Aid

On reaching the victim, the rescuer should immediately evaluate him for life-threatening problems.

The first consideration is establishment and maintenance of an open airway. If the patient is not breathing, a simple head tilt may open his airway and start respiration; or, it may be necessary to insert an S tube or oropharyngeal airway. If an open airway has been ensured but the patient is still not breathing, he must be ventilated without delay. The most effective method is mouth to mouth (or mouth to S tube). A bag-mask resuscitator may be used if one has been carried to the disaster site. Under no circumstances should the rescuer wait for a mechanical resuscitator before starting artificial ventilation.

When the patient is breathing, the rescuer's attention can be directed to controlling any serious bleeding that poses a threat to life. Until help is available, bleeding must be controlled by direct pressure, or, if limbs have been severed, by a tourniquet. If the patient has gone into cardiac arrest, CPR must be started and continued during revival attempts.

The rescuer providing the initial care should check carefully for signs of spinal injury, especially in the patient's neck. If such signs are present (as determined by examining the patient or from the mechanism of injury), the rescuer should try to stabilize the patient's head until additional rescuers arrive to assist in immobilization procedures.

Disentanglement

Once rescuers have gained access to him, the patient should be disentangled from wreckage, debris, and so forth. This activity is carried on while emergency care is being rendered.

Disentanglement, like gaining access to the patient, may be simple or highly complex. It may involve no more than cutting a patient's shoe away to release his trapped foot; or, it may require the removal of a great deal of material that surrounds him. Rescuers should be thoroughly familiar with the tools used in the disentanglement phase of rescue work, including hand tools, hydraulic rescue tools, power saws and acetylene equipment.

Preparation for Removal

When wreckage and other obstacles have been removed from around the patient and he is accessible, he can be prepared for removal. If the patient has been severely injured, extensive preparations are required. In this case, the patient should be "packaged as a unit."

The purpose of *packaging* is to minimize the danger of further damage to existing injuries. It involves procedures such as applying splints to fractured limbs, dressing and bandaging soft-tissue injuries and stabilizing impaled objects. Packaging also includes immobilization of the patient on a long or short spine board if there is evidence of spinal injury. In some cases, this procedure is more important than all other emergency care measures (except the control of life-threatening problems), because of the danger of worsening the spinal injury. If immobilization is necessary, fractures and lesser injuries can be treated after the patient is removed from the wreckage. The rescuer should be alert for changes in respiration while preparing the patient for removal.

Removal

Removal, too, may be either quite simple or very complex. The rescuer may only be required to walk with the patient, or he may have to raise a patient to another deck with ropes and a basket stretcher.

A number of techniques may be employed to remove patients. Several are described in the last part of this section. Which one is used in a particular situation will depend on the situation and the equipment available. Mechanical devices, however, are only as good as the people using them. Rescue personnel must have mechanical aptitude and knowledge, and those characteristics must be supported with ingenuity and a large measure of common sense.

Removing Victims from Electrical Hazards. The rescuer who finds a victim in the vicinity of live electrical equipment or wiring should immediately call for assistance and support. The engineering officer should be notified, so that electrical power in the area can be shut down before rescue begins. Then a *danger zone* should be established. The danger zone is the area around the accident that may be hazardous to both spectators and rescue personnel.

Rescue personnel should carefully check the area for exposed wires. If any wires are down or have been displaced, a close visual check should be made to determine what the wires are touching. The downed conductors may be contacting the deck, pools of water or other conducting substances. Unless the wires are touching the victim, are close enough to present an immediate hazard or are lying in water, it is probably safe to approach the victim and carry out emergency care procedures. However, the rescuer should not approach the victim if the deck or area surrounding the victim may be energized. If the victim is conscious and able to respond, he can be warned to remain in position until the area is electrically secure.

Energized wires should be handled only with the proper safety equipment. It would be foolhardy for rescuers to attempt to handle wires while wearing ordinary fireman's gloves and rubber boots. If rescuers must move wires, they should wear special lineman's gloves and use a tool called a "hot stick." Even with this equipment, they must work carefully and deliberately.

An alternative tool for removing an energized wire is a weighted rope. It is preferable to use a rope made of a synthetic fiber; otherwise, a good quality rope without metal strands, and absolutely dry, may be substituted. Dryness is a necessity because a wet rope can conduct electricity as well as a wire. The rope should be 0.64 cm (¼ inch) in diameter and 30.48 m (100 feet) long, with a weight of about 230 gm (½ pound) attached to each end. One end of the weighted rope should be thrown over the wire, and the other end flipped under the wire. A rescuer should then take hold of both ends of the rope, and carefully pull the wire free, making sure that the wire does not whip during the procedure. No attempt should be made to cut the wire, since it may whip during the cutting, or arc and seize the cutting tool. When the electrical hazard has been removed, the patient can be treated for his injuries. If the patient has contacted the live wire, the rescuer should check for signs of life and begin CPR immediately, if necessary.

Electrical energy affects the body in two ways. The heart is usually stimulated by a minute electrical current, which causes the muscle to contract and the heart to "beat." When an outside electrical current passes through the body, the natural heart stimulation is interrupted. The normal heartbeat is altered, and the shocked victim generally goes into cardiac arrest.

Electricity also destroys body tissue. In high-voltage, high-amperage accidents, the tissue damage may be massive. In fact, large chunks of tissue may be burned away, leaving a gaping wound that extends inward to bone or to vital organs. A patient who has contacted a high-voltage source usually has two burned areas—one at the point where he contacted the electrical source, and one at the point where the current passed from his body to the ground.

Removing a Victim with a Neck Injury from Deep Water. One rescuer should approach the patient from the head. He should place one arm under the patient's body so that the patient's head rests on the rescuer's arm and the patient's chest is supported by the rescuer's hand. The rescuer should then place his other arm over the patient's head and back, splinting the patient's head and neck between the rescuer's arms. In one movement, the patient should be rolled over, with his head and neck supported between the rescuer's arms. The rescuer can, at this point, begin mouth-to-mouth resuscitation, if necessary. As in the case of other patients with neck injuries, the head should be tilted all the way back.

During this operation, another rescuer should enter the water with a backboard, plank or other similar support. While the first rescuer holds the patient's head in a stable position, the second should slide the support device under the patient's body. The rescuers should exercise caution during this move. Most rigid devices are very buoyant and can slip loose and cause serious injury to the patient.

While the first rescuer continues to hold the patient's head in a stable position, the second should fasten a cervical collar, neck roll or other immobilizing device around the patient's neck for support. Additional support can be provided by rolled wet towels placed firmly against the patient's head. The board (or other rigid support) may then be floated and removed by other rescuers with a minimum of patient movement.

Emergency One-Man Carries

The Blanket Drag. To use the blanket drag, first gather half a blanket lengthwise in pleats. Place the pleated portion against the side of the patient (Fig. 14.29). Smooth the other half of the blanket away from the patient.

Extend the patient's arm, on the side away from the blanket, over his head in a straight line.

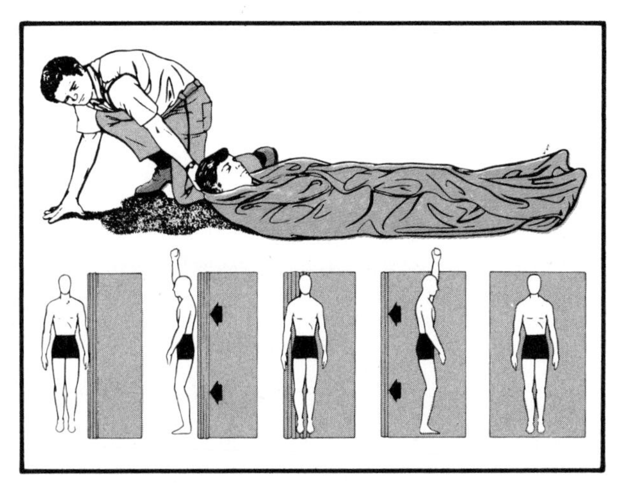

Figure 14.29. The one-man blanket drag. The steps in getting the patient onto the blanket are shown at the bottom, from left to right.

The outstretched arm will provide a cushion for the patient's head and allow his body to be rolled quite easily.

Now roll the patient onto his side, maintaining his body in as straight a line as possible. While holding the patient on his side with one hand, push the pleated portion of the blanket against the patient's back. Roll the patient back onto the blanket, on his back.

To spread the blanket, extend the patient's other arm in a straight line over his head, and roll him in the opposite direction. Now smooth out the pleats, and return the patient to the blanket, on his back. With his arms at his sides, wrap him snugly in the blanket.

The Clothes Drag. Firmly grasp the patient's shirt or coat collar so that his head is resting on your forearm. Pull him to safety, keeping his head as close to the deck as possible, and keeping his body in a straight line. Make sure that the collar is not pulled so tightly around his neck that it creates an airway obstruction.

A patient can be moved down inclined ladders by the clothes drag, with a shift in the position of your hands. When you are ready to descend the ladder, place your hands under the patient's shoulders, with your palms up. Cradle his head in your arms, and slide him as close to the plane of the ladder as possible.

The Fireman's Drag. To move a patient by the fireman's drag, place him on his back, with his arms above his head. Tie his hands together with a piece of rope, a cravat, a piece of sheeting or some similar material. Straddle the patient's body, and pass your head through the patient's trussed arms. By raising the upper part of your body, you can lift the patient's shoulders just clear of the deck. Then you can crawl on your hands and knees to safety, dragging the patient along with you.

The Fireman's Carry. This technique is not used a great deal except in dire emergencies, since the patient's entire weight is on the rescuer, tending to unbalance him. Balance and coordination are very important in the fireman's carry.

Place the patient on his back with his knees flexed. Grasp the patient by his wrists, with his palms down. Place your feet against his, and pull him forward and upward at the same time (Fig. 14.30). As you continue to pull the patient forward, crouch so that you can duck under his raised arm. Allow the patient to fall on your shoulders and, as you feel his weight, return to a standing position.

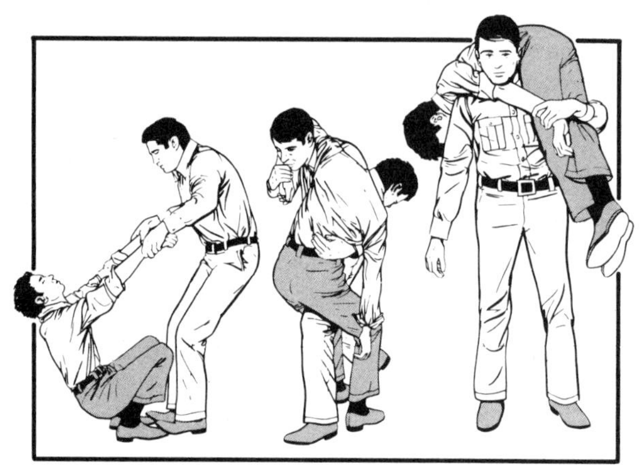

Figure 14.30. The fireman's carry. The entire lifting procedure must be performed in one continuous motion.

It is important that this sequence of movements be accomplished in an unbroken sweep. If you stop during the raise, the patient's dead weight may be too much to handle; it may then be necessary to lower him and start over again.

The Pack-Strap Carry. If the patient is conscious, and existing or suspected injuries will not be worsened by the movement, assist him to a standing position. While standing in front of the patient and supporting him, turn your back to him. Lift the patient's arms over your shoulders, and cross them over your chest. Make sure his arms are straight and his armpits are directly over your shoulders. With the patient resting on your back in this fashion, bend forward and hump him well up onto your back. By keeping the patient's arms straight and crossed over your chest, you will be able to keep him riding high on your back. Hold both his wrists with one hand, keeping the other hand free to open doors or push past obstructions.

The Rope Sling and Long Spine Board. A very effective tool for dragging a patient from beneath wreckage is the rope sling. The sling is made from 7.6-cm (3-inch) manila rope fashioned into a loop with a long splice. The loop should be approximately 1.83 meters (6 feet) in diameter. Two steel rings joined together serve to shorten the loop so that it will not slide over the patient's head.

In operation, the sling is slipped over the patient's chest and under his arms. The rings are pushed up as close to his head as possible. A long spine board is positioned at the patient's head, so that he can be moved directly onto it. The rescuer exerts a slow, steady pull on the rope, keeping it close to the deck to maintain the patient's spine in as straight a line as possible. The patient is pulled onto the long board, which then serves as a litter.

A rescuer can move a patient of almost any size in this manner. Naturally, the rope-sling method cannot be used until all wreckage is lifted from the patient; nor should it be used when the patient has chest injuries. In the event that a rope sling is not available, one of the 2.74-meter (9-foot) straps used with the spine board will make an effective substitute. The strap should be tied together behind the patient's head with a cravat.

Emergency Two-Man Carries

The two-man carries described below are considered "emergency techniques." Like the one-man carries, they are designed primarily to move sick or injured patients from hostile environments. Rescuers should remember, however, that neither of these methods is suitable for moving patients with spinal injuries.

The Two-Man Seat Carry. In this technique, the rescuers carry the patient in a seat fashioned from their arms. Both conscious and unconscious patients may be transported by the two-man seat carry.

The rescuers kneel, one on either side of the patient, near his hips (Fig. 14.31). With the arm nearest the patient's head, each man helps to raise the patient to a sitting position. When the patient is sitting, each rescuer grasps the other's upper arm with his hand, so that their arms are locked behind the patient's back. Then each rescuer slips his free hand under one of the patient's thighs and grasps the wrist of his partner. The rescuers rise slowly together. When they are standing, they adjust their arms to make a comfortable and secure seat. If the patient is conscious, he can place his arms around their necks for security.

Figure 14.31. The two-man seat carry. 1. Kneel on either side. 2. Raise the patient to a sitting position. 3. Grasp each other's arms. 4. Rise slowly.

The Two-Man Extremities Carry. The patient is placed on his back, with his legs spread apart and his knees bent. One rescuer positions himself at the patient's head, while the other stands between the patient's legs, facing his head. The rescuer at the patient's feet grasps the patient by the wrists and pulls him to a sitting position (Fig. 14.32). As the patient's upper body is raised from the deck, the other rescuer can assist by lifting his shoulders.

As the patient reaches the sitting position, the rescuer at his head drops to one knee, supports the patient's back against his other leg and passes his arms around the torso in "bear-hug" style. The rescuer at the patient's feet turns, positions himself between the patient's flexed legs, and passes his hands under the patient's knees from the outside. As soon as he is in position, the man at the feet gives the command to rise. Both rescuers stand, making sure to lift with their backs and not with their legs. The patient can then be carried, chair fashion, to safety.

The Standard Two-Man Pickup

To use the two-man pickup, both rescuers position themselves at one side of the patient. The man at the patient's head cradles the head and shoulders with one arm, and passes his other arm under the patient's body at about the belt line (Fig. 14.33). The other rescuer grasps the patient's legs under the knees, and passes his other hand over the midsection of the patient so that he can grasp his partner's hand. The rescuers lock hands and lift the patient as a unit.

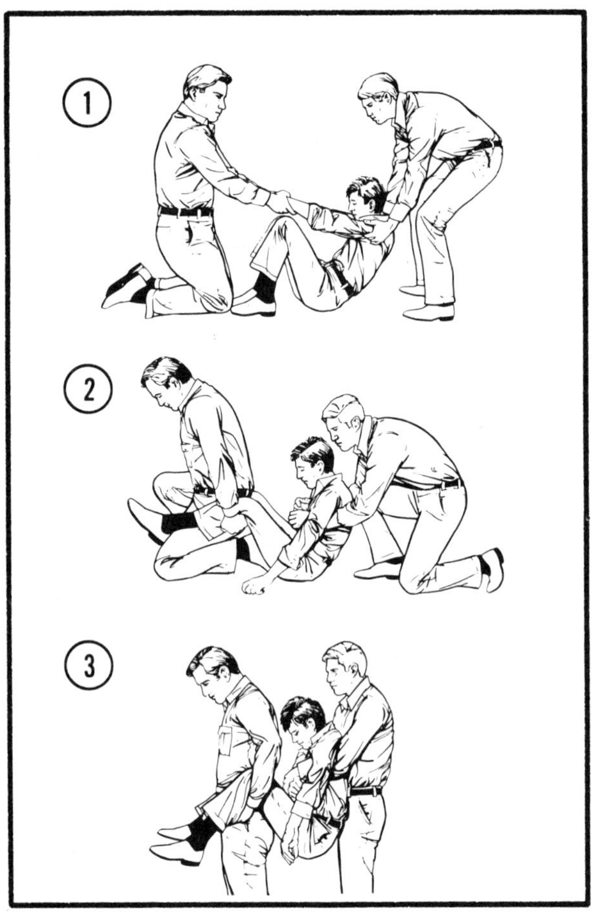

Figure 14.32. The two-man extremities carry. 1. Raise the patient to a sitting position. 2. Grasp legs and chest. 3. Rise.

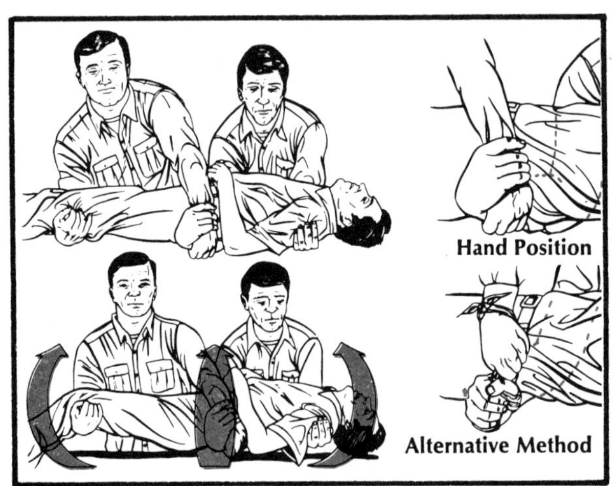

Figure 14.33. The two-man pickup. Note that the rescuers' inner hands are locked together.

Carries for Patients with Spinal Injuries

With the exception of the blanket drag and the clothes drag, none of the lifts and carries described so far should be used for patients with spinal injuries. Even the blanket and clothes drags provide only the barest minimum of support and should be used only in the most extreme emergency.

The next two techniques to be described are among the most commonly used methods for lifting and carrying patients with spinal injuries. Both are quite effective, since they immobilize the spine through the use of a long spine board. Since these procedures must be performed carefully, and in a certain sequence, they should not be used to remove a patient from a hostile environment where speed is required. They should only be used where no danger is involved and the patient's condition has been stabilized.

The Four-Man Log Roll. The log roll is most effective when a minimum of four rescuers are used to roll the patient. A fifth rescuer should be available to move the spine board. Three of the rescuers roll the patient as a unit, while the fourth maintains constant traction on the head and neck.

If the patient is found on his back, the log roll is accomplished as follows. One rescuer positions himself at the head of the patient and applies gentle traction to the head and neck (Fig. 14.34). He remains in this position and continues to apply traction until the patient is firmly secured to the board and ready for transportation. Again, a cervical collar or similar device will aid the rescuer in maintaining traction. (The rescuer at the patient's head will be referred to as the "head man.")

When traction has been applied, another rescuer raises the patient's arm (on the side to which he is to be rolled) over the patient's head. This will prevent the arm from obstructing the rolling movement. Then the three rescuers take up positions in a straight line along the patient's side, all kneeling on the same knee.

The rescuer at the patient's shoulder (the "top man") places one hand on the patient's further shoulder and passes his other hand over the patient's arm so that he can grasp the body just above the belt line. The "center man" grasps the patient's body just below the buttocks, about at midthigh. The "bottom man" places one hand behind the patient's knees and the other hand on the patient's leg, just below the calf.

When the head man is satisfied that proper traction is being applied and that the other rescuers are ready, he gives the signal for his partners to roll the patient toward them. It is important for each rescuer to coordinate his movements with the others, so that the patient's body is moved as a unit.

While the patient is held carefully in the rolled position, another rescuer slides the long spine board next to the patient. He positions it so that the patient's head and feet will be on the board when he is rolled back onto it. This fifth man then

Figures 14.34. The four-man log roll for a patient with a spinal injury. 1. Positions at the start. 2. Grasping the patient. 3. Placing the patient on the board. 4. Immobilization.

places pads where there may be spaces when the patient is placed on the board: under the neck, behind the small of the back, under the knees and behind the ankles. The pads, which will support areas of the body that do not contact the board, may be made from rolled towels, bandaging materials or multitrauma dressings.

On a signal from the head man, the rescuers carefully roll the patient onto the board, ensuring that he is moved as a unit. The fifth man can adjust the pads as the patient is lowered. When the patient is again on his back, the fifth man returns the patient's outstretched arm to his side.

The patient should be secured to the board with snugly applied straps at the chest, thighs and knees. Movement of the head should be prevented by securing it with a wide cravat applied over the forehead and passed through the slots in the side of the board. The head man must be careful to coordinate his movements with the actions of the others. He must maintain the traction until the patient is rigidly immobilized on the board.

In some accident situations, the patient is found face down; because of his injuries, it may be best to transport him in that position. The four-man log roll may be used for a patient in the prone position; the same procedure is followed, with the exception of padding the spaces.

The Straddle Slide. As in all cases where injury to the spine is suspected, one rescuer moves directly to the patient's head and applies traction. In this case, however, he does not kneel. Instead, he bends at the waist and spreads his feet wide enough to allow a spine board to pass between them. A second rescuer straddles the patient (facing the head man); he places his hands under the patient's arms, just below his shoulders (Fig. 14.35). A third rescuer also straddles the patient, placing his hands at the patient's waist. A fourth man positions the board lengthwise at the patient's head. His job is to slide the board under the patient when the other rescuers lift him slightly.

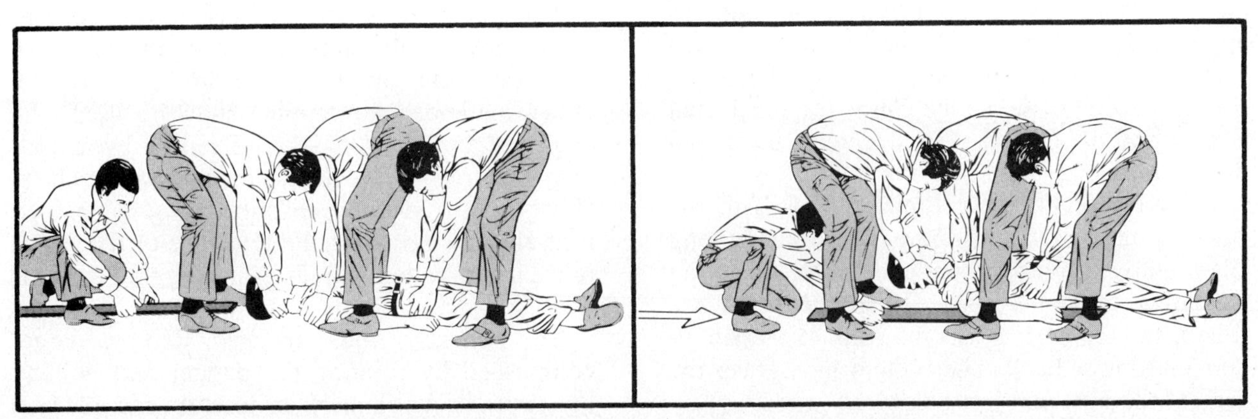

Figure 14.35. The straddle slide. The spine board is slid under the patient from the head.

At a signal from the head man, the other rescuers lift the patient just enough to allow the board to pass under his body. The board should be slid in one smooth and unbroken movement. If the patient's upper body is lifted high enough for the board to pass under, the buttocks and legs offer little resistance to the smooth board. When the spine board is completely under the patient, the rescuers lower him carefully and then strap him firmly in place. Spaces should be padded as in the log roll technique.

The Traction Blanket Lift. In some cases a long spine board may not be immediately available. Blankets and D-ring stretchers can provide the means for moving a patient with a suspected spinal-cord injury. It should be remembered, though, that the traction blanket technique is not nearly as safe as methods that make use of a long spine board or similar device. It should be used only when the patient must be removed from a hostile environment and rigid devices are not available.

As in the techniques described previously, one rescuer immediately stations himself at the patient's head and applies gentle traction. A cervical collar or improvised collar will help to maintain the head in the desired neutral position. While the head man holds the patient's head stable, two others pleat a blanket in folds of 30.5–45.7 cm (12–18 inches). Then they position the blanket so that the bottom fold is at the patient's shoulders and the top fold is under the knees of the rescuer who is applying traction (Fig. 14.36).

When the blanket is in place, four rescuers kneel next to the patient, two on each side. Each of the top men places one hand flat under the patient's shoulders and the other hand in the patient's armpit. As the top men lift the patient's shoulders slightly (the head man moving the head accordingly), the bottom men grasp the bottom fold of the blanket. They start drawing it under the patient's body in a smooth and continuous motion. As long as the top men keep the patient's upper body slightly raised, the bottom men will be able to draw the blanket under the length of the body without difficulty. Since the head man is kneeling on the blanket, it cannot be pulled down too far.

After the blanket is completely unfolded, the rescuers on each side of the patient roll the long edge tightly against the patient's sides, following the contours of his body. When the edges are rolled, the top men grasp the rolls at the shoulders and lower back. The bottom men grasp the rolls at the hips and just below the knees.

At a signal from the head man, the top and bottom men (who should be on both knees at this point) lean back. Using their back muscles and the weight of their upper bodies, they lift the patient from the deck. As the patient is lifted, another rescuer should be sliding a D-ring stretcher (or similar rigid device) under the patient from his feet toward his head. When the stretcher is in position, the rescuers maneuver the patient carefully onto the stretcher. The blanket can be unrolled and folded over the patient.

Like the log roll, the traction-blanket lift can also be used for patients found lying face down.

The Short Spine Board. The short spine board must be used when a seated patient with a suspected neck injury is packaged for removal from wreckage. Two rescuers are required.

One rescuer takes a position at the side of the patient or behind him. He immediately applies gentle traction to the patient's head and continues to do so until the patient is firmly affixed to the spine board. The second rescuer carefully secures a cervical collar around the patient's neck, to help maintain the head in a neutral position.

After applying the collar, the second rescuer positions a short spine board behind the patient's back, making sure that he does not move the patient any more than absolutely necessary. In some spaces with low headroom, it may be necessary to put the board in lengthwise and then rotate it into the proper position.

The second rescuer secures the patient's head to the board with either a special head-and-chin strap or with wraps of self-adhering bandage (which is especially well suited for the job). The bandaging material should be well secured over the patient's forehead and around the chin.

When the head is firmly fixed in place, the rescuers fasten the patient's torso to the board with two 2.74-meter (9-foot) straps. Generally the straps are passed through the upper handholds, behind the board, out the lower handholds on the opposite side, around the thighs from outside to inside, and finally under and over the thighs to the chest buckle, as close to the groin as possible. Other methods of strapping may be used; the choice of a method may be influenced by the patient's injuries, especially if they involve the chest. However, the method described should be used if at all possible, since the positions of the straps prevent the patient's body from sagging as he is lifted.

The patient is now packaged. Removal is accomplished by rotating the patient and holding him upright until another rescuer can place a

Figure 14.36. The traction blanket lift.

long spine board under the patient. When the long board is in place, the patient can be lowered onto the board and slid from the accident site. The long board then serves as a litter. If the straps tend to keep the patient's legs slightly bent, they can be adjusted as soon as he is securely on the long board and away from the wreckage.

Lifting and Moving Devices

Such devices as long and short spine boards have been mentioned in this section, but not described. Most of the standard devices used for the rescue and transport of patients are shown in Figure 14.37. They are described in the remainder of this chapter, along with some devices that may be improvised quickly.

Spine Boards. The long spine board is generally constructed of 1.91-cm (¾-inch) best exterior grade plywood, 183 cm (72 inches) long by 45.7 cm (18 inches) wide. Handholds and strap-holes, located along the long edges of the board are fashioned so that they do not present sharp or rough edges. The short edges should be tapered so that they can be slid easily under a patient. The board should be equipped with runners to

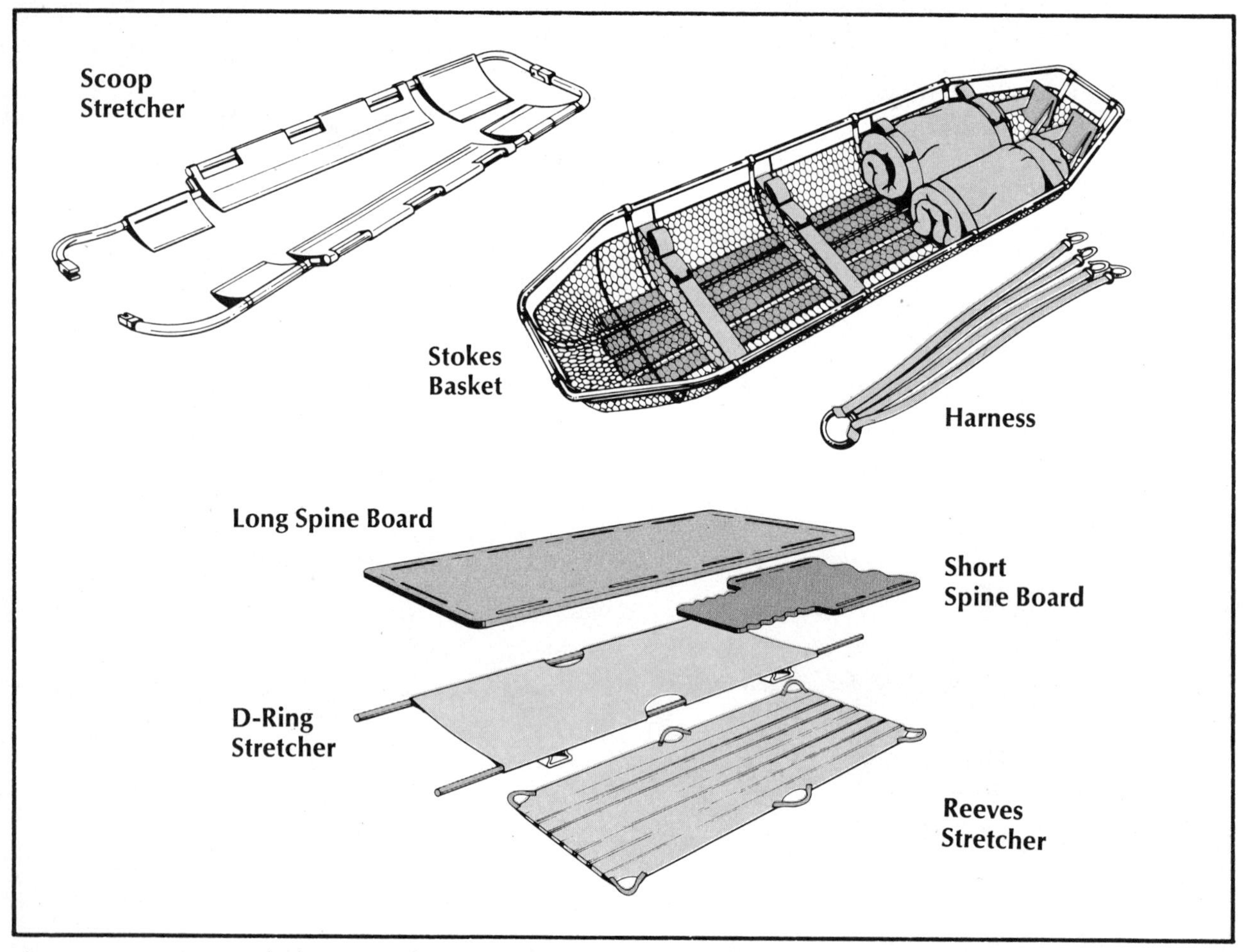

Figure 14.37. Some standard rescue and patient transportation devices.

reduce the friction between the board and the deck.

The short spine board is used mostly for the removal of sitting patients. It is usually 81.3–86.4 cm (32–34 inches) in overall length, by 45.7 cm (18 inches) wide. It too is provided with handholds and strapholes; these openings are spaced to coincide with those in the long boards. The headpiece of the short board is 20.3 cm (8 inches) wide by 30.5 cm (12 inches) long. It is notched so that the material used to hold the patient's head in place will not slip during transportation.

Spine boards should be sanded smooth and varnished or highly waxed so that they can be slid under patients and cleaned with ease.

Split-Frame or Scoop Stretchers. Several types of stretchers marketed in the past few years are especially useful for lifting and transporting sick or injured patients with a minimum of body movement. These devices are made on the split-frame or scoop principle. They are strong, well constructed and easily maintained, and they support the patient well.

In operation, the split-frame or scoop stretcher is separted along its long axis (Fig. 14.37). The two frame halves are slid under the patient from either side, mated and locked together. To keep from pinching the patient or his clothing, the rescuer should carefully lift the patient by his clothing as the halves are joined and locked. When the patient is secured with straps, he is ready to be picked up.

A disadvantage of these devices is that both sides of the patient must be accessible. Another is that the stretcher cannot be slid under the patient in the manner of the long spine board. However, the advantages of split-frame stretchers far outweigh the disadvantages, and they should be included as part of the ship's rescue equipment.

D-Ring Stretcher. The D-ring or army stretcher (Fig. 14.37) is quite common. Every ship should carry a number of these stretchers, for use in disasters or multiple victim accidents. The D-ring stretcher is useful when patients must be removed either by lowering ropes or by ladders. Rescue personnel should be thoroughly familiar with the

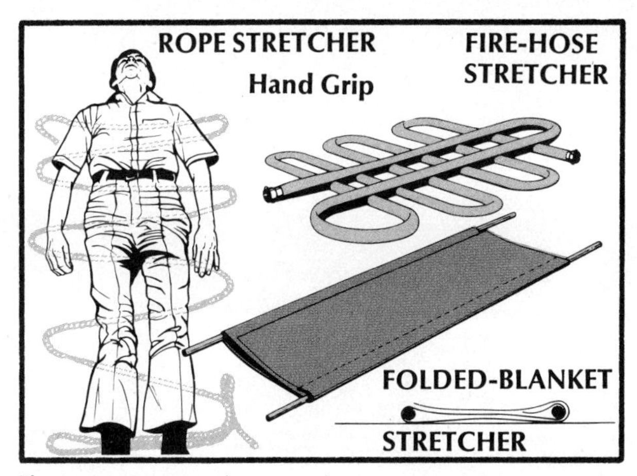

Figure 14.38. Stretchers may be improvised from rope, fire hose or a blanket and two poles.

methods of lashing and lowering these versatile stretchers.

Stokes Basket. Like the D-ring stretcher, the Stokes basket (Fig. 14.37) is a versatile piece of rescue equipment. It is useful for removing patients from heights or over rubble. Its construction offers a great deal more protection to the patient than other litters. There are several techniques for lashing and lowering Stokes baskets. Many emergency squads have developed wire or rope harnesses that allow the basket to be lowered with a single line.

Improvised Litters. Rescue personnel are often required to improvise litters when they are faced with the problem of transporting a large number of disaster victims without standard litters. A good makeshift litter can be fashioned from a blanket and two poles.

To make a litter, first spread a blanket flat on the deck. Place one of the poles across the short dimension, about one-third of the distance from one end. Fold the blanket over the pole. About 45.7 cm (18 inches) from the first pole, place the second pole across the first fold. Approximately 15.24 cm (9 inches) of the first fold should extend past this second pole; it will be rolled back over the pole when the second fold is made (Fig. 14.38). Fold the blanket back over both poles to complete the stretcher. Although the blanket seems loose, the patient's weight will "lock" it in place. However, the poles can be easily slipped from the folded blanket when the patient is placed on a cot or when he reaches a medical facility.

Another improvised stretcher can be fashioned from a 15.24-meter (50-foot) section of rope or fire hose. The method of looping and folding the rope and hose is shown in Figure 14.38. When this type of litter is used, all the loops must be held as the patient is lifted; otherwise the litter will not hold him.

BIBLIOGRAPHY

Bergeron, J. D., *Self-Instructional Workbook for Emergency Care,* The Robert J. Brady Co., Bowie, Md.

Grant, H. and Murray, R., *Emergency Care,* The Robert J. Brady Co., Bowie, Md.

Huszar, R., *Emergency Cardiac Care,* The Robert J. Brady Co., Bowie, Md.

The American Red Cross, "Cardiopulmonary Resuscitation"

———, "First Aid for Foreign Body Obstruction of the Airway"

U.S. Coast Guard, "Methods of Artificial Respiration," CG 139.

U.S. Public Health Service, "Artificial Respiration," HEW.

U.S. Navy, "Standard First Aid Training Course," NAVPERS 10081-B.

Breathing Apparatus

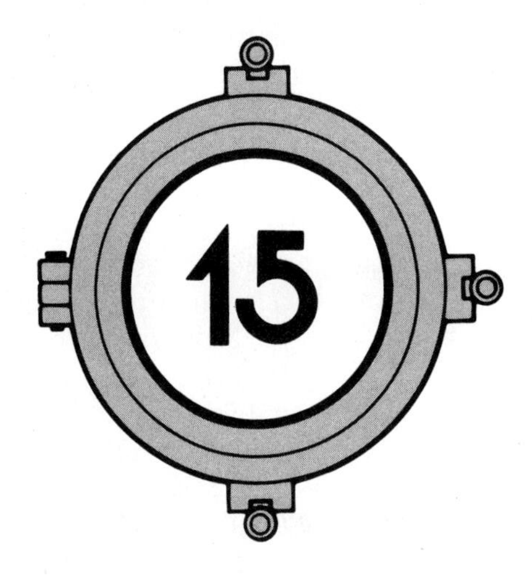

* The material in this chapter has been adapted from Faria, L: Protective Breathing Apparatus. Bowie, Md, Robert J. Brady Co, 1975.

Although the air encountered at a fire is hot, contaminated by smoke and toxic gases, and deficient in oxygen, crewmen must enter this hostile environment to fight the fire. Their problem is simple, direct and urgent—they must breathe. The equipment discussed in this chapter is designed to enable seamen to enter such a hostile environment with some degree of protection for the respiratory system.

Breathing apparatus is available in several types. Each type is effective if used properly, and each has certain advantages and disadvantages. However, no breathing apparatus provides complete protection against poisonous gases that are absorbed through the skin. Crewmen operating in atmospheres containing such poisons must wear special protective clothing (see Chapter 16).

Respiratory protection devices must be carried on every U.S. flag vessel. The specific requirements vary with the type and size of ship. Freight and tank vessels over 1016 metric tons (1000 gross tons) must carry at least two breathing devices. Passenger ships up to 10,160 metric tons (10,000 gross tons) must carry two breathing devices; those from 10,160 to 20,320 metric tons (10,000 to 20,000 gross tons) must carry three; and those over 20,320 metric tons (20,000 gross tons) must carry four. A spare charge must be provided for each breathing device in which charges are used. The tools necessary for making each device operational must also be provided. These requirements are, of course, minimums. Many ship owners and masters equip their vessels with one breathing device for each seaman who may be expected to enter a fire area, and one for the officer who will lead them. They also ensure that at least two spare charges are carried for each device.

Breathing apparatus must be stowed in convenient, accessible locations, as determined by the master. One unit should be stowed near the pilothouse; the others, outside and adjacent to the machinery space entrance. Required spare charges and tools must be stowed with the apparatus. The container for each unit must be marked to identify its contents. If the container is stored in a locker, the locker too must be so marked.

Breathing apparatus must be properly maintained, and crewmen (especially the emergency squad) must be trained in its use. Training should include:

- Instruction on the capabilities and limitations of each type of device carried on board
- Instruction on the selection of the proper type of device, depending on the hazards
- Handling of the equipment, donning of the facepiece and testing of the facepiece-to-face seal
- Drills simulating the emergency use of the equipment
- Instruction and practice in stowing the equipment.

THE STANDARD FACEPIECE

A breathing apparatus is a device that provides the user with breathing protection. It includes a facepiece, body harness and equipment that supplies air or oxygen. The facepiece is an assembly that fits onto the face of the person using the breathing apparatus, forming a tight seal to the face and transmitting air or oxygen to the user.

The standard facepieces shown in Figures 15.1

and 15.2 are used with most of the breathing apparatus covered in this chapter. Special types of facepieces will be discussed along with the equipment to which they apply.

Construction

The basic part of the facepiece is the mask. It is made of oil resistant rubber, silicone, neoprene or plastic resin. Most facepieces include a head harness with five or six adjustable straps, a flexible inhalation tube, an exhalation valve and a wide-view lens (Fig. 15.1). Some models also include a nose cup or a speaking diaphragm. The facepiece used with oxygen-generating equipment has an exhalation tube and an inhalation tube, each with a mica disk-type valve for airflow control (Fig. 15.2).

Head Harness. The function of the head harness is to hold the facepiece in the proper position on the face, with just enough pressure to prevent leakage around the edge of the mask. Before the facepiece is stowed, all harness straps should be fully extended, with the tab ends against the buckles. This helps ensure that the facepiece can be donned quickly in an emergency.

Flexible Tubes. The flexible inhalation tube carries fresh air or oxygen to the facepiece. In the facepiece with dual hose, the exhalation tube returns exhaled breath from the facepiece to the canister. The airflow through these tubes is controlled by the inhalation and exhalation valves. Like the facepiece, the flexible tubes are made of oil resistant rubber, neoprene or plastic resin.

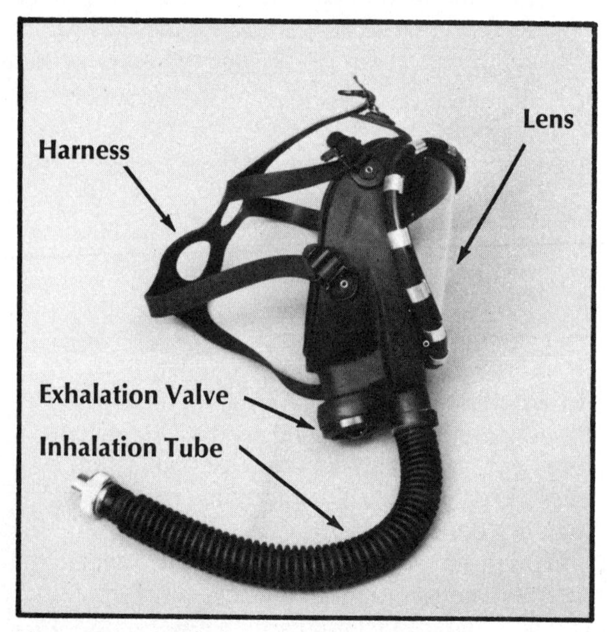

Figure 15.1. Standard single hose facepiece.

Figure 15.2. Standard dual hose facepiece.

In use, the tubes must be kept *free* and *unkinked* for the proper flow of air. All unnecessary strain on these tubes should be avoided. If they become tangled in any way, they must be freed carefully. They must not be pulled free.

Exhalation Valve. The exhalation valve on a single hose facepiece is a simple one-way valve. It consists of a thin disk of rubber, neoprene or plastic resin, secured in the center of the facepiece. It may be contained in a hard plastic mount located at the front of the chin area. The exhalation valve, commonly referred to as the "flutter valve," releases exhaled breath from the facepiece.

Lens. The facepiece may be supplied with a dual lens (Fig. 15.2) or a full-view single lens (Fig. 15.1). In some cases, the single lens is available as an optional item at additional cost. The lens gives the wearer a wide range of vision. It is made of a plastic base resin and is attached to the mask with a removable frame or metal ring.

It must be protected from scratches as much as possible, in use and during handling and packing.

Nose Cup. The nose cup is an optional removable piece that fits into the exhalation valve. It is designed to reduce fogging of the lens.

Speaking Diaphragm. The speaking diaphragm projects the wearer's voice from the facepiece with little or no distortion. It is located directly in front of the wearer's mouth and is similar in appearance to the exhalation valve.

Pressure-Relief Valve. The facepieces used with canister- and cylinder-type breathing apparatus include a combination pressure-relief and saliva valve. The valve is located in the cross tube that connects the inhalation and exhalation tubes. It automatically relieves pressure within the facepiece. By pressing a spring-loaded button, the wearer may also utilize the valve to get rid of saliva and to exhaust exhaled air to the outside.

Use and Maintenance

The donning, stowing and maintenance of the facepiece all affect its efficiency in use. For example, poorly stowed equipment is difficult to put on. Poorly maintained equipment could cause difficulties in achieving an uncontaminated atmosphere within the facepiece. Poorly donned equipment will simply not protect the wearer effectively.

Donning. When the facepiece is put on properly, the chin straps are below the ears. The harness pad is at the back of the head, as close to the neck as possible. The side straps are above the ears. The mask portion is snug but not tight. A mask that fits too tightly is very uncomfortable and could possibly interfere with the user's circulation. A mask that fits too loosely does not seal properly; it may allow contaminated air to enter the facepiece. Long hair, sideburns and beards that prevent the outer edge of the facepiece from contacting the skin may also cause leakage.

Two factors are important when the facepiece is to be put on. First, the wearer must obtain the proper seal by adjusting the harness. Second, time is precious when breathing apparatus is needed; every second counts.

After much testing, the following donning method has been proved most effective for both five-strap and six-strap facepieces. For the facepiece to be donned as recommended, the harness must be fully extended and pulled over the front of the lens. The tab end of each strap must be up against the buckle. If this was not done when the facepiece was stowed, it must precede the first step of the donning procedure.

1. Hold the facepiece at the bottom with one hand (Fig. 15.3). Place your chin in the pocket at the bottom of the mask, and fit the mask to your face.
2. Put your other hand between the mask and the harness. Your palm should be on the lens, and your fingers and thumb should be fully extended and spread.
3. In one smooth motion, push the harness over the top of your head. Push with the back of your hand and your fingers. Keep your fingers spread and extended as the harness slips into place.
4. Tighten the chin straps by gently pulling them out and back. This places the harness pad at the back of the head close to the neck. For the proper fit and seal, tighten the straps from the bottom up.
5. Tighten the side straps as described in step 4.
6. Tighten the top straps last, again as described in step 4. When steps 4–6 are completed in the proper order, the harness should fit tightly against the back of the head (see step 6 in Figure 15.3).
7. Test the facepiece for leakage as follows: For demand-type breathing apparatus, block the end of the inhalation tube with the palm of your hand while trying to inhale. If the facepiece is properly fitted, it will collapse against your face. For oxygen-generating or oxygen-rebreathing equipment, grasp both tubes while trying to inhale. Again, a properly fitted facepiece will collapse against your face.

Removal. The facepiece should be removed as follows:

1. Disconnect the inhalation tube from the supply of air or oxygen (demand-type breathing apparatus only).
2. With the tips of your fingers, release the self-locking buckles on the facepiece harness. This allows the straps to slide to the limits of the buckles, so there is no unnecessary strain on the straps.
3. Grasp the mask portion at the chin. Pull it away from your face and up over your head.
4. Fully extend all harness straps that are not already extended. If the facepiece is to be stowed, pull the harness over the front of

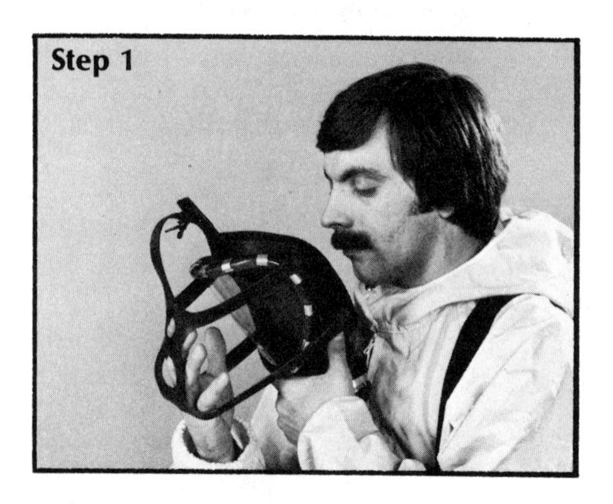

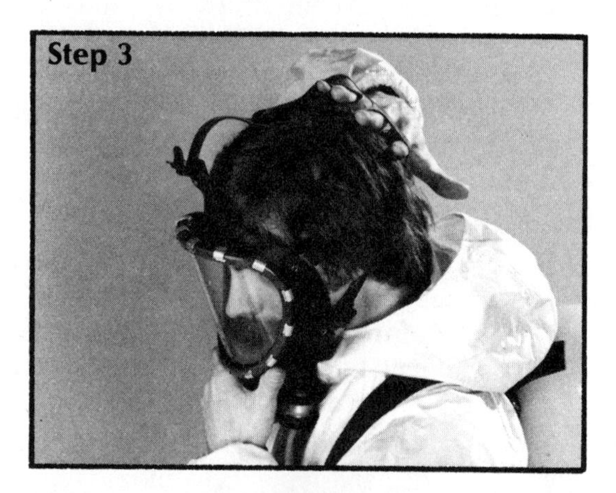

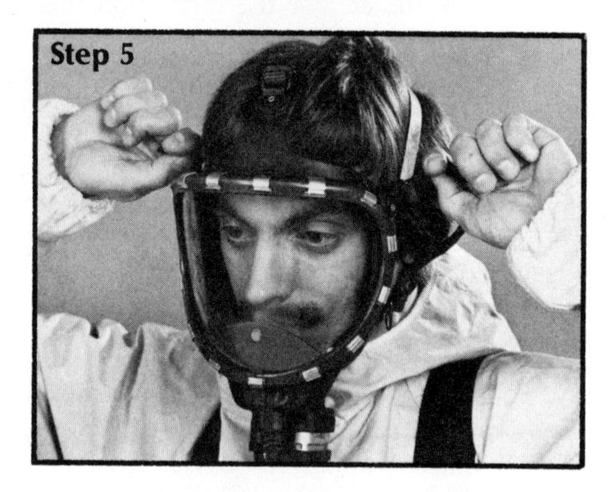

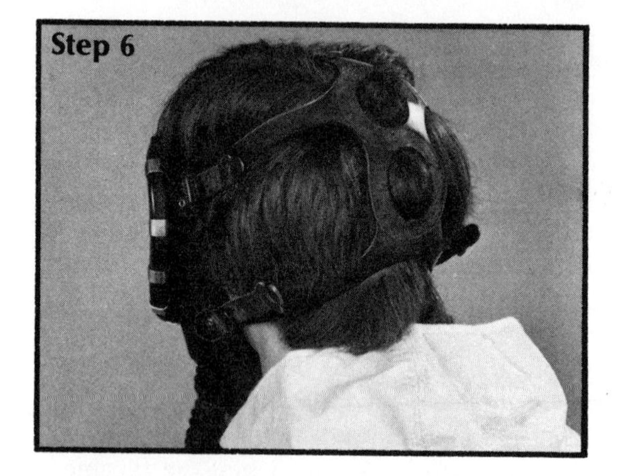

Step 1. Hold the facepiece at the bottom with one hand and place your chin in the pocket at the bottom of the mask.

Step 2. Put your other hand between the mask and the harness.

Step 3. Push the harness over the top of your head.

Step 4. Tighten the chin straps from the bottom up.

Step 5. Tighten the side straps.

Step 6. Tighten the top straps. The harness should fit tightly against the back of the head.

Step 7. Test for leakage (see discussion in text).

Figure 15.3. Donning the facepiece.

the mask before placing the unit in the carrying case. The facepiece should, however, be cleaned before it is stowed.

Maintenance. To ensure safe operation when the facepiece is needed, it must be maintained properly after every use. Cleanliness is also important. A dirty facepiece can spread colds and other respiratory diseases from one wearer to another; at the very least, it could be unpleasant to wear.

The equipment required for maintenance is

- A pail of warm water, not exceeding 38°C (100°F) in temperature, containing some mild disinfectant (such as those advertised for household or hospital use)
- A pail of clean water, not exceeding 38°C (100°F) in temperature, for rinsing
- A sponge and a soft, lintfree cloth for washing and drying.

The following maintenance procedure is illustrated in Figure 15.4.

1. Rinse the facepiece with plain water, in a bucket, under a spigot or with a hose, to remove any loose dirt, salt particles and foreign material. This initial rinse keeps the disinfectant solution clean and up to strength longer, so that several shipboard units may be cleaned with the same solution.
2. Scrub the mask, inside and out, with a sponge that is well saturated with disinfectant solution. Clean the lens with a soft cloth or sponge; never use abrasive materials on the lens.
3. Hold the facepiece by the harness, and submerge the inhalation tube and the exhalation valve in the disinfectant solution. After a few moments, remove them from the pail. Allow the excess solution to drain.
4. Remove the protective cap from the exhalation valve. With a corner of the sponge, gently lift and clean under the edge of the rubber valve. This will remove any foreign particles, which could cause a leak when the mask is next used.
5. Replace the protective cap on the exhalation valve. Completely submerge the facepiece in clear water to rinse it. Allow the excess water to drain off the facepiece.
6. Dry the entire facepiece with the clean, lintfree cloth. During the drying, check each part for damage and wear. Carefully inspect the harness and lens for tears and cracks. Inspect the inhalation tube by gently stretching the tube and looking for cracks.

Restowing. Proper restowing of the facepiece in its carrying case ensures that it is ready for its next use. The restowing procedure, shown in Figure 15.5, is as follows:

1. Check that all harness straps are extended to the tab at the buckle.
2. Pull the harness over the front of the mask, so the facepiece is ready for donning.
3. Place the facepiece in the carrying case as shown in step 3 of Figure 5. Make sure the inhalation tube is curled correctly and is not pinched or kinked. Also make sure that the lid will not touch the inhalation tube when the container is closed.

TYPES OF BREATHING APPARATUS

The types of breathing apparatus approved for use aboard ship can be divided into three groups:

1. Self-contained breathing apparatus (SCBA). These devices provide air or oxygen to the user, who wears the entire device. The user is thus completely mobile. However, the device can supply air or oxygen for only a limited amount of time. There are two kinds of SCBAs:
   a. Oxygen breathing apparatus (OBA). These devices provide oxygen chemically.
   b. Demand units. These devices provide air or oxygen from a supply carried by the user.
2. Hose masks (fresh air breathing apparatus). Here, the user wears a facepiece that is connected to a pump through a long hose. Air is pumped to the user, whose mobility is limited by the length and weight of the hose. However, the device can be used for extended periods of time.
3. Gas masks. These devices filter contaminants from air that is to be breathed. They can be used only in atmospheres that contain enough oxygen to support life.

Oxygen breathing apparatus must not be used in any atmosphere that contains, has contained or is suspected of containing flammable or combustible liquids or gases. Thus, they may not be used in cofferdams fouled by fuel oil. They may, however, be used in machinery spaces on tank vessels, where the required hose mask might not

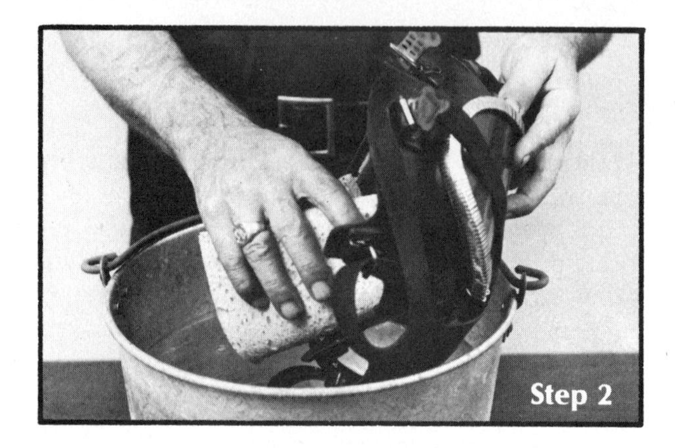

Step 1. Rinse the facepiece under a hose or a faucet.
Step 2. Scrub the mask, inside and out, with a soft sponge that is well saturated with disinfectant solution.
Step 3. Submerge the inhalation tube and the exhalation valve in the disinfectant solution.
Step 4. Clean under the edge of the rubber exhalation valve.
Step 5. Submerge the facepiece in clear water to rinse.
Step 6. Dry the facepiece with a clean, lintfree cloth and inspect the harness, lens, and inhalation tube for damage and wear.

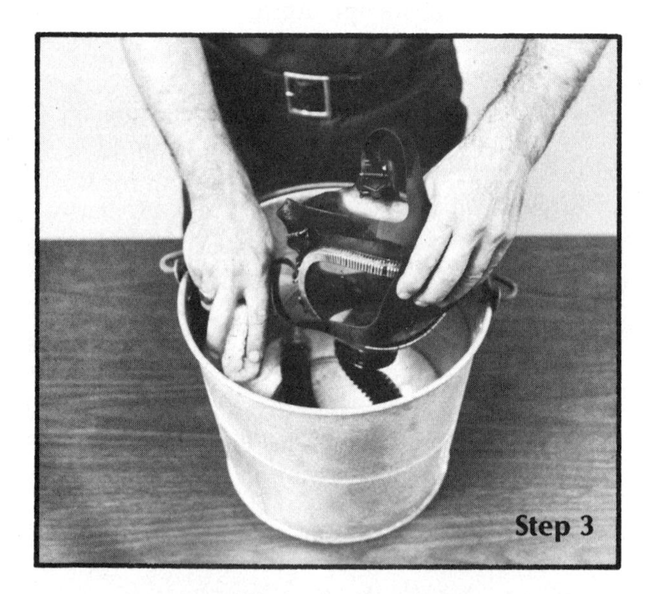

Figure 15.4. Maintenance of the facepiece.

be able to reach all parts of the space. In this case, the apparatus' container must be marked "FOR ENGINE ROOM USE ONLY," and the wearer must also use a lifeline.

The use of fresh air breathing apparatus is limited mainly by hose length. When the hose is longer than 40.2 cm (132 ft), the pump may not be able to supply enough air to the user. Demand-type apparatus consist of a facepiece, regenerator, breathing bag, inhalation tube, exhalation tube, relief valve, high pressure oxygen cylinder, high pressure reducing valve and pressure gauge, cylinder control valve, and bypass valve. A bumper plate and a spring-loaded admission valve are located in the breathing bag (Fig. 15.6).

Demand-type apparatus may be used on all vessels. They may be used instead of oxygen breathing apparatus in the machinery spaces of

Step. 1. Extend all the harness straps to the tab at the buckle.

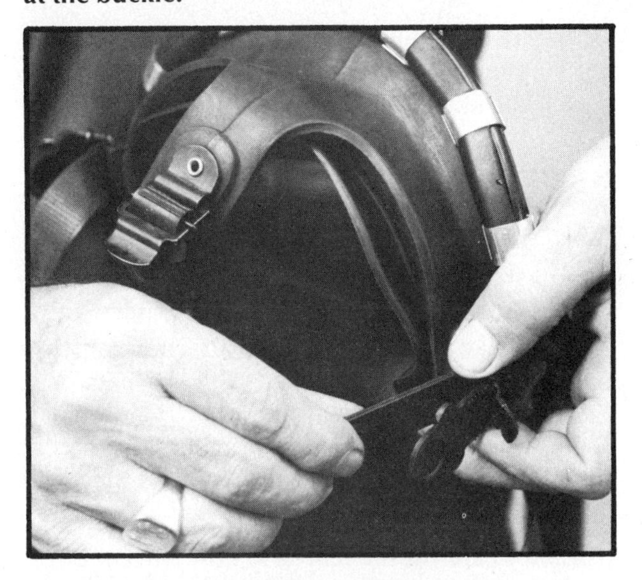

Step 2. Reverse the harness over the lens.

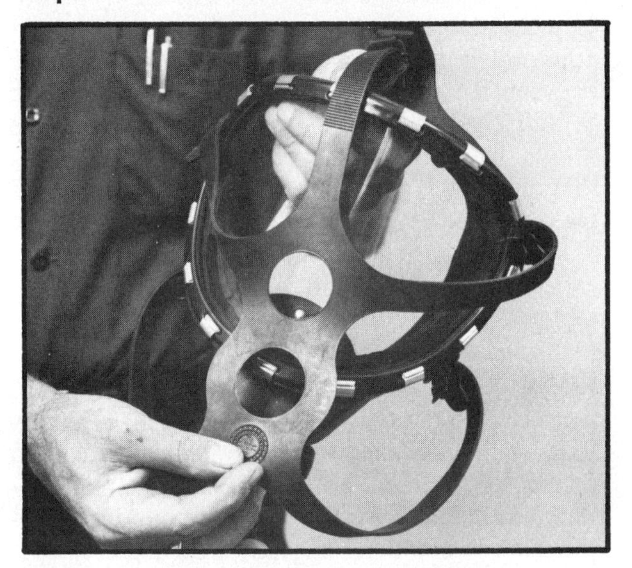

Step. 3. Place the facepiece in the carrying case.

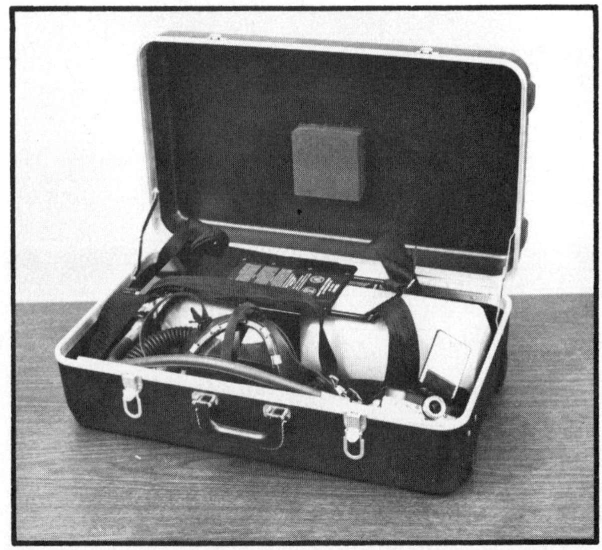

Figure 15.5. Stowing the facepiece.

tank vessels. However, they may not be used in place of the hose masks required on those vessels.

The reducing valve reduces the pressure of oxygen leaving the cylinder and entering the breathing bag. The pressure is reduced from about 125–135 atmospheres to about 20.7 kilopascals (3 psi). A safety whistle in the unit warns the wearer if the reducing valve fails when the pressure increases to about 48.3 kilopascals (7 psi). Fogging of the lens warns the wearer that oxygen is entering the bag at less than 20.7 kilopascals (3 psi) or the cardoxide is used up. In either case, the wearer must close the cylinder valve, open the bypass valve and immediately retreat to safety.

The oxygen cylinder is protected by a combination plug-rupture disk called a *safety cap*. The disk bursts if the oxygen in the cylinder reaches a temperature of 94°C (201°F). This releases the excess pressure in the cylinder.

The oxygen pressure gauge registers in atmospheres of pressure, from 0 to 150. A red zone on the dial shows pressures of 15 atmospheres. When the gauge needle enters the red zone, the wearer must retreat to safety. This is the only signal that the oxygen supply is nearly depleted—the unit does not include a timer. The cardoxide in the regenerator will be almost depleted when the pressure gauge reads in the red zone.

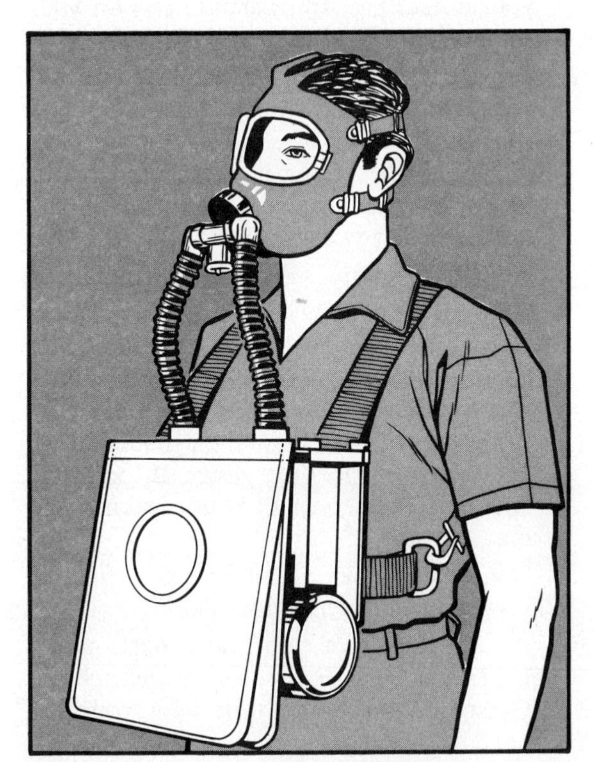

Figure 15.6. Oxygen-cylinder type OBA.

Operating Cycle

The operating cycle is started when the wearer exhales a few breaths of outside air through the exhalation valve. The warm, moist exhaled air travels through the exhalation tube and into the regenerator. There, the exhaled air reacts with the cardoxide. The CO_2 is absorbed, and heat is released. The warmed breath (now without CO_2) enters the finned cooler, where some of its heat is dissipated to the outside air. It then enters the breathing bag, which expands.

At this point, the wearer inhales air from the breathing bag through the inhalation tube. The bag collapses, causing the bumper plate to bump the admission valve off its seat. This admits a measured amount of makeup oxygen into the bag from the oxygen cylinder (through the reducing valve). The oxygen, at about 20.7 kilopascals (3 psi), is inhaled along with the regenerated breath.

Donning and Use

The body harness consists of two web straps that cross at the back, which position the unit on the wearer's chest. A thin web strap that is placed around the wearer's lower back helps stabilize the unit. A metal ring is provided for attaching a lifeline to the harness.

The donning procedure is as follows:

1. Place your head through the upper opening in the large web straps, so the unit is on your chest. The straps should rest on your shoulders.
2. Bring each snap hook around underneath your armpits, and attach it to the upper ring on the side where the strap begins. (Some people prefer to attach the hooks to the D ring on the bottom, where the thin web belt is connected.)
3. Adjust the straps so the unit is balanced comfortably on your chest.
4. Check the unit by opening the pressure-gauge valve and the cylinder valve until the pressure gauge registers the full cylinder pressure. Then close the cylinder valve, and watch the pressure gauge. If the gauge does not drop, the unit is not leaking and may be used.
5. Don the facepiece and check it for the proper fit as described earlier.
6. Place a finger of your right hand under the facepiece mask, near your right cheek. Grasp the inhalation tube with your left hand collapsing the inhalation tube shut, take a deep breath of outside air. Remove your finger from the mask, and exhale into the unit.
7. Repeat step 6 several times, until the breathing bag is fully inflated. This will start the absorption of CO_2 from the exhaled breath by the cardoxide.
8. When you are sure the unit is working properly, open the oxygen-cylinder valve.
9. Make sure your tender attaches a lifeline to the ring on the back of the harness. Also ensure that you and your tender fully understand and agree upon a set of lifeline signals. (A recommended set of signals is given in Table 15.1.)

The wearer of the unit may now enter the contaminated area. As he enters it, he should read the oxygen pressure gauge. He should then proceed to the furthest part of the contaminated area and read the gauge again. The difference between the two readings is the oxygen pressure he will need to leave the contaminated area. He should leave the area when the gauge registers that difference, or the needle reaches the red zone. He must also retreat from the contaminated area if the whistle sounds, if the lenses fog up or if he experiences any discomfort or breathing difficulties.

Recharging

To recharge the unit, a fully charged oxygen cylinder is first installed in its metal strap. It is then connected to the line leading to the pressure-reducing valve. Then the connection to the bypass line (to the breathing bag) must be made. A special wrench is provided for this purpose.

Table 15.1. Lifeline Signals between OBA Wearer and Tender

Tender to Wearer	
Pulls on line	*Meaning*
1	Are you all right?
2	Advance.
3	Back out.
4	Come out immediately.
Wearer to Tender	
Pulls on line	*Meaning*
1	I am all right.
2	I am going ahead.
3	Take up my slack.
4	Send help.

The hex-head plug on the regenerator is then removed with the same wrench. The unit is then turned over, and the old cardoxide is shaken out of the unit if not already empty. Finally, the regenerator is filled with fresh cardoxide, and the hex-head plug is reinstalled.

The oxygen cylinder must be changed whenever the cardoxide is changed, and vice versa. They are sized to operate for the same length of time. The 0.45-kg (1-lb) cardoxide charge and its oxygen cylinder will provide protection for about 30 minutes; the 0.91-kg (2-lb) charge and its cylinder, for about an hour.

SELF-GENERATING (CANISTER) TYPE OBA

The self-generating, or canister, type OBA is also a self-contained breathing apparatus. In this unit, the wearer's exhaled breath reacts with chemicals in a canister to produce oxygen. This oxygen is then breathed by the wearer.

Construction

The canister-type unit consists basically of five parts: a facepiece with an inhalation tube, an exhalation tube, and a pressure relief valve; a breathing bag; a canister holder and canister; a manual timer; and a breast plate with attached body harness. It is stored in a suitcase-type container with room for three canisters. Complete operating instructions are displayed inside the cover of the case.

The canister (Fig. 15.7) contains chemicals that react with moisture in the wearer's exhaled breath

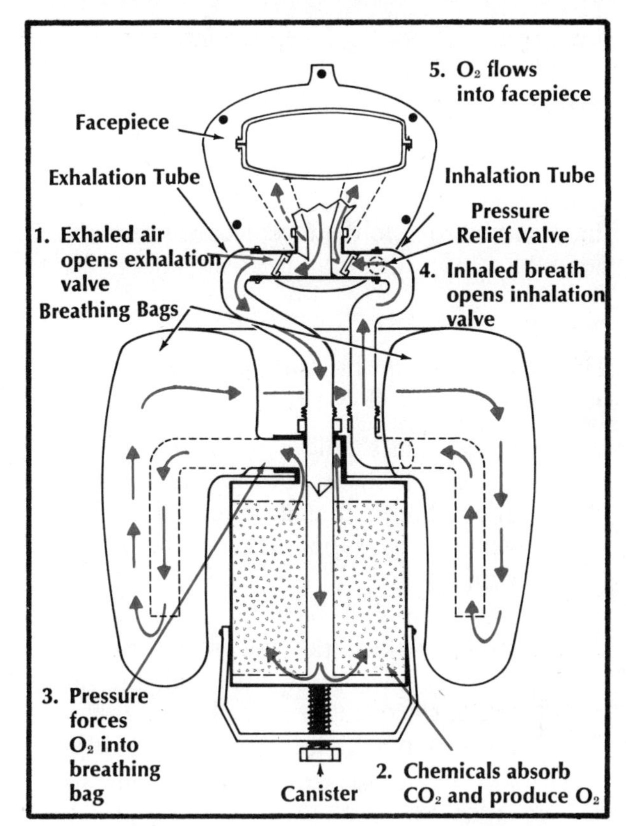

Figure 15.8. The numbers show the sequence of events during one operating cycle of the canister-type OBA. The arrows show the flow of exhaled breath and inhaled oxygen.

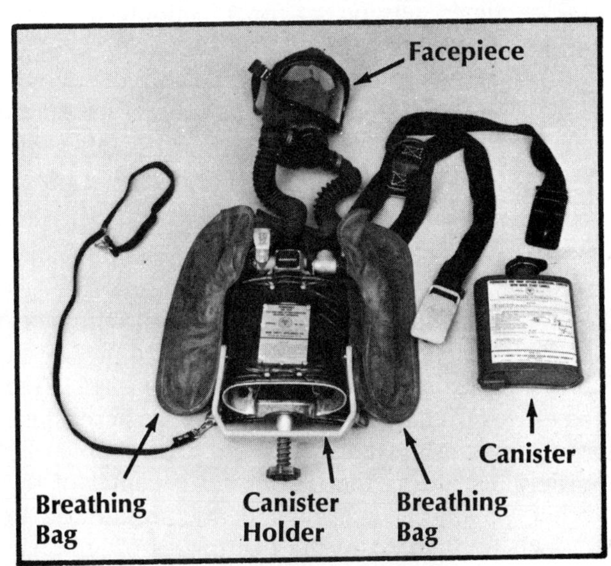

Figure 15.7. Self-generating canister-type OBA.

to produce oxygen. These chemicals also absorb carbon dioxide from the exhaled breath. If the unit is used for a short time and then removed, a new canister must be inserted before the next use. The chemicals in the canister continue to react even after the facepiece is removed and there is no accurate way of measuring the time left before the chemicals are used up. The breathing bag holds and cools the oxygen supplied by the canister and is made of reinforced neoprene.

The manual timer is set when the equipment is put into operation. It gives an audible alarm to warn the operator when the canister is nearly expended. The timer is no more than a clock; it does not indicate the condition of the canister. It should always be set to allow the wearer enough time to leave the contaminated area after the alarm sounds.

The body harness is a series of web straps that position and stabilize the apparatus. The breast plate holds the canister and protects the wearer from the heat generated by the unit.

Operating Cycle

The operating cycle of the canister-type unit is shown in Figure 15.8. The wearer's exhaled breath passes from the facepiece into the exhalation tube and then into the canister. Moisture and carbon dioxide are absorbed by the chemicals in the canister. They produce oxygen, which passes from the canister to the breathing bag. When the wearer inhales, the oxygen moves from the breathing bag to the facepiece via the inhalation tube.

Donning

The wearer can don the canister-type OBA without assistance as follows (Fig. 15.9):

1. Grasp one shoulder strap in each hand, and lift the harness over your head. This allows the equipment to rest on your chest while it is supported by the shoulder straps.
2. Reach around back to locate the side straps. Attach the side straps to the D rings on the breast plate with the hooks provided, one at a time. Then tighten the harness so it fits securely and comfortably.
3. Put the waist strap around your neck, attach the hooks at the D ring, and tighten the strap.
4. Remove a canister from the carrying case. (There are two types of canisters: self-start and manual start. You must know the type of canister you are using, for steps 9 and 10. The self-start canister has a small metal box at the bottom.)
5. To mount either type of canister, first remove the protective cap from the top to expose a thin copper seal.
6. Swing the canister retaining bail forward, and hold it with one hand. Now insert the canister in the holder, with the label facing outward, away from your body.
7. Swing the retaining bail down under the canister, and tighten the retainer (a heavy screw with a pad and handwheel) by turning it clockwise. This secures the canister in the holder and forms a seal between the canister and the central casting. The point of the central casting punctures the copper seal.
8. If you do not know which type of canister you have inserted, check the canister type to determine the correct starting action. Then don the facepiece as described earlier.
9. Start a self-start canister as follows: Locate the small triangular metal tab on the metal box at the bottom of the canister. Grasp the tab with the thumb and index finger of your right hand, and pull it downward. The small metal box will come away from the canister, exposing a lanyard. Grasp the lanyard with your index finger and thumb, and pull it straight out away from your body. *Do not pull down on the lanyard.* The correct action will activate the chemicals in the canister, filling the breathing bag with oxygen. If the lanyard breaks and does not activate the self-starter, use the manual-start procedure in step 10.
10. Start a manual-start canister in a safe, uncontaminated area by inserting one or two fingers under the facepiece, and stretching it away from your face. With the other hand, grasp the inhalation and exhalation tubes and squeeze them tightly. Then inhale. Now release the tubes, remove your fingers from under the mask, and exhale. Repeat this procedure several times, to inflate the breathing bag. This will start the chemical action in the canister. *Do not overinflate the breathing bag!* It should be firm but not rock hard.
11. Test the facepiece for leakage by squeezing the inhalation and exhalation tubes while inhaling. If the facepiece is properly fitted, it will collapse against your face.
12. Set the timer by turning the knob clockwise. On older units, the timer is set for 30 minutes. This allows the wearer 15 minutes to leave the contaminated area after the alarm sounds. On new units, the timer may be set for 45 minutes or less. The control should be turned to the extreme clockwise position and then reset to the desired time interval. This ensures that the alarm will sound for a full 8–10 seconds.

If the lenses fog up, any part of the unit malfunctions or the wearer experiences any discomfort or difficulty in breathing, he must immediately retreat to safety. One cause of difficulty in breathing is an overinflated breathing bag. If the bag is overinflated, it will seem very hard. This problem can be corrected, *in a safe area,* by briefly depressing the button in the center of the relief valve. The bag should not be allowed to deflate completely during this process. If the bag becomes underinflated, the user must repeat step 10 above.

Step 1. Lean forward from the waist with feet spread wide apart.

Step 2. Loosen the retaining screw.

Step 3. Swing the retaining ball forward and let the canister drop to the deck.

Step 4. Puncture the can several times with pick end of a fire ax.

Step 5. Submerge the canister in a pail of water. A violent boiling action will take place

Step 6. After the boiling action has stopped, empty the water into a drain or over the side of the ship. The canister can now be discarded.

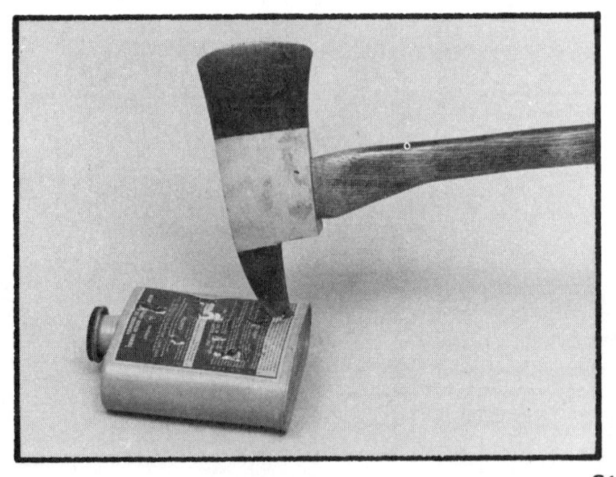

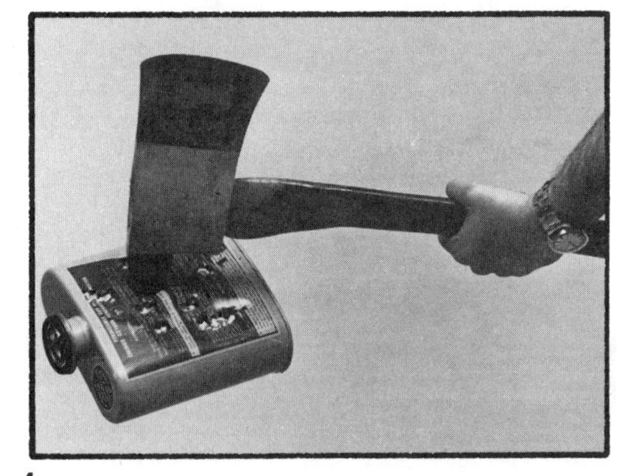

Step 4

Figure 15.10. Removing and disposing of an expended canister.

Removing the Canister

The removal and disposal of an expended canister are very hazardous operations that must be performed to avoid injury. The procedure (and the required precautions) are as follows (Fig. 15.10):

1. Spread your feet wide apart, and lean forward from the waist. (The chemical action that takes place in the canister generates sufficient heat to burn bare skin. For this reason, you must not touch the expended canister.)
2. Loosen the retaining screw by turning the handwheel counterclockwise.
3. Swing the retaining bail forward, and let the canister drop to the deck. It must not be tossed (or allowed to fall) into the bilge, or anyplace where oil, water, snow, ice, grease or other contaminants can enter the hole in the copper seal. Organic material may cause a violent reaction. Water and substances containing water will cause a rapid chemical action in the canister, creating more pressure than can be released through the small neck opening. This pressure could cause an explosion that would produce flying fragments and injure anyone in the vicinity.
4. Puncture the expended canister several times, front and back, with the pike end of a fireaxe.

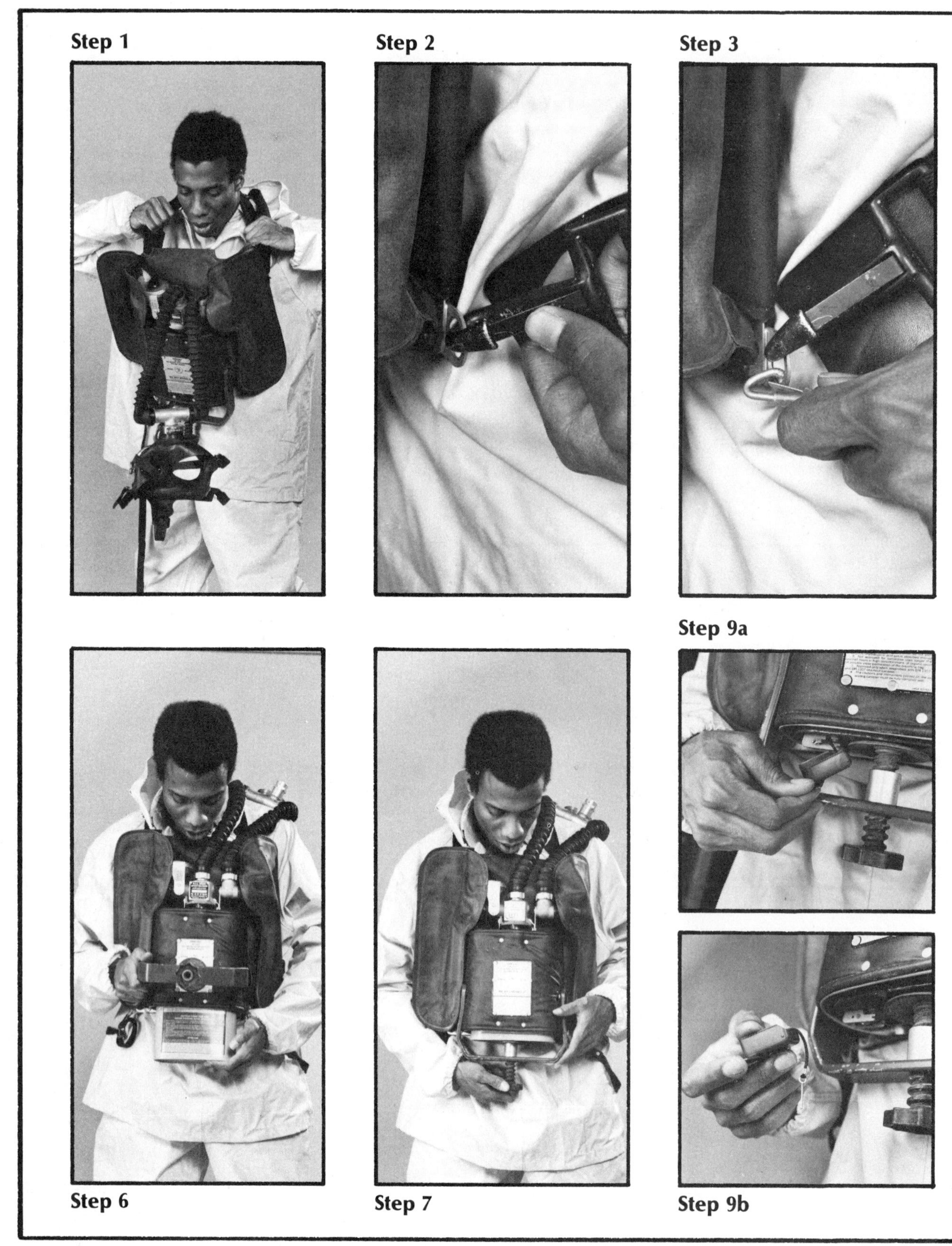

Figure 15.9. The procedure for donning a canister-type OBA.

Step 4

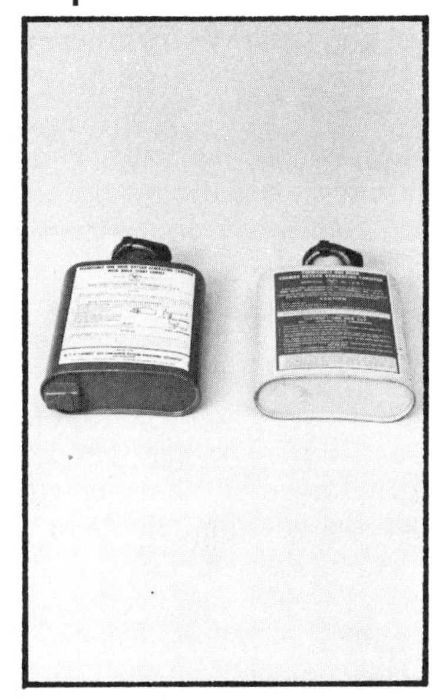

Step 5a

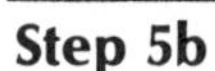

Step 5b

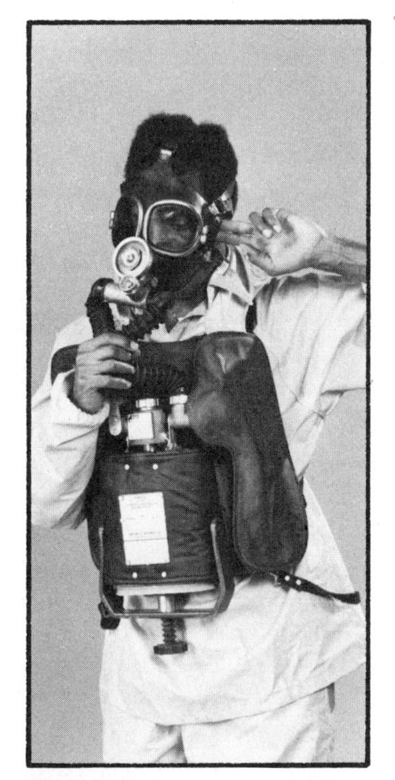

Step 10

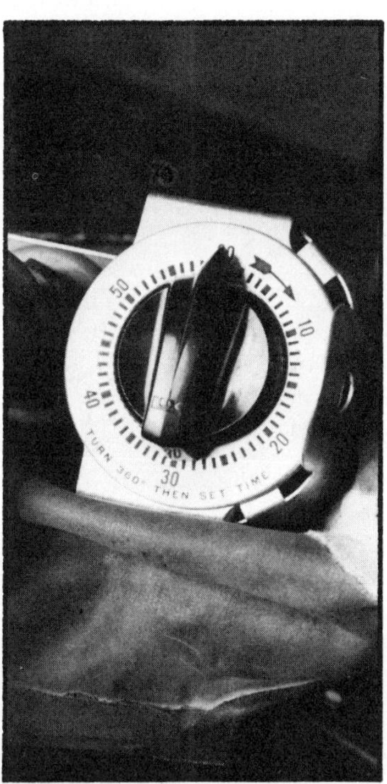

Step 12

Step 1. Grasp the shoulder straps in each hand and put harness over your head.

Step 2. Attach the side straps to the "D" rings on the breast plate.

Step 3. Attach waist harness hooks at the "D" rings.

Step 4. Remove a self-start or manual start canister from the carrying case.

Step 5. Remove the protective cap from the canister to expose a thin copper seal.

Step 6. Insert the canister in the holder.

Step 7. Swing the retaining bail under the canister and tighten the retainer.

Step 8. Check the canister and determine the correct starting action: manual or self-start.

Step 9. To activate the chemicals in the self-start canister, remove the metal box and pull the lanyard straight forward and away from your body.

Step 10. For a manual start, on inhalation, pull facepiece away from face and crimp tubes. On exhalation, release facepiece and tubes. Repeat.

Step 11. Test the mask for leakage.

Step 12. Set the timer.

5. Fill a pail with clean water, deep enough to completely submerge the canister. Gently drop the canister into the water. A violent chemical reaction will take place. However, the pressure cannot build up if the canister has been properly punctured, so there is no danger of an explosion.
6. After the boiling has stopped, empty the water (which is now caustic) into a drain or over the side of the ship. Rinse the pail thoroughly, and discard the canister.

Maintenance

The oxygen-generating apparatus must be maintained carefully. Worn or damaged parts must be replaced by the manufacturer or his representative. Periodic inspection and after-use maintenance should be performed faithfully by those who use the equipment, according to the following procedure.

1. Clean the facepiece as described in Figure 15.4. Be especially careful to dry all the equipment thoroughly.
2. Check the inhalation and exhalation valves periodically for corrosion; have them replaced if necessary.
3. Test the alarm bell to ensure proper operation.
4. Inspect the breathing bag for signs of damage and wear.
5. Inspect the canister holder and retaining bail and screw for damage, wear and proper operation. Check the central casting plunger that breaks the seal and seals the canister into the system. This plunger operates by moving in and out about 0.64 cm (¼ in.). A spring holds the plunger out. When the canister is inserted and tightened down by the bail screw, the plunger is depressed against the spring. This action ensures a tight seal. If the plunger does not work properly, it must be repaired or replaced; *it should never be lubricated.*

Safety Precautions

Certain precautions must be taken when the oxygen-generating apparatus is used. The user must be careful not to damage the breathing bag on nails, broken glass or other sharp objects. When it is necessary to operate the relief valve, he must do so carefully, so as not to deflate the breathing bag too far.

The instructions on the canister must be followed to the letter. Foreign material, especially petroleum products, must be kept from entering an opened canister. The chemical in the canister is caustic; it must not come in contact with the skin.

The apparatus must not be stowed with a canister already inserted. After one use, regardless of how short, the canister must be discarded as described. For older units without the self-start action, three fresh canisters should always be kept in readiness, *with their caps intact,* in the storage case. For newer units with the self-start action, two fresh canisters may be kept in the case.

Advantages and Disadvantages

The greatest advantage of the oxygen-generating apparatus is its staying time. The canister produces sufficient oxygen for comfortable breathing up to 45 minutes. In addition, this unit is much lighter than other self-contained units. Thus, it is advantageous for use in large contaminated spaces where ventilation may be difficult; where it is difficult to locate the fire or the source of contamination; and wherever an uninterrupted operating time of up to 45 minutes is required.

Among the disadvantages of the canister-type apparatus are these:

- Approximately 2 minutes is required to start a manual-start canister and get the equipment into operation.
- If the relief valve is not operated properly, the breathing bag may lose its oxygen. The wearer must then return to an uncontaminated area to restart the unit.
- The bulkiness of the unit and its location on the wearer's chest may reduce maneuverability and the ability to work freely.
- The heat produced by the canister, the possibility of explosion if the canister is not disposed of properly and the explosive reaction if petroleum products are introduced into the canister opening make the unit hazardous if not used properly.
- The unit is not easily used for buddy breathing in rescue work.
- The apparatus cannot be used in an atmosphere that has contained or is suspected of containing flammable or combustible liquids or gases.
- When the alarm bell sounds, it rings once and stops. Owing to noise or some other distraction, the wearer may not hear the alarm.

SELF-CONTAINED, DEMAND-TYPE BREATHING APPARATUS

Demand-type breathing apparatus is being used increasingly aboard merchant ships. Its popularity

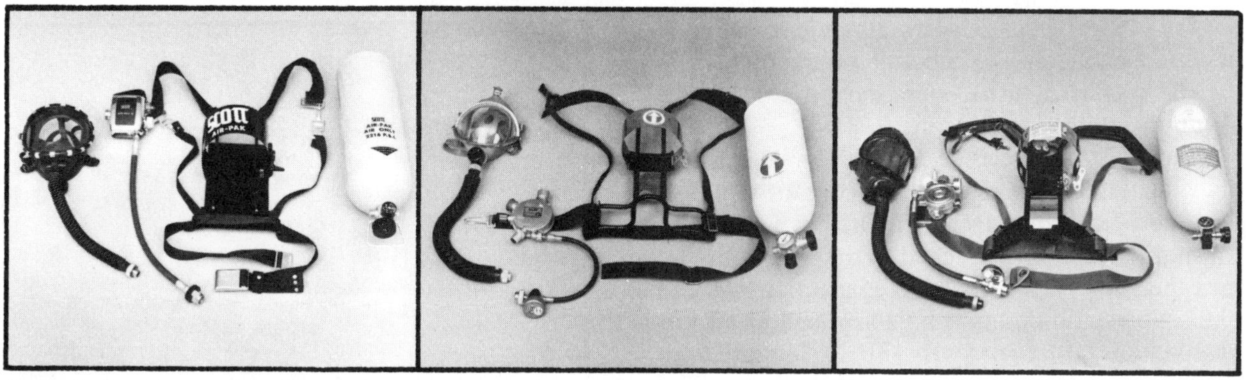

Figure 15.11. Three self-contained, demand-type breathing units.

stems from its convenience, the fact that it supplies the user with cool fresh air, the speed with which it can be put into service and its versatility. Figure 15.11 shows three self-contained, demand-type units produced by three different manufacturers. The units are dissimilar enough so that their components are not interchangeable—except for the air cylinders.

The demand-type apparatus gets its name from the functioning of the regulator, which controls the flow of air to the facepiece. The regulator supplies air "on demand"; i.e., it supplies the user with air when he needs it and in the amount his respiratory system requires. It thus supplies different users with air at different rates, depending on their "demand." *Note:* Newer model demand-type breathing apparatus are being supplied with a positive flow to the facepiece. The slight pressure in the facepiece prevents contaminated air from entering the facepiece and getting into the respiratory tract. This positive air pressure lessens the critical nature of the facepiece fit against the user's face.

Construction

The self-contained, demand-type apparatus consists of four assemblies: the facepiece with inhalation tube, exhalation valve, head harness and wide-vision lens; the regulator with pressure gauge, valves, high-pressure hose and alarm bell; the air cylinder with valve and pressure gauge; and the backpack or sling pack with adjustable harness. (Some manufacturers consider the high-pressure hose and alarm to be a separate assembly.)

Facepiece. The facepiece used is the standard full-face type discussed earlier in this chapter.

Regulator. Figure 15.12 is a schematic diagram of a demand-type breathing apparatus. Air from the supply cylinder passes through the high-pressure hose and a preset pressure-reducing valve in the regulator. The admission valve is normally closed. However, when the user inhales, he produces a partial vacuum on one side of the admission valve. This opens the valve, allowing air to

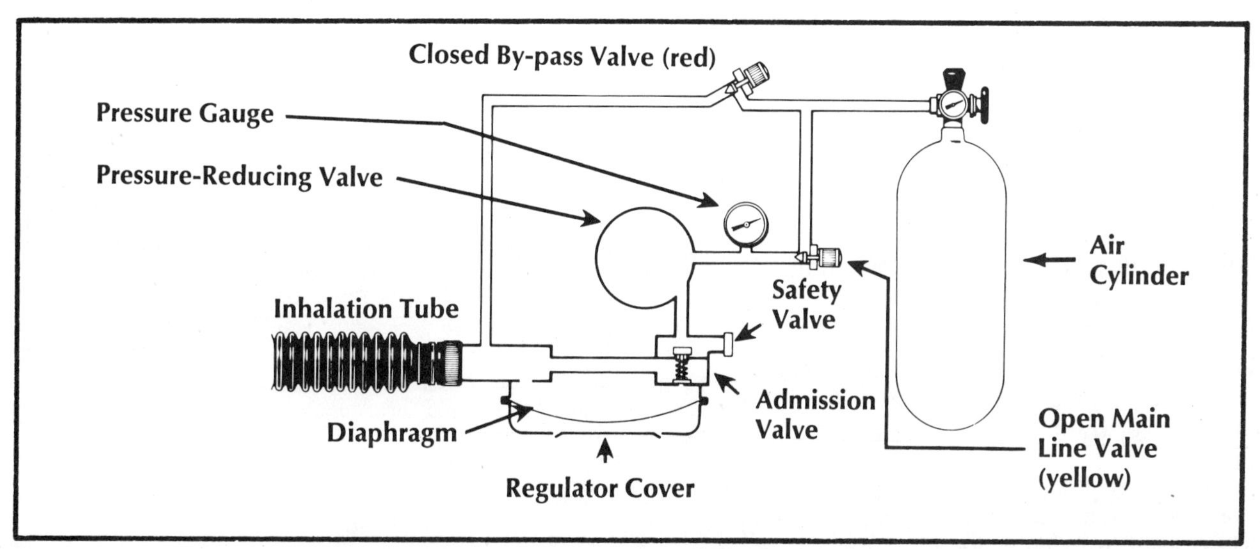

Figure 15.12. Schematic diagram of the self-contained, demand-type breathing apparatus.

pass into the facepiece. The amount of air supplied depends on the amount of vacuum produced, which in turn depends on the user's air requirements.

Figure 15.13 shows four commercially available regulators. The regulator in Figure 15.13A does not show an alarm bell. In this model, the low-air alarm bell is attached to the high-pressure hose near the threaded tank connection. (Some older models do not have an alarm bell at all.) In the regulator in Figure 15.13B, a low-pressure alarm bell is located in the regulator case. The low-pressure alarm bell for the regulator in Figure 15.13C is located near the tank connection on the high-pressure hose. The regulator in Figure 15.13D has a low-pressure alarm bell attached to the high-pressure hose. Older models of this regulator were equipped with a reserve valve. The reserve-valve lever is placed in the "Start" position when the equipment is donned. When the cylinder pressure falls to approximately 3450 kilopascals (500 psi), breathing becomes difficult. At this time the wearer must move the reserve lever to the "Reserve" position. This allows the wearer 4–5 minutes of reserve air with which to leave the contaminated area. An alarm bell kit can be installed on this older regulator model.

Air Cylinder. The air cylinder includes a pressure gauge and a control valve. On most cylinders the threaded hose connection is a standard size. Cylinders are rated according to breathing duration, which depends on the size and pressure of the cylinder. There are four standard sizes. United States Coast Guard regulations require an air supply sufficient for at least 10 minutes of normal breathing. The IMCO code for tank ships requires a cylinder capacity of 1200 psi (42 ft^3) of air. This should be sufficient to provide breathing protection for approximately 30 minutes.

Backpack or Sling Pack. The backpack or sling pack and the harness are designed to hold the unit securely and comfortably on the wearer. They differ slightly according to the manufacturer, but all makes are donned in about the same way. However, backpack units are donned and stowed differently from sling-pack units.

Backpack Unit

The backpack unit is the most commonly used demand-type breathing apparatus. Its air supply has a longer duration than that of the sling-pack unit.

Donning. When a backpack unit has been properly stowed in its carrying case, it can be donned

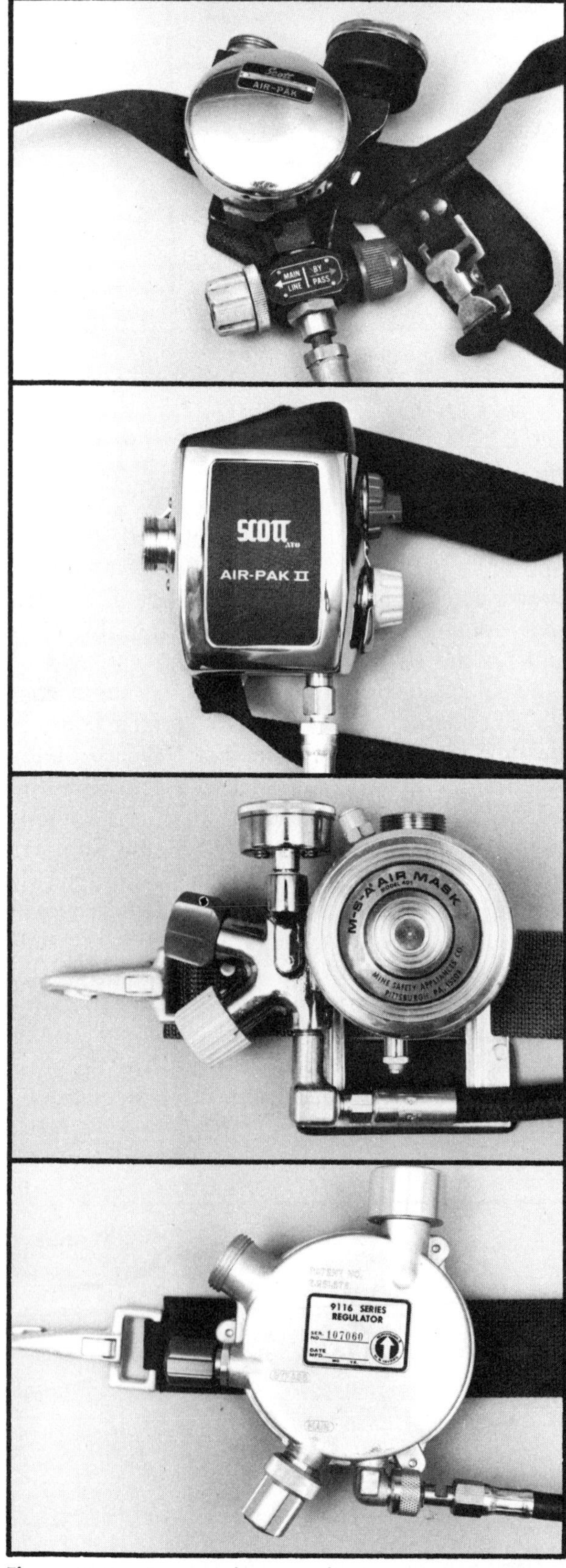

Figure 15.13. Four regulators, each of a somewhat different configuration.

by the user without assistance. The unit should be stowed with the tank down, backpack up and harness straps fully extended (Fig. 15.14). The high-pressure air hose should be lying along the front of the case, with the regulator at the front right-hand corner. The harness take-up straps must be attached to the chest straps. One should be to the left of the regulator, and the other should be attached to the metal buckle on the right chest strap. The waist straps should be rolled or folded neatly between the backpack and the cylinder valve. The facepiece should be placed between the air cylinder and the high-pressure air hose.

When the unit has been stowed as described, it is donned in this way (Fig. 15.15):

1. Take a crouched position at the right end of the open case. With one hand, grasp the cylinder valve handle, and stand the cylinder and backpack on end. Check that the main-line valve (usually a yellow knob) is opened and locked in the open position. Check that the bypass valve (a red knob) is closed.
2. Check the *cylinder* gauge to be sure the cylinder is full. Then open the cylinder valve three turns. Now check the *regulator* gauge; it should read within 1380 kilopascals (200 psi), of the cylinder gauge. If the difference is more than 1380 kilopascals (200 psi), assume the lower reading is correct. At the first opportunity, check the gauges for accuracy and make any necessary repairs.
3. Grasp the backpack with one hand on either side, making certain that the harness straps are resting on the backs of your hands or arms. Now, from the crouched position, lift the unit over your head. Allow the harness to drop into position over your arms.
4. After the harness has cleared your arms, lean forward, still in the crouched position. Lower the unit to your back. While still in this position, fasten the chest buckle.
5. Stand, but lean slightly forward to balance the cylinder on your back. Then grasp the two underarm adjusting strap tabs. Pull the tabs downward to adjust the straps. To get the equipment as high on your back as possible, bounce the cylinder by moving your back and legs; at the same time, pull the tabs to position the cylinder.
6. Locate both ends of the waist harness, hook the buckle, and tighten the strap. Once this is done the equipment is secure, and you may stand erect.
7. Remove the facepiece from the case, and don it as described earlier. The donning of the facepiece should be practiced and mastered before this equipment is used.
8. Insert the quick connect coupling of the inhalation tube at the regulator, and tighten it down. To conserve air, this step should be performed just before you enter the contaminated area.

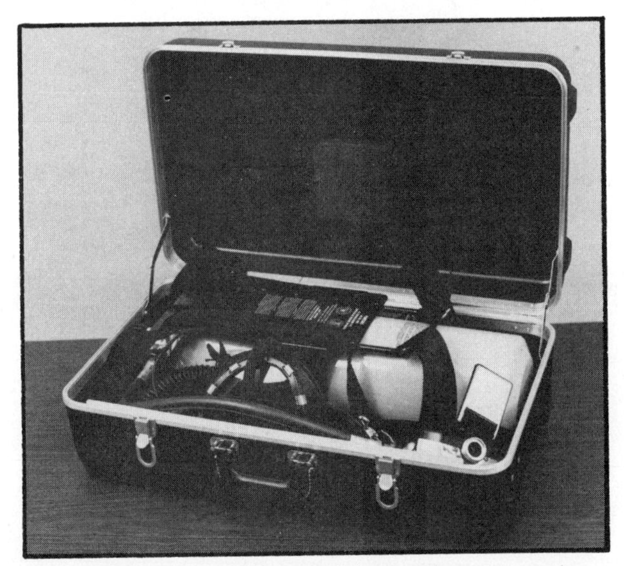

Figure 15.14. A properly stored backpack unit.

The user's breathing should now feel and remain normal. If the unit does not supply sufficient air automatically, the main-line valve (yellow knob) should again be checked to ensure that it is fully opened and locked. The bypass valve (red knob) must be closed at all times; it is opened only if the regulator malfunctions. Then the air flows directly to the facepiece, bypassing the regulator. When the bypass valve must be opened, the main-line valve should be closed.

Removal and Restowing. The backpack unit should be removed as follows (Fig. 15.16):

1. Disconnect the inhalation tube from the regulator.
2. With the tips of your fingers, release the self-locking buckles on the facepiece harness. Remove the facepiece as described earlier.
3. Make sure the facepiece harness straps are fully extended. Pull the harness over the front of the facepiece, and place the facepiece in the carrying case.
4. Unbuckle the backpack waist belt, and extend the belt fully.

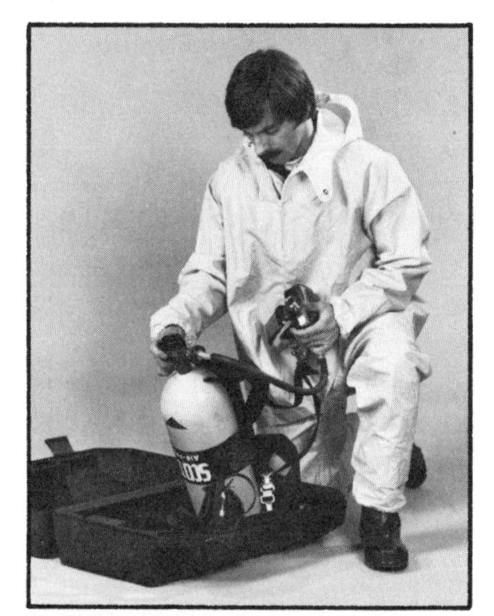

Step 1

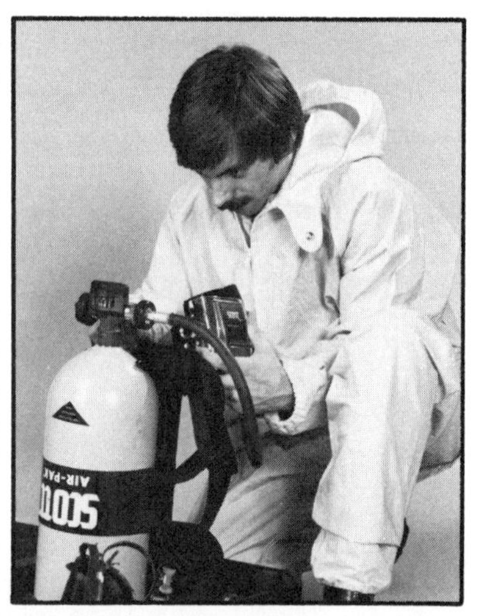

Step 2

Step 1. Taking a position at the right end of the open case, grasp the cylinder valve handle with one hand and stand the cylinder and backpack on end.

Step 2. Check the cylinder gauge, open the valve, and compare the regulator gauge with the cylinder gauge (it should be within 200 psi).

Step 3. Lift the unit over your head, allowing the harness to drop down over your arms.

Step 4. Lower the unit onto your back and fasten the chest buckle.

Step 5. Bounce the cylinder into position on your back and pull the underarm strap tabs to secure its position.

Step 6. Hook the waist harness buckle and tighten the strap.

Step 7. Don the facepiece.

Step 8. Tighten down the quick connect coupling of the inhalation tube at the regulator.

Step 3

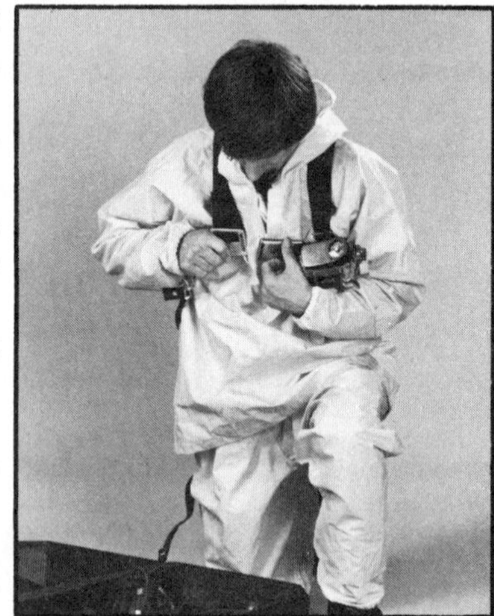

Step 4

Step 5

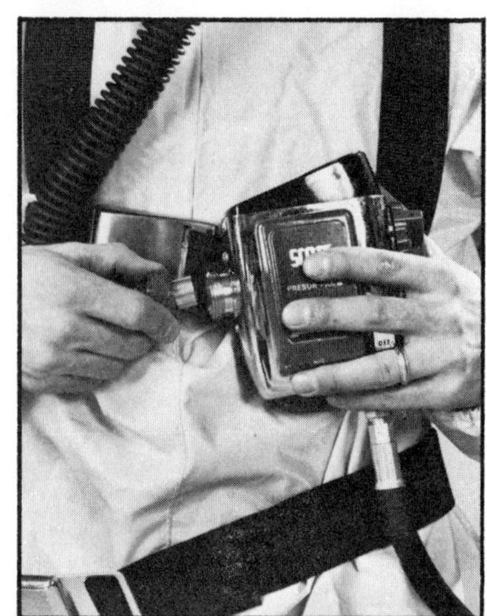

Step 8

Figure 15.15. Proper stowage of the backpack unit (tank down, backpack up, harness straps fully extended) allows one crewman to don the unit without assistance.

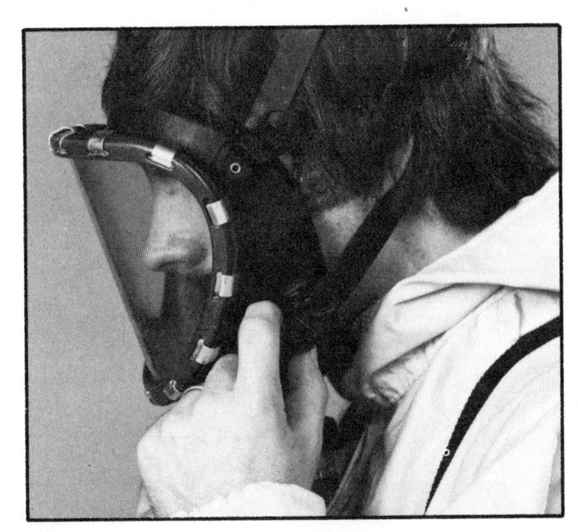

Step 2

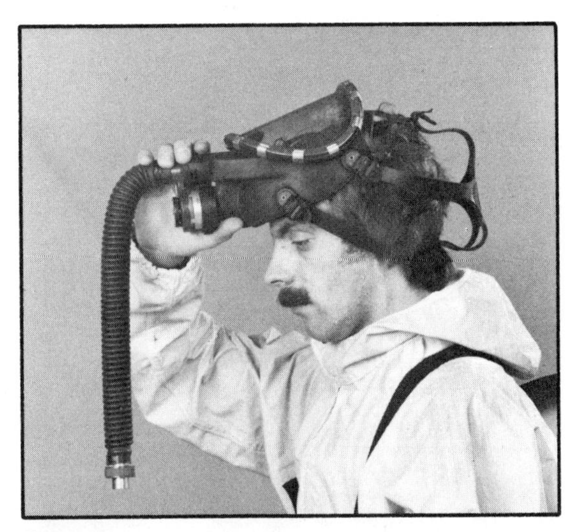

Step 2

Step 7

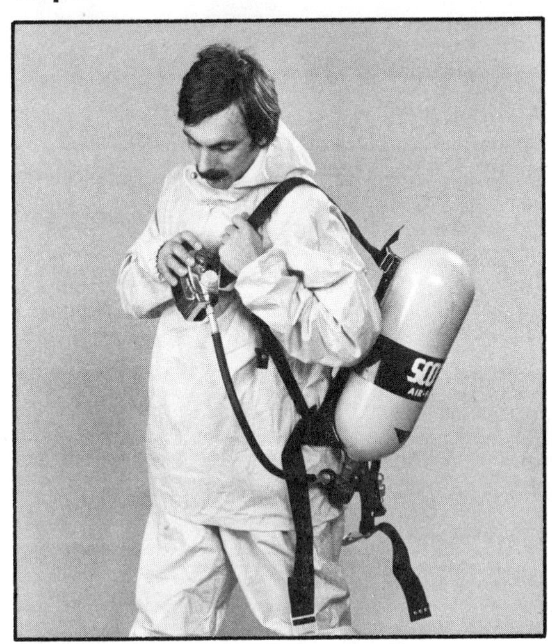

Step 8

Step 1. Disconnect the inhalation tube from the regulator.
Step 2. Release the self-locking buckles on the facepiece harness. Remove the facepiece.
Step 3. Pull the harness over the front of the facepiece and place the facepiece in the carrying case.
Step 4. Unbuckle the waist belt.
Step 5. Release and hold the underarm strap buckles.
Step 6. Disconnect the chest buckle.
Step 7. Hold the body harness and regulator in your left hand and slip your right arm out of the harness.
Step 8. Grasp the harness and regulator in your right hand and remove the unit from your left arm.
Step 9. Close the valve on the air cylinder and stow the equipment in the carrying case as detailed in the text.

Figure 15.16. Removing the backpack unit.

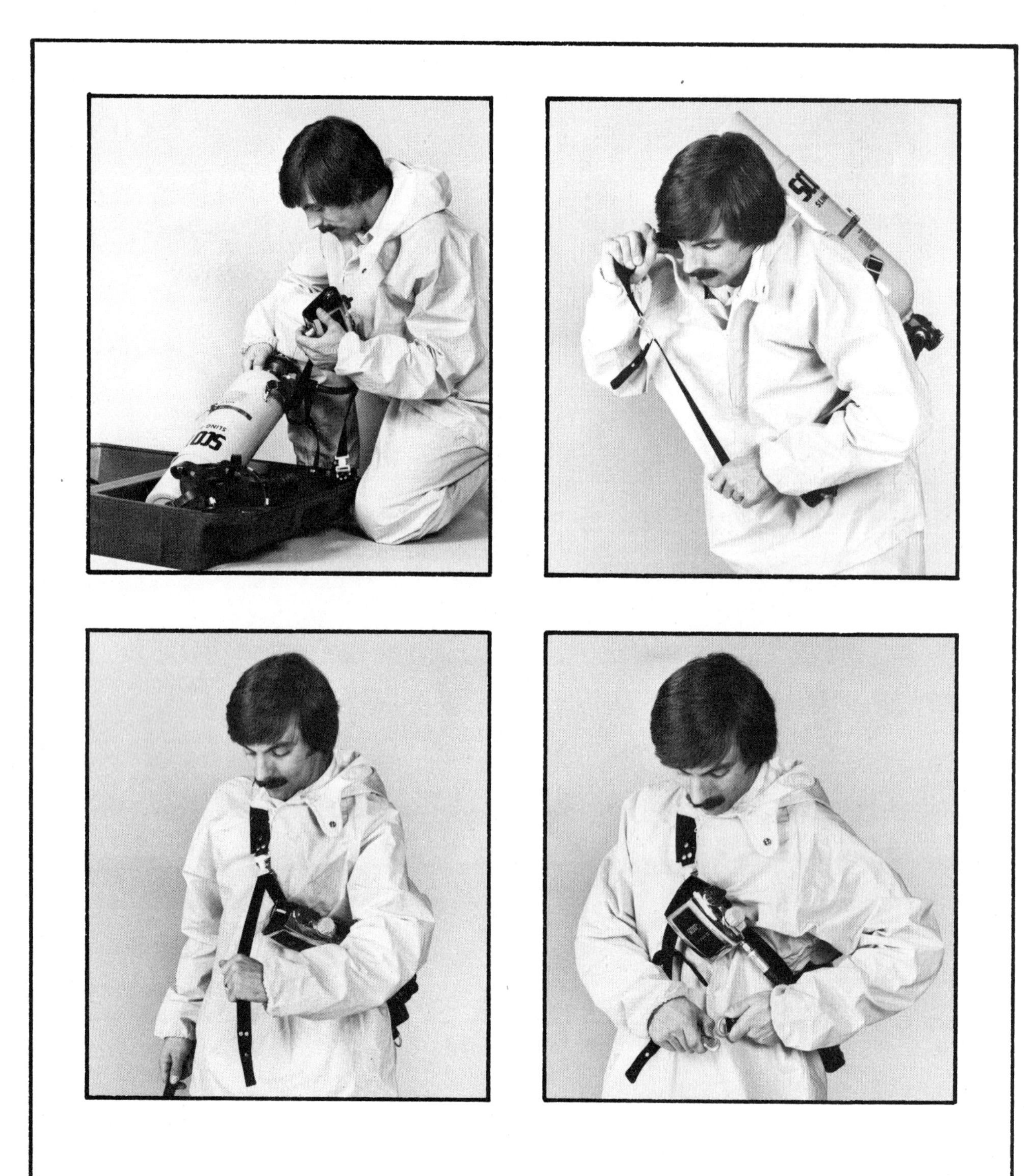

For step-by-step instructions for donning the unit, see page 341.

Figure 15.17. Donning the sling-pack unit.

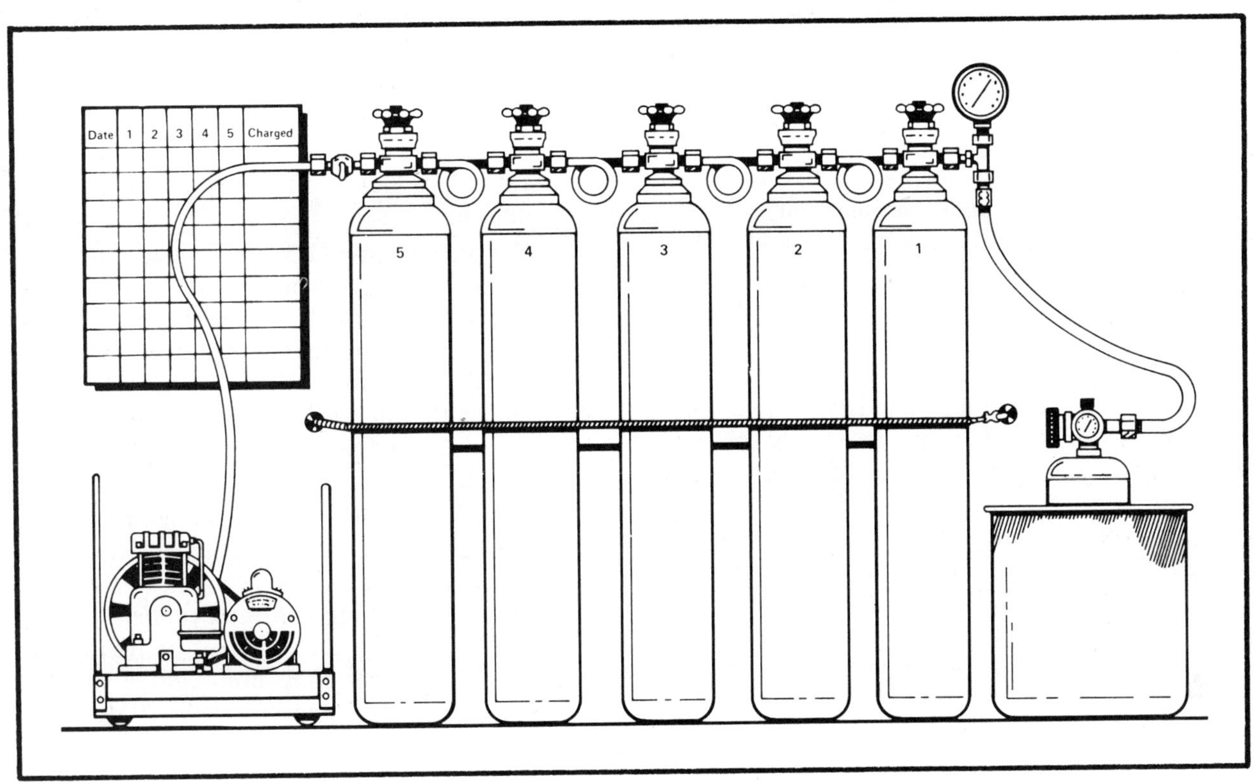

Figure 15.18. Cascaded air tanks for refilling breathing apparatus cylinders.

5. With your thumb and index finger, release and hold the underarm strap buckles, and extend them fully.
6. Disconnect the chest buckle.
7. Get a firm grip on the body harness and the regulator with your left hand, at the point where they are attached. Slip your right arm out of the harness as if you were removing a vest.
8. Grasp the harness with your right hand, above and as close to the regulator as possible. Then remove the equipment from your left shoulder and arm. By removing the equipment this way, you will keep the regulator from striking nearby objects, which could damage it.
9. Close the valve on the air cylinder. Remove the air pressure from the regulator by cracking the bypass valve open momentarily.

The unit should be thoroughly cleaned, and the air cylinder should be replaced immediately with a full cylinder. These procedures will be described shortly. However, it may be necessary to restow the equipment before it is cleaned and its cylinder is replaced. It should then be stowed in its case as described above. The case should be marked or tagged "Empty Cylinder."

Sling-Pack Unit

The sling-pack unit is generally stowed in a case. However, it is donned as follows no matter how it has been stowed (Fig. 15.17).

1. Lay the facepiece aside, in a clean, dry place.
2. Grasp the shoulder strap with your right hand. The air cylinder should be to your left, and the regulator to your right.
3. In one motion, swing the unit onto your back while putting your left arm through the harness. Carry the shoulder strap over your head, and place it on your right shoulder.
4. Pull the strap, to take up the slack.
5. Clip the waist straps together; tighten them by pulling the strap end to your right.
6. Don the facepiece as described previously.

The sling-pack unit is removed by reversing these steps. Before the unit is stowed, it should be cleaned and its cylinder should be replaced.

Minipack Unit

The minipack unit (Fig. 15.18) is a small cylinder-supplied, demand-type breathing apparatus. It is most often used with aluminum and asbestos firefighting and proximity suits. It is worn under the suit with a sling-type harness, or carried in a pocket built into the suit for this purpose. It is used with a half mask (respirator type) that covers the nose and mouth only, and is held onto the face with a lightweight elastic harness.

This unit is not meant for the usual shipboard duties. It is used for quick "hit-and-run" operations, such as shutting down tank valves in flammable-liquid fires or possibly rescuing victims trapped by a flammable-liquid fire.

Changing Air Cylinders

When the alarm bell on a demand-type breathing apparatus sounds, a 4–5-minute supply of air (approximately 3450 kilopascals (500 psi)) remains in the cylinder. If several crewmen equipped with breathing apparatus are working together, it may be difficult to tell whose alarm bell is sounding. A crewman who believes his bell is sounding should put his hand on the bell. If it is his alarm bell, the sound will be deadened, and he will feel the vibration of the bell. He should immediately leave the contaminated area to replace his air cylinder.

A second crewman should help change the air cylinder on a backpack or sling-pack unit while the equipment is being worn. The exchange of cylinders should be performed carefully.

1. When you are outside the contaminated area, remove your facepiece and locate a full air cylinder. Spare cylinders are usually stowed with the apparatus. It is important that you locate a full cylinder. To avoid confusion with used cylinders, hold onto the full cylinder until it is placed into your unit.
2. Someone should be available to assist you in changing the cylinder. Take advantage of the cylinder change time to rest. Kneel on one knee, with your back to your helper, while he makes the change. Hold the full cylinder on the ground in front of you.
3. The helper closes the cylinder valve and disconnects the high-pressure hose coupling from the used cylinder. If a wrench is required, it should be kept in the cylinder storage compartment, on a length of light chain.
4. The helper must support the used cylinder with his left hand while he releases the cylinder clamp.
5. The helper next removes the empty cylinder from your pack and places it on the ground by his feet.
6. Now place the full cylinder on your shoulder. The helper then—and only then—takes the full cylinder and places it directly into your pack.
7. When the cylinder is in the proper position, the helper locks it in place with the locking device.
8. The helper now checks the opening of the cylinder valve to ensure that it is free of foreign material. If it is dirty, he releases a short burst of air to clear it. When he is certain the valve opening is clear, he attaches the high-pressure hose to the cylinder outlet. Again, a wrench may be required.
9. You or the helper may now open the cylinder valve. Then check the pressure gauge on your regulator, while the helper checks the pressure gauge on the cylinder. Owing to the age of the equipment or its design, the two gauges may not register exactly the same. A difference of 1380 kilopascals (200 psi) is acceptable.

Maintenance

Self-contained, demand-type breathing apparatus must be carefully maintained. Any part of the unit that fails should be replaced or repaired by the manufacturer or his authorized representative. The equipment should be inspected periodically as recommended by the manufacturer. After each use, the wearer should clean the apparatus and replace the used air cylinder with a full one. If the unit must be stowed with an empty or used cylinder, the carrying case must be so tagged or marked.

Cleaning the Apparatus

1. Clean the facepiece as described earlier.
2. Wipe down the entire unit, including the harness straps, with a sponge soaked in a mild disinfectant solution or a mild soap-and-water solution. This will remove any loose particles and help deodorize the equipment.
3. Turn the carrying case upside down, to shake out loose particles. Wipe down the entire case, inside and out, with a sponge and disinfectant solution.
4. The following check should be made of the regulator parts:
    a. Inspect the threaded fittings for damaged threads and obstructions.
    b. Inspect the gauge for visible damage, such as dents or a cracked lens.

c. Inspect the main-line valve (yellow knob) to be sure it is fully open and locked (if a locking device is provided).
d. Inspect the bypass valve (red knob) to be sure it is closed tightly.
e. Inspect the alarm bell by first opening the cylinder valve to put air pressure on the regulator. Then close the cylinder valve, and breathe the air pressure off the regulator slowly. The alarm should sound when you have reduced the pressure in the regulator to approximately 3450 kilopascals (500 psi).

5. Inspect the harness for signs of wear and damage. A worn or damaged harness or buckle could break the next time the equipment is used, endangering the wearer.
6. Wipe the case and the unit dry with a lint-free cloth. Restow a backpack unit as described earlier in this chapter.

Refilling Air Cylinders. Every vessel that is required to carry the demand-type breathing apparatus must also carry a spare air cylinder. Some vessels may have a recharge system of air tanks shown in Figure 15.19 called a *manifold* or *cascade* system. Each tank should be numbered. A chart should be hung near the tanks, recording the air pressure in each tank, each date on which the cascade system was used and the number of cylinders refilled. (A sample chart is shown in Table 15.2.) This is very important for the proper use of the cascade system.

The pack cylinders for the breathing apparatus are filled from the large tanks in the following manner.

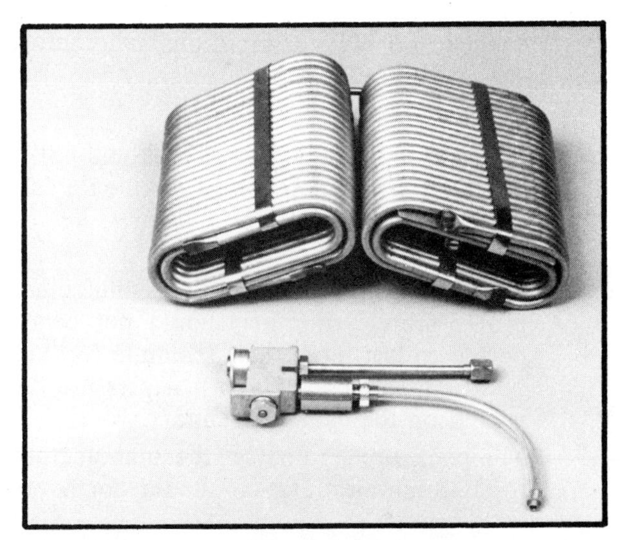

Figure 15.19. Air modules and flow regulation unit for a module-supplied, demand-type apparatus.

Table 15.2. Cascade System Air-Pressure and Usage Chart

	Cascade Tanks					*Pack Cylinders Charged*
Date	*1*	*2*	*3*	*4*	*5*	
4/20	2400	2400	2400	2400	2400	—
4/22	1700	2250	2400	2400	2400	5
4/29	1500	1925	2400	2400	2400	6
5/1	1200	1675	2100	2400	2400	6
5/8	800	1250	1600	1950	2400	8
5/20	300	825	1000	1450	1975	6
Charged	X	X	X	—	—	—
5/21	2400	2400	2400	1450	1975	—
5/23	2400	2400	2400	1025	1825	3
5/30	1700	2275	2400	700	1450	7
6/4	1150	1775	2275	275	900	6
Charged	X	—	—	X	X	—
6/5	2400	1775	2275	2400	2400	—

1. Check the cascade system record chart to find the air pressure in each tank.
2. Connect the charging hose to the cylinder to be charged.
3. Check the pressure of the cylinder to be charged. Open the valve on this cylinder. Then open the valve on the cascade tank with the least air pressure that is greater than the pressure in the cylinder to be filled.
4. Release air into the pack cylinder slowly, to keep from heating it excessively. Placing the cylinder in a container of cold water helps keep the cylinder cool. When the cylinder and tank pressures have equalized, close the valve on the cascade tank. Open the valve on the tank with the next highest pressure. Continue this procedure until the pack cylinder is filled to the desired pressure.
5. Repeat steps 2–4 for any other pack cylinders that are to be filled.
6. After all the pack cylinders have been filled, record the pressures remaining in the cascade tanks on the chart. If the cascade system is equipped with a compressor, refill all the cascade tanks to their maximum pressure. Mark the chart "full" or "recharged."

Safety Precautions

As with all emergency equipment, the most effective safety procedure for demand-type breathing apparatus is training, followed by constant practice. However, crewmen should take certain precautions when using this breathing equipment. When used properly, a demand-type unit will protect the wearer in any situation requiring respiratory protection equipment except underwater search.

Demand-type equipment should not be used after running or strenuous work. The air will be used up rapidly, and the wearer may feel that the unit is not giving him all the air he needs.

Before donning the equipment, the user should check the pressure gauges on the air cylinder and the regulator. As noted above, they should register within 1380 kilopascals (200 psi) of each other. The backpack or sling-pack harness should be tight; owing to the weight of the unit, a loose pack can cause injury.

Before entering a contaminated area, the wearer should check the facepiece for the proper seal. He should also check to see if his unit has an alarm bell. If it does not, he must check the regulator gauge frequently while he is in the contaminated area. When the gauge reads 3450 kilopascals (500 psi), he must leave the area immediately.

Whenever possible, crewmen wearing breathing apparatus should work in pairs. In all cases, a lifeline must be tied to the firefighter using the demand-type breathing apparatus, especially in a compartment with large open areas. When a lifeline is used, someone should monitor the line, using prearranged signals (Table 15.1). The weight of the unit changes the wearer's center of gravity, making it easier to become unbalanced and fall (especially backward). Wearers must be aware of this possibility when climbing ladders, working near the edge of a deck opening and in other precarious positions.

If it is necessary to operate the bypass valve (red knob), it should be opened slowly and only enough so the wearer may breathe comfortably. If it is opened quickly and too far, the rush of air could shift the facepiece, cause a leak and waste valuable air. The main-line valve (yellow knob) should be closed when the bypass valve is opened.

If a unit runs out of air in a smoke-filled compartment, the wearer should disconnect the breathing tube from the regulator, push its end into his shirt or coat through a front opening, and continue to breathe through the mask. The fabric may filter the air somewhat, and the facepiece will protect his face from the extreme heat. He should, of course, retreat to safety immediately.

Self-contained, demand-type breathing apparatus should never be stowed with pressure on the regulator. To relieve this pressure, the person stowing the unit should hold the threaded connection of the regulator between his thumb and index finger. He should then place his mouth over his thumb and finger, and breathe the pressure off the regulator.

Advantages and Disadvantages

The major advantage of the self-contained, demand-type breathing apparatus is the speed with which it can be donned and put into operation. When the equipment is properly stowed in its carrying case, a well-trained seaman can be ready for work in 45 seconds. The unit can be donned and started in smoke, and the facepiece can be cleared afterward. However, it is far safer to don the equipment in an uncontaminated atmosphere and check it according to prescribed procedures, before assuming that it will function properly in a hostile atmosphere.

Since the bulk of the equipment is on the wearer's back, it does not limit his arm movements. The wearer can use all hand tools, handle hose and operate nozzles without interference from his breathing equipment. Some regulators have a place where a second facepiece may be connected, for use in rescue work.

There are two major disadvantages to self-contained, demand-type breathing equipment: the operating time limitation and problems due to its size and weight.

The operating times for air cylinders are based on the normal breathing rate of an average person. However, during firefighting and rescue operations, air is used up more quickly than usual, because of the exertion, the psychological effect of wearing the breathing apparatus and the extreme heat. For this reason, more severe guidelines should be used:

- Backpack units rated by the manufacturer for 30 or 45-minute duration should not be expected to last more than 1 minute for each 690 kilopascals (100 psi) of pressure registered on the cylinder gauge.
- Sling-pack units rated by the manufacturer for 15-minute duration should not be expected to last more than 1 minute for each 1380 kilopascals (200 psi) of pressure registered on the cylinder gauge.
- Minipack units rated by the manufacturer for 6–8-minute duration should not be ex-

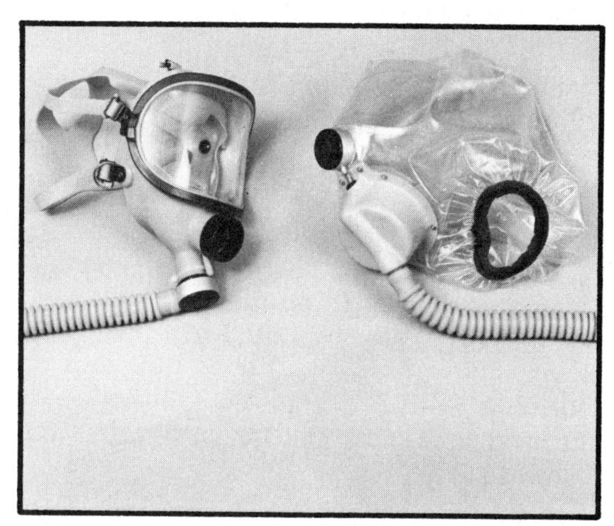

Figure 15.20. Standard facepiece *(left)* and polyurethane hood facepiece *(right)*.

pected to last longer than 1 minute for each 2760 kilopascals (400 psi) of pressure registered on the tank gauge.

These figures are, of course, averages. Some wearers may exceed these times by several minutes. However, to maintain a margin of safety, crewmen should not expect more than the average from their breathing equipment.

The second disadvantage results from the size and weight of the apparatus. The backpack equipment, which is the most popular, is quite bulky and weighs over 13.6 kg (30 lb). The bulkiness makes it difficult for the wearer to work in confined spaces. The weight adds to the physical strain on the wearer.

AIR-MODULE-SUPPLIED DEMAND-TYPE BREATHING APPARATUS

Recently, a manufacturer of demand-type breathing apparatus introduced a model that is supplied with air by modules. The apparatus consists of an air supply unit, facepiece, carrying case and harness.

Air Supply Unit

The air supply unit consists of an air module, which is made of small diameter, stainless steel tubing pressurized to 37,920 kilopascals (5500 psi), and a flow regulation unit. The latter includes a control valve, safety disk, fill valve, pressure gauge and pressure-reducing regulator (Fig. 15.20). The pressure-reducing regulator is threaded into a start-valve assembly. It reduces the pressure of air leaving the module, from 37,920 to 483 kilopascals (5500 to 70 psi).

The flow regulation unit (the main-line regulator) is located within the coils of the air module. A bypass valve, including a second pressure-reducing regulator, is also housed within the air module.

Facepiece

Two types of facepieces can be used with the air module pack. The first type is of the conventional design discussed earlier in this chapter; it is constructed of yellow silicone.

The second type is a soft polyurethane hood that seals around the wearer's neck (Fig. 15.21). The hood type has a hard polycarbonate view plate. It has a service temperature range of from −40° C to 121° C (−40° F to 250° F) with no loss of physical properties. At 177° C (350° F) there is a 50% loss of physical properties; melting begins at 218° C (425° F). The hood contains an aspirator–absorber that reduces carbon dioxide levels in the facepiece, directs incoming air over the nose and mouth during inhalation and provides a controlled supply of air to the facepiece. Exhaled breath is pulled out of the facepiece by the aspirator and is pumped through the carbon dioxide absorber. It then flows back into the facepiece. Compressed air from the air modules replaces the oxygen used in breathing, powers the aspirator and cools the air coming out of the absorber.

The facepieces have a quick-connect coupling assembly at the end of the air supply hose (Fig. 15.22). The assembly contains a quarter turn shutoff valve and an audible alarm whistle. The audible alarm has two modes. During main-line operation, when the pressure in the air modules

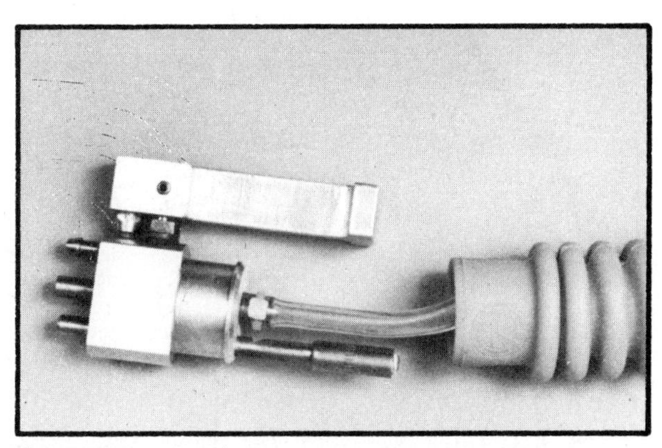

Figure 15.21. Quick-connect coupling assembly for module-supplied apparatus.

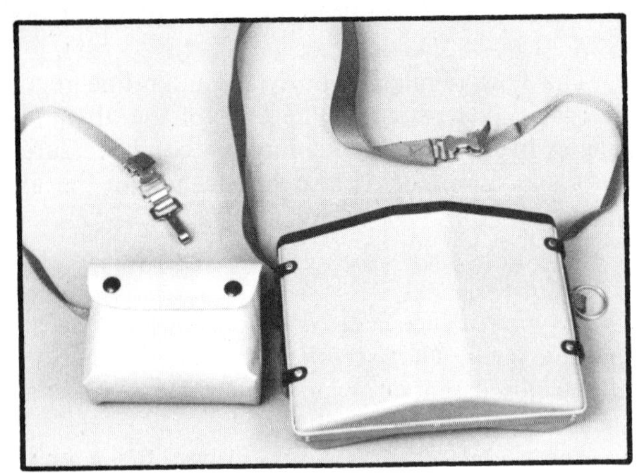

Figure 15.22. Harness and carrying case for the module-supplied apparatus.

drops to 25% of the normal operating pressure, the whistle sounds whenever the wearer exhales; it is silent during inhalations. This allows the wearer to distinguish his alarm from others in the area by breathing rapidly or by holding his breath. During bypass-system operation, the whistle sounds continuously. It is recommended that the wearer leave the hazardous area as soon as the bypass system is activated.

Figure 15.23. The module-supplied, demand-type apparatus.

Harness and Carrying Case

The harness is an adjustable, sling-type assembly with an adjustable waist strap (Fig. 15.23). The air supply unit is carried within a lightweight, high impact case. The start ring, bypass ring, pressure gauge and quick-connect coupling outlet are located along the top cover of the carrying case. The adjustable harness permits the air supply unit to be carried on the user's front, right side or back.

Donning

The module-supplied unit should be donned as follows (Fig. 15.24):

1. Lift the shoulder harness over your head, and place it on your left shoulder. The air pack should be resting on your right side.
2. Tighten the shoulder strap by pulling down on the adjustment strap.
3. Attach the snap hook on the waist strap to the ring on the carrying case, and pull the end to adjust it.
4. Don the facepiece. The hood type should be grasped by the elasticized collar and pulled over the head. The standard type should be donned as described earlier.
5. Attach the quick-connect coupling, open the quarter turn shutoff valve and pull the main-line start ring. The red start ring operates the bypass regulator, which should be used only in an emergency.

Recharging

The air modules can be recharged with a booster charging station (available from the manufacturer). The booster station must be coupled to a cascade system or a compressor. The cascade system or compressor supplies an air pressure between 4140 and 13,790 kilopascals (600 and 2000 psi). The booster station, which is simple to operate, boosts this pressure to 37,920 kilopascals (5500 psi).

FRESH-AIR HOSE MASK

Hose-mask type protective breathing apparatus is required on all tank vessels. In this type of apparatus, a length of hose connects the facepiece to an electrically driven pump or a hand-operated blower. The pump or blower supplies fresh air to the facepiece.

Construction

The fresh-air hose-mask apparatus in Figure 15.25 consists of a facepiece with breathing tubes

Figure 15.24. Crank-driven fresh-air hose mask.

and an exhalation valve; an air hose with body harness; and a manual blower with a hand crank. The unit is stored in a suitcase-type container, completely assembled except for the hand crank. The crank must be placed in the pump through a hole in the side of the carrying case.

The facepiece is of the standard full-face type, although some models have two inhalation tubes.

The wire-reinforced air hose comes in 7.62-m (25-ft) lengths, with threaded connections on both ends. One end attaches to the blower, and the other end attaches to the body harness. The hose is connected to the harness, rather than the facepiece, to protect the wearer in case the hose becomes entangled. If the hose were connected directly to the facepiece, a snagged hose could pull the facepiece from the wearer's face.

The blower is a small centrifugal or displacement pump with one air-intake connection and supply connections for one or two facepieces. The pump is operated by a hand crank as illustrated in Figure 15.25. Electrically driven pumps are available for some units. At least two men are required to tend a fresh-air hose mask. One additional man must operate the hand crank or supervise the motor-driven pump supplying air to one or two men wearing the equipment.

Figure 15.25. Typical approved gas mask.

Operation

Perhaps the most important step in putting a hose-mask apparatus into operation is finding a good location for the blower or pump. The location must be close enough to the contaminated area to allow the wearer to enter it with the hose. At the same time, the air surrounding the blower or pump must be free of contamination, since the wearer will be breathing air pumped from that location. Here is the procedure:

1. Select the appropriate location. It should be close to the contaminated area, to allow the wearer as much air hose as possible for his work; upwind of the contaminated area, so the wind will not spread contamination into the pump area; away from other sources of contamination such as operating engines; away from areas where dust or any other substance, liquid or solid, could enter the intake opening of the air pump; and well forward of the smoke stack.
2. Open the carrying case and remove the facepiece, harness and hose. The hose should be faked to ensure against tangling.
3. Attach a lifeline, as long as the air hose, to the D ring on the harness.
4. Install the hand crank in the pump.
5. The wearer now dons the harness, in the same manner as a vest or jacket. The buckles should be in front, and the hose connection and D ring at the wearer's back. The facepiece is then passed over the wearer's head, from back to front. If the facepiece has two inhalation tubes, one tube should rest on each shoulder as the facepiece is brought forward.
6. The wearer dons the facepiece, as described earlier in this chapter. The pump must be started when the facepiece is tightened, and it must be operated until the facepiece is removed.
7. Adjust the rotational speed of the blower to satisfy the wearer before he enters the contaminated area.
8. Make sure that the lifeline signals (Table 15.1) are fully understood by the wearer and tenders before the contaminated area is entered.

Maintenance

Whether or not it is used regularly, the fresh-air hose mask must be checked periodically for proper operation and signs of wear. Most important, after each use and before the equipment is stowed, the following maintenance procedures should be performed. The disinfectant solution used to clean the hose mask is also used for the case.

1. Clean and dry the facepiece, head harness, inhalation tubes and exhalation valve. Check these components for damage and wear as described in the section on facepieces.
2. Thoroughly inspect the air hose for damage. Wash and dry it before restowing.
3. Inspect all threaded connectors for damaged threads and for missing or damaged washers or gaskets.
4. Clean and lubricate the air pump according to the manufacturer's instructions.
5. Clean the case, inside and out. This helps keep the equipment clean after it is stowed in the case.
6. Inspect the hand crank for damage and wear. Stow the crank in its proper place. A misplaced crank makes the apparatus useless.

Safety Precautions

If at all possible, at least two crewmen should enter the contaminated area together, wearing similar breathing equipment. This will allow one to support the other if a problem arises. A lifeline that is the same length as the air hose should be attached to each wearer of a hose mask. This is especially important if a crewman must work alone in the contaminated area.

The wearer should never remove his fresh-air hose mask in the contaminated area. He must remember to leave the area by the route he used to enter it to keep his hose and lifeline untangled.

The pump or blower must be operated in an area that is well away from the contaminated air.

Advantages and Disadvantages

The major advantages of the fresh-air hose mask are its light weight and the unlimited air supply it provides. The wearer may work as long as necessary to complete his assigned tasks.

Among the disadvantages of the hose mask are the need for personnel to operate the blower and the restrictions due to the long length of hose. The hose limits the wearer's movements and may make breathing difficult. In certain compartments, it could become jammed or tangled on doors, cargo or machinery. In addition, the wearer must leave a compartment by the route through which he entered. This could be a problem if the wearer had to retreat quickly from a space that was involved with fire. Finally, the blower or pump must be located as close as possible to the contaminated area yet in an area that is itself free of contaminants.

GAS MASKS

Gas masks have been used over the years to provide protection against certain gases or vapors. These masks, often referred to as *filter, canister* or *all-service* masks, are air-purifying devices. They are designed only to remove specific contaminants from air that contains sufficient oxygen for breathing. Gas masks are not approved for firefighting, but they are approved for use aboard U.S. merchant ships for protection against toxic refrigerant vapors. An approved self-contained breathing apparatus may be substituted for a required gas mask.

Construction

The filter mask (Fig. 15.26) consists basically of four parts: a facepiece with an inhalation tube, exhalation valve, and speaking diaphragm; an external check valve; a canister; and a harness (in which the canister is held) with adjustable neck and body straps. An in-line timer is provided on carbon monoxide masks and "all-service" masks.

Operating Cycle

Inhaled air is first drawn through the timer (when provided), where the airflow operates a nutating disk that moves the timer dial needle. It then flows through the canister, which contains chemicals that remove or neutralize the contaminants. The air is then drawn through the inhalation tube and into the facepiece, where it passes over the lenses before it is taken into the lungs. Exhaled air leaves the facepiece through an exhalation valve. The external check valve prevents exhaled air from passing through the canister.

Donning

To don the gas mask, proceed as follows:

1. Remove the equipment from its case. Remove the bottom seal (if present), and adjust the neck strap for size.
2. Grasp the neck harness in one hand, and the facepiece in the other. Place the neck harness over your head, around the back of your neck.
3. Don the facepiece as described earlier.
4. Test the mask for leaks as follows: First place the palm of your hand over the opening in the bottom of the canister; inhale, and hold your breath for 10 seconds. If there are no leaks, the facepiece will collapse partially. Then exhale, and note whether air blows out of the sides of the facepiece. If it does not, the external check valve is functioning properly. Do not use the mask unless it passes both these tests.

Limitations

Filter masks are simple and compact. However, they are useless in atmospheres that do not contain enough oxygen to support life. They may not be used in atmospheres that contain more than 3% smoke, dust, mist or ammonia, or more than 2% carbon monoxide, acid vapor or organic vapor.

The canister is reliable for up to 5 years from the date of manufacture if the seal is unbroken, but only 1 year after the seal is broken. One canister can provide up to 2 hours protection in atmospheres containing the maximum concentrations of toxic gases given above. A flame safety lamp must always be used with the gas mask.

BIBLIOGRAPHY

Maryland Fire and Rescue Institute, *Basic Fireman's Training Course,* pp. 306–315. University of Maryland, College Park, Md, 1969.

Ohio Trade and Education Service, *Fire Training Manual,* pp. 317–341. State Department of Education, Columbus, Ohio, 1977.

Miscellaneous Fire Safety Equipment

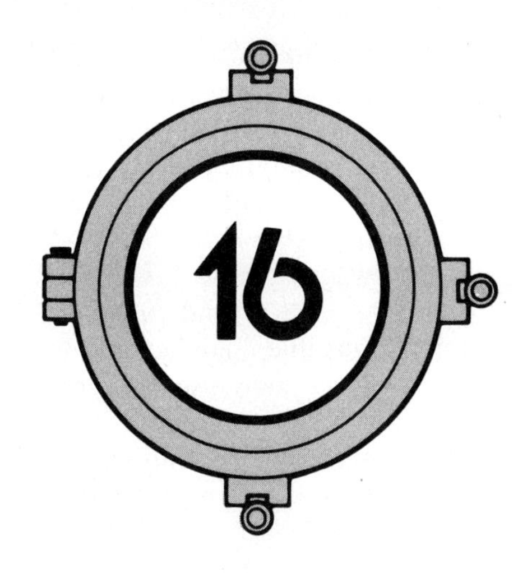

The equipment discussed in this final chapter is not used to detect or fight fire, but rather to protect personnel in the event of a fire. For example, *construction features* such as bulkheads are installed on vessels for strength, to enable the hull to withstand the forces of the sea. At the same time, they are designed to retard the spread of heat, and thus of fire, through the vessel. Another category of fire safety equipment might be called *portable devices*. Such devices (e.g., the oxygen indicator) are used to determine whether the atmosphere in a space is safe. A third category, *personal equipment*, includes equipment that is worn by crewmen during firefighting operations.

BULKHEADS AND DECKS

Bulkheads and decks divide a vessel into a number of separate divisions. Heat or flame must penetrate through these subdivision bulkheads if the fire is to spread from an involved space to other spaces. There are three means by which a fire might penetrate a bulkhead or deck:

- By igniting the bulkhead. The burning bulkhead or deck would then spread flames to combustible materials in the space adjoining the involved space.
- Through openings in the bulkhead, which would allow heat, flame and hot combustion products to travel to uninvolved spaces.
- By the conduction of heat through the bulkhead to nearby combustible materials.

The U.S. Coast Guard is constantly seeking ways in which to prevent the extension of fire by these means.

Structural Fire Protection

Bulkheads and decks must be constructed of approved noncombustible materials. A noncombustible material is one that will not burn or support combustion. A number of noncombustible materials are known, but only a few with suitable properties have been approved for use in ship construction. And even these materials cannot withstand an intense fire for an extended period of time.

For example, the strength of steel makes it an ideal shipbuilding material, and it is an approved noncombustible material. But although steel is noncombustible, it is affected by heat. The heat of an intense fire can cause exposed steel decks and bulkheads to warp, buckle or separate (fail) completely.

The extent to which a noncombustible subdivision bulkhead will be affected by heat depends on the temperature and the exposure time, as well as on the dimensions of the subdivision. Bulkheads and decks are therefore rated as to their ability to withstand heat in a standard fire test. Regulations specify where bulkheads with certain ratings may be located within the vessel.

A class A bulkhead is one that will resist the passage of flame and smoke for 1 hour when subjected to temperatures up to 927°C (1700°F). Since the subdivision is noncombustible, it will not ignite. It will also resist buckling and warping sufficiently to confine the fire and the combustion products to the involved space for at least the 1-hour period.

Class A bulkheads must be made of steel and are the only class of subdivision that may be used as main bulkheads and decks.

A Class B bulkhead is one that will resist the passage of flame and smoke for 30 minutes when subjected to temperatures up to 843°C (1550°F). A class C bulkhead is essentially unrated; it is not expected to resist flame or smoke for any length of time.

Openings in Subdivision Bulkheads

Fire can travel through any opening that will pass heat, hot combustion products or flames. No matter how well a bulkhead or deck resists flames, an opening in the bulkhead is an invitation for fire to spread.

Doorways, hatches, ductwork and accommodations for wires and pipes are all openings in noncombustible bulkheads. They all serve a purpose, but they also can permit fire to extend from one space to another. For this reason, such openings should be constructed so that they do not destroy the fire resistance of the bulkhead in which they are located. For example, suppose a ventilation duct passes through a class B bulkhead. Then the duct and its opening should be constructed according to the class B standard. They should be able to resist the passage of flame and smoke for at least 30 minutes when subjected to a temperature of 843° (1550°F).

Watertight doors (discussed in the next section) are constructed in this way. Although they do not add to the strength of a bulkhead, they do not reduce its strength or tightness. However, a door will resist the spread of fire when it is closed tightly. The watertight door may be opened while at sea if required during the normal course of ship operations.

Conduction of Heat Through Subdivision Bulkheads

Steel is a very good conductor of heat; aluminum is an even better conductor. Both metals, when used as bulkheads, can conduct enough heat into an uninvolved space to ignite nearby combustible materials. In the early stages of a fire, conduction can be a much more dangerous source of fire spread than bulkhead failure. (For this reason, the protection of exposures was stressed in the chapters of Part II.) Combustible materials should be moved away from bulkheads that separate involved spaces from uninvolved spaces. Hot bulkheads and decks should be cooled with water fog.

The materials most liable to be ignited by conducted heat are combustible paneling, furring and reefer insulation that are installed in direct contact with bulkheads. (*See* SS *Hanseatic* in Chapter 3.) Present regulations require, with few exceptions, that all vessels of 4064 metric tons (4000 gross tons) or more contracted after January 1, 1962, have noncombustible sheathing, furring and holding pieces. In addition, passenger vessels are required to be subdivided into main vertical zones for the purpose of fire control. These zones shall not generally exceed 131 feet. The bulkheads forming these zones shall be fire resisting. Classes A and B shall have insulation to prevent a temperature rise of more than 121°C (250°F) on the unexposed side of the bulkhead for up to 1 hour, depending on its location within the vessel. The details are specified in 46 CFR 72.05–10.

Neither of these requirements prevents fires, but they do restrict its extension—particularly in concealed or inaccessible locations—giving crewmen added time to reach and attack the seat of the fire. However, even on passenger ships, fire control bulkheads need not be located around cargo spaces, except where they abut certain types of spaces. Instead, cargo spaces are generally protected by fixed fire detecting and fire extinguishing systems.

DOORS

Doors are, of course, installed to allow access to compartments and passageways. Although they are not designed specifically for use in fighting shipboard fires, closed doors will help restrict the spread of fire from space to space. Some doors are provided with remote closing mechanisms. If the smoke and heat of a fire prevent crewmen from approaching a door, it may be closed from a remote location.

Watertight Doors

A watertight door is, as its name implies, designed to prevent the movement of water through the doorway. Generally, the fire retarding capabilities of a watertight door match those of the bulkhead in which it is installed.

Classifications. In terms of operation, there are three classes of watertight doors:

- Class 1: manually operated hinged doors
- Class 2: manually operated (with hydraulic assist) sliding doors
- Class 3: manually and power-operating sliding doors.

All three classes of doors must be capable of being closed with the ship listed 15° to either port or starboard.

Class 1 Doors. Class 1 watertight doors are constructed of steel. They are hinged, and must be swung open or closed manually. When a class 1 door is closed, a knife edge on the door fits against a rubber gasket on the bulkhead. The door is secured in the closed position by hinged levers called *dogs*. There are usually six dogs; when they are hand tightened, they cause the gasket and knife edge to form a watertight seal.

A class 1 door should be undogged as indicated in Figure 16.1. First the dog nearest the upper hinge should be released; then the dog nearest the lower hinge; and then the center dog on the hinge side of the door. (The hinges are attached through slotted or elongated openings.) Then the dogs on the side opposite the hinges should be released in the same order—upper, then lower, and center dog last.

Class 1 doors are used for all exterior deckhouse openings on weather-deck levels. Their use in these locations provides protection against inclement weather and heavy seas. They may also be used during and after firefighting operations, as openings for venting heat and smoke to the outside.

Class 2 Doors. Class 2 watertight doors are steel sliding doors used below the waterline. Some are operated manually, by turning a wheel that moves the door via a set of gears. However, most class 2 doors are operated by a manual system with hydraulic assist. A rotary hand pump produces the hydraulic pressure that opens or closes the door. A class 2 door must be capable of operation from either side of its bulkhead and must

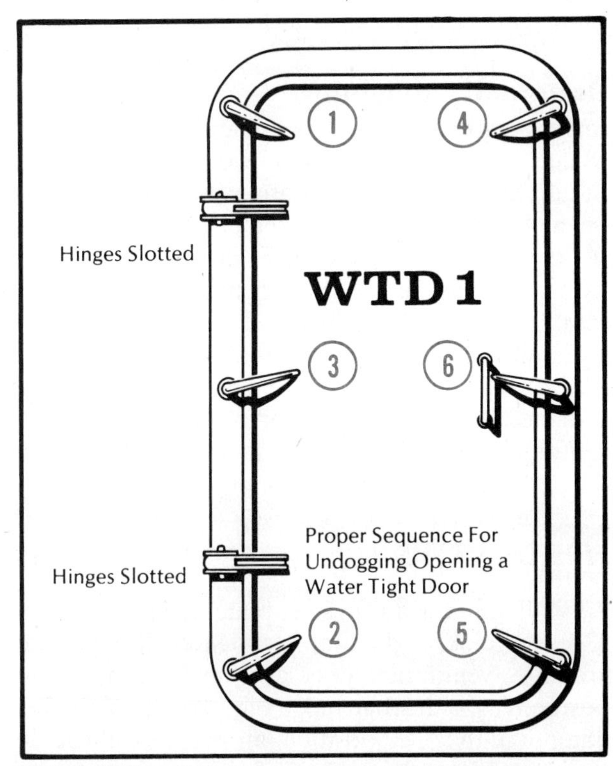

Figure 16.1. The numbers show the proper sequence for releasing the dogs on a watertight class 1 door.

be able to close in 90 seconds or less when the vessel is not listing.

A second means for closing (not opening) the door must be provided from an accessible position above the bulkhead deck. This is usually a mechanical means; a wheel valve is turned to operate gears that slide the door closed. A door position indicator must be installed at the remote closing location, so that anyone attempting to close the door can easily determine its position.

Class 3 Doors. The class 3 watertight door (Fig. 16.2) is a sliding steel door that may be operated by either an electric hydraulic system or a manual hydraulic system. In the former, a switch activates an electric motor that drives the hydraulic opening and closing mechanism. The manual hydraulic system is similar to that installed on class 2 doors. Both systems must be capable of operation from both sides of the bulkhead and must be able to close the door in 90 seconds or less when the ship is in an upright position.

A manual hydraulic operating system is also provided at a remote location, usually a deck above the door. As for class 2 doors, the remote mechanism is used only to close the door. A door position indicator must be installed at the remote closing location.

On passenger vessels, class 3 doors must be capable of being closed from a central location on the bridge. The doors must also be capable of closing automatically if they are opened at the bulkhead after being closed from the bridge. When a door control is activated on the bridge, a warning signal at the door must sound a minimum time interval of 20 seconds is provided from the time of the signal until the door reaches the closed position. Also, there must be at least a 1 second warning signal before the door moves into the clear opening.

Ships fitted with more than one class 3 door can be equipped with a central control station (Fig. 16.3). The doors can be operated simultaneously or separately from the control station. Their positions are monitored, via electric circuits, on a lighted display board. Display boards are usually located on the bridge. They allow the positions of the ship's watertight doors to be evaluated quickly during a fire, to determine if CO_2 flooding systems can be employed.

Testing. Manually operated doors should be tested to ensure that they can be opened easily, that they close properly and that all the dogs operate freely. The seal can be tested by putting

Figure 16.2. A horizontal watertight class A (and class 3) door separating the engine room from the shaft alley. (Courtesy Walz and Krenzer Inc.)

chalk on the knife edge, closing the door and dogging it down. Chalk marks will show on the entire rubber gasket if the door closes properly and the gasket is in good shape. If chalk marks skip any part of the gasket, it should be adjusted or replaced. The Coast Guard requires that all watertight doors be hose tested in the closed position during installation.

The testing of hydraulic doors is complex and requires particular mechanical skills and knowledge. These doors should be tested according to the manufacturer's recommendations.

Fume Doors

Fumetight (gastight) doors are constructed of metal. They swing open and shut on hinges and are dogged down manually to form a gastight seal. They are almost identical to class 1 watertight doors but are of lighter construction.

Fumetight doors and their fittings must pass more exacting tests for tightness than watertight doors. They are installed in bulkheads surrounding spaces that may contain poisonous or toxic fumes, such as battery rooms, refrigerated cargo spaces and paint lockers. Openings into such spaces (for pipes or wiring) must also be fumetight. Ducting is used to direct fumes vertically from the space to a safe discharge point.

Doors and Firefighting

In brief, a charged hoseline should be available whenever a closed door is to be opened. The door should be felt with the bare hands before it is opened. If it is cool, it may be opened cautiously. If the door is hot, it should be cooled thoroughly with water fog before it is opened—and again it should be opened cautiously. The door should be reclosed quickly if the fire that is found cannot be controlled with the extinguishing equipment at hand. (*See* Chapter 10 for a discussion of the techniques for opening doors and using doors during firefighting operations.

FIRE DAMPERS

A fire damper is a thin steel plate at least 3.2 mm (1/8 in.) thick, and suitably stiffened. It is placed within a ventilation duct and held in the open position by a fusible link (Fig. 16.4). With the damper in the open position, air may flow through the duct. When the air in the duct reaches a temperature of about 74°C (165°F), or 100°C

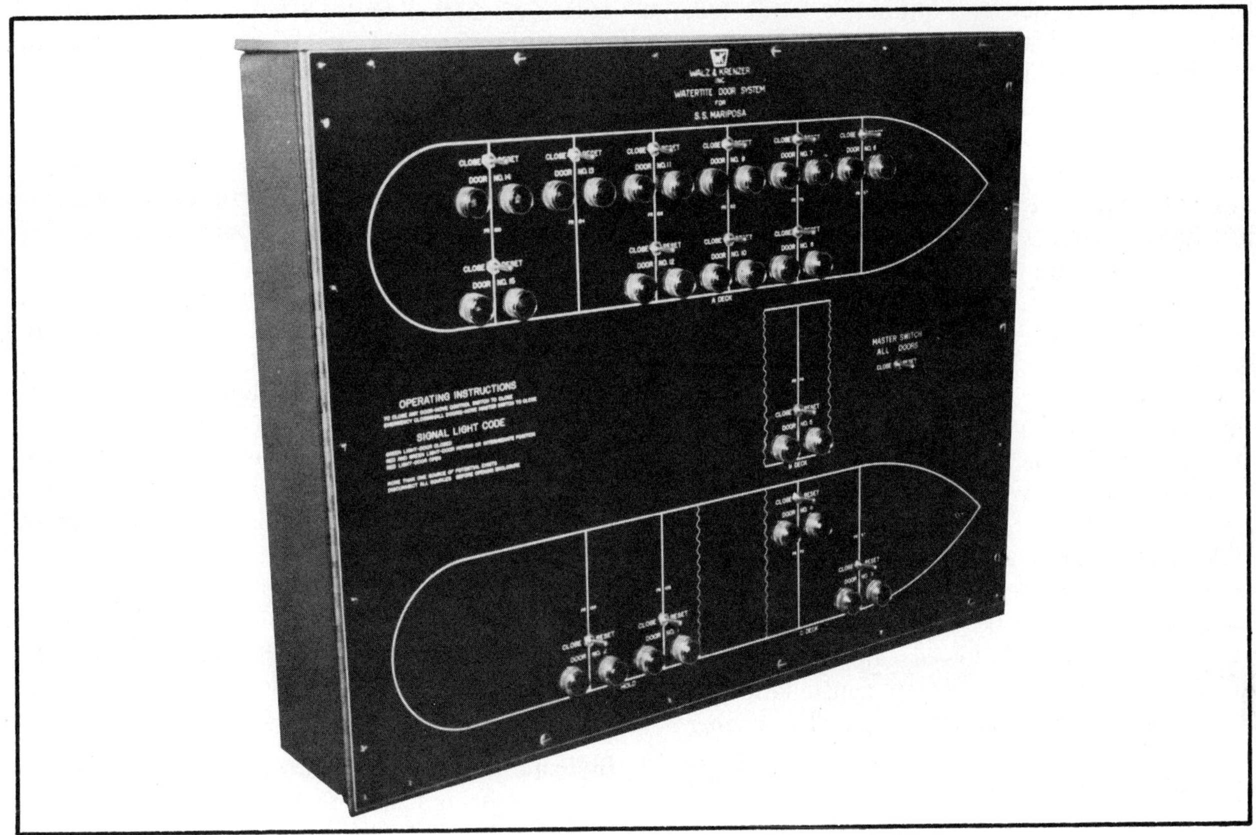

Figure 16.3. Central control station for watertight doors. (Courtesy Walz and Krenzer Inc.)

(212°F) in hot areas such as galleys, the fusible link melts, allowing the damper to close. A visible indicator on the outside of the duct shows whether the damper is closed or open.

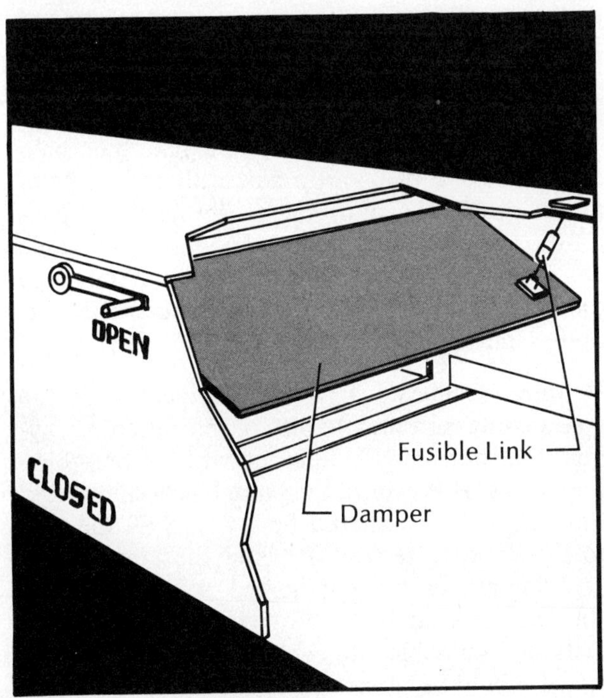

Figure 16.4. Typical fire damper.

Dampers can also be closed manually. They must be capable of manual operation from both sides of the bulkhead through which the duct passes.

Fire dampers will not prevent fires, but they can help stop fire from spreading. They do this in two ways: First, they reduce or shut off the supply of air to the fire. This reduces the rate at which the fire intensifies and thus reduces the heat buildup. Second, they block heat, smoke and flame, so that these combustion products do not spread the fire through the ducting and into uninvolved spaces.

On passenger ships, all ventilation systems must have fire dampers, but not all dampers must be automatic. However, automatic dampers are required in ventilation ducts that pass through main bulkheads. On some vessels, ventilation system motors can be shut down from the bridge or from the CO_2 room. With the ventilation fans shut down and the dampers closed, the travel of fire through the ducts is slowed considerably.

FLAME SAFETY LAMP

Air normally contains about 21% oxygen. A concentration of about 16% is considered sufficient to maintain human life. At concentrations of 15% or less, muscular coordination is affected. Lower concentrations of oxygen in breathed air will affect judgment and body functioning and may result in unconsciousness. Death may result from the breathing of air with less than 6% oxygen, even for a few minutes. Thus, crewmen must not enter a fire area without breathing apparatus until its atmosphere has been tested and found to contain sufficient oxygen.

The flame safety lamp is a portable device that is used to detect oxygen deficiencies in confined spaces. The lamp in Figure 16.5 is approved by the Coast Guard for use on cargo and passenger ships. The lamp uses naphtha as a fuel for its flame. Changes in the flame size and brightness indicate the relative amount of oxygen in the atmosphere being tested.

The maintenance of flame safety lamps and the preparation of lamps for use vary with the model and manufacturer. The manufacturer's instructions should be followed carefully. In particular, instructions regarding the installation of asbestos washers and the cleaning and replacing of wire gauzes must be followed to the letter. Wire gauzes are the main safety feature of the lamp; they prevent the flame from igniting flammable gases that may be present in the atmosphere being tested. For this reason, they must be in perfect condition.

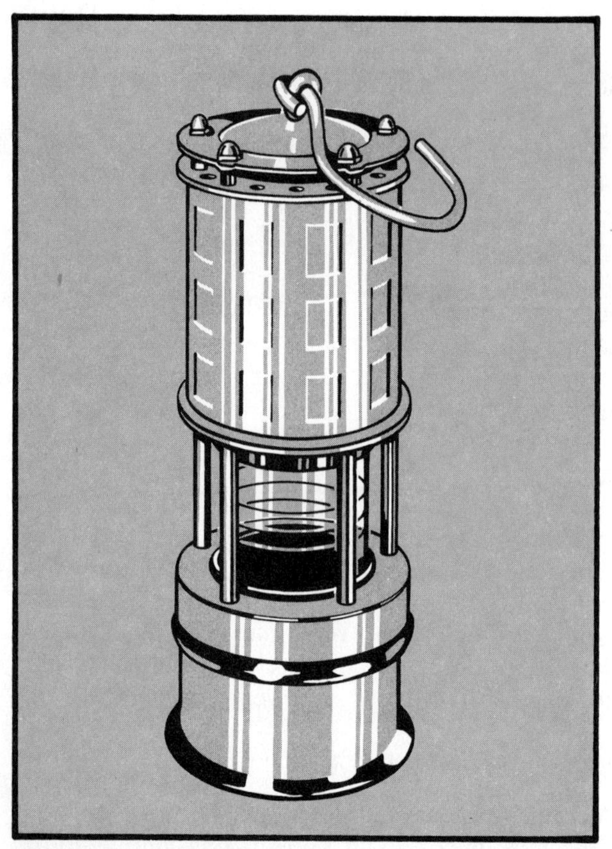

Figure 16.5. Flame safety lamp approved only for detecting oxygen deficiency. (Courtesy Koehler Manufacturing Co.)

Using the Lamp

The lamp wick must be ignited (in an uncontaminated area, away from the compartment to be tested) and the flame size adjusted according to the manufacturer's instructions. Then the lamp must be allowed to warm for the specified period (at least 5 minutes).

The space to be tested should be ventilated before it is tested. The lamp should be vertical, whether it is being carried or lowered into the compartment. It should be advanced into the compartment *slowly*. If the lamp is to be carried into a space, it should be held well ahead of the crewman who is carrying it. The crewman entering the space should wear a self-contained breathing apparatus or a filter mask. This is especially important if the atmosphere within the space could possibly contain toxic gas.

Indications. If the flame continues to burn in the space being tested, there is enough oxygen in the space to support life.

If the flame slowly decreases in size and flickers or goes out, the atmosphere is deficient in oxygen. The flame is extinguished by a concentration of oxygen below 16%. This is not enough oxygen to support life.

If the flame "pops" the atmosphere is explosive. The lamp must then be withdrawn *slowly* from the space. (In case of reignition, rapid movement could force the flame through the gauze and cause an explosion.) The lamp should be flushed thoroughly in a safe atmosphere.

If the flame gets brighter, or a pale blue halo appears above the orange flame, there is a flammable gas in the atmosphere. If the flame continues to burn, the concentration of flammable gas is below the lower explosive limit. If the flame dies out after brightening, the concentration is above the upper explosive limit.

Required Actions. If the lamp indicates that the space contains a breathable concentration of oxygen, the space may be entered without breathing apparatus. However, if the lamp has been lowered into the space, it should be lowered all the way to the deck, to test all levels of the atmosphere. If a flammable gas is indicated at a lower level, the lamp should be withdrawn slowly.

If a flammable atmosphere is indicated, the space should be ventilated after the lamp is slowly withdrawn. The atmosphere should then be retested—first with a combustible-gas indicator and then with the flame safety lamp. If the lamp shows an oxygen deficiency, the space should be ventilated and retested.

Precautions

Even though it will indicate the presence of flammable gas, the flame safety lamp should not be used in any space that has contained or is suspected of containing flammable or combustible gases or liquids. It should not be used in cofferdams fouled by fuel oil or in atmospheres that may contain hydrogen or acetylene gas. If possible, the space should first be tested with a combustible-gas indicator.

The lamp should be checked for defects such as a chipped globe, broken seals or gaskets and damaged gauzes before it is used. A defective lamp should not be used.

No attempt should be made to relight the lamp in the compartment being tested.

The flame safety lamp will not indicate the presence of carbon monoxide. It may not indicate accurately after an explosion or fire, if excessive amounts of any combustion products are present. For this reason, the space should be well ventilated before it is tested with the lamp.

OXYGEN INDICATOR

The oxygen indicator is an instrument that measures the amount of oxygen in the atmosphere of a confined space. The device consists of a case with a meter, an aspirating bulb and a long rubber tube. The end of the tube is placed in the atmosphere to be tested; samples of the atmosphere are drawn into the case by squeezing the bulb. The percentage of oxygen in the sample is indicated by the meter needle.

Using the Oxygen Indicator

The instrument should be maintained and calibrated according to the manufacturer's instructions, often located on the device itself. The instrument should be stowed in the upright position when it is not in use. The batteries (for instruments that use them) should be stowed separately. If the meter needle cannot be set to zero, the batteries are weak. All the batteries should be replaced at the same time.

The rubber tube should be fed slowly into the space whose atmosphere is being tested. The cotton filter should be in place in the end of the tube. The case of the instrument should be level or nearly level for accurate readings. Liquids should not be drawn into the instrument, as they destroy its accuracy; the instrument must then be flushed before further use.

When a sample of air is drawn into the instrument, the meter needle will move back and forth, and then settle at a reading. The meter should therefore not be read immediately; about 10 seconds should be allowed for the meter to stabilize. All levels and all parts of the space should be tested. After each reading, the instrument should be purged by squeezing the aspirating bulb five or six times, with the tube in fresh air.

A concentration of 16%–21% oxygen throughout the tested space will sustain life. A concentration of 15% oxygen or less is considered inert (will not support combustion).

Limitations

Exposure to certain gases will affect the accuracy of the oxygen indicator. For example, an exposure of 10 minutes or more to CO_2 will cause the meter to register an incorrect high oxygen concentration. Exposure to CO_2 or flue gas should be brief; the instrument should be flushed with fresh air after such exposure. High concentrations of sulphur dioxide, fluorine, chlorine, bromine, iodine and oxides of nitrogen will interfere with the operation of the device. Strongly acidic gases may damage the instrument enough so that it requires an overhaul before it can be used again.

PORTABLE COMBUSTIBLE-GAS INDICATOR

Combustible-gas indicators detect and register concentrations of dangerous gases in the air in confined spaces. Most shipboard explosions occur when flammable or combustible gases are ignited in enclosed or partly enclosed spaces. It is thus extremely important to test the atmospheres of such spaces as tanks after cleaning, or holds after a fire.

The portable combustible-gas indicator, or explosimeter, is similar in appearance to the oxygen indicator. It consists of a case with a meter, an aspirating bulb and a long rubber tube. The open end of the tube is placed in the atmosphere to be tested. A sample of that atmosphere is drawn into the case by squeezing the aspirating bulb. The meter needle registers the presence of combustible gas as a percentage of the lower explosive limit.

There are various types of combustible-gas indicators, including a small instrument about the size of a flashlight. This model gives visible and audible indications of gas concentrations. Manufacturers provide instructions for the calibration, use and maintenance of their instruments; their instructions should be followed carefully. All combustible-gas indicators are battery operated. The batteries should not be stowed with the instrument. Instead, a set of fresh batteries should be installed in the unit whenever it is to be used.

Using the Combustible-Gas Indicator

Before the instrument is used to test a space for combustible gas, it should be purged. This is done by squeezing the aspirator bulb five or six times, plus one additional time for each 1.52 m (5 ft) of tubing. The open end of the tube should be in fresh (or at least uncontaminated) air during the purging.

As samples are drawn into the instrument, they are burned within the case. Thus, the atmosphere being tested must contain sufficient oxygen to support combustion. The atmosphere must therefore be tested with an oxygen indicator before it is tested with a combustible-gas indicator. The case itself must remain outside the atmosphere that is being tested.

Once the instrument has been calibrated and purged, the open end of the tube is placed in the space to be tested (Fig. 16.6). The bulb is squeezed to draw a sample of the atmosphere into the case. The heat generated during the burning of the sample is translated into a meter reading through a Wheatstone bridge (a device for measuring electrical resistance). As noted above, the meter indicates the concentration of flammable gases as a percentage of the lower explosive limit (LEL). Because of the length of the sampling hose and the way the device operates, several seconds must elapse before the meter shows a reading. The meter scale is red at and above 60% of the LEL. These high concentrations are dangerous—too close to the explosive range to be safe for crewmen.

If the meter needle moves to the extreme right side of the scale and stays there, the atmosphere is explosive. If the meter moves rapidly across the scale and then drops near or below zero, the concentration of flammable gas may be above the upper explosive limit (UEL). In this case, the instrument should be flushed out with fresh air and the atmosphere retested. In fact, each reading should be rechecked at least once, to ensure its accuracy. If possible, the instrument should be flushed with fresh air between readings.

All levels of the space should be tested. Many flammable gases are heavier than air and will contaminate only the lowest parts of a compartment. The sampling tube should not, however, be allowed to contact any liquid as it is lowered into a space. The liquid will cause a false read-

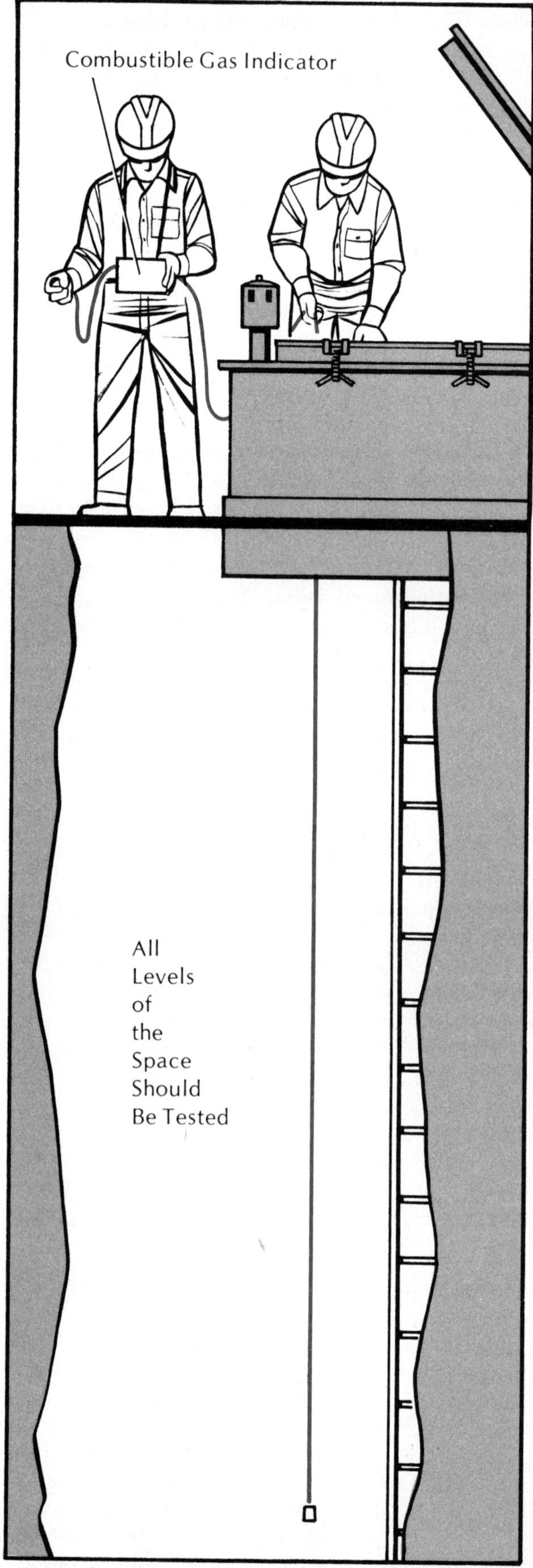

Figure 16.6. The sampling tube is inserted into the space to be tested for combustible gas. The instrument itself remains outside the space.

ing. If necessary, a probe can be used to determine how much freeboard there is above a liquid surface; the tube may then be inserted only far enough to sample the atmosphere above the liquid.

The combustible-gas indicator should be purged with fresh air after each use.

Limitations

Continued sampling of combustible-gas concentrations above the UEL may burn out the testing mechanism. This is indicated when the meter needle moves to the extreme right and cannot be adjusted to zero.

Hot vapors may affect the indicator reading if they condense inside the instrument. An inhibitor filter is required for the testing of atmospheres that may contain leaded gasoline fumes.

Many indicators will show incorrect readings if the sampled atmosphere contains less than 10% or more than 25% oxygen. Certain indicators are designed for specific contaminants. For example, different instruments must be used to test for oxyhydrogen and for oxyacetylene.

The presence of hydrogen sulfide gas or of silanes, silicates or other compounds containing silicone in the sampled atmosphere may cause serious problems. Some of these materials rapidly "poison" the detector element, so that it does not function properly. When it is suspected that such materials are present in the atmosphere being tested, the instrument must be checked frequently (at least after each five tests). Some manufacturers produce calibration kits for this purpose.

COMBINATION COMBUSTIBLE-GAS AND OXYGEN INDICATOR

The instrument in Figure 16.7 measures the concentrations of both combustible gas and oxygen. Each is indicated on a separate meter. A sample of the atmosphere is drawn into the instrument by a battery-operated pump. The sample passes through two separate sections, one for combustible gas and the other for oxygen. Each section operates in about the same way as the comparable single-purpose indicator.

As is usual with this type of instrument, the indicator should be purged and calibrated before each use according to the manufacturer's instructions. The combustible-gas portion will not measure the percentage of combustible vapor in steam because of the lack of sufficient oxygen. It will not indicate the presence of explosive or combustible mists or sprays formed from lubrication oils, or the explosive dust produced by grain or coal.

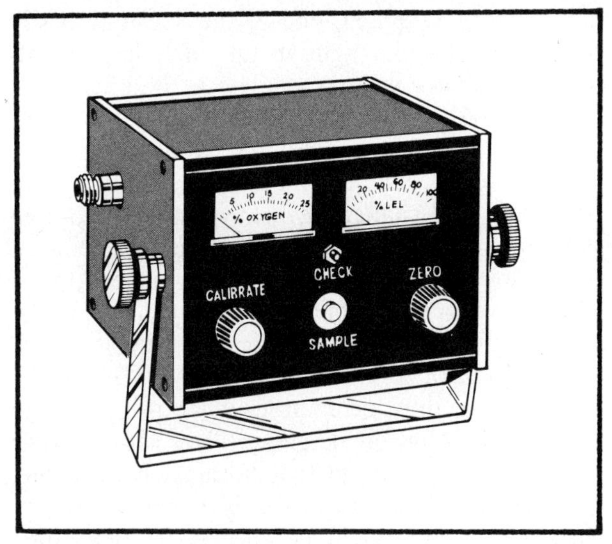

Figure 16.7. Combination portable combustible gas and oxygen indicator. (Courtesy Mine Safety Appliances Company)

FIREAXE

The pikehead axe (Fig. 16.8) is a versatile, portable firefighting tool. Every vessel is required to carry two of these fireaxes on international voyages.

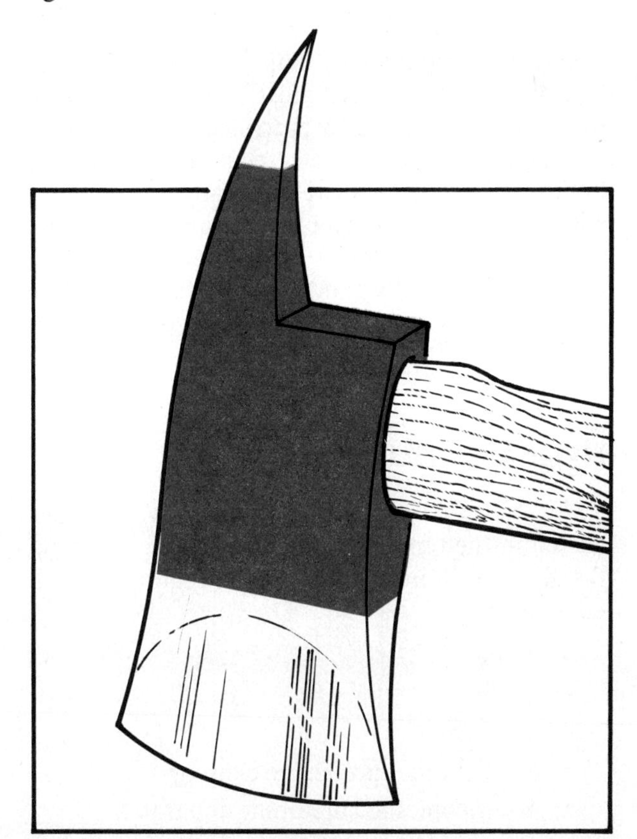

Figure 16.8. Pike head fireaxe.

The pike (pointed) end of the axe may easily be driven through light metal, including metal-clad fire doors and some class C bulkheads. It can be used to make openings quickly, to check for smoke or fire extension. It is also useful for tearing apart mattresses and upholstered furniture and for shattering heavy glass (including tempered glass) when necessary. The broad end of the axe can be used to pry open hinged doors, to remove paneling and sheathing to expose recesses and voids (avenues of fire travel), or to chock doors open.

Crewmen must be cautious when using axes to force a door or break glass. They should wear gloves and other protective clothing, if available. A door should be forced only when necessary. First, the door should be checked to determine whether it is unlocked. If it is not, there may be time to obtain a key, especially if the fire is a minor one and lives are not in danger. On the other hand, when a door must be forced, this should be done without hesitation.

Axes should be inspected periodically and sharpened, cleaned or repaired as necessary. The blade and pike ends should be kept sharp and free of burrs. The handle should be tight in the axe head and free of splits and splinters. An occasional light oiling will keep the head from rusting.

KEYS

Emergency equipment that is stowed in foot lockers or locker spaces must be accessible at all times. If a storage area is locked, the key must be placed in a receptacle secured to a nearby bulkhead. The receptacle is usually a small box with a glass front. A hammer is provided for use in breaking the glass in an emergency. The receptacle is generally painted red, with information regarding the key stenciled on the bulkhead. A key for the CO_2 room must also be available, or the room may not be locked.

FIREMAN'S OUTFIT

Three types of protective clothing will be discussed in the remainder of this chapter. The first, the fireman's outfit, is shown in Figure 16.9. It consists of

- Boots
- Gloves
- A helmet
- A set of outer protective clothing
- A self-contained breathing apparatus
- A lifeline
- An approved flashlight
- A flame safety lamp
- A fireaxe.

The boots and gloves must be made of rubber or a similar nonconducting material. The helmet must provide effective protection against impact. The outer protective clothing must protect the wearer's skin from the heat of a fire and from scalding by steam.

At least two fireman's outfits must be carried on every U.S. flag vessel on an international voyage. However, in the event of a serious fire, the two required outfits would not be sufficient to protect all the crewmen involved in firefighting and rescue operations. Recognizing this fact, many ship owners provide additional outfits or additional sets of breathing apparatus and protective clothing.

PROXIMITY SUIT

An approach suit, or proximity suit, consists of:

- Jumper-type pants that cover the legs and upper part of the body, including the arms

16.9

Figure 16.9. Fireman's protective outfit. A lifeline, flashlight, flame safety lamp and fireaxe must also be carried as part of the outfit. (Courtesy C. J. Hendry Co.)

Figure 16.10. The proximity suit protects the wearer against high heat but not against direct contact with flames. (Courtesy Globe Firefighting Suits)

- A hood (with a transparent heat-reflecting vision shield) that covers the entire head, shoulders and upper part of the body
- Heavy gloves
- Special coverings for the feet.

The outer surface of the suit is covered with a highly reflective material. (The suit reflects as much as 90% of the radiant heat.)

When properly donned, the proximity suit encases the wearer in a heat resistant envelope (Fig. 16.10). It may be used to approach close to a fire, but *it is not designed to protect the wearer during direct contact with flames.* A self-contained breathing apparatus must be worn under the proximity suit. Otherwise, the intense heat near the fire can damage the wearer's respiratory tract.

Proximity suits are used in fighting flammable-liquid and LPG fires, which generate tremendous amounts of heat. They allow firefighters to approach close enough to attack the fire effectively. If a cooling shield of water is used during the approach, the wearer is reasonably safe. The modern proximity suit does not require "wetting down" before the approach to the fire, as earlier asbestos suits did.

ENTRY SUIT

The entry suit (Fig. 16.11) consists of boots, trousers, coat and hood. Each of these is constructed of nine layers of fiberglass insulating material separated by aluminized heat-reflecting glass fabric. The outermost layer is aluminized fiberglass. The vision shield is of a special heat-reflecting material and is sealed into the hood.

Drawstrings and snaps on the suit provide an airtight seal around the wearer. The hood is attached to the coat with straps when the suit is donned, so that it cannot be accidently removed. An air pack (demand-type breathing apparatus) is worn under the entry suit. The suit, but not the air pack, is stowed in a suitcase.

The entry suit is adjustable. Designated wearers should be familiar with the donning procedure and should practice it until they can don the suit and the air pack quickly. After each use, the suit should be cleaned (especially of oil). Tears should be repaired with the repair kit provided, or according to the manufacturer's instructions, before the suit is restowed. A full air cylinder should be placed in the air pack.

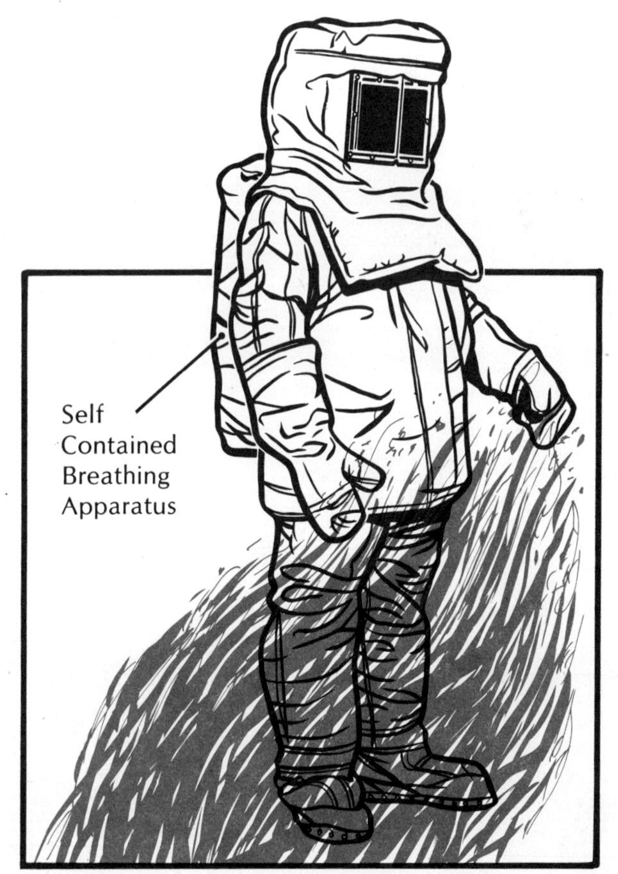

Figure 16.11. The entry suit allows the wearer to enter flames for a short time—the shorter the better. (Courtesy Fyrepel)

The entry suit will protect the wearer from direct contact with flames up to a temperature of 815.5°C (1500°F) for a short time. It may be used to enter flames for rescue, to close a fuel valve and for similar emergency tasks. However, *the wearer cannot linger in the flames;* he must move in, do what is necessary and move out quickly. A crewman wearing an entry suit can be baked like a potato in aluminum foil if he assumes the suit will provide unlimited protection against flames.

CONCLUSION

The ships of the U.S. maritime service range from older, smaller break-bulk cargo vessels to supertankers and vessels of unique design. They carry their cargos to and from all parts of the world, usually with speed and efficiency. These ships are durable; when safely navigated and properly maintained, they can serve their owners and crews for long periods of time. But they are also vulnerable to neglect and carelessness.

Throughout this manual, the need for fire prevention has been stressed; it is hoped that the information presented will help prevent destructive fires. However, if fire does occur at sea, it must be extinguished or controlled; there is no other alternative. The methods of detecting fire have been thoroughly described, along with the ship's firefighting capabilities, which are perhaps limited only by the training and capability of the crew.

If you are a crewman who will be confronted by fire at sea, the authors and all who have contributed to this text pray its information will be instrumental in your survival and the survival of your ship.

BIBLIOGRAPHY

Marine Officers Handbook, Edward A. Turpin and William A. MacEwen, Cornell Maritime Press, Inc., Cambridge, Maryland 1965.

Damage Controlman, U.S. Navy Training Manual, U.S. Navy Training Publication Center, Washington, D.C. 1964

Fire Fighting-Ship, Bureau of Ships Manual, U.S. Navy, Washington, D.C.

IFSTA Air Crash Rescue
Oklahoma State University
Stillwater, Okla.

GLOSSARY

adapter a hose coupling device for connecting hoses of the nominal size, but which have different type threads.

air foam see *mechanical foam.*

air foam nozzle (mechanical foam nozzle) a special pick-up tube or nozzle incorporating a foam maker to aspirate air into the solution to produce air foam.

air line mask a face mask where the air is supplied through an air hose attached to a blower outside of the contaminated space or area.

all-purpose nozzle (combination) a mechanical device that fits on the end of a hose that controls the water pressure inside the hose three ways by operating a single valve. The three positions of the valve are: *1)* FWD—off, *2)* vertical—HV/LV fog and *3)* back—solid stream.

applicator a special pipe or nozzle attachment that fits into the all-purpose nozzle high velocity outlet. Applicators used aboard ship are 4′, 10′ and 12′ lengths and are equipped to change high velocity fog into low velocity fog. The 4′ and 10′ applicators fit the standard 1½″ nozzles and the 4′ has a 60° curve and the 10′ has a 90° curve on the outlet end. The 12′ applicator fits the standard 2½″ nozzle and has a 90° curve at the outlet end.

aqueous film forming foam (AFFF) a fluorocarbon surfactant that acts as an effective vapor securing agent due to its effect on the surface tension of the water. Its physical properties enable it to float and spread across surfaces of a hydrocarbon fuel with more density than protein foam.

arcing pure electricty jumping across a gap in a circuit. The intense heat at the arc may ignite any nearby combustible material or may fuse the metal of the conductor.

automatic alarm an alarm usually activated by thermostats, sprinkler valves or other automatic devices that activate electrical circuits to the control station located on the bridge.

automatic sprinkler system a device that fulfills both the functions of a fire detecting system and a fire extinguishing system; the water is held back normally with a fixed temperature seal in the sprinkler head, which melts or shatters at a predetermined temperature.

backup man the man positioned directly behind the nozzleman; he takes up the weight of the hose and absorbs some of the nozzle reaction so the nozzle can be manipulated without undue strain.

bleve (pronounced "blevey") a boiling liquid–expanding vapor explosion; failure of a liquefied flammable gas container caused by fire exposure.

blitz attack firefighters hit the fire with everything at their disposal.

body harness a series of web straps on the protective breathing apparatus that position and stabilize the apparatus.

boilover occurs when the heat from a fire in a tank travels down to the bottom of the tank causing water that is already there to boil and push part of the tank's contents over the side.

breast plate that part of the protective breathing apparatus that holds the canister and protects the wearer from the heat generated by the unit.

breathing apparatus a device that provides the user with breathing protection; it includes a facepiece, body harness and equipment that supplies air or oxygen.

carbon dioxide (CO_2) a heavy, colorless, odorless, asphyxiating gas, that does not normally support combustion. It is one and one-half times heavier than air and when directed at the base of a fire its action is to dilute the fuel vapors to a lean mixture to extinguish the fire. Normally carried on board in 15 lb portable extinguishers and 50 or 100 cylinders in the installed system.

chain breaking a method of fire extinguishment that disrupts the chemical process that sustains the fire; an attack on the chain reaction side of the fire tetrahedron.

chain reaction a series of events, each of which causes or influences its succeeding event. For example, the burning vapor from a fire produces heat which releases and ignites more vapor; the additional vapor burns, producing more heat, which releases and ignites still more vapor; and so forth.

check valve a valve that permits a flow in one direction only and will close to prevent a flow in the opposite direction.

chemical foam foam formed by mixing an alkali with an acid in water.

class A fire a fire involving common combustible materials which can be extinguished by the use of water or water solutions. Materials in this category include wood and wood-based materials, cloth, paper, rubber and certain plastics.

class B fire a fire involving flammable or combustible liquids, flammable gases, greases and similar products. Extinguishment is accomplished by cutting off the supply of oxygen to the fire or by preventing flammable vapors from being given off.

class C fire a fire involving energized electrical equipment, conductors or appliances. Nonconducting extinguishing agents must be used for the protection of firefighters.

class D fire a fire involving combustible metals, for example, sodium, potassium, magnesium, titanium and aluminum. Extinguishment is accomplished through the use of heat-absorbing extinguishing agents such as certain dry powders that do not react with the burning metals.

combination combustible gas and oxygen indicator an instrument that measures the concentrations of both combustible gas and oxygen; each is indicated on a separate meter.

combination nozzle *see all-purpose nozzle.*

combustible gas indicator an instrument used to determine whether the atmosphere of a particular area is flammable; also called an explosimeter.

combustion see *fire.*

compressed gas a gas that, at normal temperatures, is entirely in the gaseous state under pressure in its container.

conduction the transfer of heat through a solid body.

convection the transfer of heat through the motion of heated matter, that is, through the motion of smoke, hot air, heated gases produced by the fire and flying embers.

convection cycle the pattern in which convected heat moves. As the hot air and gases rise from the fire, they begin to cool; as they do, they drop down to be reheated and rise again.

cooling a method of fire extinguishment that reduces the temperature of the fuel below its ignition temperature; a direct attack on the heat side of a fire tetrahedron (also see *fire tetrahedron).*

cryogenic gas a gas that is liquefied in its container at a temperature far below normal temperatures, and at low-to-moderate pressures.

demand breathing apparatus a type of self-contained breathing apparatus that provides air or oxygen from a supply carried by the user.

dry chemical a mixture of chemicals in powder form that has fire extinguishing properties.

dry powder an extinguishing agent developed to control and extinguish fires in combustible metals (class D fires).

dry system an automatic sprinkling system that has air under pressure throughout installed piping in areas that might be subjected to freezing temperatures. The operation of one or more sprinkler heads release the air pressure and activate the control valve allowing water to flow into the system.

electric fire sensor system a device capable of lighting a panel in the wheelhouse when it detects fire in a certain area of the ship.

entry suit protective clothing designed to protect the wearer from direct contact with flames for a short time.

exhalation valve a simple one-way valve on a single-hose facepiece, consisting of a thin disk of rubber, neoprene or plastic resin secured in the center of the facepiece and designed to release exhaled breath; also called a flutter valve.

explosimeter see *combustible gas indicator.*

explosive range flammable range; the range of the mixture of air and flammable gas or flammable vapor of liquids that must be present in the proper proportions for the mixture to be ignited. The range has upper and lower limits; any mixture above the *upper explosive limit* (UEL) or below the *lower explosive limit* (LEL) will not burn.

exposures combustible materials that may be ignited by flames or radiated heat from the fire.

extinguisher normally portable equipment approved for use on certain types and classes of fires.

extinguishing agent a substance that will put out a fire and is available as a solid, liquid or gas.

facepiece an assembly that fits onto the face of the person using the breathing apparatus, forming a tight seal to the face and transmitting air or oxygen to the user.

fire a chemical reaction known as rapid oxidation that produces heat and light in the form of flames, gases and smoke.

fire detector a device that gives a warning when fire occurs in the area protected by the device; it senses and sends a signal in response to heat, smoke, flame or any indication of fire.

fire extinguisher a self-contained unit, portable or semiportable, consisting of a supply of the extinguishing agent, an expellant gas (if the apparatus is not pressurized) and a hose with a nozzle.

fire extinguishing system a means of putting out fires consisting of a supply of the extinguishing agent, an actuation device (manual or automatic), and the piping, valves and nozzles necessary to apply the agent.

fire gases the hot gases produced by burning materials.

fire line automatic system the system used to detect fire in open spaces and to activate alarms and/or firefighting equipment automatically, for example, a pneumatic tube fire detector.

fire-main system a system that supplies water to all areas of the vessel; it is composed of the fire pumps, piping (main and branch lines), control valves, hose and nozzles.

fire point the temperature at which a liquid fuel sustains combustion.

fire station consists basically of a fire hydrant (water outlet) with valve and associated hose and nozzles.

fire tetrahedron a solid figure with four triangular sides illustrating how the chain reaction sequence interacts with heat, fuel and oxygen to support and sustain a fire.

fire triangle a three-sided figure illustrating the three essential components of fire: fuel (to vaporize and burn), oxygen (to combine with fuel vapor), and heat (to raise the temperature of the fuel vapor to its ignition temperature).

flame safety lamp an instrument used to test for oxygen deficiency; if there is enough oxygen in the surrounding atmosphere to keep the flame burning, there is enough oxygen to support life.

flammable range see *explosive range.*

flashover the ignition of combustibles in an area heated by convection, radiation or a combination of the two. The action may be a sudden ignition in a particular location followed by rapid spread or a "flash" of the entire area.

flash point the temperature at which a liquid fuel gives off sufficient vapor to form an ignitable mixture near its surface.

flexible tubes the part of the facepiece designed to carry fresh air or oxygen from the canister to the facepiece and, in the facepiece with a dual hose, to return exhaled breath from the facepiece to the canister.

flutter valve see *exhalation valve.*

foam a blanket of bubbles that extinguishes fire mainly by smothering. The blanket prevents flammable vapors from leaving the surface of the fire and prevents oxygen from reaching the fuel. The water in the foam also has a cooling effect.

foam concentrate liquids of 3% or 6% concentrations that are mixed with water to produce mechanical foam.

foam generators devices for mixing chemical foam powders with a stream of water to produce foam. Pressure type foam generators are closed devices containing the necessary chemicals with provision for admission of water when foam is needed.

foam proportioner a device that regulates the amount of foam concentrate and water to form a foam solution.

foam solutions the result of mixing foam concentrates with water.

fog (spray) streams a method of projecting a stream of water in which a specifically designed nozzle causes the water to leave the nozzle in small droplets, thereby increasing the water's heat absorption efficiency.

fresh-air breathing apparatus a hose mask; a facepiece connected to a pump by a long hose through which air is pumped to the user. Mobility is limited by the length and weight of the hose.

fuel any combustible material adding to the magnitude or intensity of a fire; one of the essential sides of the fire triangle.

fumes a smoke, vapor or gas given off by a fire which could be irritating, offensive or dangerous to the fire fighter.

gas a substance that has no shape of its own but which will take the shape of, and fill the volume of its container.

gas free an area, tank or system previously used to carry inflammable or poisonous liquids that has been entirely cleared of such liquids and certified by a chemist as clear of any danger.

gasket a sealing ring necessary to make a watertight connection between female and male hose couplings.

gas mask a device that filters contaminants from air that is to be breathed; it can only be used in an atmosphere that contains enough oxygen to support life.

goosenecking directing a stream of water over the vessel's side, perpendicular to the water surface.

GPM the initials for "Gallons Per Minute" and is a measure of water flow through the fire main system.

halogenated extinguishing agents Halon; made up of carbon and one or more of the halogen elements: fluorine, chlorine, bromine and iodine.

Halon see *halogenated extinguishing agents.*

hazard a condition of fire potential defined by arrangement, size, type of fuel and other factors which form a special threat of ignition or difficulty of extinguishment. A "fire hazard" refers specifically to fire seriousness potential and a "life hazard" to danger of loss of life from fire.

head harness that part of the mask designed to hold the facepiece in the proper position on the face, with just enough pressure to prevent leakage around the edge of the mask.

heat temperature above the normal atmospheric temperature, as produced by the burning or oxidation process; one of the essential sides of the fire triangle; often referred to as "ignition temperature" in fire fighting instructions.

heat transfer the movement and dispersion of heat from a fire area to the outside atmosphere. An example of heat transfer would be fire fighting water being converted into steam and expanding its volume, thus creating a slight pressure and carrying the heat and heated water vapor to the outside atmosphere Also see *connection, conduction,* and *radiation.*

high-expansion foam a foam that expands in ratios of over 100 : 1 when mixed with water; it is designed for fires in confined spaces.

high pressure fog (high velocity fog) produced when using the all purpose nozzle with the handle in mid-position. It is a high capacity jet spray produced at very high pressure and discharged through small holes of a cage type sprayer tip.

hose a flexible tube used to carry fluid from a source to an outlet. Standard shipboard fire hoses are 1½″ or 2½″ in diameter. They are normally 50 feet in length, with a female coupling installed on one end and a male coupling on the other.

hose jackets the covering over the inside liner of a hose. It is a woven jacket (or jackets) of cotton or synthetic fibers.

Index

Italicized page numbers indicate illustrations.

NOTES

NOTES

NOTES

NOTES

NOTES

NOTES

NOTES

NOTES

NOTES

NOTES